T0324148

Performance Management for the Oil, Gas, and Process Industries

Performance Management for the Oil, Gas, and Process Industries

A Systems Approach

Robert Bruce Hey

Gulf Professional Publishing is an imprint of Elsevier
50 Hampshire Street, 5th Floor, Cambridge, MA 02139, United States
The Boulevard, Langford Lane, Kidlington, Oxford, OX5 1GB, United Kingdom

British Library Cataloguing-in-Publication Data
A catalogue record for this book is available from the British Library

Library of Congress Cataloging-in-Publication Data
A catalog record for this book is available from the Library of Congress

ISBN: 978-0-12-810446-0

For information on all Gulf Professional Publishing publications
visit our website at https://www.elsevier.com/books-and-journals

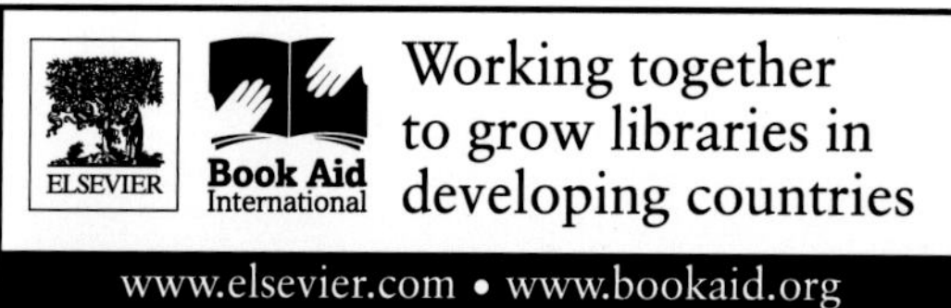

Publisher: Joe Hayton
Senior Acquisition Editor: Katie Hammon
Senior Editorial Project Manager: Kattie Washington
Production Project Manager: Kiruthika Govindaraju
Designer: Mark Rogers

Typeset by TNQ Books and Journals

This book is dedicated to Keeran and his fellow millennial engineers, the future of the developing world.

Contents

PART 2 GOVERNANCE AND PERFORMANCE

PART 10 CONCLUSION

Preface

This book provides a practical guide to those who manage the type of process plants that change bulk materials, chemically or physically, into salable products. These include oil and gas, petrochemical, aluminum, cement, paper, steel, and power plants and are generally referred to as the process industry.

The process industry is coming under severe pressure to cut costs. The construction of larger integrated units and the application of increasingly stringent environmental policies have put a number of older operators out of business, and short-term cost cutting has also been proven to be highly detrimental to the industry. The optimization of operating costs to ensure the integrity of the physical assets to produce at design capacity for the life of the investment is therefore crucial for survival of the business.

There is a gap in the market for a readable book on the whole business cycle and the use of techniques to undertake **step/episodic/breakthrough** improvement in the performance of the company. The methodology for this step improvement requires a focus on the "value adder" or "revenue generator" **Core System and** the **company direction statement** – vision, mission etc.

This book describes an approach which assures **significant sustainable improvements** in business and operational performance of process operations with **special emphasis on the hydrocarbon industry**.

The book will enable the reader to:

- Apply a "**Systems approach**" to managing the company
- Utilize the **best practice principles of good governance** for long-term performance enhancement
- Identify the most **significant performance indicators** for overall business improvement
- Set appropriate and realistic **short and long-term targets**
- Apply **strategies** to ensure that targets are met in agreed time frames
- Use appropriate **reporting tools** to influence **effective decision-making**.

This book is intended for all managers in the oil, gas and process industry, who will possibly have experience in some aspects of the book, but not necessarily in all. It's intended as a reference work-book, where the reader can skip to whichever **part** that is relevant to his/her immediate needs.

It is hoped that oil and gas and process industry management students will find the information useful as an added text for their studies.

The book is divided into **parts** that focus on different aspects of performance management which need to come together to ensure step improvement of the business. The **Parts** are:

INTRODUCTION

Why a **Systems approach**, using a **Business Unit** concept to improve performance, is essential for the success of the company.

PART 1: SYSTEMS

A Systems approach to managing the company.

PART 2: GOVERNANCE AND PERFORMANCE

Foundations for control and the constraints within which the company has to work.

PART 3: RISK AND PERFORMANCE

Requirement of company-wide risk management as part of performance management.

PART 4: PERFORMANCE INDICATOR SELECTION

Selection of those performance indicators that have the most effective influence over the performance of the company.

PART 5: ASSET PERFORMANCE MANAGEMENT

Various assets of the company and the performance of these assets.

PART 6: BENCHMARKING

Comparing the performance of the company with that of its competitors.

PART 7: ASSESSMENT, STRATEGIES, AND REPORTING

Performance assessment using suitable tools.

PART 8: BUSINESS OVERSIGHT

Performance management framework for monitoring, reviewing, and taking corrective action.

PART 9: OIL AND GAS: SPECIFIC ISSUES

Hydrocarbon accounting and issues relating to refineries and gas plants.

Each **part** consists of a number of **chapters**. In addition, each **chapter** has real life examples and anecdotes of personal experiences to clarify the chapter contents. A reference list is appended at the end of each part.

Appendices include a glossary of terms, listings of KPIs, and checklists.

Note on the use of words in this book:

The words "**System**," "**Management System**," "**Business System**," and **Business Management System** are synonymous. These words, as they pertain to this book, are defined in Chapter 4, Management Systems Determination and Requirements. A **System** must **contain** all of the **key elements** described in Chapter 4. The word "**Process"** is also defined as used in this book in Chapter 4, Management Systems and their Requirements.

The words "**KPIs**," "**key performance indicators**," and "**performance indicators**" are synonymous. They are used generically in all parts of the business. They could be applied to a **Process**, **System**, or the complete **Business Unit**.

Step, episodic, breakthrough, or **transformational change** occurs when there is paradigm shift in approach to a problem or issue (a **need**). It may or may not require a formal project management approach to change management.

Other technical words and acronyms are defined in Appendix A, Glossary of Terms.

Acknowledgments

I wish to thank the following for their contributions and support in getting this book to press.

First, I thank those who have contributed to aspects of the book. These include Professor Marios Katsioloudes in Qatar, Dr. Jim Thomson in the United Kingdom, Keith Turner in Canada, Dennis Murray in Australia, Randy Sonmor in Azerbaijan, KhaiZhen Foo in Malaysia, and Rudy de Beer and Nikko Hamp in South Africa.

Second, I thank those who have shared their experiences. They include Rob Burchell, who expounded the phrase "opportunity favors the prepared mind"; Rob Duncan for his health and safety experiences; and Dirk Bode for his benchmarking advice.

Third, I am grateful to those who have influenced me the most in my career. The names that come to mind are Bill Weckesser (City of Cape Town Electricity Generation Department), Dave Bradley (City of Cape Town Mechanical Engineering Department), Dr. Duncan McLean (ICI Fibers), Shra Smeets (Cape Portland Cement), Johan Lubbe (Chevron Cape Town Refinery), Mark Wessel (Bahrain Petroleum Company), and Ahmed Al Mulla (Qatar Petroleum Company).

Fourth, I am very grateful to the editors at Elsevier, including Kattie Washington, Katie Hammon, Kiruthika Govidaraju and Fiona Geraghty, who believed the book had potential.

Last, I thank my family, who have supported me throughout my career. My grandfather, "JC" Camp, allowed his grandchildren to experiment in his vast workshop. He had an engineering supply company—Campco. My uncle, Rod Camp, actively encouraged me as a student. He retired as chief engineer for Mobil Oil Southern Africa. My father, Bob Hey, who died at the early age of 49, set the standard of building one's own home, which all three sons achieved. My mother, Barbara, worked to support me as a full-time university student living at home. She has written a number of books on herbs. My brothers, Jimmy and Doug, taught me how to use my hands. Jimmy recently retired as a forensic tribologist for Petronas Malaysia. Doug owns an electrical switchgear company. My niece, Nikki, a fine arts graduate, developed the "before and after" Illustration. My wife, Sheree, provided support and encouragement throughout the long writing process.

CHAPTER

WHAT IS PERFORMANCE MANAGEMENT? 1

1.1 INTRODUCTION

Quote: "The only constant is change."

I regard performance improvement as a natural part of the evolution of man. Man harnessed fire and invented the wheel in the early stages of his development, and these led to leaps of progress that have ensured man's continuing advancement.

Similarly, management theories have evolved with time. For example, Demming, Juran, and others initiated various structured approaches for product quality control after World War 2, with a focus on statistical process control and continuous improvement. This culminated in the establishment of International Standards Organization (ISO) 9001 for "product" Quality Management Systems (QMS) in 1987.

In parallel, other management improvement concepts were developing, including Business Process Reengineering, where processes were redesigned, and which involved step or episodic improvement. In addition, Business Process Mapping evolved into Business Process Management, and the Balanced Scorecard helped to align business activities with strategy, thus focusing on top Performance Indicators.

The parallel approach to performance management is discussed further in Box 1.1.

BOX 1.1 DISCUSSION ON PARALLEL APPROACHES TO PERFORMANCE MANAGEMENT

The disparate and, often disconnected activities that can be described as 'quality management' or 'total quality management' have often created confusion among those to whom quality programs have been 'done'. In many cases, a collection of techniques has been presented to an unsuspecting workforce in the form of 'quality training'. Often, the training was entirely unconnected to the work that those being trained were engaged in, or worse, they thought that it was! In other cases, one particular aspect has been focused on, say 'top management commitment' – with the consequence that a 'leadership' program has been prescribed for unconvinced and uncommitted 'top management'. 'Communication' followed close behind, when these other panaceas failed. 'Empowerment' became a focus as the missing ingredient, particularly when the issues of service quality rather than product quality came into vogue.

In parallel, there were the various 'gurus', all of whom had their own approach to 'making it all happen'. Deming had his 14 points, whilst Crosby had his 14 step procedure. Peters went 'in search of excellence' while Camp introduced 'benchmarking', and Hammer developed 'business process reengineering'. Many of these approaches were pursued with vigor reminiscent of the Crusades – and in some cases with an equal proportion of bigotry, with one group of followers deriding and belittling the work and contribution of the other or others.

From the paper "Knowledge Creation and Advancement of Organizational Excellence." Authored by Rick L. Edgeman.[1]

Performance Management for the Oil, Gas, and Process Industries. http://dx.doi.org/10.1016/B978-0-12-810446-0.00001-3

Many of these concepts started to merge with the introduction of ISO 9001:2000, which applied a Systems approach to managing the company. Various ISO standards were subsequently established for Environmental Management, Health and Safety Management, Asset Management, Energy Management, etc., using the same approach as ISO 9001:2000. Company internal audits, which have traditionally focused solely on the finances of the company, changed to risk-based auditing (which includes a Systems approach).

The Systems approach focuses on the Core System, which is the "value adder" and "revenue generator" of the business, with support from Strategic and Support Systems.

It is important to note here that, for our purposes, Processes are subordinate to Systems. The reasons for this are explained in Chapter 4, Management Systems Determination and Requirements (some businesses use the word "process" as a generic term to encompass all Systems and Processes).

The Systems approach, using the concept of **a** Business Unit, is evergreen and embeds both risk and opportunity management and improvements (step and continuous). Fig. 1.1 shows the Evolution of Performance Management and the merging of a number of theories.

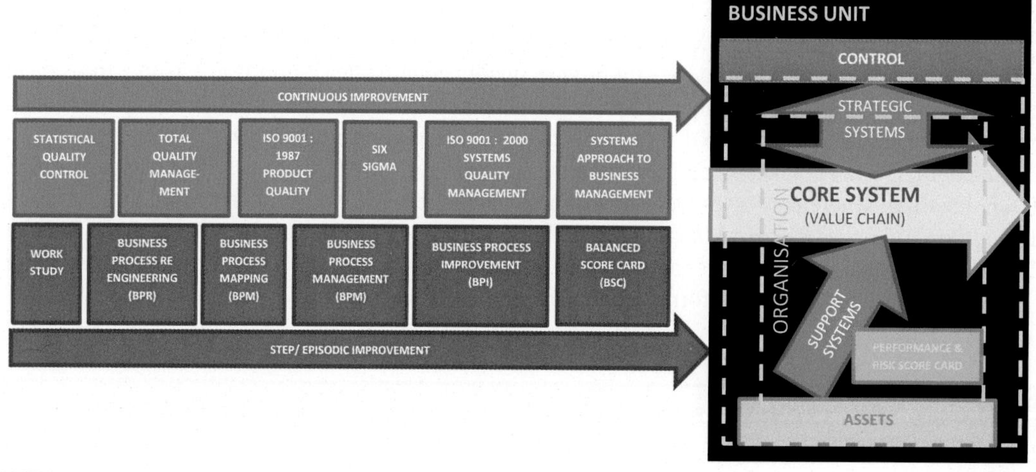

FIGURE 1.1

The Evolution of Performance Management.

The approach described in this book is not necessarily the only approach to performance management. Nevertheless, whichever route is taken, it should always focus on long-term value addition in a safe environment, within a formal control framework, and have a structured mechanism for continual improvement.

Enthusiasm for continual improvement should be embedded in the psyche of all staff (a Behavior-Based Safety Program ensures enthusiasm.)

1.2 PERFORMANCE MANAGEMENT THEORY

Performance management incorporates the identification of needs (the driver phase), the end state to be achieved to satisfy these needs, the means to achieve the end state, and how the outcomes are measured.

In the driver phase, an influence to change is defined through identified **risks and opportunities**, and **performance results**. These determine the need for change. This determines where we are.

The end state is defined by the company **vision**, which is elaborated through **goals** that are quantified by **objectives**. This is where we want to be.

The means of achieving the goals and objectives are through **strategies**, which are enabled through **tactics**. This is the means to achieve the vision.

Strategies and tactics generate **measurable outcomes**, which result in benefits to the company (value realization). Benefits can be defined by the competitiveness, efficiency, and effectiveness of the company. Measurable outcomes monitor progress toward achieving the vision.

The previous is summarized in Fig. 1.2.

1.3 WHY HAVE A STRUCTURED APPROACH TO PERFORMANCE MANAGEMENT?

This question provokes three more key questions:

1. Where are we?
2. Where do we want to be?
3. How do we get there?

These questions are addressed in parts of the book as follows:

1. Where are we?
   Part 1, Systems, discusses the requirement of focusing on the Core System of the business to maximize value addition.
   Part 2, Governance and Performance, discusses the soundness of the foundations for the existence of the company.
   Part 3, Risk and Performance, discusses current risks and opportunities (performance gaps), and probable future occurrence of risks and opportunities, to ensure performance improvement.
   Part 4, Performance Indicator Selection, discusses the correct selection of indicators to determine optimum improvement.
   Part 5, Asset Performance Management, discusses the performance requirements of different types of assets and the restraints on their application.
2. Where do we want to be?
   Part 6, Benchmarking, discusses how to determine where we are relative to the competition and we then determine, based on the industry leaders' (pacesetters) performance, where we would like to be in the future.
3. How do we get there?
   Once we have determined where we would like to be in the future, we need to ensure it is embedded in our Company Vision (see Chapter 10: Alignment). This is then used as the focus for all strategies and plans. These are discussed in Part 7, Assessment, Strategies, and Reporting. Strategies and plans have to be endorsed and monitored. This is discussed in Part 8, Business Oversight.

> What gets measured and rewarded gets done.

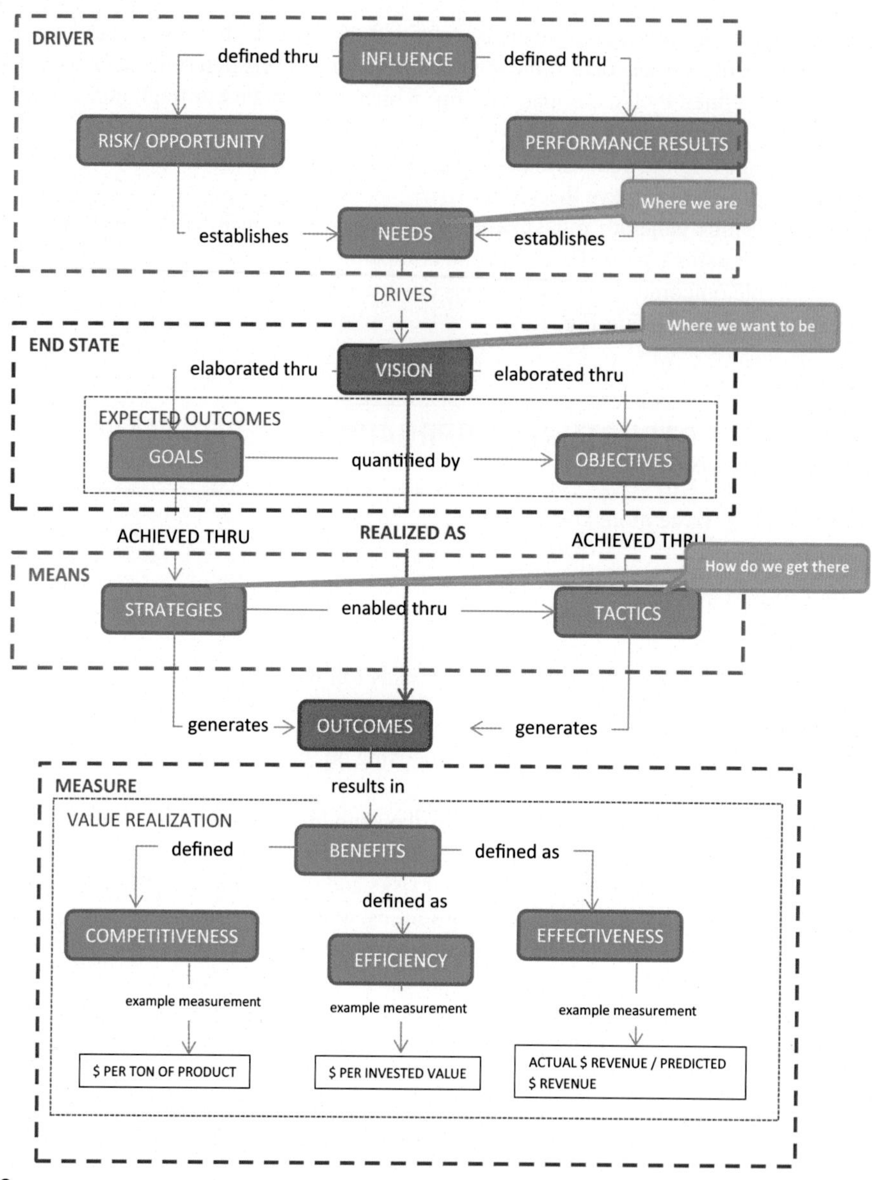

FIGURE 1.2

Performance Management theory.

Process Industry Performance Management acronyms and terms are kept to a minimum. Generic terms are used wherever possible.

Appendix A, Glossary of Terms explains the terms used in the Oil, Gas, and Process Industry.

1.4 PERFORMANCE MANAGEMENT STRUCTURE

Performance Management is a structured process for improvement. **It entails managing the overall performance of the whole Company**. It is imperative that the concept of continuous improvement is embedded in every aspect of the Company.

As a result of Strategic Planning, direction is given by the Company Board in the form of Company **Objectives** and **Strategies**. The outcome is the Business Plan and Budget, approved by the Board. Business Units work to the approved Business Plan and Budget, and their progress is monitored to ensure achievement of agreed targets.

Performance Management, as part of the Business Cycle (Strategy Review and development of Business Plans), is discussed in Chapter 5, The Business Cycle.

Assessment of Performance (Appraisal) is discussed in Part 7, Assessment, Strategies, and Reporting.

Management Review is discussed in Part 8, Business Oversight.

Key Performance Indicators (KPIs) are embedded in Management Systems, which must, in turn, align vertically with the Company direction statement and horizontally with the customers' needs. This is discussed in detail in Chapter 10, Alignment.

The "scorecard" is a performance reporting mechanism, which is discussed in Chapter 31, Reporting.

The Performance Management data flow is depicted simply in Fig. 1.3.

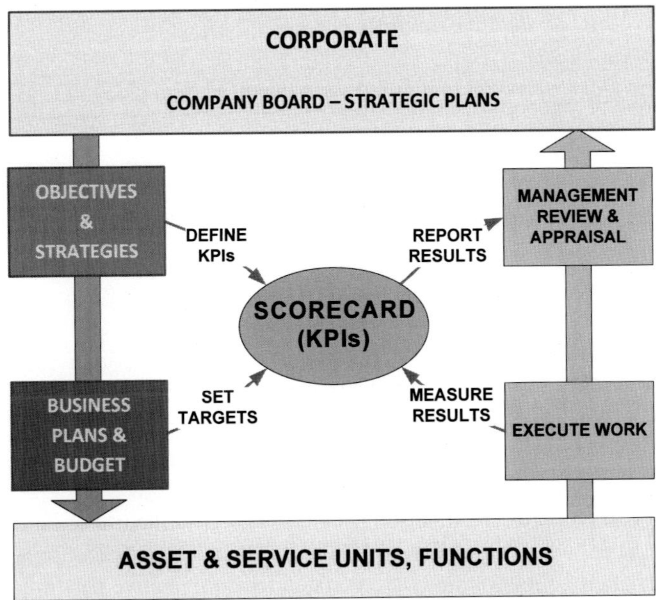

FIGURE 1.3

Performance Management data flow.

1.5 PERFORMANCE IMPROVEMENT PROCESS

Performance Management must include improvement processes. An improvement process is continuous, i.e., never-ending. Specific improvements could be continuous (possibly in small steps), or they could be big steps. Big steps require a structured change management process to succeed. Chapter 30, Strategies and Actions, discusses the change management process.

It is important that the fundamental "plan-do-check-act" cycle be instilled in all aspects of the business. Various comparative performance analysis approaches can be adopted, some within the Company and others outside the Company. For example, besides benchmarking, an external comparative performance analysis could compare against industry standards using, for instance, a review or audit process.

INTERNAL COMPARATIVE PERFORMANCE ANALYSIS EXAMPLE

A corporate-wide SAP maintenance module gives access to comparative statistics from different operations within the Company, where common targets can be set. For instance: percentage of emergency work orders as part of total work orders. This is a very useful leading KPI for Asset Integrity; see Chapter 22, Physical Asset Performance Management, for details.

EXTERNAL COMPARATIVE PERFORMANCE ANALYSIS EXAMPLE

A refinery undertakes the following:

Benchmarking (every 2 years) of the following:

- *Refinery complex*
- *Laboratory*
- *Fluidic Catalytic Cracker: detailed analysis*
- *Reliability and Maintenance*
- *Advanced Process Control (APC) and Safety Instrument System (SIS)*

Reviews

- *Quantitative Risk Analysis: carried out before initial startup and major plant modifications*
- *Insurance Review: carried out annually to determine the insurance premium*

Audits

- *Systems*
- *Certification*

HIGH-LEVEL IMPROVEMENT PROCESS

The high-level improvement process is as follows:

1. The current Risk and Opportunity situation is determined with the use of KPIs and other Metrics at all levels within the business (see Part 4: Performance Indicator Selection).
2. Relevant current data required for study are collated for comparative performance analysis.
3. Comparative performance analysis is carried out (see Part 6: Benchmarking).

4. The results of the comparative performance analysis are assessed, and those factors affecting the high-level performance of the business are used to refine the Company Direction Statement and supporting documentation (see Part 7: Assessment, Strategies, and Reporting).
5. Action plans are developed, agreed by the actioners, and implemented (also Part 7: Assessment, Strategies, and Reporting).
6. The loop is closed by returning to reassessment of the current situation (1–3 year cycle).

After repeated cycles, "world-class" performance may be achieved. It is a slow process, which takes committed leadership and dedication to the improvement process.

This is depicted simply in Fig. 1.4.

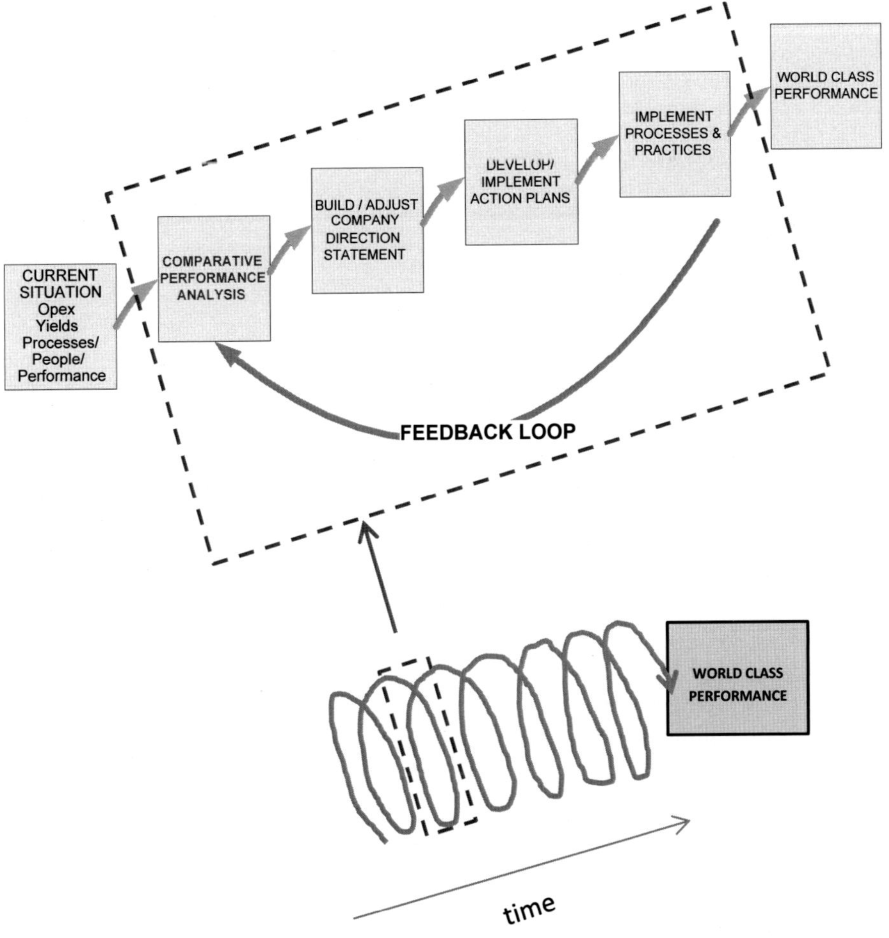

FIGURE 1.4

Improvement process.

Note that risk mitigation and improvement opportunities do not always arise from comparative performance analysis. For instance, repeated mechanical failures increase costs and reduce availability of the plant. Box 1.2 relates a situation where there was a need to improve, but those involved were not given the necessary incentive to do so.

BOX 1.2 NO INCENTIVE TO IMPROVE

When I started my career in a municipal coal-fired power station, I noticed that drives on the coal chain-grates of the eight boilers failed regularly, but there was no incentive to undertake Root Cause Analysis (RCA) and permanently fix the recurring problem.

Coal chain-grates are much like big caterpillar tracks, which support the bed of coal while it burns. As they travel toward the back of the furnace, fresh coal feeds onto the grate and ash fails off the other end. If the grate stops, steam generation stops and the turbogenerators stop producing electricity. If a failure occurred during a time of high demand, the Municipality was obliged to buy electricity at a premium from the National Power Company.

I later discovered that the fitters were "called out" on double pay each time there was a failure. Added to this, maintenance management was not responsible for controlling overtime costs. Clearly, it was in the interests of the staff **not** to find a permanent solution to the problem, since they stood to profit each time the drives failed.

Nevertheless, management could not be persuaded that they could reduce imported power and therefore save money and improve the plant availability by finding the root cause of the problem. To the best of my knowledge, this situation continued until the power station was decommissioned in the 1990s.

1.6 SUMMARY

Performance management is a structured process that has been designed for the improvement of a company. Specific improvements could be continual (possibly in small steps), or they could be in a number of big steps.

Managing performance to stay competitive is crucial to the success of the Company. For real improvement, the Company needs to improve at a higher rate than the competition. Thus "big" improvement steps are required.

The general business process entails setting objectives, planning and executing the work, reviewing and closing the loop, and back to adjusting/setting objectives. Monitoring of performance is generally achieved with the use of scorecards.

The improvement process entails comparing and assessing the current situation, adjusting the direction statement, developing implementation action plans, implementing processes and practices, and finally closing the loop by comparison and assessment of the new current situation. Eventually this leads to world-class performance (pacesetter status).

A **Systems** approach enables:

1. choice of the "right" performance indicators;
2. comparative performance analysis; and
3. **step** improvement.

This leads to the following questions:

1. Where are we?
2. Where do we want to be?
3. How do we get there?

The clues to answering these three critical questions are revealed as you proceed through the book.

REFERENCE

1. Edgeman RL, et al. Knowledge creation and advancement of organizational excellence. *Qual Eng* **12**(4) http://asq.org/qic/display-item/index.pl?item=14599.

PART

SYSTEMS

1

OVERVIEW

WHY IS A SYSTEMS APPROACH SO EFFECTIVE?

A Systems approach eliminates silos by reducing barriers and duplication of effort presented by silos.

A Systems approach is all-embracing. Its main focus is on the "Value Addition" of a Core System and is contained within a "Black Box".

A "Black Box" is "a clearly defined modular entity" and is referred to as a "Business Unit" in this text.

"Value Addition" creates the wealth which ensures continuation of the business or enterprise.

The Core System is the system that physically produces the "Value Addition" and is invested in primary assets – physical, human and financial.

The "Black Box" model helps us manage complex Systems & Processes and their interrelationships, which we may not fully understand.

THE OBJECTIVE

The objective of Part 1 is to lay the groundwork for building a business framework in which elements of performance & risk can be measured & assessed, and clear steps for improvement can be implemented.

PICTORIAL VIEW

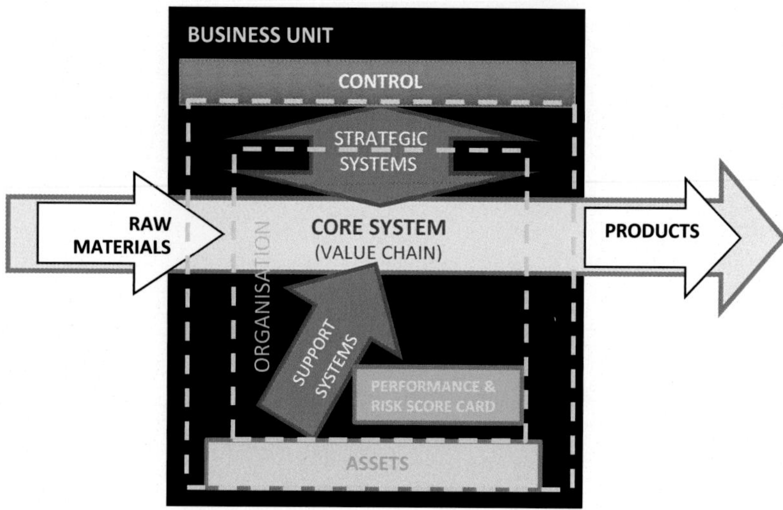

OUTLINE OF PART 1

- Modules, comprising "Business Units", are described. These Business Units come together to form a Company or Group of Companies within a Corporate Governance Framework.
- With the focus on Core (Business) Systems, definitions, as used in this book, are explained using an analogy.
- The benefits of a Systems approach are detailed.
- The relationship between Systems and Processes is clearly defined.
- To qualify as a System, key elements are described.
- Examples of different Business Systems and their underlying computer systems are described.
- Modeling is demonstrated as the basis of accurate performance measurement and assessment. Models focus on different aspects of the Business Unit and don't always align with Systems.
- Strategic Systems are shown to focus on the Business Cycle where high level business targets are set and achieved.
- Business Process Management (BPM) is explained as a tool to streamline Processes within Systems, thus improving the competitiveness and efficiency of the business.

CHAPTER

THE BUSINESS UNIT

2

2.1 INTRODUCTION

An organization can be viewed as a combination of Business Units, aligned to realize the overall company's Direction Statement. Each Business Unit could be a complex production facility. Thus the concept of a "Black Box" is applied.

DEFINITION OF THE "BLACK BOX" CONCEPT

A "Black Box" is a clearly defined modular entity that helps us manage complex systems, processes, and their interrelationships, which we may not fully understand.[1]

The concept of the "Black Box" is very simple. It is a modular entity around a **Core** "Value Addition" **System**. This "Black Box" could encompass an entire refinery, a petrochemical plant, or even a service provider. Each "product complex" or "service" is therefore regarded as a **Business Unit**.

The Business Unit could be a stand-alone business, or it could be part of a bigger enterprise or group of companies. Integrating and extending the value chain, using a series of Business Units, will boost the "Value Addition" of the bigger enterprise or group of companies.

A Business Unit, as defined in this book, is thus based on the concept of a "Black Box."

2.2 ADVANTAGE OF THE BUSINESS UNIT APPROACH USING THE BLACK BOX CONCEPT

The **advantage** of the Business Unit approach is that a complex business can be simplified into "profit centers." Relative contribution to "Value Addition" and the Direction Statement of the company can then be assessed. Those Business Units that do not make a contribution to the company need to be disposed of. Noncore business, not related to Core business activities, should also be disposed of.

Box 2.1 gives an example of a company choosing to divest from noncore Business Units so as to focus on its Core business.

Performance Management for the Oil, Gas, and Process Industries. http://dx.doi.org/10.1016/B978-0-12-810446-0.00002-5

BOX 2.1 FOCUS ON CORE BUSINESS: NEWS ARTICLE IN BUSINESS TIMES MALAYSIA, THURSDAY, MAY 7, 2015

Felda Global Ventures Holdings Bhd to Sell Noncore Businesses

Open Tender: Company Divesting Felda Prodata Systems, Felda Travel, and Felda Properties by End of Next Month

Felda Global Ventures Holdings Bhd (FGV) plans to hive off three noncore businesses by next month to focus on becoming a fully integrated plantation player in palm oil, sugar, and rubber, says its chief.

The company is planning to sell Felda Prodata Systems, Felda Travel, and Felda Properties via an open tender.

Any activity that does not contribute to "value addition" within the Business Unit should be discarded or, alternatively, the relevant service could be outsourced or contracted out.

Box 2.2 gives an example of an activity that does not contribute to "value addition" of the company.

BOX 2.2 DISPOSAL OF THE COMPANY SCHOOL

Bahrain Petroleum Company (Bapco; the National Oil Company of Bahrain) formed a school, funded by the company, for the children of company employees.

In the 1990s, the company reassessed its competitive position and decided to focus on its **Core Business**: refining oil. It handed the school over to the St. Christopher's Group of Schools, who already managed other schools in Bahrain.

In place of free schooling from the company school, staff were given an education allowance so that they could send their children to a school of their choice. The upshot was a win–win solution for both company and parents.

2.3 THE SYSTEMS APPROACH USING THE BUSINESS UNIT

The basic Business Unit is applicable to any "Value Addition" asset or service.

The Business Unit consists of:

1. a Governance Framework, which includes **Strategic, Core** (Value Chain), and **Support Business Systems**; and
2. the application of Financial, Human, and Physical **Assets** to produce **Salable Products** from **Raw Materials** or, alternatively, to offer a Service.

Fig. 2.1 depicts the Business Unit as a Black Box.

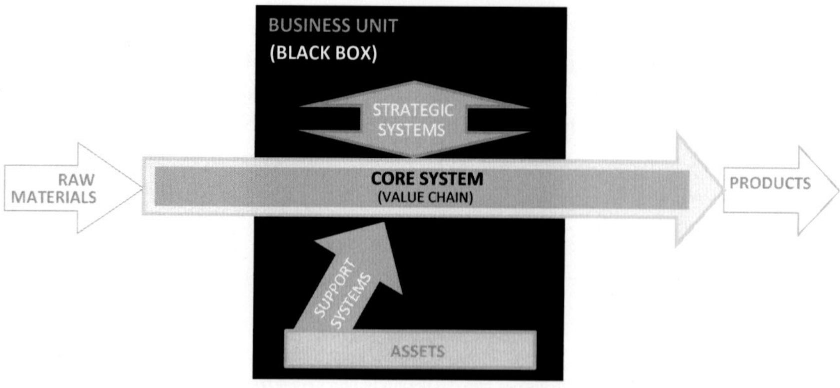

FIGURE 2.1

The Business Unit as a Black Box.

The Systems approach is expanded in Chapter 4, Management Systems Determination and Requirements, and Chapter 8, Governance Framework.

An **Asset Business Unit** would consist of the entire Process Plant, with its Primary Process Units as the Core System, converting bulk raw materials chemically or physically to salable products.

A **Service Business Unit** would consist of the "value addition" of the services provided.

The Business Unit concept can also be applied, in modular form, to the whole company, or group of companies, which is configured as discussed in Section 2.6.

2.4 BOUNDARIES AND CONTENT OF THE BUSINESS UNIT

The Business Unit is defined by its boundaries and what enters and exits these boundaries.

BOUNDARY TYPES

Boundaries could be defined by the following:

Product

The Product boundary of a Business Unit generally applies to an Asset Business Unit producing, for example, transport fuels, petrochemicals, heating fuels, or electricity. These boundaries are determined by the fiscal meters for raw materials in and products out, regardless of where these meters are located.

Examples

- *An integrated gas plant with associated pipelines with fiscal meters at the customer's site*
- *A power plant with electricity meters at the customers' sites*

Physical Asset

The Physical Asset boundary of an Asset Business Unit is the boundary fence around the plant, which produces the Salable Products.

Examples

- *A refinery with fiscal meters at the boundary fence*
- *An aluminum smelter with alumina feed conveyor into the plant to aluminum ingots out of the plant*
- *A cement plant with limestone from the quarry to bagged and bulk cement out of the plant*

Geographical

A Geographical boundary could be applied to all activities of the Business Unit within a certain territory.

Examples

- *An exploration block*
- *A municipality*
- *A country*

Discipline

This boundary could be applied to a service that the Business Unit supplies.

Examples

- *The supply of Inspection Services*
- *The supply of Turnaround Management Services*

Legal

The legal boundaries of Business Units could be described by any of the previous and documented in one of the following:

1. Mandate: for directly managed Business Units
2. Joint Venture Agreement (JVA): for Joint Venture Business Units
3. Memorandum and Articles of Association: for Subsidiary Business Units
4. Production Sharing Agreement (PSA): for PSA Affiliate Business Units
5. Government Directive: for State Service Units

Legal boundaries are expanded in Chapter 32, Business Relationships.

INPUTS AND OUTPUTS

Inputs are Raw Materials or a "Request for Services." Both raw materials and services are provided through a procurement contract.

Resources (People and Finance) are provided to ensure "Value Addition." These are provided by the Corporate Functions with the approval of the Company Board of Directors.

Outputs are Products (salable and waste) or "Service Completed." Products are disposed of through a sales agreement, and completed Services are as per the procurement contract.

CONTENTS

The contents consist of the Physical Assets required to undertake the value addition as well as other Assets that make this possible: primarily Information, Human Capital, Financial, and Intangible. These are discussed in detail in Part 5, Asset Performance Management.

Note: People and Finance can be viewed as both assets and resources.

People Assets who cannot be readily replaced are regarded as Human Capital.

Example: An operator with extensive knowledge of the type of process, and particularly the plant he works on, is clearly an asset and needs to be retained. An operator may initially be recruited as a resource and could, in time, become an asset.

However, a human resource might also be considered a commodity that can be readily replaced.

Example: Certified welders are usually regarded as a resource, as they can be recruited, utilized immediately, and then released.

Finance is an asset when applied to a particular investment. However, it is regarded as a resource when it is money sitting in the bank and readily available.

Example: A bank sets funds aside for a project, which requires funding at regular intervals. When the money transfers to the project, and is used for project goods and services, it moves from being a resource to being an asset. Once the project is commissioned, the total spent on the project is registered in the company Asset Register.

2.5 BUSINESS UNIT TYPES AND RELATIONSHIPS

The modular Systems approach differentiates between production and service activities as follows:

ASSET BUSINESS UNITS

Asset Business Units are the **Producing/"Value Addition" units** of the business.

SERVICE BUSINESS UNITS

Corporate **Service Business Units** (including **Functions**) supply services in varying degrees to the previously mentioned **Asset Business Units.**

Functions assist the Managing Director and Board by providing functional direction, support, and leadership for the Group of Companies and provide services to the Business Units and also to other Functions.

Business Unit relationships are modeled as per Fig. 2.2.

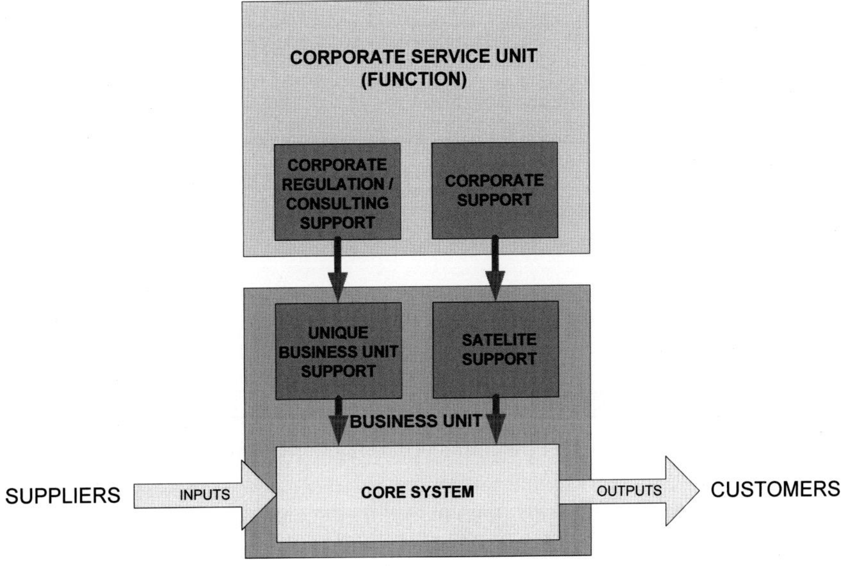

FIGURE 2.2

Business Unit relationships.

Unique Business Unit Support could require a Maintenance Business System with access to a Corporate Maintenance Consulting Group. Satellite Support could necessitate a local representative of the Corporate Training Function.

Control over Subsidiaries and Joint Ventures are determined by the relationships decided on by inter-company agreements (for example, JVAs, Operating and Maintenance agreements, etc.) Chapter 32, Business Relationships, discusses this in more detail.

Examples of Corporate **Service Business Units (Functions)** are as follows:

- Strategic Planning, Risk, and Performance
- Audit
- Finance
- Information Technology
- Human Relations and Training
- Legal and Corporate Affairs
- Procurement
- Health, Safety, Environment (HSE) and Integrity
- Capital Projects

Examples of **Asset Business Units** are as follows:

- Upstream Oil and Gas Production
  - Onshore
  - Offshore
- Refining Complex
- Gas Processing Complexes
  - Flowing Gas/Natural Gas liquefaction (NGL)
  - Liquefied Natural Gas (LNG)
  - Gas to Liquid (GTL)
- Petrochemical Complexes
- Power Plant
- Aluminum Smelter
- Cement Plant
- Paper Mill
- Steel Mill

2.6 INTERNATIONAL STANDARDS ORGANIZATION CERTIFICATION[2–11]

The Business Unit approach facilitates the application of International Standards Organization (ISO) certification.

If the business is certified to ISO 9001, then the following eight **Principles** apply:

1. Customer focus
2. Leadership
3. Involvement of people

4. **Process approach**
5. **Systems approach to management**
6. Continual improvement
7. Factual approach to decision-making
8. Mutually beneficial supplier relationships

To obtain ISO certification, it is best to start with an **Integrated Management System (IMS)** for the **Business Unit**. In doing so, appropriate Business Systems need to be identified and developed before looking at the requirements for certification to a particular standard. By starting **with** an IMS, future add-ons, such as ISO 50001, "Energy Management," and ISO 55001, "Asset Management," are easily accommodated. Starting **without** an IMS could create difficulties should the company decide to certify to other ISO standards at a later stage. Box 2.3 gives examples of the two approaches.

The development of an **IMS** is described in Chapter 4, Management Systems Determination and Requirements.

BOX 2.3 INTERNATIONAL STANDARDS ORGANIZATION (ISO) 9001 CERTIFICATION: TWO APPROACHES

Using the Existing Establishment

Bahrain Petroleum Company (Bapco) was under pressure from their biggest customer, the US Navy, to obtain International Standards Organization (ISO) 9001 so that the US Navy could downsize their quality oversight team in Bahrain. At the time, ISO 9001 focused solely on product quality, and, in accordance with the requirements, Bapco collated a set of procedures within the bounds of their existing organization structure. In 1994, the company became the first refinery in the Arabian Gulf and in Chevron Oil Company (their Joint Venture Partner) to obtain ISO 9001 for product quality. Bapco then wished to obtain certification for other ISO standards, which was complicated by the silo effect of their organization's structure. As a result, it was not until 2009, 15 years later, that Bapco obtained certification to ISO 14001, "Environmental Management."

Comment: Establishing an Integrated Management System (IMS) first would have greatly facilitated the ISO 14001 certification and reduced the time to obtain this.

Using a Systems Approach

After Bapco received ISO 9001 certification in 1994, other Chevron refineries were under pressure to comply. The management of Chevron Cape Town Refinery decided to embark on establishing an IMS, based on a **Systems** approach to satisfy the requirements of ISO 9001 and any future compliance requirements, including ISO 14001, "Environmental Management," and API 750 (the predecessor of OHSAS 18001/ISO 45001 Occupational Health and Safety Management). Based on its experience of certification for the previously mentioned standards, Det Norske Veritas (DNV) was chosen as the initial Accreditation Certifying Authority (see Chapter 4: Management Systems Determination and Requirements, for discussion on management system products offered by DNV and others).

Chevron Cape Town Refinery received certification to ISO 9001 in May 1998.

A primary outcome of the implementation of a **Systems** approach was that the ISO embedded these ideas in the revised ISO 9001:2000. This made it easier to adopt other standards modeled in the same way.

The ISO certification process in a complex organization is facilitated by adopting a modular approach using the Business Unit concept.

Each complex or site that contains Process Units could be regarded as individual Business Units.

Example

A Company owns and directly manages the following Asset Business Units:

- *Upstream Oil and Gas Production*
  - *Onshore*
  - *Offshore*
- *Refining Complex*
- *Gas Processing Complexes*
  - *NGL*

The Corporate Support Business Units (Functions) are the following:

- *Strategic Planning, Risk, and Performance*
- *Audit*
- *Finance*
- *Information Technology*
- *Human Relations and Training*
- *Legal and Corporate Affairs*
- *Procurement*
- *HSE*
- *Capital Projects*

Each Asset Business Unit that manufactures products can, initially, apply for stand-alone certification to ISO 9001, "Quality Management," ISO 14001, "Environmental Management," and ISO 45001 (formerly OHSAS 18001), "Occupational Health and Safety Management." To this end, it is preferred that these Business Units each individually establish an IMS. Each Business Unit will have a program for certification, and thus certification will probably be obtained at different times.

At the same time, Corporate Service Units can be grouped appropriately for certification to ISO 9001. The expansion of ISO 14001 and ISO 45001 to the Corporate Service Units is dependent on the need for this certification.

Once certification has been obtained for each Business Unit, application can then be made for Group certification. Compliance with other ISO standards can follow, as the IMS will easily accommodate them.

An application of the previously mentioned example is described in Box 2.4.

BOX 2.4 CERTIFICATION OF A COMPLEX BUSINESS: QATAR PETROLEUM

Qatar Petroleum (QP) consists of the following directly managed Asset Business Units:

1. Onshore Oil and Gas Production
2. Offshore Oil and Gas Production
3. Natural Gas Liquefaction (NGL) Plant
4. Oil Refinery

These and various QP subsidiaries and Joint Ventures (JVs) are supported by Support Business Units: Finance; Legal; IT; Procurement; Health, Safety, Environment (HSE); HR; Training; Project Engineering; Industrial Cities.

QP applied for Certification of each Business Unit individually, starting with the Oil Refinery, and then applied for group certification to ISO 90001.

Onshore Oil and Gas Production and Offshore Oil and Gas Production have subsequently obtained ISO 14001, "Environmental Management," while at the time of writing, certification of other Business Units to ISO 14001, "Environmental Management," and ISO 45000 "Occupational Health and Safety Management," is in progress.

2.7 INTEGRATION

Integration of Business Units to extend the supply chain gives "added value" and competitive advantage.

In order to reduce operating costs, the Process Industry is moving toward integration, especially in National Oil and Gas Companies.

For example, crude goes straight from the wells to the refineries in pipelines. Refineries directly feed Petrochemical Plants, which are housed in the same industrial complex. Field Gas produces condensate, which is also fed to the Refinery. The resulting dry gas is used in LNG, NGL, or GTL plants, still all within the same industrial complex, with common side products of ethane, propane, and butane, which are either fed to the local petrochemical plants or exported. Low-grade gas is fed to a Cement Plant for clinker production in rotary kilns and Combined Cycle Power Plants, which can attain high generation efficiencies. Part of this power generated could be used in electric arc furnaces for the production of Aluminum (the gas for Aluminum smelting is sometimes referred to as Gas to Solids).

Fig. 2.3 depicts a basic hydrocarbon integration scenario.

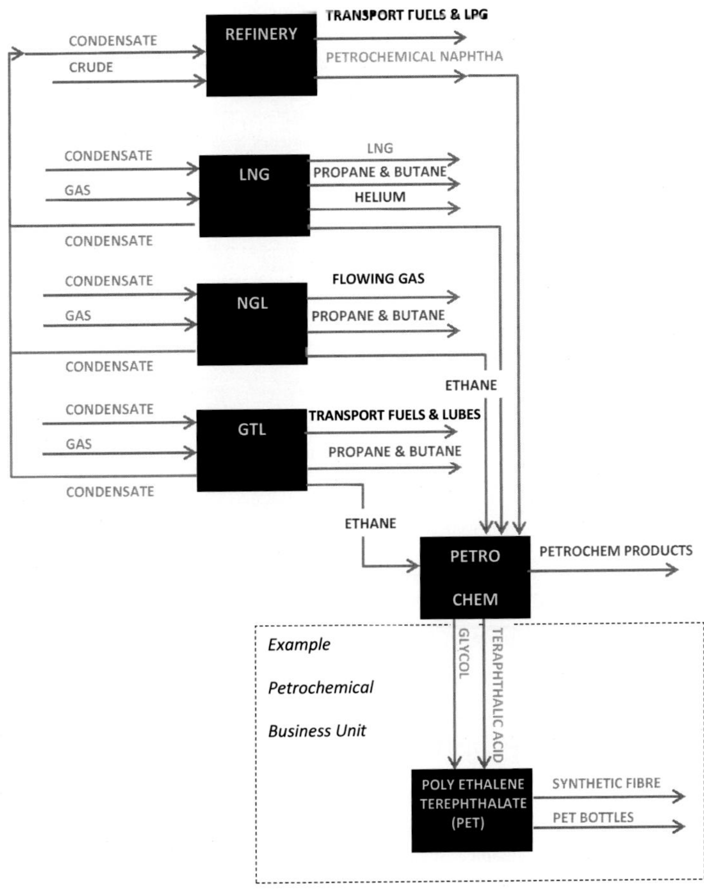

FIGURE 2.3

Hydrocarbon integration.

Fig. 2.4 shows the complete petrochemical flow from natural gas and crude oil feedstocks to basic products, intermediate products, and final products.

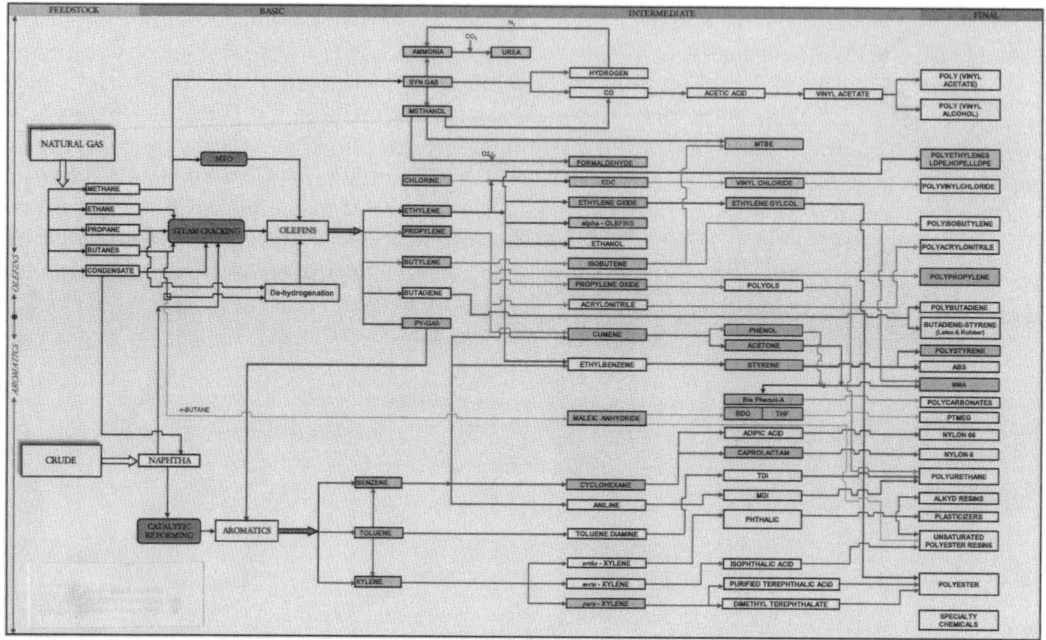

FIGURE 2.4

Petrochemical flowchart.

In order to extend their supply chain, some LNG companies are considering buying shares in pipeline and gas-fired power companies in their client countries.

An example of the integration of Business Units to extend the supply chain is related in Box 2.5.

BOX 2.5 SUPPLY CHAIN INTEGRATION

Malaysia's Pengerang Integrated Petroleum Complex (PIPC)

PIPC is the catalyst that will transform Johor into a new regional oil and gas hub in the coming years and position the state as the Rotterdam of Asia.

Pengerang Integrated Petroleum Complex (PIPC) comprises Refinery And Petrochemical Integration Development, undertaken by Petronas, and its associated facilities, including a co-generation plant, Liquefied Natural Gas (LNG) regasification terminal, air separation unit, raw water supply, as well as other ancillary facilities.

RAPID integrates the refinery and gas plants with various petrochemical plants, using their outputs to produce numerous products including well-known consumables, such as polythene, PVC, synthetic fiber, etc.

An example of integration of disparate Business Units for increased Return on Investment (RoI) is given in Box 2.6.

BOX 2.6 IMPROVED RETURN ON INVESTMENT FROM THE SALE OF A "WASTE PRODUCT"

Saldanha Steel and Pretoria Portland Cement Sales Agreement

South Africa's exports had a major boost with the installation of a railway, exclusively for very long iron ore trains, from the large iron ore mines at Sishen in the interior to the coast. To add value to the exported product, a steel mill was built at Saldanha Bay.

The main by-product of steelmaking is slag. This waste product normally has no value, but with a minor investment on-site by the local cement producer, the slag was blended with cement to produce a marketable product, "Slagment." This product is marketed in competition to Ordinary Portland Cement.

Increased RoI could also be obtained by the availability of new technology and the changing market for products.

Example:

The increase in use of diesel for cars in Europe has resulted in an increased demand for diesel, with a relative reduction in the demand for gasoline. However, only a certain percentage of diesel can be extracted from a barrel of crude oil, and this actually results in an excess of gasoline. Now, with the development of new technology and a relatively minor investment, excess gasoline can be used to produce high-value petrochemicals.[12]

2.8 SUMMARY

A "Black Box" approach is a simple way of managing the performance of a complex business. Essentially, it aims to modularize the Business Unit around each "value addition" System (Core Business System), be it a stand-alone company or a part of a bigger enterprise or group of companies. All aspects of the Business Unit are contained, with clear interfacing of inputs and outputs.

Business Units could either be Asset Business Units supplying the products of the company as per the Company Direction Statement (Vision, Mission, etc.), or they could be Service Business Units that support the Asset Business Units.

The Systems approach descends from the Business Unit down to Core, Strategic, and Support Business Systems.

The bigger enterprise or group of companies would consist of a group of Business Units as follows:

1. Parent Company Business Unit consisting of Strategic, Core, and Support Business Systems
2. Asset Business Units carrying out value addition of the products of the business with Strategic, Core, and Support Business Systems
3. Support Business Units supporting the Asset Business Units with Core and Support Business Systems

Integration of Business Units to extend the supply chain will boost the value addition of the bigger enterprise or group of companies, giving competitive advantage. Minor additions to an investment to maximize value addition may also enhance RoI.

The Systems approach is discussed in detail in Chapter 4, Management Systems Determination and Requirements.

The framework for control of the Business Unit is discussed in Chapter 8, Governance Framework.

Businesses, from the legal perspective, are discussed in Chapter 32, Business Relationships.

REFERENCES

1. Krogerus M, Tschappeler R. *The decision book*. Profile Books; 2012.
2. ISO 9001:2008. *Quality management.*
3. ISO 29001:2003. *Quality management system for the oil and natural gas industry.*
4. ISO 14001:2004. *Environmental management systems.*
5. ISO 31000:2009. *Risk management principles & guidelines.*
6. ISO 31010:2009. *Risk management – risk assessment techniques.*
7. ISO 45001:2016. *Occupational health & safety management systems.*
8. ISO 50001:2011. *Energy management systems.*
9. ISO 55000:2014. *Asset management – overview principles & terminology.*
10. ISO 55001:2014. *Asset management – requirements.*
11. ISO 55002:2014. *Asset management – guidelines for application of ISO 55001.*
12. Gentry JC. *Refining/petrochemical integration – a new paradigm.* GTC white paper May 2015. Fuel consumption favors diesel over gasoline http://www.digitalrefining.com/article/1001081,Refining_petrochemical_integration______FCC_gasoline_to_petrochemicals.html#.VwDK785OJdg.

CHAPTER

MODELS

3

3.1 INTRODUCTION: WHAT ARE MODELS AND WHY DO WE NEED THEM?

Models depict the interaction between components of the Business Unit. They can be dynamic (live) showing updates by the minute, hour or day, or they could be static, where data is downloaded into them on, say, a monthly basis. They are used to facilitate accurate measurement, without which performance cannot be measured. They are also used to compare performance and undertake "what if" analyses to determine what interaction there would be between components of the Business Unit, if a certain scenario is entered into the model.

The limitations of modeling are determined by the scope of the model. Exclusions from the model have to be noted and their impact weighed before decisions are made based on model data.

3.2 THE SCOPE OF MODELS

Modeling is undertaken on different parts of the Business Unit. The following are the common models used.

BUSINESS MODELS

Business Models are generally based on Enterprise Resource Management (ERM) software and business intelligence platforms and integration tools, such as those offered by SAP and Oracle. This could be regarded as live modeling. Daily production, health and safety, and environment statistics could be summarized on a dashboard for management. Monthly reports would also include financial and maintenance statistics.

MASS AND ENERGY BALANCE MODELS

Mass and Energy Balance Models are essential for assessing losses from the raw materials processed. When an imbalance occurs, an immediate investigation can reveal an accounting/measurement error or a real loss to be reduced or eliminated. This is regarded as historical (static) modeling.

PROCESS SIMULATION MODELS

Process Simulation Models run simulations based on real and potential inputs. These could be stand alone for determining different options for operation, referred to as "what if" simulation, or could be live, giving feedback directly to the process controls for continuous process optimization, normally referred to as Advanced Process Control (APC). "What if" simulation determines the potential for the process and thus can be used to determine targets for Key Performance Indicators (KPIs).

Performance Management for the Oil, Gas, and Process Industries. http://dx.doi.org/10.1016/B978-0-12-810446-0.00003-7

UTILITY MODELS

Utility Models focus on the efficiency of utilities. In the case of gas plants and refineries, the utilities are totally integrated into the process, and so a total energy balance is required as a cross-check on the mass balance.

OPERATING EXPENSE MODELS

Operating Expense Modeling, using a standard template, is a powerful tool for comparative cost studies. It can be used for budget review and/or benchmarking and is generally carried out on an annual basis.

BENCHMARKING MODELS

Benchmarking Models are used for comparing "apples" with "apples" and thus only focus on aspects of the Business Unit that can be compared, while other aspects are excluded.

The scopes of models within the Business Unit are shown in Fig. 3.1 and are depicted in red.

Various models are described in more detail in the following paragraphs.

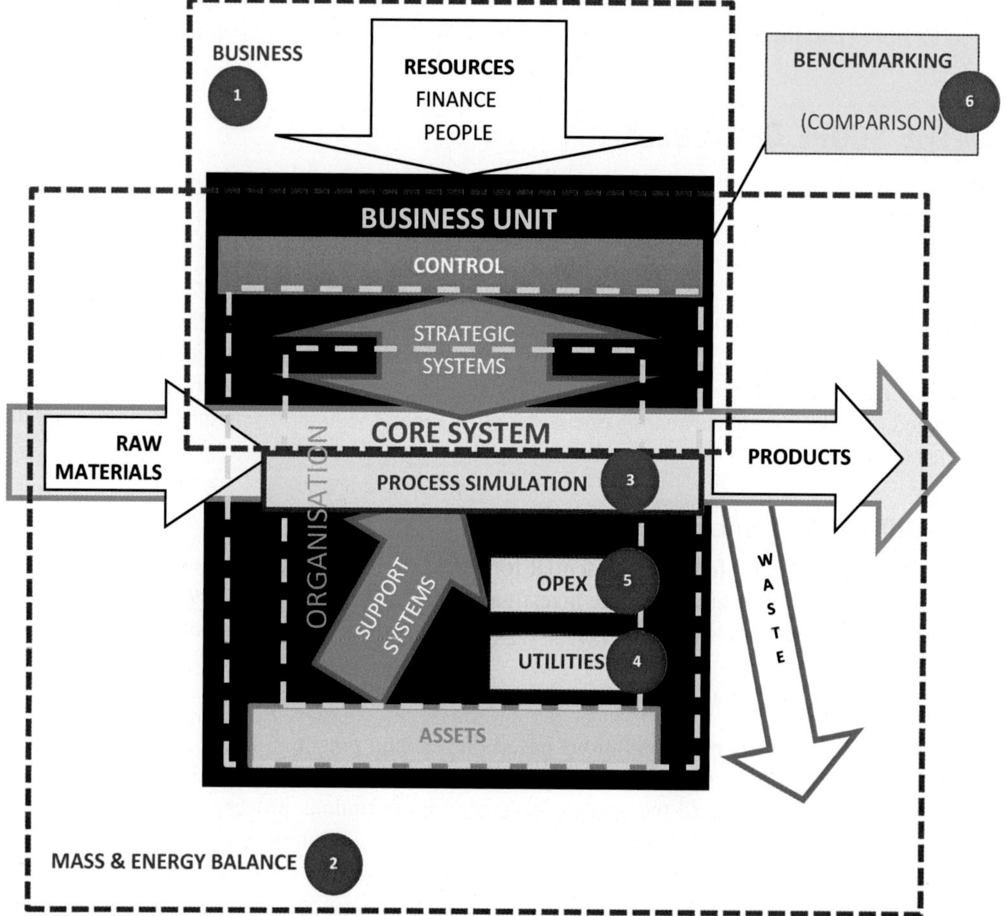

FIGURE 3.1

Model scopes [depicted in red].

3.3 BUSINESS MODELS

These involve ERM modules and business intelligence platforms and integration tools.

Example:

A ***financial module*** *will include budgets and expenditure, and* ***materials modules*** *will track stock prices, levels, and consumption. A* ***maintenance module*** *will draw down on stocks and expenditure on materials and people. Accurate performance indicators are thus drawn from these three modules and possibly rolled up into a high-level reporting tool linked to Enterprise Risk Management Systems. These tend to be consolidated data reporting tools showing the data in formats for easy decision-making.*

Forrester undertake a comparative study of Business Performance Software every quarter.[1]

Example:

The Forrester Wave Business Performance Solutions Q4 2009 Report put IBM Cognos, Oracle, SAP, and SAS in the lead.

Box 3.1 outlines the SAP Business Models.[2,3]

BOX 3.1 EXAMPLE BUSINESS MODELS: SAP

SAP Strategy Management: Supported Business Processes and Software Functions

- Communication: transform written plans into living documents that can be used with employees to define, discuss, share, and update goals.
- Collaboration: motivate employees and encourage greater collaboration by making performance-relevant data available in a contextually appropriate, personalized way.
- Strategy management: deploy resources more efficiently by understanding interdependencies, "below horizon" objectives, risks, and the importance of initiatives to strategic goals.

SAP: Business Planning and Consolidation Application: Supported Business Processes and Software Functions

- Business planning and budgeting: streamline and automate strategic planning and budgeting by using a collaborative top-down and bottom-up approach.
- Forecasting: produce more accurate plans and budgets, create rolling forecasts, and incorporate real-time actuals with historical data analysis.
- Predictive analysis: receive automatic alerts about potentially at-risk Key Performance Indicators (KPIs) and recommended actions, plus one-click access to explanations for variances and root causes.
- Reporting and analysis: gain one-click access to up-to-date production and management reporting, financial and operational analysis, and multidimensional analysis.
- Consolidation: centralize all performance related data, shave weeks off consolidation processes, and ensure compliance.

3.4 MASS AND ENERGY BALANCE MODELS

Mass and Energy Balance Models are required for balancing the material "books" in a similar way to balancing financial books. Specific models are discussed in this paragraph.

REFINERY MODELS

All measurement of input and output are required from meters: mass, volume, and density. In some cases, such as "coke on catalyst," the value is calculated based on other measurements.

The key is to capture all hydrocarbons into and out of the Business Unit, as well as internal consumption.

Fig. 3.2 is an example of a simple picture for identification of tags for input and output of a refinery mass balance model. This model includes feed to, and return from, a Benzene Plant located next to the refinery. It also identifies "coke on catalyst" produced in a Fluidic Catalyst Cracking Unit. The energy value of the "coke on catalyst" is the boundary of the model. However, the energy from the "coke on catalyst" is sent to a Waste Heat Boiler, which feeds a steam turbine that generates 12 MW of power. The conversion of energy to electrical power would only be identified in the energy component of a Benchmarking Model (See Section 3.8: Benchmarking Models).

When the mass and energy are balanced in one model and the same outer limits set for both, then data input errors can be quickly identified and immediately addressed.

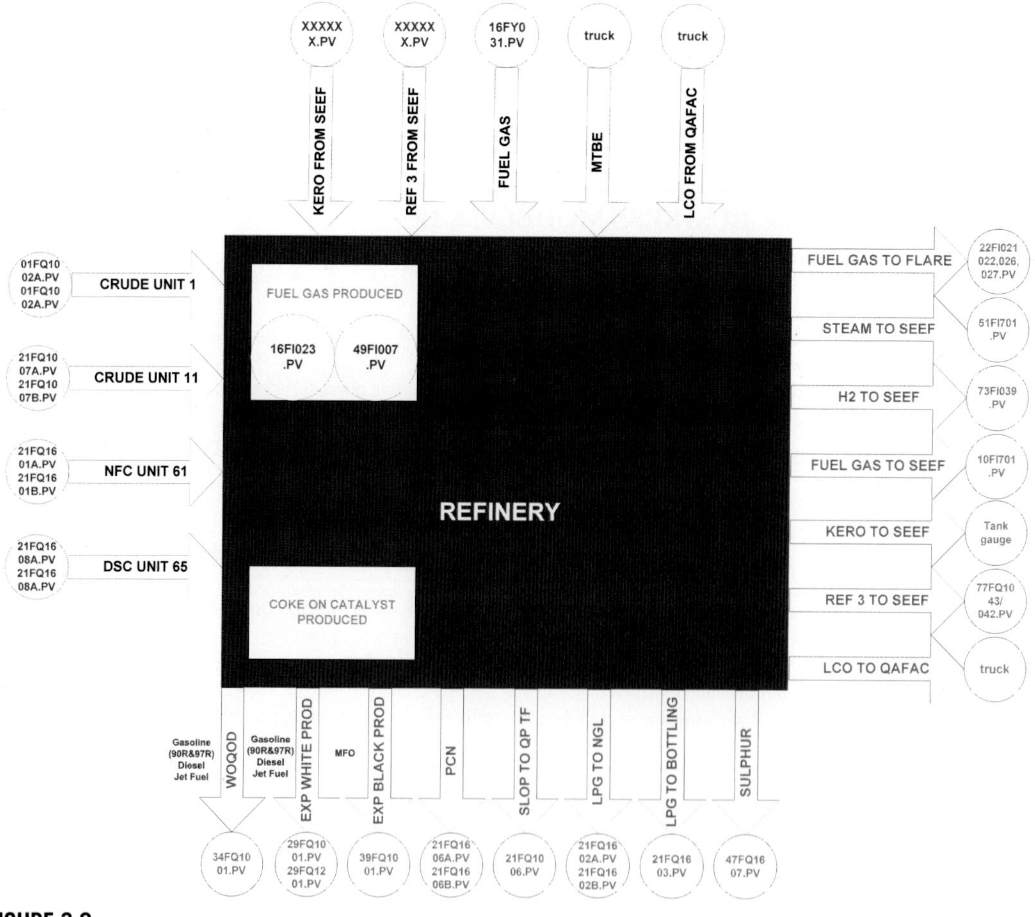

FIGURE 3.2

Refinery mass balance model example.

GAS PLANT MODELS

In gas plants, the production of energy for internal use is significant, and so, mass **and** energy needs to be balanced in the same period. Mass units are used for royalty calculations and mass balance, whereas energy units are used for sales, where the product is priced in $/million BTUs (see Appendix A: Glossary of Terms). This requires both mass and energy balance.

The import of electricity, fuel, and steam relative to self-generation of energy also needs to be assessed. Section 17.7, Energy, discusses the use of a reference gas for comparison.

All energy into the Business Unit, consumed internally, lost, and produced (whether salable or not) needs to be accounted for. Fig. 3.3 depicts this balance.

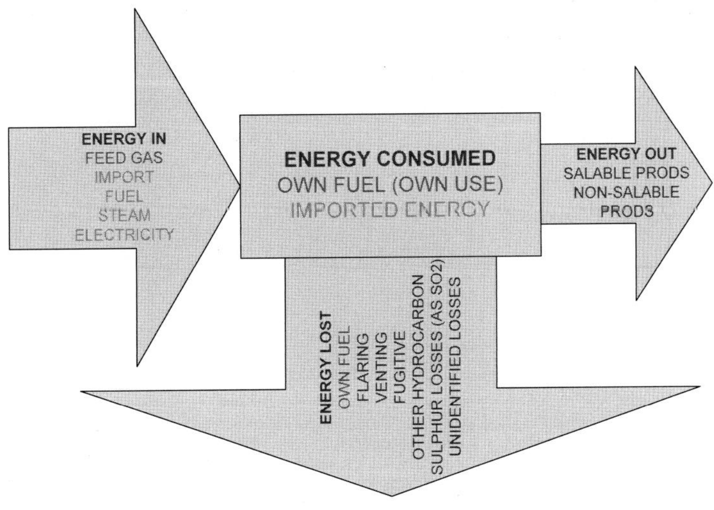

FIGURE 3.3

Energy balance components.

Fig. 3.4 shows a simple picture for identification of tags for input and output of a Mass/Energy Balance model for a Liquefied Natural Gas (LNG) plant. The grouping of process units is used for internal balancing as follows:

1. Offshore and preseparation
2. Condensate stabilization
3. Gas treating
4. Sulfur recovery
5. Ethane fractionation/treating
6. Propane fractionation/treating
7. Butane fractionation/treating
8. Plant condensate fractionation
9. Gas liquefaction
10. Helium extraction

11. LNG storage and loading
12. Utilities
13. Off-sites

Measurement takes place where **blue circles** are shown.

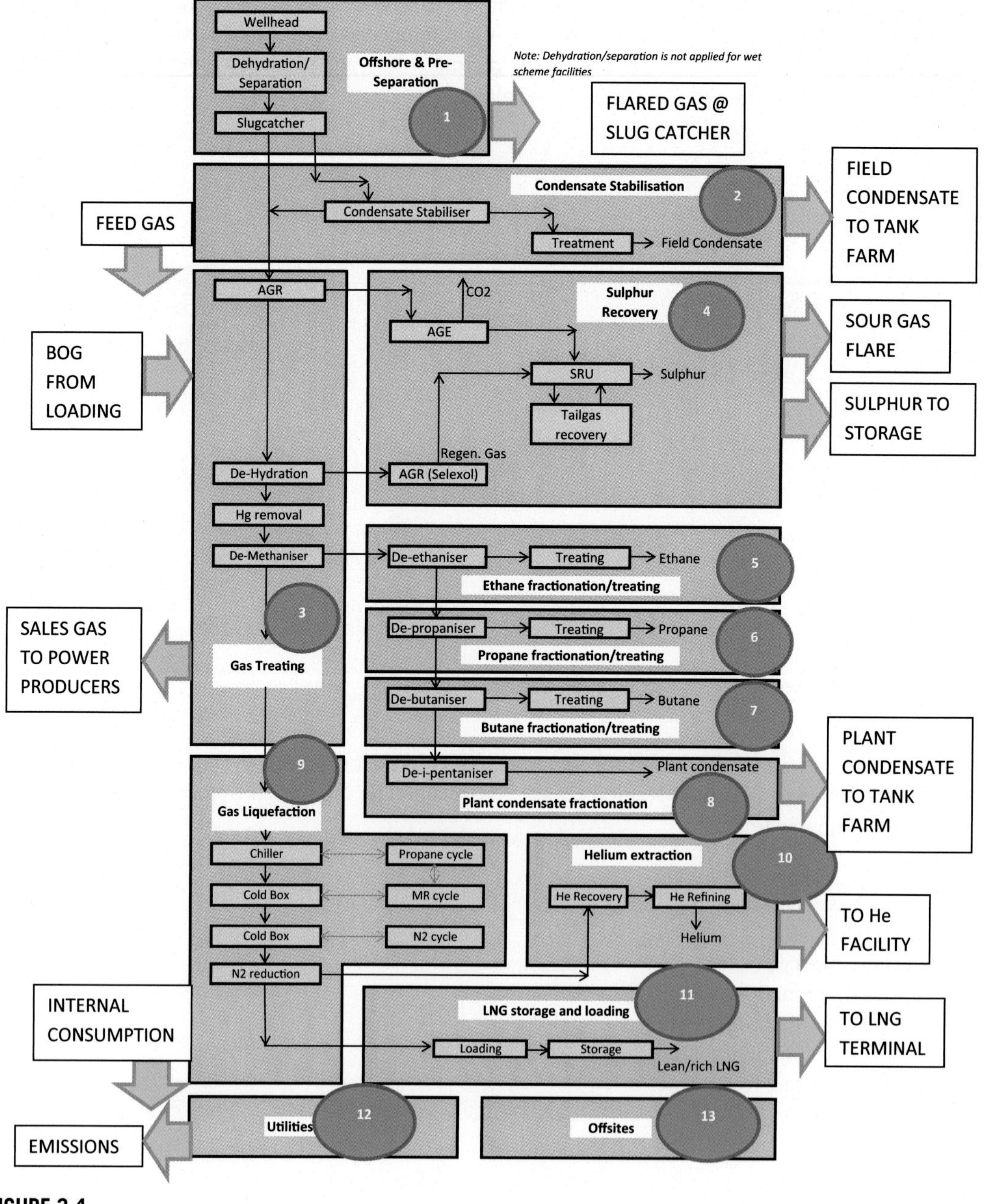

FIGURE 3.4

Liquefied Natural Gas (LNG) gas plant mass/energy balance model.

The inputs and outputs would be readings from mass and volume flow meters, density meters, and gas chromatographs. A component balance can then also be determined. Component balance is a useful tool for Root Cause Analysis of imbalances. Chapter 36, Hydrocarbon Accounting, gives further details and examples.

The key is to capture all hydrocarbons into and out of the Business Unit, as well as internal consumption.

Note that the wellhead gas flow is normally back-calculated from feed gas meters into the gas treating unit and the flared gas at the slug catcher and wellhead.

Specific gas plant models, listed as follows, are discussed in detail in Chapter 36, Hydrocarbon Accounting.

Sigmafine[4]

This is used to identify problems with hydrocarbon balancing and helps to address them.

Spreadsheet Mass Balance Template

This does mass and component balance of the Business Unit using measured process data. Monthly reporting and analysis is carried out.

Spreadsheet Energy Efficiency Model

This evaluates energy efficiency of the Business Unit compared to design, using measured process data. Quarterly evaluation is normally carried out.

Shell Energy Model[5]

KPIs, such as "Plant Specific Power" and " Plant Thermal Efficiency," are determined from the Shell Energy Model depicted in Fig. 3.5.

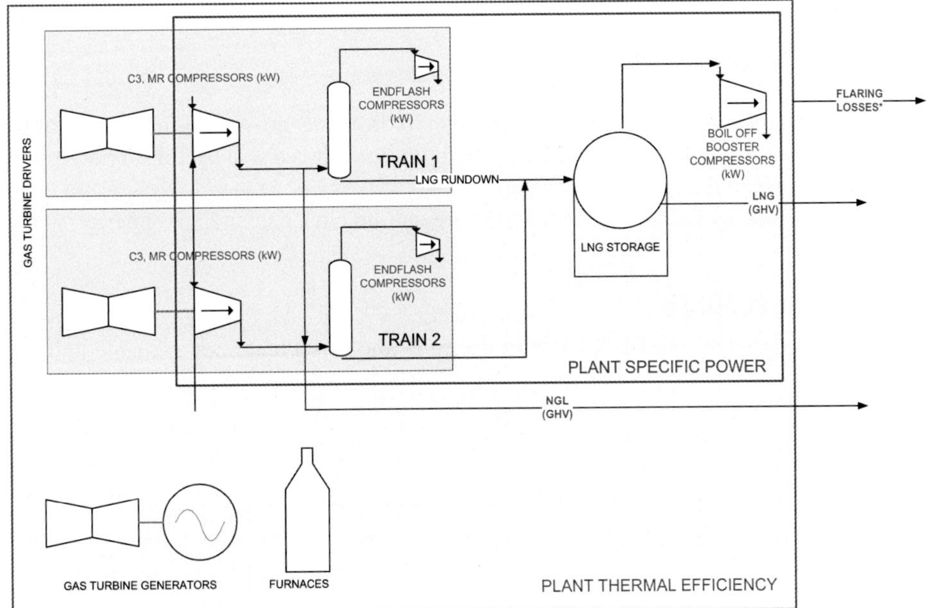

FIGURE 3.5

Plant specific power and plant thermal efficiency model.

INDUSTRY-WIDE ENERGY MODEL

Sankey diagrams map all producers and consumers of electricity, gas, etc., in an industrial complex or country. Thickness of the line relates to quantity of energy and color relates to type of energy. These can be used to optimize energy production and consumption.

Sankey diagrams for all countries are available on the International Energy Agency website.[6]

Examples of the value of Mass/Energy models are discussed in Chapter 36, Hydrocarbon Accounting.

3.5 PROCESS SIMULATION MODELS

REFINERY

The following are some examples.

Linear Programming

Models, such as PIMS,[7] from Aspentech are used.

Aspen PIMS optimizes feedstock evaluation, product slate production, and plant design and operations for enhanced efficiency and profitability.

These are used for planning optimum production from a particular crude slate.

Process Simulation

Popular simulation models are as follows:

Aspen HYSYS[8]

Aspen HYSYS is a comprehensive process modeling tool used by the world's leading oil and gas producers, refineries, and engineering companies for process simulation and process optimization in design and operations.

Petro-SIM From KBC[9]

This models the complete refinery in order to simulate, for instance, gross margin $/barrel for particular changes in process conditions based on specific configurations. It is also a useful support tool for engineering major changes to the operating envelope.

The services offered by KBC using Petro-SIM are outlined in Chapter 37, Refineries.

PETROCHEMICAL MODELS

These tend to be proprietary models owned by licensors and as such are not available unless one is a licensee.

LICENSOR MODELS

Licensor models are the patented models of process inventors and developers. Hydrocarbon Processing[10] publishes a full list of petrochemical, gas, and refining processes on a regular basis.

Examples:

- *CLAUS/SCOT Sulfur Plant Process*
- *UOP fluidic catalytic cracking Technology*

3.6 UTILITY MODELS

Utility models don't always capture all energy producers and consumers in a complex. Gas plants are a case in point, where producers and consumers of energy are in both the producing units and utility areas. Fig. 3.6 depicts the conventional view of a separate utilities area and a view of a gas plant with a separate utilities area and utilities fully integrated into the process.

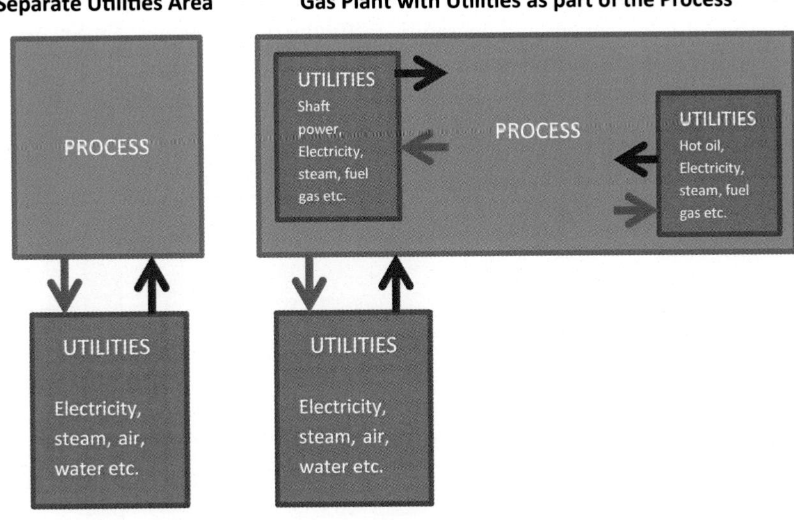

FIGURE 3.6

Utility model comparison.

The following are some examples:

ENERGY MANAGEMENT INFORMATION SYSTEM FROM SHELL GLOBAL SOLUTIONS

This is used for real-time monitoring of site key energy equipment performance.

ASPEN UTILITIES[11]

Aspen Utilities is a model-based, equation-oriented simulation and optimization software tool. It optimizes the purchase, supply, and use of fuel, steam, and power at a refinery, based on process unit energy demands and system constraints caused by equipment or environmental regulations. The software analyses conditions such as supply contract variability, alternative fuel options, optimum loading of steam boiler equipment, motor versus turbine driver decisions, and importing versus exporting of steam, fuel, and power. The software performs the following functions:

- Facilitates optimal planning of utility equipment
- Assists in optimal operation of the utility plant and associated equipment
- Provides real-time information on site-wide energy performance, utility costs, and revenues
- Provides real-time information for use in prioritizing maintenance tasks

An example of the application of this software is outlined in Chapter 37, Refineries.

PROSTEAM FROM KBC[12]

KBC's **ProSteam** application is specifically designed to model steam, fuel, and power systems to help identify both system-wide and equipment-specific operational cost improvements. It aids development of comprehensive utility investment and operating cost reduction strategies. ProSteam builds increasingly complex models from electronic steam tables. This is depicted in Fig. 3.7.

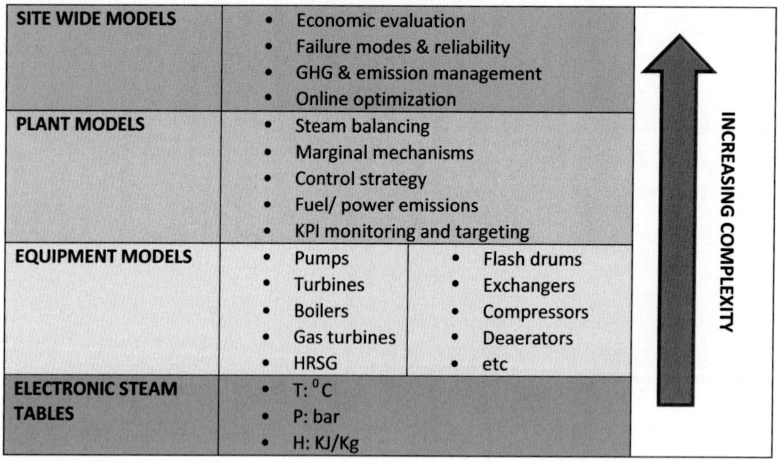

SITE WIDE MODELS	• Economic evaluation • Failure modes & reliability • GHG & emission management • Online optimization	
PLANT MODELS	• Steam balancing • Marginal mechanisms • Control strategy • Fuel/ power emissions • KPI monitoring and targeting	
EQUIPMENT MODELS	• Pumps • Turbines • Boilers • Gas turbines • HRSG	• Flash drums • Exchangers • Compressors • Deaerators • etc
ELECTRONIC STEAM TABLES	• T: ^{0}C • P: bar • H: KJ/Kg	

INCREASING COMPLEXITY

FIGURE 3.7

ProSteam model.

3.7 OPERATING EXPENSE MODELS

SPREADSHEET OPEX TEMPLATE

The Spreadsheet Opex Template is used to evaluate the primary fixed costs (people and maintenance) using Replacement Value or Normal Shift Positions as a common denominator. The template is very similar to the Philip Townsend Associates International (PTAI) data input template. Evaluation should take place on a biannual or annual basis.

This is depicted as in Fig. 3.8.

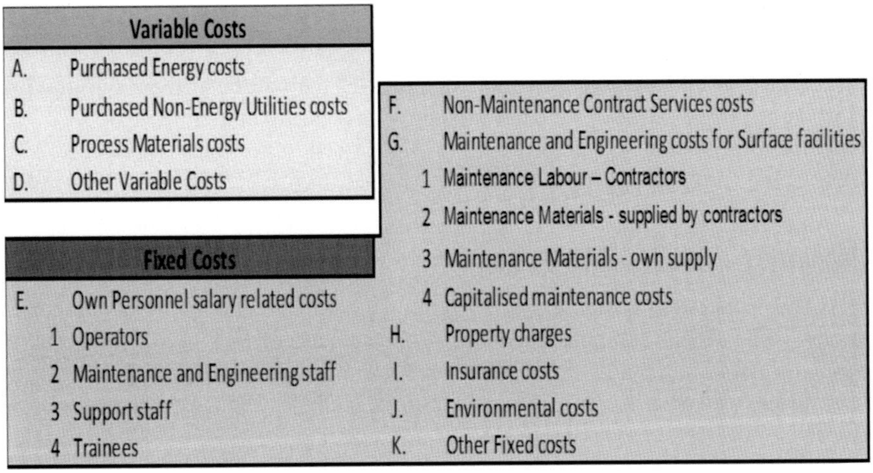

Variable Costs	
A.	Purchased Energy costs
B.	Purchased Non-Energy Utilities costs
C.	Process Materials costs
D.	Other Variable Costs

Fixed Costs	
E.	Own Personnel salary related costs
1	Operators
2	Maintenance and Engineering staff
3	Support staff
4	Trainees
F.	Non-Maintenance Contract Services costs
G.	Maintenance and Engineering costs for Surface facilities
1	Maintenance Labour – Contractors
2	Maintenance Materials - supplied by contractors
3	Maintenance Materials - own supply
4	Capitalised maintenance costs
H.	Property charges
I.	Insurance costs
J.	Environmental costs
K.	Other Fixed costs

FIGURE 3.8

Opex model.

An application of the spreadsheet is discussed in BOX 25.4, Example of Cost Comparison Using Replacement Value as Denominator.

This is a powerful tool during the review of the company's Business/Work Program and Budgets and for setting annual expenditure targets. However, maintenance and operations target setting has to be done in conjunction with the setting of other targets, such as Utilization and Availability.

Opex modeling is discussed further in Chapter 21, Finance.

3.8 BENCHMARKING MODELS

The primary measurements for comparison are focused around **product and finance as follows**:

1. Refineries: gross margin and net margin in $/barrel
2. Gas plants: $/ton or $/million BTUs
3. Petrochemical plants: $/ton
4. Power plants: $/MWH

REFINERY

Profile II From Solomon Associates

The specific refinery model is built for ongoing evaluation of performance of benchmarked indicators. Key elements are mass and energy balance. Mass input is downloaded from tags indicated, for example, in Fig. 3.2. Monthly reports are generated for profit ($/bbl) and other KPIs. It has limited simulation capability.

The model is constructed using the unique features of the particular refinery.

PETROCHEMICAL MODELS

Consultants, such as PTAI and Solomon Associates, have benchmarking models for most petrochemical industries. The consultant enters the unique features of the particular plant into their model, and then the client submits annual data to the consultant for processing in their models.

The requirement for a Standard Reference Gas is discussed in Section 17.7, Energy.

3.9 SUMMARY

The Business Unit depicts the complete Value Addition System. To attempt to model the performance of the Business Unit, various computer modeling tools are used. However, it is important to note that each modeling tool has limitations.

Business models tend to focus on high-level performance by extracting and consolidating information from various sources.

Mass and energy balance models rely on fiscal metering and are therefore the most reliable for accounting purposes. Nonetheless, manual and/or automatic cross-checks are required.

Process simulation models reflect the operation of the physical assets to assess value addition and are useful tools to predict the behavior of the process plant. These could be fed back into the process Distributed Control System, as is done with APC. Licensors generally run their proprietary simulation model in-house and then give the results to the client company, whereas companies tend to run simulations on generic models.

Utility models reflect the energy consumption of the plant, and recommendations for improvement in energy consumption are then derived.

Operating expense spreadsheets are the basic input sheets for submission to benchmarking consultants. However, they can also be used as a standard template to summarize submissions of the Business Plan and Budget to shareholders who are then able to compare the operating costs of their shareholdings in various similar ventures.

Benchmarking models are not normally accessible to clients. However, Solomon's Profile II can be used to monitor performance in between benchmarking submissions.

The capability limits of modeling need to be taken into account for decision-making.

REFERENCES

1. Business Performance Solutions. *The Forrester Wave: Q4*. 2009. https://www.forrester.com/report/The+Forrester+Wave+Business+Performance+Solutions+Q4+2009/-/E-RES48391. Comparative study of performance software.
2. Enterprise-wide financial control SAP business planning and consolidation application. www.sap.com/solutions/performancemanagement. Underlying finance computer system model.
3. SAP strategy management. www.sap.com/solutions/performancemanagement.
4. Sigmafine from OSI. www.osisoft.com. Mass balance model.
5. Wilson J, van der Wal G. *Leading the pack for efficiency and environmental performance at a base load LNG plant*. Thailand. GasTech; 2008. Shell plant thermal efficiency and plant specific power model.
6. International Energy Agency (IEA). www.iea.org/Sankey.

7. PIMS from Aspentech. http://www.aspentech.com/products/aspen-pims/.
8. HYSYS from Aspentech. http://www.aspentech.com/products/aspen-hysys/.
9. Petrosim from KBC. www.kbcat.com.
10. Process handbooks hydrocarbon processing. http://www.hydrocarbonprocessing.com/ProcessHandbooks.html.
11. Aspen utilities from Aspentech. http://www.aspentech.com/products/aspen-utilities-operations.aspx.
12. ProSteam™ KBC energy and utilities software for modelling steam and power. http://www.kbcat.com/energy-utilities-software/prosteam.

CHAPTER

MANAGEMENT SYSTEMS DETERMINATION AND REQUIREMENTS

4

4.1 INTRODUCTION

The elimination of functional silos is of prime importance since silos tend to duplicate effort, and can lead to empire building, especially when management's compensation relates to staff numbers.

The value addition of the product or service needs to be seamless. Accordingly, the focus of all staff must be two-pronged, both satisfying the customer needs and aligning with the direction statement of the company. Departments therefore need to align with Systems as far as possible. Performance measurement and continual improvement are required to be part of daily life.

This chapter describes an "open" approach for tailor-making your own Management System using "first principles" to ensure total buy-in from all employees of the company.

This chapter further describes how Systems and Processes combine with a focus on the Core Management System as the "value addition" part of the Business Unit. Three primary groups of Management Systems (Core, Strategic, and Support) are outlined and the benefits of a Systems approach explained.

Note on terminology:

> The words **"System," "Management System," Business System, and Business Management System** are defined in Section 4.3, and are synonymous. A **System** must contain all of the key elements described in Section 4.5.
> However, **"system"** with a lowercase "s" is generic and does not necessarily comply with the previously mentioned definition. For example, it could refer to a computer system.

The "before and after" are depicted in Fig. 4.1. The "before and after" is discussed further in Chapter 8, Framework.

Performance Management for the Oil, Gas, and Process Industries. http://dx.doi.org/10.1016/B978-0-12-810446-0.00004-9

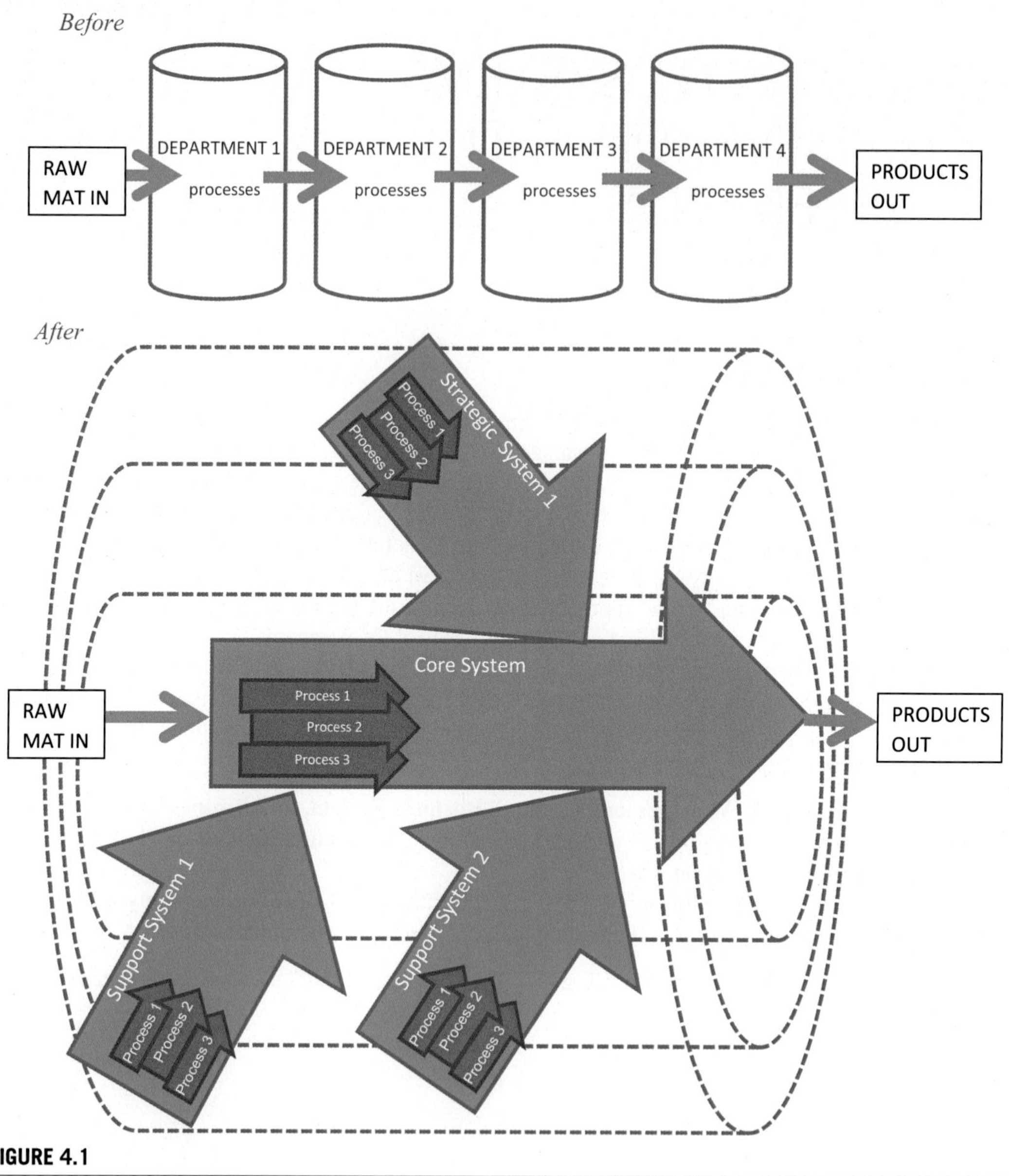

FIGURE 4.1

Functional silos versus Systems.

4.2 ANALOGY AND AN EXAMPLE OF A SYSTEM

Many organizations use the terms systems and processes interchangeably/synonymously or have different definitions. For the purposes of clarifying the approach taken in this book, I will use the analogy of a bus to describe the differences between Management Systems and Processes.

A bus has the primary ***Objective*** of transporting passengers from A to B, much like a ***Management System*** having an ***Objective.*** The bus or ***Management System*** consists of parts or ***Processes,*** which work together to achieve the ***System Objective***. Each part (engine, transmission, wheels, etc.) or ***Process*** is critical for the overall functioning of the ***System*** in that the transport of passengers cannot occur without the interaction of the parts. The bus requires a driver ***(System Owner)*** to steer the bus along the chosen route and ensure that it arrives at point B (the ***Objective***).

Fig. 4.2 depicts this analogy.

FIGURE 4.2

Bus analogy.

A good example of a System is an Environmental Management System, which is outlined as follows:

The Environmental Management System's primary objective is to minimize the detrimental impact that the Business Unit has on the environment by promoting environmental awareness and responsibility. The System comprises of a number of concurrent and sequential processes such as:

- Monitor Environmental Performance
- Conduct Environmental Impact Assessments
- Report Environmental Performance to stakeholders
- Comply with Environmental Requirements, etc.

Together, these processes are required to ensure that the Systems' objectives are achieved. The System would be less effective if any one of the above processes were eliminated.

4.3 DEFINITIONS

MANAGEMENT SYSTEM

The Oil and Gas Producers (OGP)[1] definition is as follows:

A structured and documented set of interdependent practices, process, and procedures used by the managers and the workforce at every level in a company to plan, direct, and execute activities.

A Management System is made up of a collection of Processes, which together result in the achievement of a business objective or set of objectives. ISO 9001 defines it as a set of interrelated or interacting elements (particularly processes).

A Management System:

- exists to achieve business objectives;
- consists of a number of sequential and/or concurrent Processes;
- may span multiple functional departments; and
- has a logical beginning and end, which is not dictated by the boundaries of a department.

CORE MANAGEMENT SYSTEM

- Represents the Core or Primary Business of the organization
- Links directly to the reason for the existence of the business and thus adds value to the business
- Develops, produces, sells, and distributes an entity's products and services
- Does not necessarily follow traditional organizational or functional lines, but reflects the grouping of related business activities

STRATEGIC MANAGEMENT SYSTEMS

- Define how management sets direction, monitors the external environment, assesses strategic implications, allocates resources, and aligns strategic business objectives with System objectives (*example: Strategic Planning*)

SUPPORT MANAGEMENT SYSTEMS

- Provide appropriate support resources to the other business processes (*example: Maintenance*)

PROCESS

- A set of interacting activities, which transform inputs into outputs

4.4 BOUNDARIES OF A SYSTEM

The boundaries of a System could be based on the following:

1. Product
2. Physical asset
3. Geographical area
4. Discipline
5. Paper

Asset and Geographical Boundaries are normally the same as that of the **Business Unit** as discussed in Chapter 2, The Business Unit. However, the **Business Unit** boundary is referenced to the Asset owner, and the **Core System** boundary is referenced to the transfer of product or delivery of service. Box 4.1 outlines Liquefied Natural Gas and Refinery cases.

BOX 4.1 EXAMPLES OF ASSET AND CORE SYSTEM BOUNDARIES THAT ARE DIFFERENT

Liquefied Natural Gas Case

An integrated Liquefied Natural Gas (LNG) company is owned by a number of Asset Owners, but operated by an Operating Company.

The **Asset Business Unit** boundaries are the reservoir wellhead and the loading arms to the LNG tanker.

The **Core System** boundaries are the reservoir wellhead and the **unloading arms** in the destination country. The LNG tankers and the storage facilities in the destination country are owned by others. Royalties are based on the quantity and heating value at the **unloading arms**.

Refinery Case

A refinery delivers product to a client's depot (storage and delivery facility) via a pipeline owned by a third party.

The **Asset Business Unit** boundaries are the crude fiscal meters at the refinery fence and entry to the pipeline at the refinery fence.

The **Core System** boundaries are the crude fiscal meters at the refinery fence and the fiscal meters into the client depot at the end of the pipeline.

Examples of System boundaries are as follows:

Product Boundary

A Laboratory Management (Support) System could include inputs of product samples for analysis and outputs could be the analysis reports (in the Laboratory Information Management System).

Geographical Boundary

A Production and Storage ***Core System*** *of a Refinery could be defined from the Crude Fiscal Meters at the Refinery fence, through the crude storage tanks, process units, product storage tanks, and pipelines to the customer boundary fence where the Product Fiscal Meters are placed.*

Discipline Boundary
An Environmental Management (Support) System input could list all the environmental aspects of an asset, and outputs could be all the mitigation actions completed.
Paper Boundary
A Maintenance (Support) System input is "a maintenance request" and the output is "a completed work order."

4.5 THE KEY ELEMENTS OF A SYSTEM

These are essential for the existence of a Management System. They are as follows:

1. **Objective**
2. Scope
3. Policies, Regulations, and Standards
4. **Processes**
5. Responsibility
6. **Goals and Key Performance Indicators (KPIs)**
7. **Targets**
8. **Strategies and Plans**

These are expanded as follows.

OBJECTIVE

The reason for the existence of the System. It must be linked to higher-level business objectives, mission, and vision (the company direction statement).

SCOPE

This gives clear boundaries for the System: product, asset, geographical, discipline, paper, etc. (see Section 4.4).

POLICIES, REGULATIONS, AND STANDARDS

Compliance with Company/Government/International policies, regulations, and standards is to be assured. These are generally regarded as **external constraints.**

PROCESSES

The detailed documented contents of the System: procedures, operating manuals, work instructions, forms, records, software, etc.

OWNERSHIP (RESPONSIBILITY)

The owner should be a Division/Section Head who has the MOST influence over one or more of the following for the whole System:

1. Outputs of System
2. System Budget and Expenditure
3. Application of Company/Government/International Policies, Regulations, and Standards

The owner must be able to manage the day-to-day administration of the System. Department managers are normally Sponsors of Systems within their departments.

GOALS AND KEY PERFORMANCE INDICATORS

These must be linked to higher-level business KPIs, objectives, mission, and vision.

TARGETS

Numerical measurements of each KPI are required with target values related to time.

STRATEGIES, PLANS, AND PROGRAMS

These are required for actions that affect the KPIs.

Descriptions for each of the Key Elements of a System are shown in Table 4.1.

Table 4.1 Key Elements of a System

Element	Description
Objective	Generic statement to be in alignment with the corporate direction statement
Scope	Clear boundaries of the **Management System**
Policies, regulations, and standards	Guidelines and restraints (external and internal) (Including product specifications)
Processes	Procedures/software
Responsibility	Clear delegation of responsibility
Key Performance Indicators (KPIs)/Targets	What we wish to achieve and what targets have been agreed
Strategies and plans	Actions to ensure targets are achieved

Fig. 4.3 is an example of a Mind Map used to develop a Refinery Integrated Management System (IMS). It identifies all the Key Elements of a System.

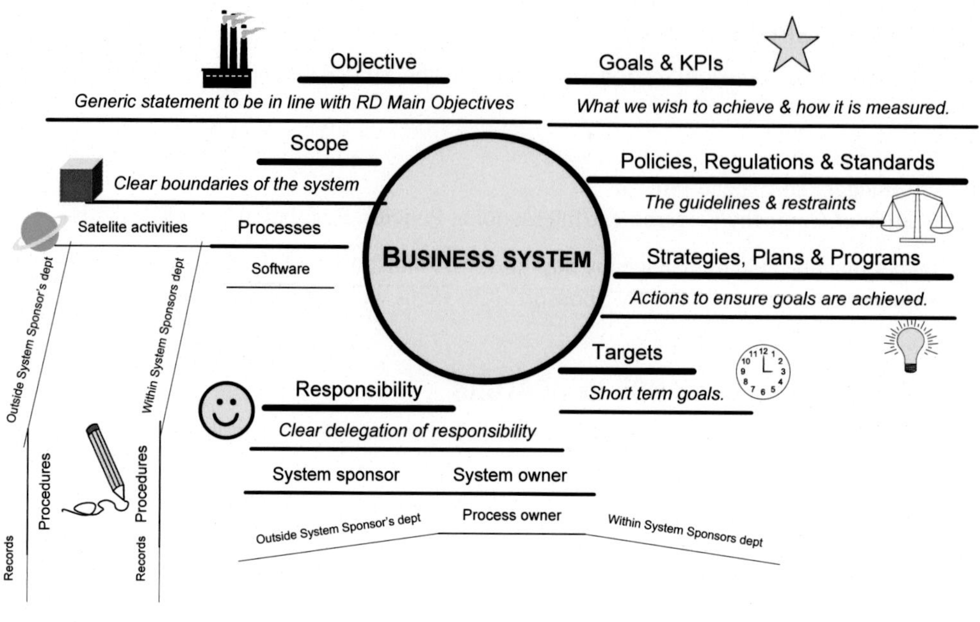

FIGURE 4.3

Business Management System mind map example.

The Systems and Processes of a Business Unit could be hosted on the Company intranet. Box 4.2, shows an example of the Elements of the System as displayed on an intranet system.

BOX 4.2 CORE MANAGEMENT SYSTEM INTRANET FRONT PAGE (SEE SCREEN VIEW BELOW)

- **Name:** Production Planning Management System
- **Objective:** as shown on System Front Page
- **Key Performance Indicators (KPIs)/Targets:** see left of System Front Page, "*PP Goals and KPIs*"
- **Scope:** as shown on System Front Page
- **Policies, Regulations, and Standards:** shown on top right of Integrated Management System (IMS) Homepage, "*Product Specs*": see Box 4.5
- **Strategies and Plans:** see left of System Front Page, "*Continual Improvement*"
- **Processes:** see bottom of System Front Page: Procedures and Forms
- **Responsibilities:** as shown on System Front Page

BOX 4.2 CORE MANAGEMENT SYSTEM INTRANET FRONT PAGE—cont'd

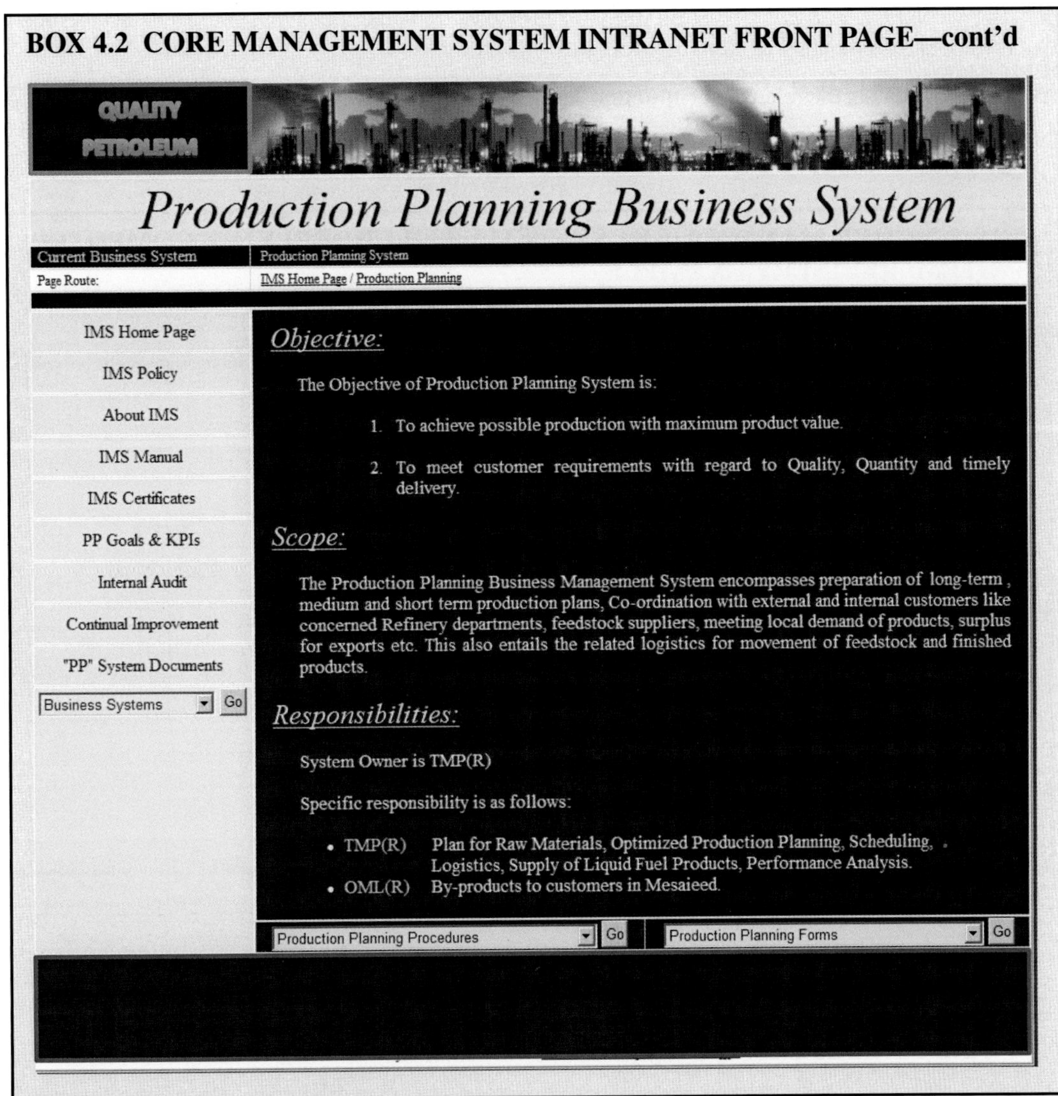

Box 4.2 shows one of two Core Management Systems chosen by the implementation team. In creating the Core Systems, the logic was as follows:

1. The first Core System, **"Production Planning"** (shown in the screenshot in Box 4.2), entailed ensuring that the raw materials are available for Production and that the products are removed from the Storage tanks in a timely manner.

2. The second Core System, **"Production and Storage,"** entailed the physical refinery from crude receipt at the boundary to shipping from the refinery tanks.

   Box 4.3, depicts the related **"Production and Storage"** System example.

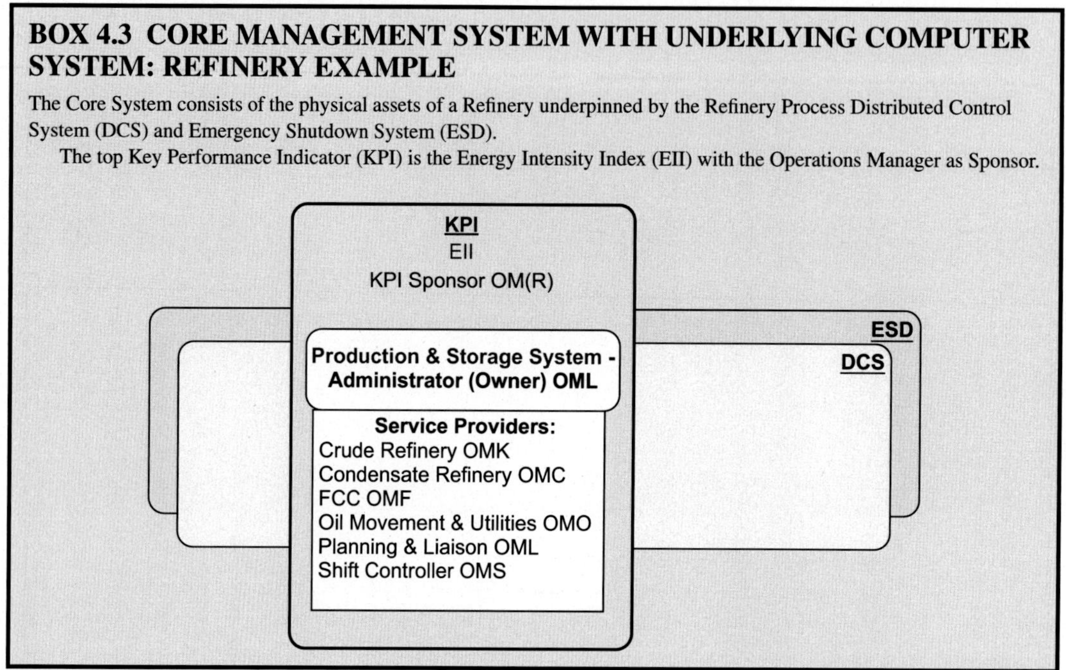

BOX 4.3 CORE MANAGEMENT SYSTEM WITH UNDERLYING COMPUTER SYSTEM: REFINERY EXAMPLE

The Core System consists of the physical assets of a Refinery underpinned by the Refinery Process Distributed Control System (DCS) and Emergency Shutdown System (ESD).

The top Key Performance Indicator (KPI) is the Energy Intensity Index (EII) with the Operations Manager as Sponsor.

4.6 DETAILED COMPARISON BETWEEN MANAGEMENT SYSTEMS AND PROCESSES

There is a clear distinction between Systems and Processes. The main differences are in the following:

1. Description
2. Objective
3. Targets

Table 4.2 details the differences between **Management Systems** and **Processes** with examples.

Table 4.2 Management System and Process comparison

Management System	**Process**
Description	
A collection of processes that together result in the achievement of a business goal as expressed in the **Business Unit's** direction statement (vision, Mission, etc.). *Examples:* • *Production and Storage System* • *Environmental Management System*	A sequence of logically grouped activities. A Process in itself cannot achieve the business goal(s) of the System, but is critical to the achievement of the goal(s). *Examples:* • *Startup and Shutdown of process unit X* • *Conduct Environmental Impact Assessments*
Objective	
Each System requires a statement of objectives. The objective statement communicates the objectives and intention(s) of the System and details guidelines to be followed in support of the objectives. *Example:* *Objective Statement: Compliance with Integrated Management System (IMS) Policy* *Objectives:* • *Supply customers with products that meet the latest applicable Product Standard* • *Minimize rework or downgrading of finished and intermediate products*	Objective statements are not mandatory for processes. The System objective statement acts as an "umbrella" for all the processes within the System.
Targets	
A System target is defined for the System as a whole and should ideally be a composite of the specific targets of the processes that constitute the System. This ensures alignment of Process targets with System targets, goals, and objectives. *Examples:* • ***Goal****: Reduce Emission* • *Key Performance Indicator (KPI): Carbon Emission Index* • ***Target****: 100* • ***Goal****: Achieve Study Quartile 2 for Europe, Africa, and Middle East in Solomon Benchmarking* • ***KPI****: Volumetric Expansion Index* • ***Target****: >90*	At a lower level, a more detailed target may be defined for each Process within a System. *Example:* • ***Goal****: Reduce particulate emission from fluidic catalytic cracking unit* • *KPI: MT/day* • ***Target****: <5*

4.7 PRIMARY GROUPS OF SYSTEMS

Systems, as part of the "Control Organization Assets Systems" Governance model described in Chapter 8, Framework, depicts **"How we do things."**

They are grouped as follows:

1. **Strategic:** How management sets direction, monitors the external environment and assesses strategic implications, allocates resources, and aligns Strategic Business Objectives with System Objectives.
2. **Core:** The Systems that develop, produce, sell, and distribute an entity's products and services. These Systems do not follow traditional organizational or functional lines, but reflect the grouping of related business activities.

3. **Support:** Management Systems that provide appropriate support resources to the other Systems.

The **Strategic Systems** are discussed as follows:

1. "Governance" is discussed in Chapter 8, Framework
2. "Strategic Planning and Budgeting" is discussed in Chapter 5, The Business Cycle
3. "Risk" is discussed in Chapter 11, Risk Management
4. "Allocation of Resources: Human Capital" is discussed in Chapter 20, Human Capital
5. "Allocation of Resources: Financial" is discussed in Chapter 21, Finance
6. "Allocation of Resources: Hydrocarbons" is discussed in Chapter 36, Hydrocarbon Accounting
7. Health Safety Environment Quality/Asset Integrity is discussed in Chapter 20, Human Capital and Chapter 22, Physical Asset Performance
8. Investment/Affiliate Oversight is discussed in Chapter 32, Business Relationships and Chapter 33, The Opportunity Life Cycle.

The "steady state" **Core Systems** are under "supply chain" where "investment" and "resource life-cycle" are "cradle to grave," intercepting the supply chain at **operate/produce/process**.

The **Support Systems** are divided into "certification," "critical," and "other." Examples are given in Section 4.9.

These can be shown in an Oil and Gas environment as per Fig. 4.4.

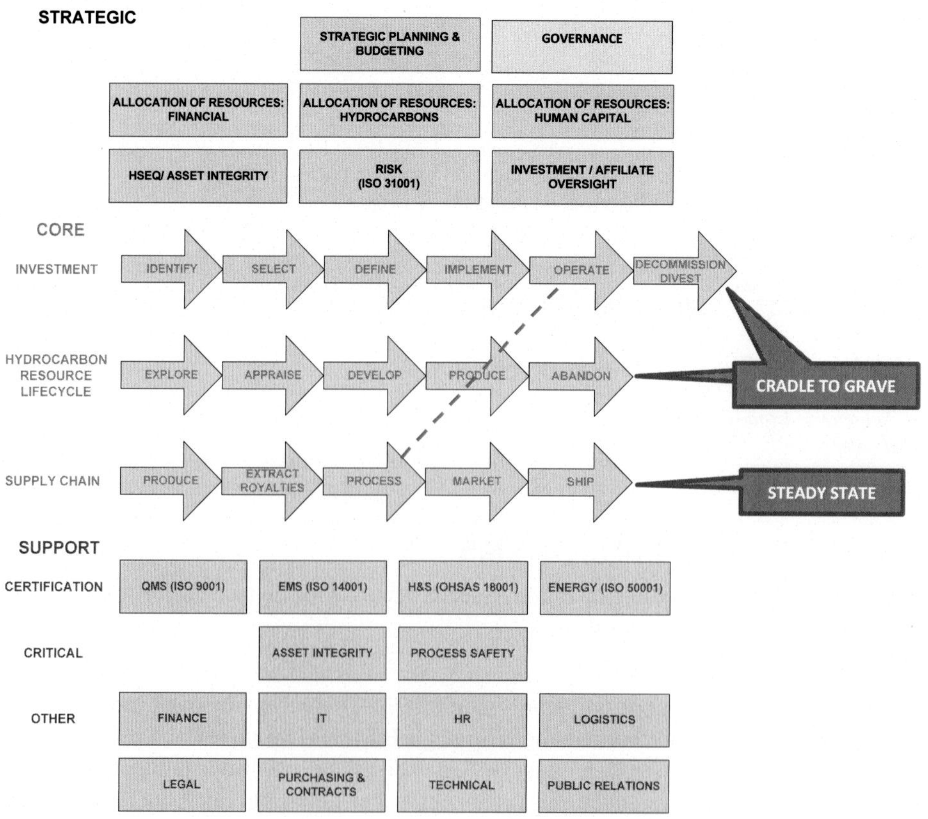

FIGURE 4.4

Strategic, Core, and Support Systems: oil and gas company example.

The Core System for the Cement industry could be depicted as shown in Fig. 4.5.

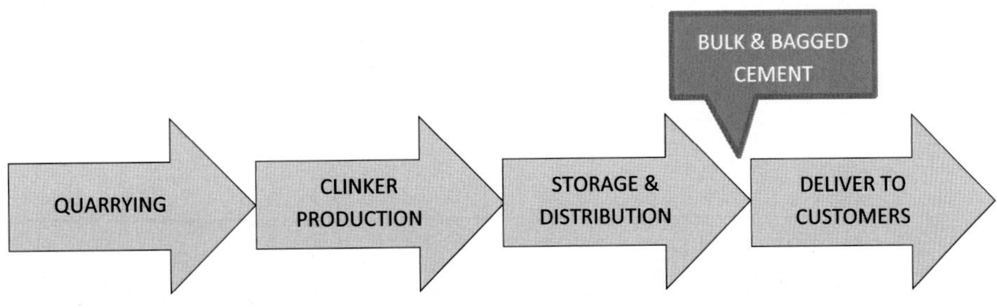

FIGURE 4.5

Cement industry Core System.

The Core System for the Mining industry could be depicted as shown in Fig. 4.6.

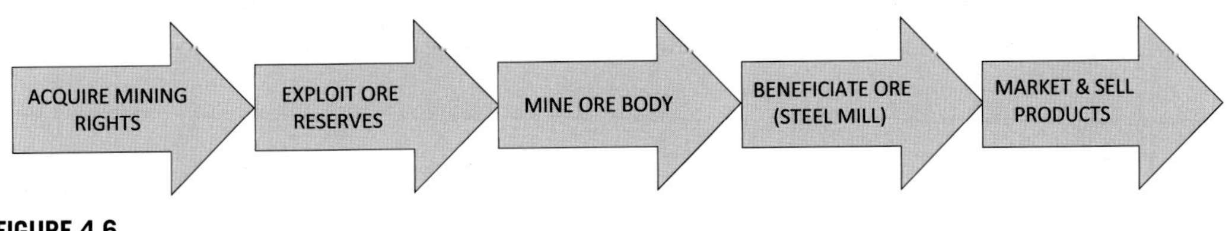

FIGURE 4.6

Mining industry Core System.

The Core System for the Aluminum industry could be depicted as shown in Fig. 4.7. However, there is a major recycling component.

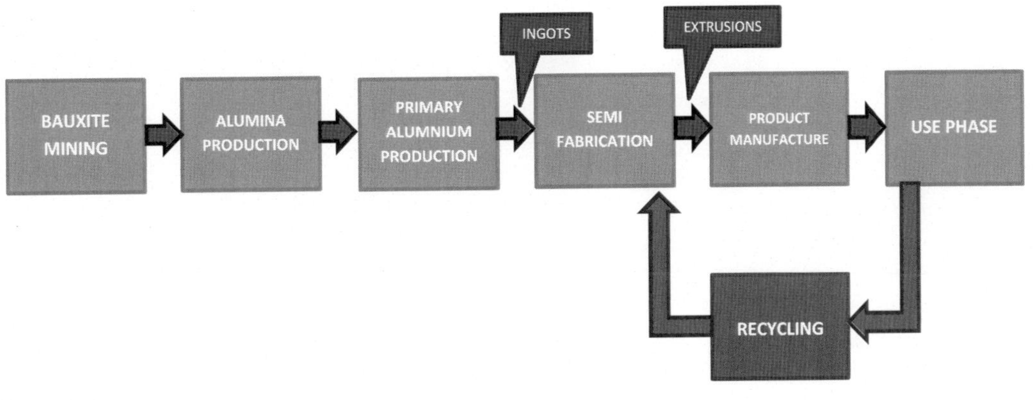

FIGURE 4.7

Aluminum production Core System.

The Strategic and Support Systems for the above are essentially the same as the Oil and Gas example.

4.8 UNDERLYING COMPUTER SYSTEMS

Most Management Systems have underlying computer systems.

An example of a Core System for a Refinery showing the top KPI, software, and responsibilities is shown in Box 4.3.

An example of Support Systems in a Department, showing ownership and support software, is shown in Box 4.4.

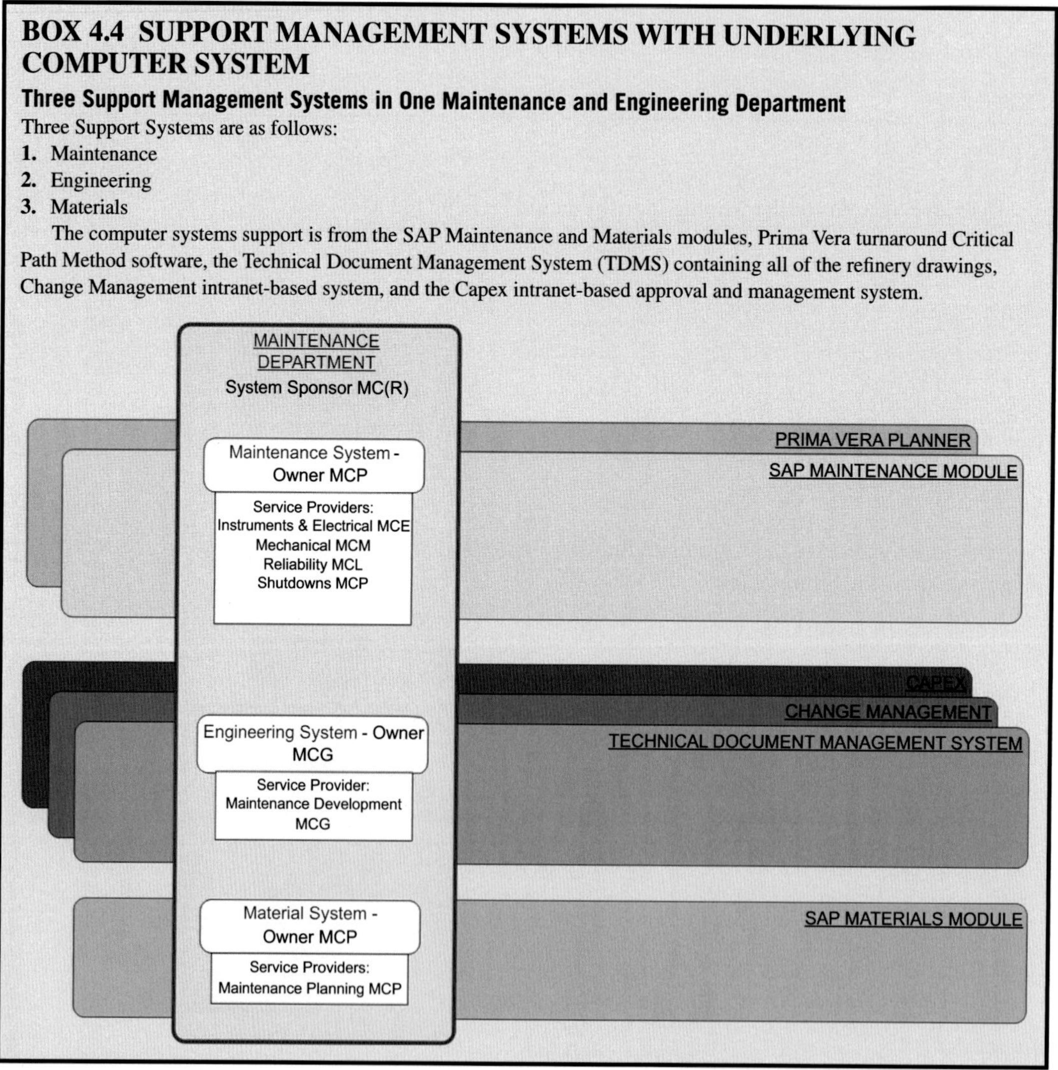

BOX 4.4 SUPPORT MANAGEMENT SYSTEMS WITH UNDERLYING COMPUTER SYSTEM

Three Support Management Systems in One Maintenance and Engineering Department

Three Support Systems are as follows:

1. Maintenance
2. Engineering
3. Materials

The computer systems support is from the SAP Maintenance and Materials modules, Prima Vera turnaround Critical Path Method software, the Technical Document Management System (TDMS) containing all of the refinery drawings, Change Management intranet-based system, and the Capex intranet-based approval and management system.

4.9 SYSTEM EXAMPLES

Some Systems are described as follows.

INTEGRATED MANAGEMENT SYSTEM (CERTIFICATION TO VARIOUS ISO STANDARDS)

The IMS should cover the whole Business Unit for compliance with the following:

- ISO 9001, Quality Management
- ISO 14001, Environmental Management
- ISO 45001, Occupational Health and Safety Management (formerly OHSAS 18001—*Occupation Health and Safety Assessment Series*)
- ISO 31001, Risk Management
- ISO 50001, Energy Management
- ISO 55001, Asset Management

SUPPORT SOFTWARE

Section 4.13, gives details of commercially available software.

Box 4.5 gives an example of an IMS for certification to various ISO standards.

BOX 4.5 REFINERY INTEGRATED MANAGEMENT SYSTEM

This is the Integrated Management System (IMS) of a 140,000 barrel-a-day oil refinery. The refinery started with obtaining certification for its laboratory to ISO/IEC 17025: 2005, "General requirements for the competence of testing and calibration laboratories," and then obtained ISO 9001, "Quality Management." At the time of writing, it was applying for certification to ISO 14001, "Environmental Management," and ISO 45001, "Occupational Health and Safety Management," It will soon apply for ISO 50001, "Energy Management."

The IMS is shown pictorially as follows.

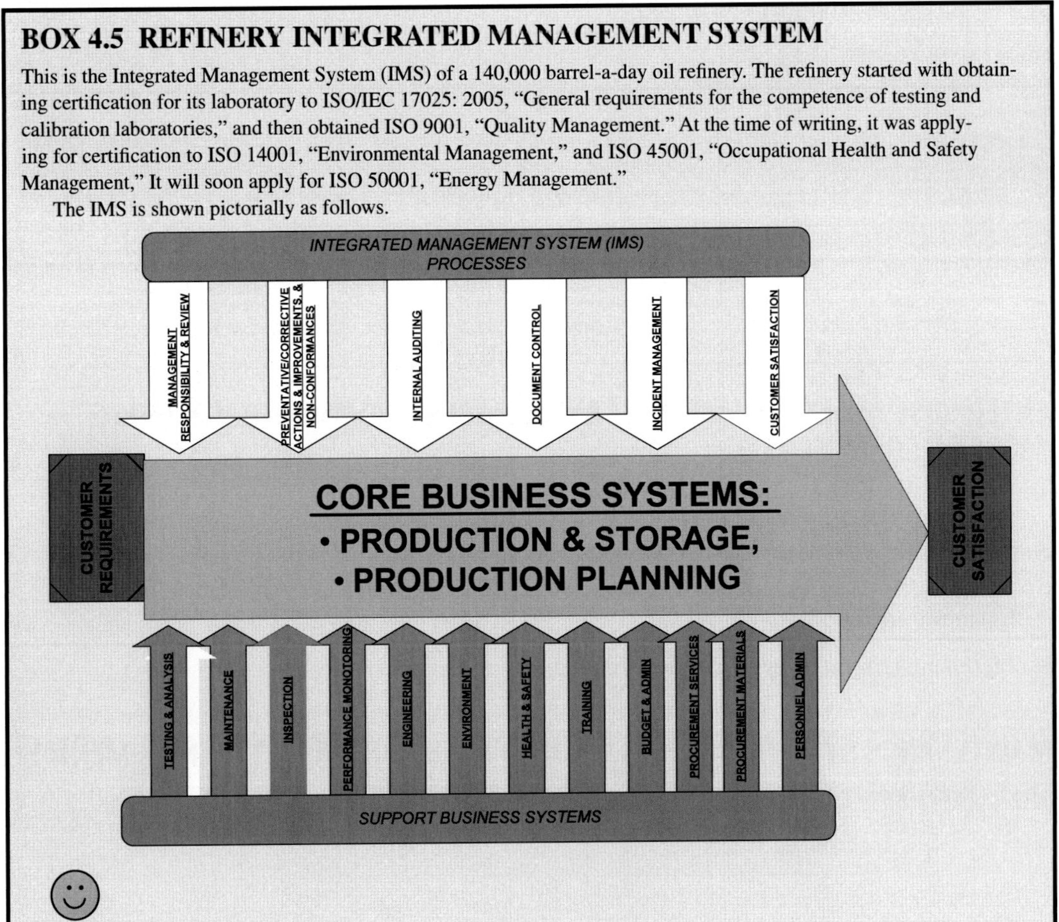

Continued

BOX 4.5 REFINERY INTEGRATED MANAGEMENT SYSTEM—cont'd

The intranet-based homepage of the Refinery IMS is as follows.

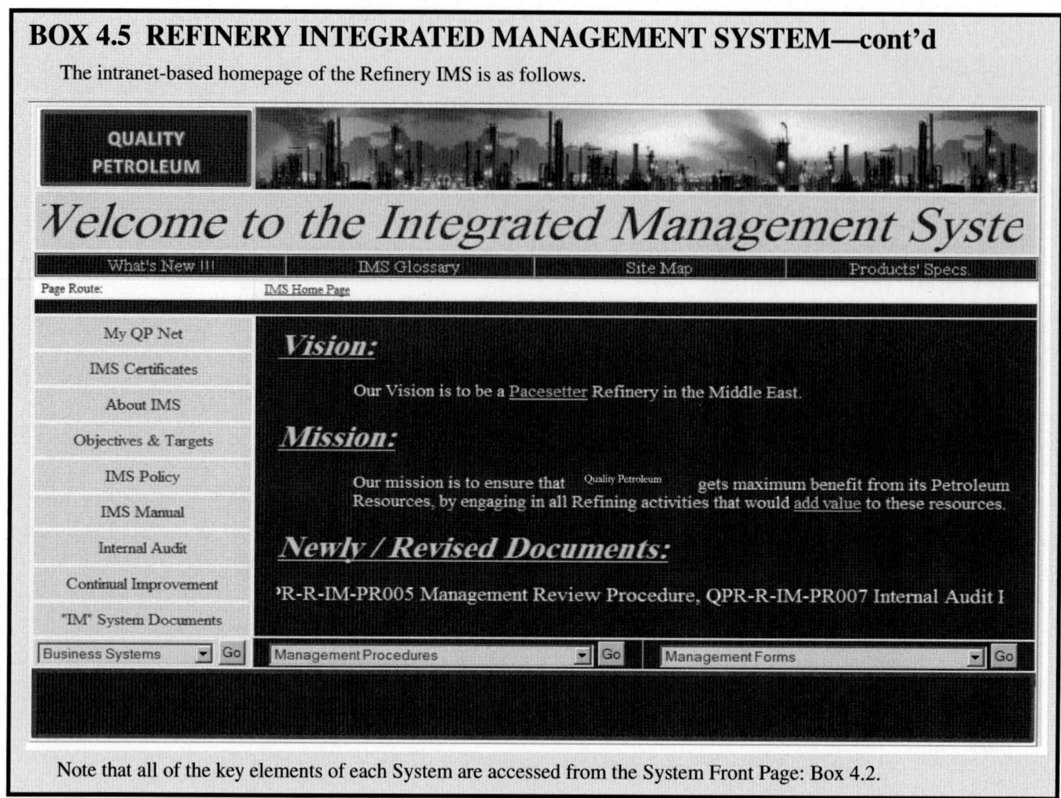

Note that all of the key elements of each System are accessed from the System Front Page: Box 4.2.

ENTERPRISE ASSET MANAGEMENT SYSTEM (CERTIFICATION TO ISO 55001, "ASSET MANAGEMENT")

The System should cover the requirements of ISO 55001.

Support Software

Requirements are discussed in detail in Section 22.9, Asset Performance Management Software. Examples are as follows:

- Meridium, Orksoft, Lloyds, Capstone, etc.

MAINTENANCE SYSTEM (CERTIFICATION TO ISO 55001 "ASSET MANAGEMENT")

The System should cover aspects of the requirements of ISO 55001.

Computerized Maintenance Management Systems (CMMS) are now applied to all instances where the work orders are the heart of the System.

Box 4.4, includes a Maintenance System.

Support Software

Examples are as follows:

- SAP Maintenance and Materials Modules, Oracle, Maximo
- Online Analytical, such as GE Bently Nevada; used to monitor the mechanical condition of rotating equipment

PROCESS SAFETY MANAGEMENT SYSTEM (CERTIFICATION TO ISO 45001, "OCCUPATIONAL HEALTH AND SAFETY MANAGEMENT," FORMERLY OHSAS 18001)

The System should cover the requirements of ISO 45001.

Process Safety (PS) Management must encompass the following:

1. Behavior-Based Safety
2. PS
3. Occupational Health and Safety

The main elements expected in the System are as follows:

- PS Leadership
- The PS Management System processes, procedures, records, etc.
- PS Culture
- Clearly defined expectations and accountabilities for PS
- Support for line management
- PS auditing
- Board monitoring
- Leading and lagging performance indicators for PS
- Aspiring to be an industry leader

Application of the system must include the following:

- Shift change cycles
- Alarm management
- Permit to work (PTW)
- Management of change
- Empowering staff

Support Software

Support software examples include the following:

- *Incident Management: SAP,* etc.
- *ePTW: Honeywell,* etc.

ENERGY MANAGEMENT SYSTEM (CERTIFICATION TO ISO 50001, "ENERGY MANAGEMENT")

The System and energy model should be in line with ISO 50001, "Energy Management."

Support Software

Software includes the following:

- Shell Global Solutions: online monitoring
- KBC: energy modeling
- Aspentech : energy modeling

EMPLOYEE PERFORMANCE MANAGEMENT SYSTEM

Primary components of the System must include management of the following:

1. Behavioral skills
2. Technical skills

Support Software

Software includes the following:

- SAP HR and Training modules, etc.
- Competency Management software

PROJECT MANAGEMENT (IN LINE WITH ISO 21500:2012, "GUIDANCE ON PROJECT MANAGEMENT," AND ANSI 99-001-2008, "PROJECT MANAGEMENT BODY OF KNOWLEDGE")

Capital Projects should be managed in line with ISO 21500:2012, ANSI 99-001-2008, "Project Management Body Of Knowledge" and, where possible, using International Oil Company (IOC) Design and Project Management Manuals and Basic Design Data Sheets (BEDS).

Support Software

Software includes the following:

- SAP Project Management Module, etc.
- Oracle Primavera, Open-Plan, etc., for Critical Path Method (CPM)

SHUTDOWN MANAGEMENT

Shutdown modeling using CPM software and links to company Enterprise Resource Management system.

Support Software

Software includes the following:

- Oracle Primavera, Open-Plan, etc., for CPM
- SAP or other Maintenance and Materials Modules

4.10 SYSTEM AND PROCESS OWNERSHIP

Each System is assigned to a System Owner who has primary responsibility for the efficient functioning of the System. The System Owner needs to have the authority to initiate and review the functioning of the System and will therefore be a Division/Section Head from the Functional Department that plays the largest or most dominant role in the functioning of the System. This is **the** key position.

Assigning Business System Owners facilitates accountability for measurement and its outcomes. Making one person accountable for the performance of a given System gives others a clear point of contact for guidance if that System needs improvement or better integration with other Systems. Staff also know who to go to for support or clarification. The System Owner should be someone who has knowledge of the expected baselines of performance or at least understands the data well enough to know when a System is not performing correctly. This responsibility can motivate the System Owner by explicitly assigning him or her with overseeing the performance of his or her group.

The System Owner owns the end-to-end System and carries the following responsibilities:

- Provide System direction by developing System vision, strategy, and objectives.
- Develop and implement System improvement initiatives.
- Define the Processes that are part of the System.
- Monitor System performance.
- Develop and manage policies and procedures related to the System.
- Ensure System adoption.

The System Sponsor is the "Champion" of the System and is normally a senior member of staff: a Department Manager or higher. Departmental Managers, as Sponsors of Systems within their Departments, would carry out the Management Reviews of the Systems for which their Divisional/Section Heads are responsible.

Each Process within the System requires a Process Owner, typically the actioner of the Process, and/or author of the procedure covering the Process.

Names are required for allocation of responsibilities. These should be consistent and minimized. Management System Job Names may be required where the same tasks are performed by a group of positions. *For example, all mechanical fitters would be titled "Fitter: mechanical.*"

Names can also be allocated to roles (normally part time activities). *An example is the name "auditor," which would be used for whoever is auditing a System, regardless of his/her position.*

The Shell approach of giving alphanumeric names to positions in the organization clearly depicts accountability and responsibility. *For example: MCM is the name for Head of Plant Maintenance, and MCM1 is a Senior Engineer in Plant Maintenance Section.*

Detailed responsibilities are shown in Table 4.3.

Table 4.3 Responsibilities

Sponsor/ Department Manager	System Owner	Process Owner	Process User
Responsible for: • management review of Systems owned by his Division/Section heads to ensure agreed actions are taken by the required actioners by agreed deadlines	*Responsible for:* • setting and meeting System goals • compiling the System manual, including performance targets • checking process documentation for processes in their Systems • coordinating System reviews • taking corrective actions on nonconformances in the System • establishing communication channels within the System	*Responsible for:* • setting and meeting Process goals • effective and efficient functioning of the Process • ensuring that all Process users are trained in the Process • taking corrective actions on nonconformances in the Process • designing the Process to satisfy Policies/ Regulations/Standards requirements • consulting Process users and customers during the course of Process design	*Responsible for:* • correct application of Process in accordance with documented procedure • checking Process design to ensure it satisfies technical and functional requirements

Box 4.6 gives a sample a questionnaire for determining Ownership of a System.

BOX 4.6 SYSTEM OWNERSHIP QUESTIONNAIRE

1. How much control do you have over the outputs of the X Management System?

None	Little	Half	Most	All

2. How much control do you have with respect to the Y Department Budget?

None	Little	Half	Most	All

3. How much control do you have with respect to the Y Department Expenditure?

None	Little	Half	Most	All

4. Are you responsible for the application of Company/Government/International Policies, Regulations, and Standards specific to Y Department?

No	Little	Half	Most	Yes

5. Are you in charge of the day-to-day administration of the X Management System?

No	Little	Half	Most	Yes

4.11 SUSTAINABILITY OF THE SYSTEMS APPROACH

AWARENESS

Since company staff come and go, it is essential that both initial and refresher training on the Systems in the Business Unit are provided.

Box 4.7, gives an example of a two-sided A4 flyer used by a refinery to explain the basics of the IMS of the Business Unit. This is issued to every staff member and displayed on all notice boards.

BOX 4.7 REFINERY INTEGRATED MANAGEMENT SYSTEM (IMS) GUIDELINE

REFINING INTEGRATED MANAGEMENT SYSTEM

REFERENCE GUIDE

(see IMS website for details)

Flowcharting Symbols used in Procedures

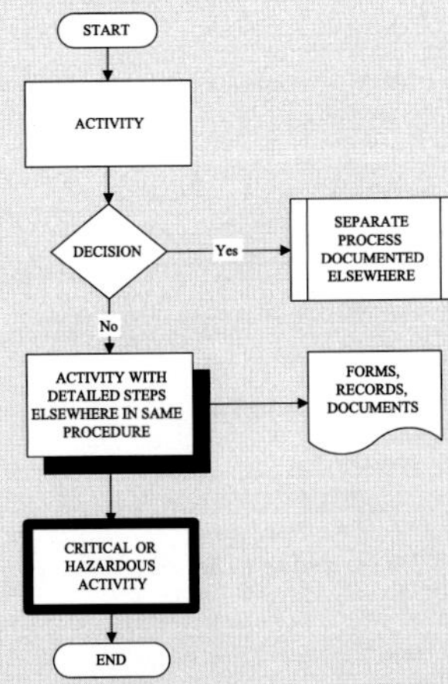

Responsibility is to be shown for each activity.

Abbreviations

- IMS - Integrated Management System
- BMS - Business Management System

Definitions

- ***Process*** – A set of interacting activities which transform inputs into outputs (example: baking a cake).
- ***Procedure*** – Key document for a specific *Process* (example: recipe for baking a cake).
- ***Process owner or custodian*** – Any SS.
- ***Business Management System (BMS)*** – Collection of *Processes*, to satisfy a business goal, which could span functional departments.
- ***System Owner*** – Division/ Section Head who has most control over administration of the *System.*
- ***System Sponsor*** – Department Manager - *System* Owner's Manager.

Standards selected for IMS

- ISO 9001 – Quality Management System
- ISO 14001 – Environmental Management System
- OHSAS 18001 – Health & Safety Management System

The IMS way of life

- **Procedures** are followed to ensure optimum Health, Safety, Environment and Quality (HSE&Q) performance.
- **Records** are kept as proof that procedures are adhered to.
- **IMS website** is used to view high level quality documents and obsolete copies are destroyed when documents are revised.
- **Internal audits** are conducted regularly to ensure system implementation & continual improvement.

Continued

BOX 4.7 REFINERY INTEGRATED MANAGEMENT SYSTEM (IMS) GUIDELINE—cont'd

Document Types

MN-Manual – Integrated Management System Manual or other System Manual which is in hard & soft copy. Business Management Systems (BMS) are on the ***IMS website*** and include the BMS's objectives, scope, responsibility, KPIs/ targets, strategies/plans/programs, & other information.

PR*-Management or Business Procedure* – A key document for a specific business process. It describes who does what, where, when and how for a particular process. It includes only information that is auditable. It provides clear actionable steps.

FC*-Flowchart* – Used to map out the steps of a process.

FM*-Form* – A template for recording information.

GI*-Guideline* – Detailed information concerning a process, supporting a procedure & providing additional information & guidance – not to be used as a working document for a process.

OP*-Operating Procedure*-A controlled procedure within a Business System.

QP*-Quality plans*–Any business plans e.g. Production plans, Budget, SD plans etc.

CL*-Check list*

GR*-Graph*

DW*-Drawing*

OC*-Organization chart*

WI*-Work Instruction* – A detailed sequence of steps. e.g. Steps for repair of a pump.

OT*-Other*–Not specified above.

Document Numbering

Quality-**T**ype-**D**irectorate-**S**ystem
Type **S**erial No

QPR-R-PS-PR001

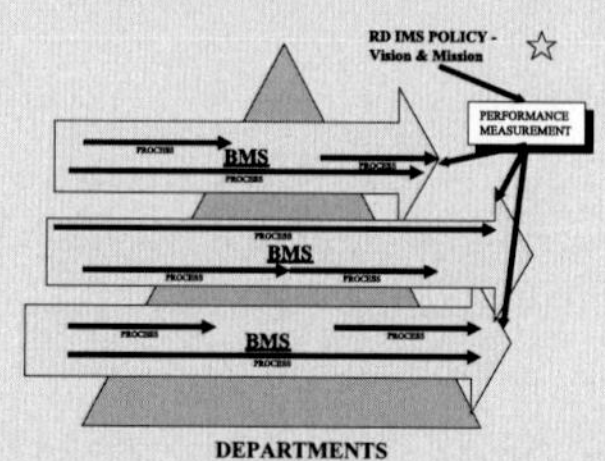

Business Management Systems

- **PS-Production & Storage** – Operation of all refinery process units and utilities
- **PP-Production Planning** – Production planning & scheduling
- **IM-Management** – Management responsibility & review; preventative/ corrective actions/ improvements, & non-conformance; incident management; internal auditing; document control
- **TA-Testing & Analysis** – All laboratory testing
- **MC-Maintenance** – Maintenance of all process equipment, tanks, pipelines, utilities-online & shutdown
- **PE-Performance Monitoring** – Performance of refinery process units
- **IN-Inspection** – All static, rotary, electrical, safety inspections
- **EN-Engineering** – All Plant Change Requests (PCRs) & Engineering Document Management
- **EV-Environment** – ISO 14001 compliance
- **HS-Health & Safety** – OHSAS 18001 compliance
- **TR-Training** – All needs & training carried out to fulfill the needs
- **BA-Budget & Admin** – All financial & admin items
- **CO-Procurement Services** – All contracts and agreements for services
- **MA-Procurement Materials** – All materials purchases
- **HR-Personnel Admin** – Reference to HR procedures & policies

Other tools to retain sustainability include regular Awareness Bulletins and gifts, such as T-shirts and mugs with the IMS logo and slogans.

MAINTENANCE OF A STANDARD COMPLIANCE TABLE

To ensure compliance with various standards, as well as making it easier for a compliance auditor, a table needs to be created showing the relevant standard and paragraph number(s) in the standard(s) related to a Process, procedure, form, or record in the Management System, which ensures compliance to the relevant standard(s).

4.12 INTEGRATING SYSTEMS IN A BUSINESS UNIT

All Systems have common attributes as follows:

1. Handling change: control of change in Systems, Processes, documentation, plant, and technology
   - incorporating a **Document Control procedure**
2. Handling failures: identification and investigation of incidents and findings
   - incorporating an Incident Management procedure
3. Preventive/corrective actions: eliminating the causes of failures
   - incorporating a procedure for logging and managing improvements and nonconformities
4. Audits and Reviews: assessment of Processes and review of Systems and documentation
   - incorporating an Internal Certification Auditing procedure
5. Management Responsibility and Review
   - incorporating a procedure on the Management Review process and the responsibilities of management
6. Customer Satisfaction
   - incorporating a procedure to monitor customer satisfaction and actioning customer concerns and complaints
7. Continuous improvement: identification of ways of improving competitiveness, efficiency, and effectiveness

The previously mentioned attributes would be covered in an IMS Manual as an integral part of the Governance Framework (see Section 8.4: Physical Picture of the Model Governance Framework).

Box 4.5 identifies the processes of an IMS for a Refinery Business Unit.

4.13 INTEGRATED MANAGEMENT SYSTEMS: SPECIFIC AND GENERIC EXAMPLES

The integration of numerous requirements for the existence of a business is complex and often untidy. This chapter has covered Systems that are subordinate to the top Company document, often referred as the Corporate "Governance Framework," which is discussed in Chapter 8, Governance Framework.

Specific

Some of the following examples use the word "processes," which could possibly align with "Systems," as described in this Chapter.

Statoil[2]

Statoil utilized ARIS from Software AG to develop their **top document** electronically (see Chapter 6: Business Process Management). This is "The Statoil Book." As shown in Box 4.8, Statoil divide their

BOX 4.8 THE STATOIL BOOK PROCESSES

Corporate Management and Planning processes					
Ambition and action	Capital value process (CVP)	Business development	Risk management	Monitoring	
Value Chain Processes					
Exploration	Petroleum technology	Drilling and well	Project development	Operation and maintenance	Marketing and supply
Support Processes					
Management System	Health, safety, and environment	Supply chain management	Finance and control	Human resources	Information technology
Technology development	Ethics and anticorruption	Legal	Corporate social responsibility	Communication	Facility management

Emergency response and business continuity
Organizational management and control
External requirements
Technical requirements
Service management

Business processes into three primary categories: Corporate management and planning, Value chain, and Support. Functional and Business Area requirements are subordinate to "The Statoil Book."

Shell[3]

Shell has a "Shell Control Framework" document, which documents a **single overall control framework** for Shell.

In support of the **"Shell Control Framework"** and relevant manuals, their management processes are divided into three primary categories: **Strategic management, Core business. and Resource management**. Section 8.6, Examples from Industry, outlines the bigger picture.

BP[4]

BP has an **Operating Management System (OMS)**. It integrates all BP Group operating standards into one consistent set of expectations, defining the BP Group requirements for how they operate their business. The OMS also provides a platform for sustainable improvement through a consistent process that assesses and prioritizes risk, and then improves local operating performance on a continuous basis.

This is only part of the bigger BP picture described in Section 8.6, Examples from Industry.

ExxonMobil[5]

ExxonMobil introduced their **Operations Integrity Management System** in 1992. It provides a set of expectations embedded into everyday work processes at all levels of the organization and addresses all aspects of managing safety, health, security, environment, and social risks at their facilities worldwide. It meets the requirements of ISO 14001:2004 and OHSAS 18001:2007.

They also have a **Controls Integrity Management System,** which provides a structured approach for assessing financial control risks, establishing procedures for mitigating concerns, monitoring conformance with standards, and reporting results to management. Aspects of this System are discussed in Section 8.6, Examples from Industry

GENERIC

Oil and Gas Producers[1]

OGP Report No. 510, "Operating Management System Framework," gives guidance on establishing a framework for an Operating Management System. The Framework comprises two interdependent components:

A. Four Fundamentals focus attention on management principles that are arguably the most important for an effective OMS: Leadership, Risk Management, Continuous Improvement, and Implementation.

B. Ten elements establish a structure to organize the various components of an OMS.

4.14 BENEFITS OF A MANAGEMENT SYSTEM APPROACH

These are as follows:

1. Alignment of vision, mission, policies, goals, and targets
2. Delegated Management System ownership
3. Customer orientation
4. Streamlined interfaces between departments
5. Consistent approach
6. Focus on performance measurement and analysis and actions to achieve targets
7. Duplication is eliminated
8. Improved utilization of resources
9. Continuous improvement is assured

These are discussed in more detail as follows.

ALIGNMENT OF VISION, MISSION, POLICIES, GOALS, AND TARGETS

This entails specific **Goals and Targets** to aim for, as to ensure continuous improvement. To be effective, these have to be in line with the corporate vision, mission, and policies.

DELEGATED MANAGEMENT SYSTEM OWNERSHIP

Roles and Responsibilities have to be clearly laid out so that actioning parties know exactly what is required of them. Their Management System is then audited to ensure that they are achieving what they have set out to do.

CUSTOMER ORIENTATION

Customer satisfaction is the key to a successful business. To ensure this, **focus** has to be **on the core business** and thus hopefully ensure a profit. Other interested parties' needs are to be identified and taken into consideration as well. The customer and other interested parties are generally referred to as stakeholders. *For example, Environmental Management Systems include the Environmental Regulation Authority as a stakeholder.*

STREAMLINED INTERFACES

All staff have to be focused on satisfying the customers' needs (internal and external). Managers then become leaders and facilitators to ensure this is done with **no interdepartmental barriers.**

CONSISTENT APPROACH

Documented Management System procedures that are **applied** ensure a consistent approach.

FOCUS ON PERFORMANCE MEASUREMENT AND ANALYSIS AND ACTIONS TO ACHIEVE TARGETS

High-level indices (KPIs) are identified on which to focus, with clear **Management System Owner** responsibility to **monitor and achieve the targets** that have been set. They all relate to improving profitability of the Business Unit.

DUPLICATION IS ELIMINATED

Clear delineation of each Management System eliminates duplication. Interface documentation is essential to refer between Systems to ensure this.

IMPROVED UTILIZATION OF RESOURCES

Focus on actions that have most effect on performance and applying financial and IT resources for most effect.

CONTINUOUS IMPROVEMENT IS ASSURED

With clear audited documentation showing activities, responsibilities, and targets in line with the customer's requirements, the process of continuous improvement goes on.

4.15 SUMMARY

The "Bus Analogy" describes the relationship between Management Systems and Processes. In short, the bus transports people from A to B, where a System gives "value addition" to a material or service.

The Systems approach is not new. For example, ISO 9001 gave clear definitions for Systems and Processes in 2000.

Differences between Systems and Processes are detailed by description, objectives, and targets, each with examples for clarity. For example, an Environmental Management System could have a Process for control of stack emissions.

The benefits of a Systems approach to managing the business are listed. These include the following:

1. Alignment of vision, mission, policies, goals, and targets
2. Delegated Management System ownership
3. Customer orientation
4. Streamlined interfaces between departments
5. Consistent approach
6. Focus on performance measurement and analysis and actions to achieve targets
7. Duplication is eliminated
8. Improved utilization of resources
9. Continuous improvement is assured

Primary Systems are **Strategic**: necessary for Corporate Governance, **Core**: the Value Chains of the business, and **Support**: required to ensure smooth achievement of Value Chain targets.

The key elements of a System include the existence of objectives, KPIs, scope, policies/regulations/standards, strategies and plans, and processes and responsibilities.

The underlying computer systems help define the Management Systems. For example, a Maintenance Management Systems has an underlying CMMS.

Systems within a Business Unit need to be consolidated into an IMS with common elements grouped to prevent duplication. The integration of Systems within the Governance Control Framework is discussed in Chapter 8, Governance Framework.

REFERENCES

1. Oil & Gas Producers (OGP) Report No 510. *Operating management system framework.* June 2014.
2. Solevågseide C, Frode Saasen J. *How statoil uses ARIS to support safe, reliable and efficient global operations.* Orlando Florida: ARIS. ProcessWorld; 2012.
3. Shell. http://www.shell.com/investors/environmental-social-and-governance/corporate-governance.html.
4. BP. http://www.bp.com/en/global/corporate/investors/governance.html.
5. Exxon-Mobil. http://corporate.exxonmobil.com/en/investors/corporate-governance/corporate-governance-guidelines/guidelines.

FURTHER READING

1. Hey RB. *Integrating quality management.* Dubai: Institute of International Research; 2000.
2. Hernaus T. *Generic process transformation model: transition to process-based organization.* Working Paper Series Paper No. 08–07. University of Zagreb Croatia; 2008.
3. *Guidelines for risk based process safety.* AIChE Center for Chemical Process Safety (CCPS) Wiley – Interscience; 2007.
4. *Guidelines for integrating process safety management, environment, safety, health & quality.* Center for Chemical Process Safety of American Institute of Chemical Engineers; 1996.
5. ISO 19011:2011. *Guidelines for auditing management systems.*

CHAPTER

THE BUSINESS CYCLE

5

5.1 INTRODUCTION

The Business Cycle is sometimes also referred to as Strategic Management and is generally split into three linked Systems that are discussed in this chapter: Strategic Planning (Strategy Formulation), Annual Planning and Budgeting (Implementation), and Performance Management (Evaluation and Control). All other Strategic Systems revolve around these three.

WHAT IS STRATEGY?

Definition[1]

Strategy is the direction and scope of an organization over the long term, which achieves advantage in a changing environment through its configuration of resources and competences with the aim of fulfilling stakeholder expectations.

Corporate-level strategy is concerned with the overall purpose and scope of an organization and how value will be added to the different parts (Business Units) of the organization.

Business Unit level strategy deals with ways to compete successfully in particular markets.

Operational strategies are concerned with how the elements of an organization effectively deliver the Corporate and Business Unit level strategies in terms of resources (raw materials, finance, and people) and processes.

WHAT IS STRATEGIC MANAGEMENT?

Definition

Strategic management is the art and science of formulating, implementing, and evaluating cross-functional decisions to enable the Company or Business Unit to achieve its objectives.

THE BUSINESS CYCLE: A BRIEF INTRODUCTION

The Cycle is completed on an annual basis.

Strategic Planning (Strategy Formulation): the direction of the business is formulated in detail for the next year and with less detail for at least the following 5 years, i.e., a total of 7 years: this year, next year plus 5 years.

Annual Planning and Budgeting (Implementation): the outcome of Strategic Planning is used as the framework for detailed planning and budgeting for the next year with a further 5-year projection of expected expenditure and planning.

Performance Management for the Oil, Gas, and Process Industries. http://dx.doi.org/10.1016/B978-0-12-810446-0.00005-0

Performance Management (Evaluation and Control): the firm budget and plan for the next year, approved by the Board of Directors (BoD), is actioned and then monitored at regular intervals (at least monthly by the leadership team and BoD, and quarterly by the shareholders).

The cycle is depicted in Fig. 5.1.

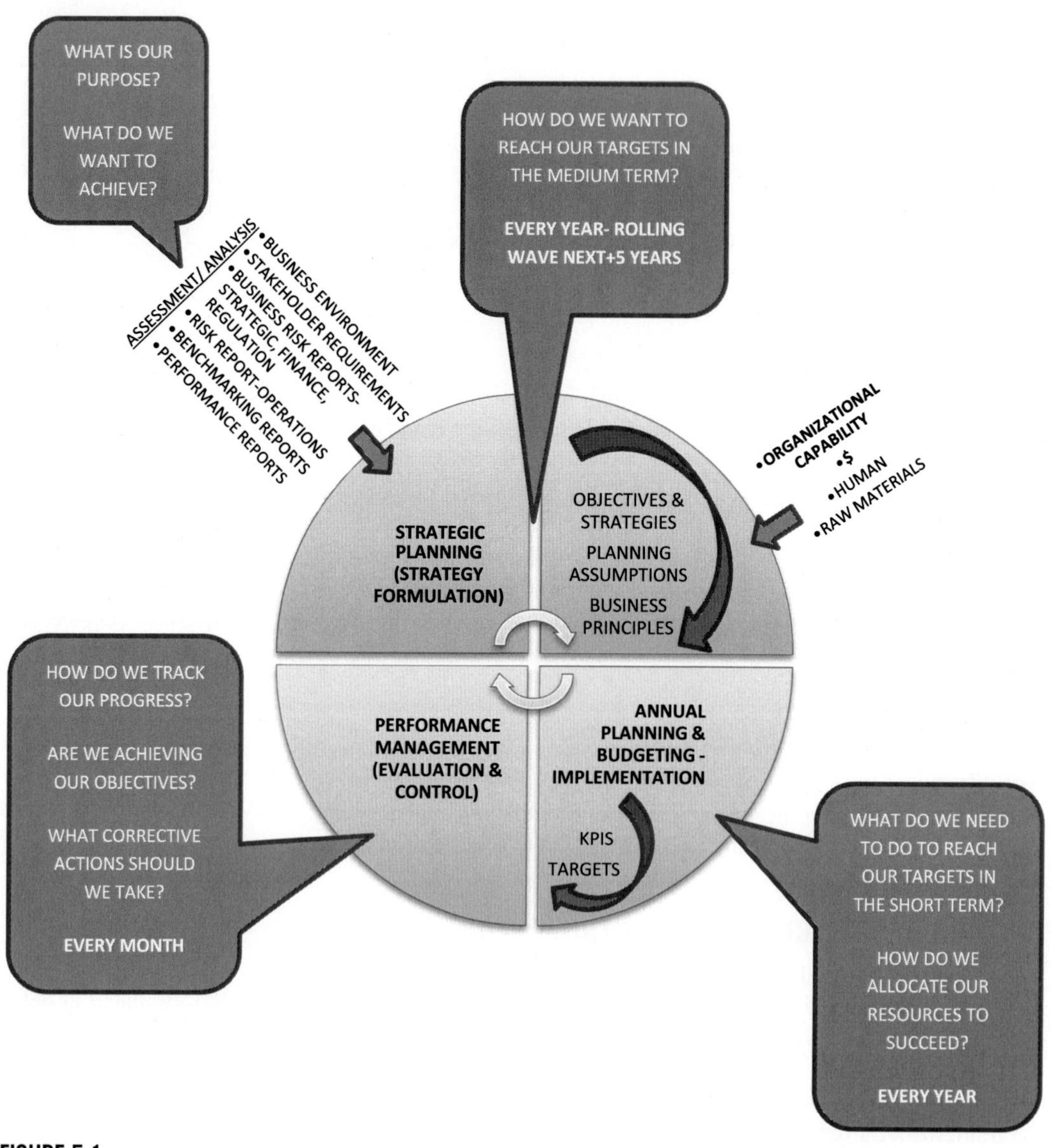

FIGURE 5.1

The business cycle basics.

5.2 STRATEGIC PLANNING (STRATEGY FORMULATION)

The **business environment** and **stakeholder requirements** are the usual starting points of strategy formulation. The business environment entails all external risks and opportunities, and the stakeholders include government, shareholders and partners, industry, suppliers/contractors, customers, public, and employees.

Analysis of the business environment and stakeholder requirements is undertaken by the leadership team and compared to the outcomes of the following:

1. Strategic studies: product demand and pricing, pricing of raw material and availability, etc.
2. High-level business risks and opportunities assessment

Subsequently, Critical Success Factors are considered and confirmed before reviewing, and possibly updating, the **principles** by which the business is governed (including vision, mission, and various policies).

The company **planning assumptions** are also **established** as a result of the analysis of the business environment and stakeholder requirements.

At the same time, an assessment of the business performance, relative to budget, targets, and the competition, is carried out.

In parallel, the organizational capability requirements, including human capital, finance, and raw materials requirements, are assessed.

Vision and mission are discussed in Chapter 10, Alignment. Adjustments to the company vision and mission should only be undertaken when there is a major change in the direction of the company.

STRATEGIC STUDIES

Strategic studies are normally undertaken by expert consultants or industry associations. Ongoing demand and pricing studies for petrochemical products are undertaken by organizations such as the following:

- IHS Chemical Market Advisory Service: Global Plastics and Polymers[2] (formerly CMAI Global Plastics and Polymers Market Advisory Service) provides a comprehensive view of world markets for six major thermoplastic resins (PE, PP, PS, EPS, PVC, and PET) and their relationships with upstream and downstream sectors.
- Strategic Business Analysis Ltd. (SBAcci)[3] is a leading consultancy to the global polyester industry and upstream feedstocks (PX, PTA/DMT, MEG, fibers, PET resin, and film). SBAcci produces a unique worldwide macroeconomic supply/demand, precise price/margin models, and reports on the global value chain, supported by individual company presentations.

On some occasions, strategic studies are needed to affirm or modify the company direction statement, as the example in Box 5.1 shows.

BOX 5.1 STRATEGIC STUDY

Defining Strategic Direction of a Newly Formed International Subsidiary of a National Oil Company.

A National Oil company decided to enter the international market and consequently employed a management consultant to recommend a suitable strategic direction for the newly formed international subsidiary.

The recommendations of the study included:

The focus is to be on extending the value chain that exists in the parent company to international markets. Ventures to include focusing on selected markets for distribution of petroleum products and the extension of the LNG value chain by investing in gas pipelines and power generation in the countries where their products are already marketed.

BUSINESS RISKS AND OPPORTUNITIES

Business risks and opportunities are determined using planning techniques, such as the following:

SWOT

SWOT, which is possibly the most popular strategic planning tool, provides a clear assessment of a situation, project or venture by looking at its Strengths, Weaknesses, Opportunities, and Threats.

PPESTT

PPESTT analysis is a model that is used to scan the external environment in which the company operates. PPESTT elements are as follows:

1. **P**olitical
2. **P**hysical environment
3. **E**conomic
4. **S**ocial
5. **T**echnological
6. **T**rade

Porter's Five Forces

Porter's Five Forces framework can aid the understanding of the driving forces of the industry within which the business operates. These are as follows:

1. Industry competitors intensity of rivalry
2. Suppliers
3. Buyers
4. New entrants into the industry
5. Substitutes

McKinsey 7S Framework

The McKinsey framework describes seven areas of a company that should be focused on when executing a strategy. These are as follows:

- Shared values
- Strategy
- Structure
- System
- Staff
- Skills
- Style

Scenario Planning

Scenarios are detailed and plausible views of how the business environment of an organization might develop in the future, based on key drivers for change (about which there are different levels of uncertainty).

Scenario Planning is used in conjunction with other models to develop alternative views of the future.

This is often used for product or raw material pricing where worst, best, and most likely case scenarios are played out to see the effect on the business strategy.

Example:

The Oil and Gas industry attempts to predict future crude and LNG prices. While in some markets these two prices are linked, in other markets they are not, giving two entirely different sets of scenarios.

Section 12.6, Scenario Planning, discusses **risk analysis** as part of Scenario Planning.

Consultants

Scenario planning consultants, such as Clem Sunter,[4] facilitate strategic conversations with executive teams of multinational companies and use facilitation techniques, which include the sketching and construction of tailor-made scenarios and the best options to cope with the challenges and risks they contain.

Strategic Workshops

Strategy workshops away from the work environment (sometimes called strategy retreats, away days, or off-sites) can be used to determine organizational strategy. They usually involve groups of executives working intensively for 1 or 2 days.

Questions that could be asked in these workshops include the following:

1. What do you think is wrong?
2. What do you think can be done?
3. If you were the manager, what would you do?

Key elements are as follows:

- Insisting on prior preparation
- Involving participants from outside the senior executive team
- Involving outside consultants as facilitators
- Breaking organizational routines
- Making an agreed list of actions
- Establishing project groups
- Circulating agreed actions
- Making visible commitment by the top management

Horizon Scanning

Horizon scanning could be part of PPESTT or Scenario planning

The example in Box 5.2 demonstrates agility when unanticipated events occur.

Strategic risks associated with particular actions are determined based on probability. However, the risks need to be within the company's "risk appetite," which is determined by the BoD (see Section 11.5: Risk Appetite).

BOX 5.2 HORIZON SCANNING

In the years prior to 2011, a large Middle Eastern LNG producer was exporting a significant part of its production to the US. However, with the relaxation of legislation on "fracking" in the US, local gas production there increased to a point where the US became a net exporter. As a result, the Middle Eastern producer lost a significant share of its market in a short period.

Tragically, as this was happening, Japan experienced a major tsunami, which caused a significant part of its nuclear power generation to shut down. Japan quickly increased power generation from LNG, not only absorbing the Middle Eastern producer's lost US market, but also paying a higher price than what the US market was paying.

Comment

Horizon Scanning could have helped foresee the downturn in the US market, but could not have predicted the "positive" spin-off of the Japanese tsunami, a "black swan" (See Box 12.4: Black Swan).

An example of a National Risk Assessment is outlined in Box 5.3.

BOX 5.3 NATIONAL RISK MANAGEMENT ANALYSIS: "MANAGING RISK THROUGH QUANTIFICATION"

LNG risks are outlined as follows:

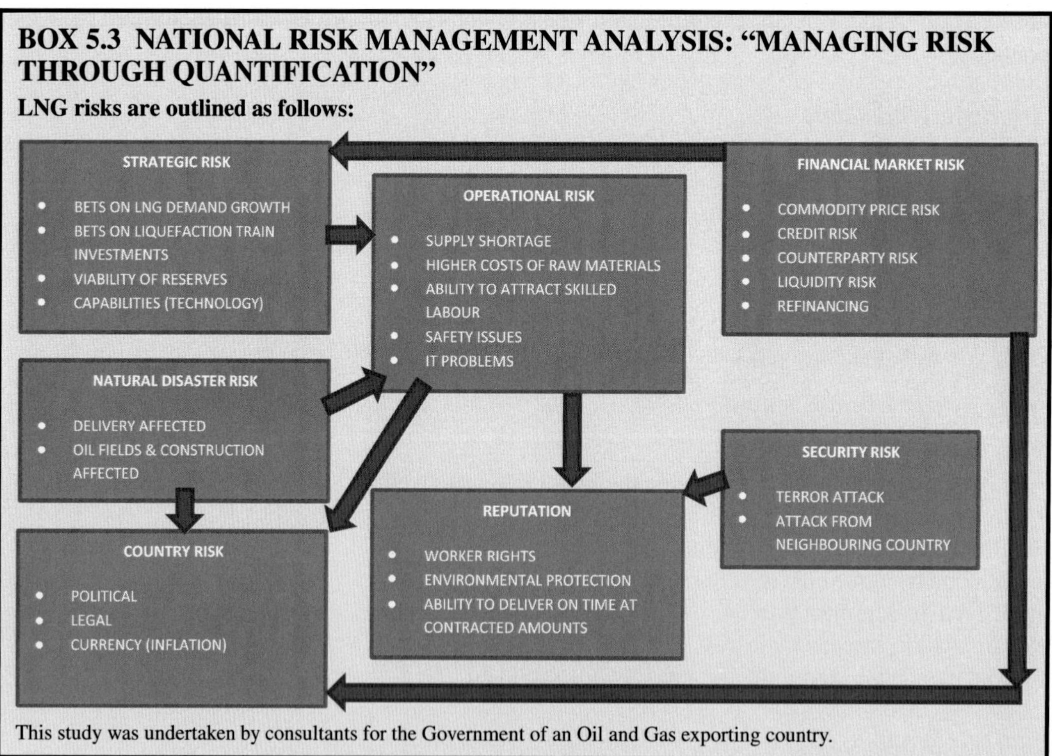

This study was undertaken by consultants for the Government of an Oil and Gas exporting country.

The determination of **objectives and strategies** for improvement (risk mitigation and opportunity promotion) are discussed in Chapter 31, Strategies and Actions.

Strategic Planning/Strategy Formulation decisions commit the Business Unit to specific products, markets, resources, and technologies over an extended period of time.

ORGANIZATIONAL CAPABILITY

Once Strategic Objectives have been defined, **Organizational Capability** in terms of **Human** and **Financial** resources are assessed and proposed for inclusion in the annual budget and plan. The availability of raw materials (**Natural** resources) also needs to be assessed.

Organizational Capability is defined as follows:

Ability and capacity of an organization expressed in terms of its:

1. Human resources: their number, quality, skills, and experience;
2. Financial resources: money and credit;
3. **Natural** resources: oil, gas, minerals, water, wind, solar, etc.;
4. Information resources: pool of knowledge, databases;
5. **Intellectual** resources: copyrights, designs, patents, etc.; and
6. **Physical and** material resources: machines, land, buildings.

The growing intensity of globalization, bringing with it increased competition for resources and concurrent economic stresses, have impacted upon the strategic importance of Organizational Capability as a prime factor for organizations to achieve and sustain competitive advantage.

Organizational Capability is discussed further in Section 19.2, Organizational Capability: The Information Asset Aspect and Section 20.2, Organizational Capability: The Human Capital Aspect.

Part 5, Asset Performance Management, discusses the previous as assets.

The following are then finalized before being used as input to the annual budgeting and planning process:

1. Company planning assumptions for the next planning period (next year and at least 5 years thereafter)
2. Affirmation of the Company Vision, Mission, and Policies
3. Clear Objectives and Strategies for the next planning period

Strategic Planning/Strategy Formulation is summarized in Fig. 5.2.

5.3 ANNUAL PLANNING AND BUDGETING (STRATEGY IMPLEMENTATION)

Strategy implementation requires the establishment of annual performance targets, use of Processes and procedures, allocation of resources (human and financial), and motivation of staff to ensure execution of the agreed Objectives and Strategies. This includes developing a Strategy-supportive culture and linking employee compensation to organizational performance.

Annual performance targets are discussed further in Chapter 30, Strategies and Actions.

Development of **Processes and procedures** are discussed Chapter 6, Business Process Mapping.

Allocation of **resources** should be reviewed in the Management Review Meetings as an agenda item as discussed in Chapter 35, Review.

Motivation of staff is discussed in Chapter 20, Human Capital.

Strategy implementation requires personal discipline, commitment, and sacrifice. Successful strategy implementation depends on a manager's ability to motivate his/her staff, and excellent interpersonal skills are especially critical.

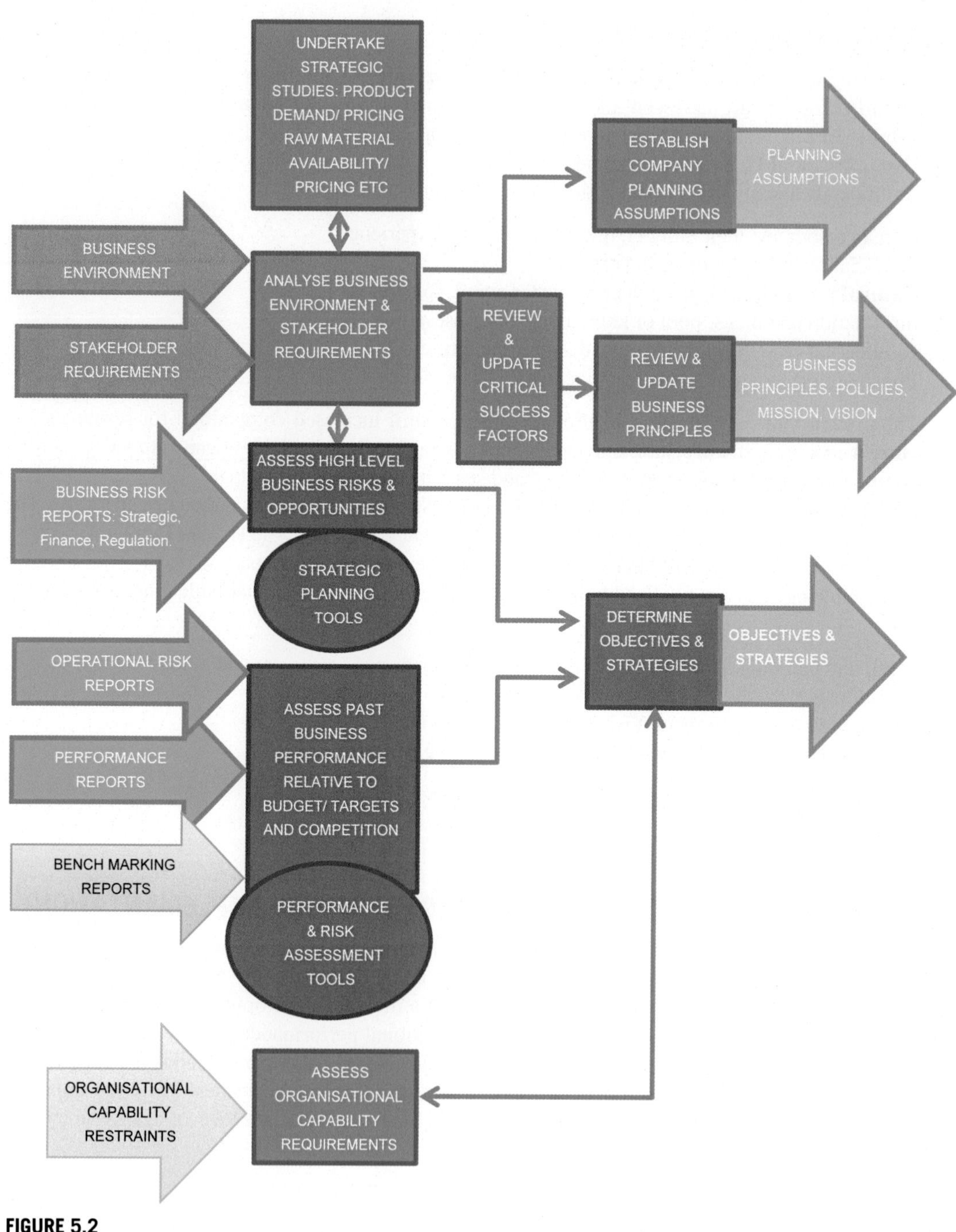

FIGURE 5.2

Strategic planning/strategy formulation.

The Annual Planning and Budgeting cycle is at the heart of Performance Management. This is generally depicted as per Fig. 5.3.

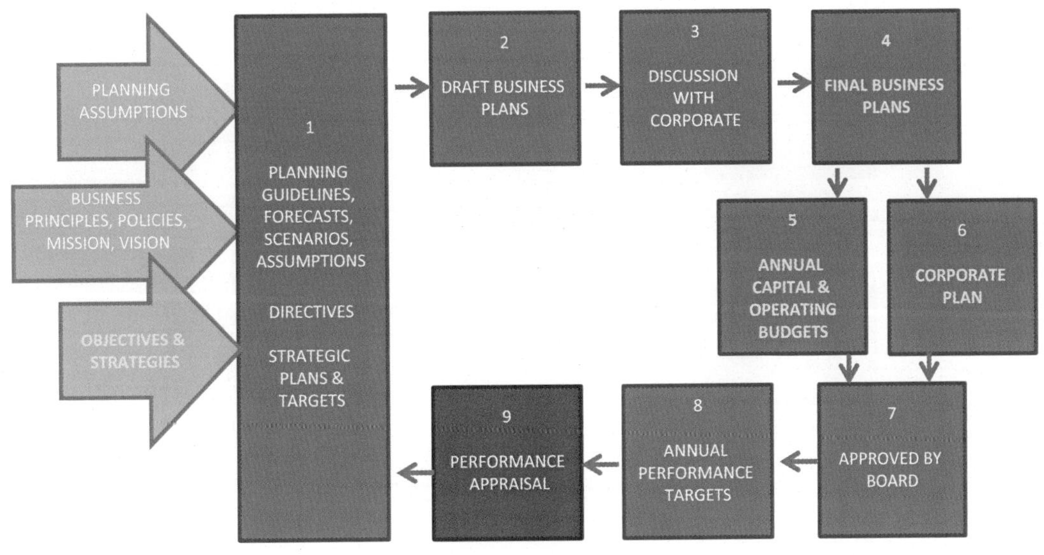

FIGURE 5.3

The annual planning and budgeting cycle.

5.4 PERFORMANCE MANAGEMENT (STRATEGY EVALUATION AND CONTROL)

Strategic evaluation is the primary means used to review strategies in light of changing internal and external factors, performance to date, and corrective actions taken. Decisions are made on adding new strategies, dropping strategies that are not working, and re-prioritizing strategies within the restraints of the approved budget and business plan.

To quote Peter Drucker:

> "...the question: 'What is our Business?' This leads to the setting of objectives, the development of strategies, and the making of today's decisions for tomorrow's results. ***This clearly must be done by a part of the organization that can see the entire business***; that can balance objectives and the needs of today against the needs of tomorrow; and that can allocate resources of men and money to key results."

PERFORMANCE SCORECARD INTERFACING

Performance is a two-way system shown in Fig. 5.4. The left side depicts an annual process, while the right is a monthly, quarterly, and annual process.

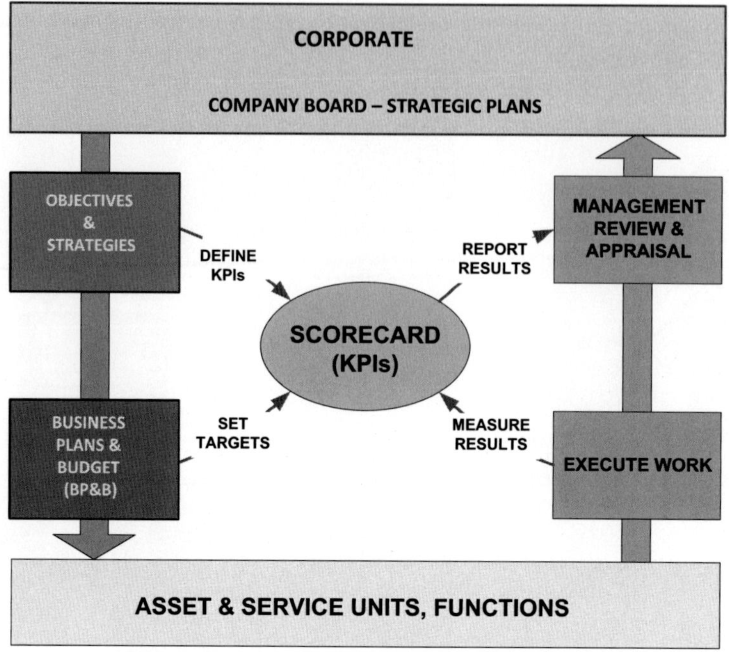

FIGURE 5.4

Performance monitoring data flow. *KPI*, Key Performance Indicator.

Scorecards need to be developed in a way that maximizes the use of **predictive** indicators for long-term integrity of the assets, **and** must include indicators mitigating high-level business risks. The "human factor" is **key** to the success of any company.

These points are discussed further in the following chapters:

- Chapter 13, The Merging of Performance and Risk
- Chapter 15, Performance Focus
- Chapter 20, Human Capital
- Chapter 22, Physical Asset Performance Management

RISK AND OPPORTUNITY KEY PERFORMANCE INDICATOR ROLL-UP

It is critical to escalate higher-level risks and opportunity Key Performance Indicators (KPIs) for review and inclusion in the Strategic Planning process to ensure mitigation of risks and closing of opportunity gaps.

Box 5.4 gives a graphical example of the risk and opportunity roll up in a National Oil Company, which has a variety of interests.

BOX 5.4 ANNUAL PLANNING AND BUDGETING AND PERFORMANCE MANAGEMENT RISK AND OPPORTUNITY KEY PERFORMANCE INDICATOR (KPI) ROLL-UP

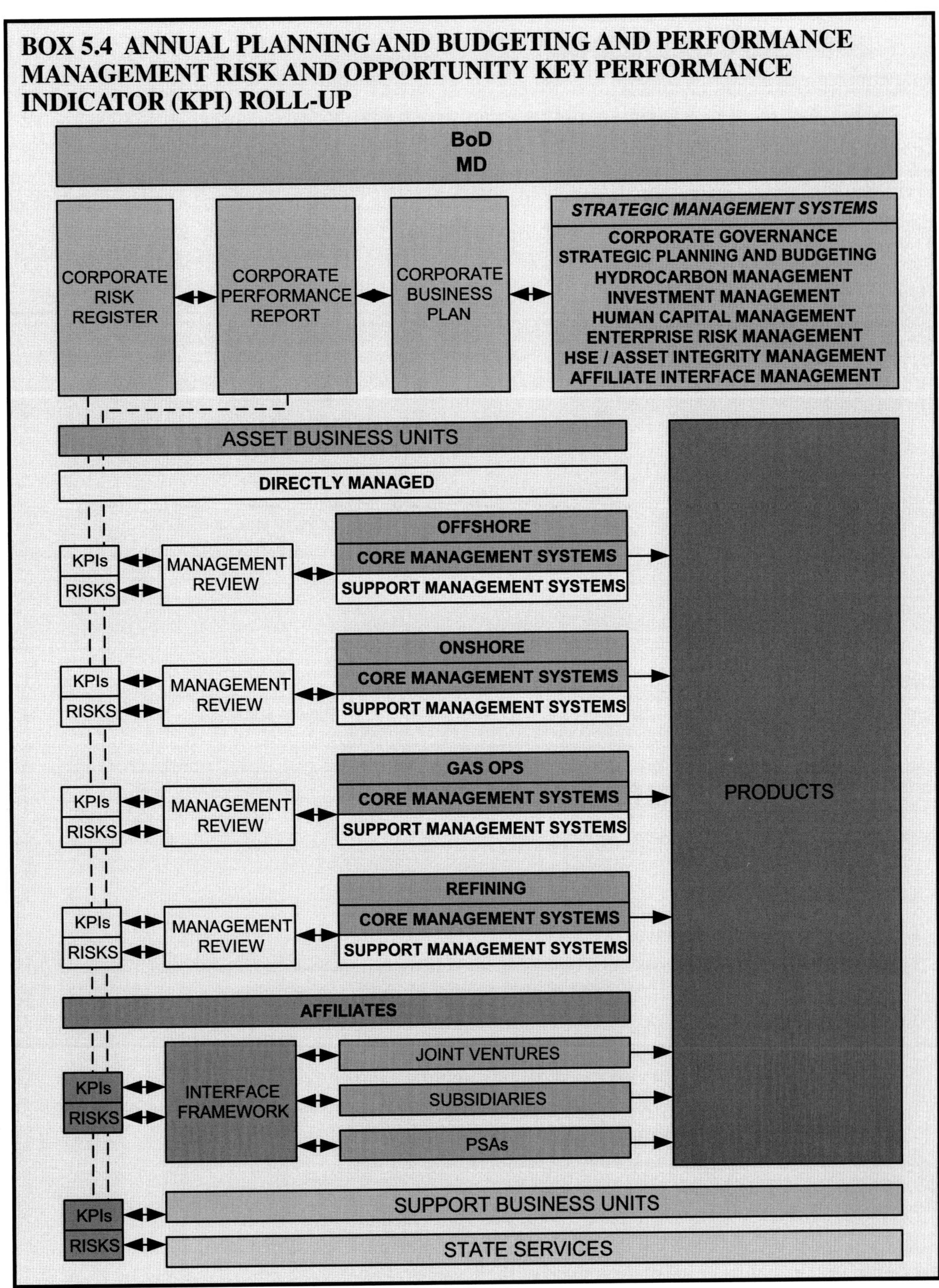

5.5 BUSINESS CYCLE TIMING

The full cycle takes place annually.

The results of the previous year's performance are assessed by management after the Company Annual Report is produced. Corporate guidelines and forecasts for the next cycle are then formulated and issued early in the second quarter. Business Unit strategic plans are formulated at midyear. Business Unit level strategic plans are discussed with corporate management in the third quarter. Business plans and budgets are developed in the fourth quarter and approved by the BoD by year-end. Fig. 5.5 depicts this cycle.

FIGURE 5.5

Business cycle timing.

5.6 BENEFITS OF STRATEGIC MANAGEMENT

Strategic management offers the following benefits:

1. Identifies, prioritizes, and mitigates risks
2. Identifies, prioritizes, and exploits opportunities
3. Provides an objective view of problems
4. Minimizes the effects of adverse conditions and changes ("chance favors the prepared mind")
5. Provides a framework for improved coordination and control of activities
6. Allows major decision to better support the established objectives
7. Allows more effective allocation of time and resources to identified opportunities
8. Reduces resources and time allocated to "fire-fighting" and corrective action
9. Improves communication and knowledge sharing between departments and divisions, and Business Unit and Corporate Management
10. Combines the individual effort into a group effort (1 + 1 = 3)
11. Clarifies individual and group responsibilities
12. Encourages a favorable attitude to change and forward thinking
13. Encourages a cooperative approach to resolving problems (Cross-Functional Teams)
14. Provides discipline to the management of the Business Unit

5.7 SUMMARY

The business cycle encompasses **Strategic Systems,** which are necessary for Corporate Governance.

Top risks and performance issues are primary inputs into the Strategic Planning System after passing through the assessment process (see Part 7: Assessment) and being elevated through the management review process (see Part 8: Oversight).

The Annual Business Plan and Budget (sometimes referred to as the Work Plan and Budget) is derived from the Strategic Planning System and its Processes.

Once the Annual Plan and Budget are approved by the BoD, business performance is monitored against this approved package and enables Line Management to take corrective action to ensure targets are achieved.

Performance monitoring also takes place against benchmark targets, although this tends to be in the longer term.

REFERENCES

1. Johnson, Scholes, Whittington. *Exploring corporate strategy*. Pearson; 2008.
2. IHS Chemical Market Advisory Service: Global Plastics & Polymers. https://www.ihs.com/products/chemical-market-plastics-polymers-global.html.
3. Strategic Business Analysis Ltd. www.sba-cci.com.
4. Sunter C. *The mind of a fox*. Tafelberg: Human & Rousseau; 2001. http://www.mindofafox.com/site/home.

FURTHER READING

1. Katsioloudes MI. *Strategic management*. Elsevier; 2006.
2. Buzan T. *Mind maps for business*. Pearson; 2014.
3. Grant RM. Strategic planning in a turbulent environment: evidence from the oil majors. *Strategic Manag J* June 2003;**24**(6):491–517. www.jstor.org/stable/20060552.
4. HBRs most reads on strategy. Product 12601. Harvard Business Review. www.hbr.org.
5. Awamleh R. *Strategic thinking & business planning*. Doha Qatar: Projacs; November 2007.
6. *The outlook for energy: a view to 2030*. ExxonMobil.
7. Van der Heijden K. *Scenarios: the art of strategic conversation*. John Wiley; 2004.

CHAPTER

BUSINESS PROCESS MANAGEMENT

6

6.1 INTRODUCTION

Quote: "If you don't have good manual processes you will have disastrous automated processes."

Deficiencies in processes are determined from a number of sources including the following:

1. **Benchmarking**, where one knows the competition is operating at a higher level of efficiency
2. **Certification Audits**, where there are either inadequate processes or employees are not following existing processes. Quality peer auditing (in line with ISO 9001 and its derivatives) determines if a documented Process is being implemented or not, and whether it needs to be improved or replaced
3. **Internal Audits**, where a risk-based "Systems" approach is taken, with a focus on the potential for corruption, and thereby possibly creating an opportunity for implementation of computerized processes

Embedding processes into computer systems removes any subjectivity and diminishes the potential for corruption.

Business Process Management (BPM) has evolved from Business Process Mapping, where a structured "project management" approach is used to ensure implementation and sustainability of the process improvement or modification. Change management is critical to the success of the implementation.

BPM is applied for development and enhancement of Processes within Systems. The classification of Systems and Processes is discussed in Chapter 4, Management Systems Determination and Requirements. The simplest processes can be dealt with by flowcharting the process, as discussed in Section 6.2. More complex processes need a formal structured BPM approach, as discussed in Sections 6.3 and 6.4.

6.2 FLOWCHARTING

Before writing any procedure for a process, the process has to be mapped with a start and end, decision points, and responsibilities.

Swim lines could be used to separate responsibilities for easy identification.

Some basic rules of flowcharting include:

- activity boxes where the activity description starts with a verb;
- a decision triangle is required where alternative paths are to be taken;

Performance Management for the Oil, Gas, and Process Industries. http://dx.doi.org/10.1016/B978-0-12-810446-0.00006-2

- activity boxes with side bars identifying a separate process to be described elsewhere;
- activity boxes with a shadow identifying activities with detailed steps elsewhere in the process and activity boxes with an enhanced outline showing critical activities; and
- elliptical start and end notation.

The previous is depicted in Fig. 6.1.

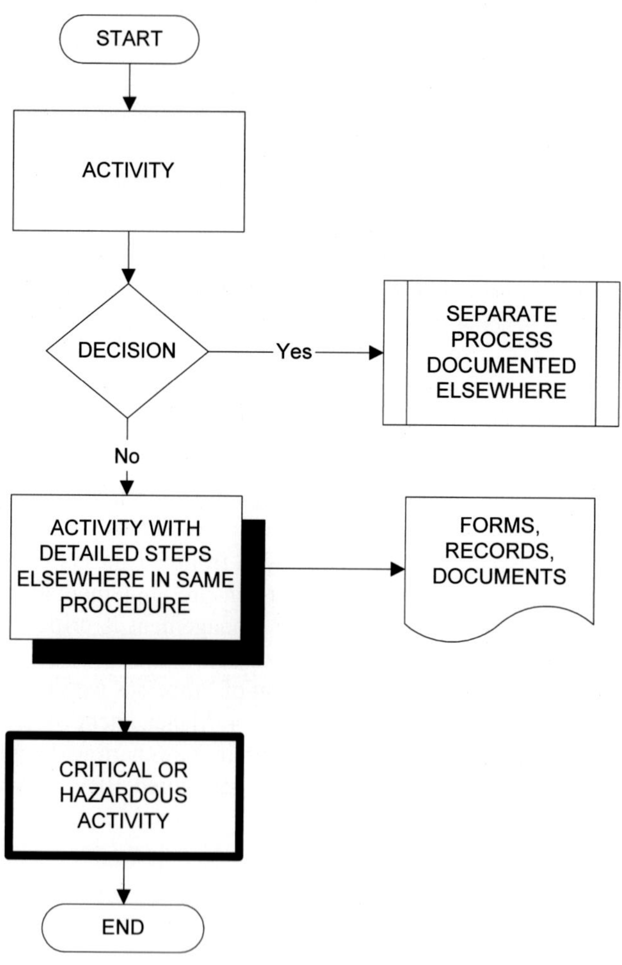

FIGURE 6.1

Flowchart with notation.

Flowcharting applied to the Annual Planning and Budgeting Process is shown in Box 6.1.

BOX 6.1 FLOWCHART: ANNUAL PLANNING AND BUDGETING PROCESS

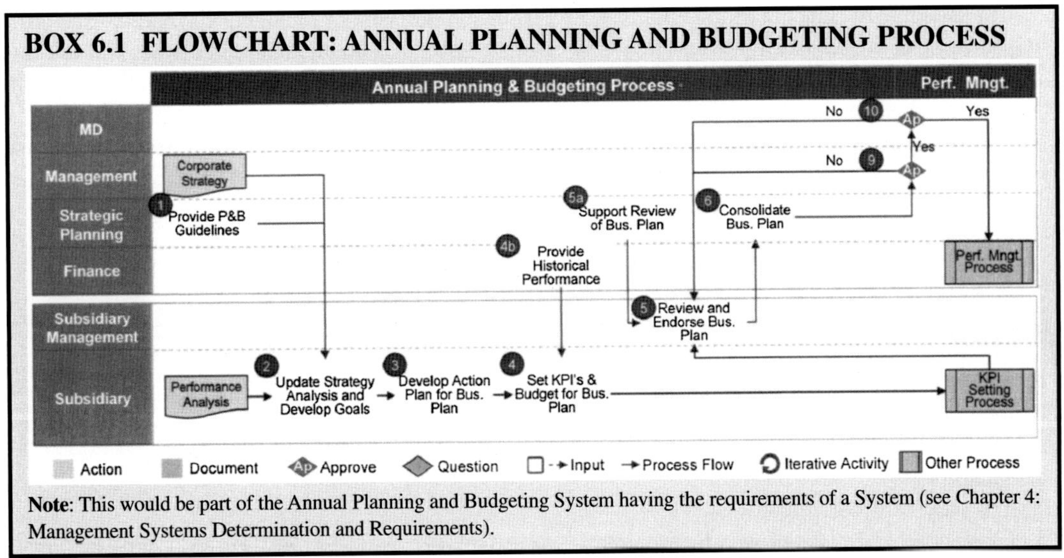

Note: This would be part of the Annual Planning and Budgeting System having the requirements of a System (see Chapter 4: Management Systems Determination and Requirements).

6.3 BUSINESS PROCESS MANAGEMENT AS A RESULT OF BENCHMARKING

One of the strategies resulting from benchmarking may be the modification or replacement of a Process. This has been labeled "business reengineering" in the past.

This option is depicted as in Fig. 6.2.

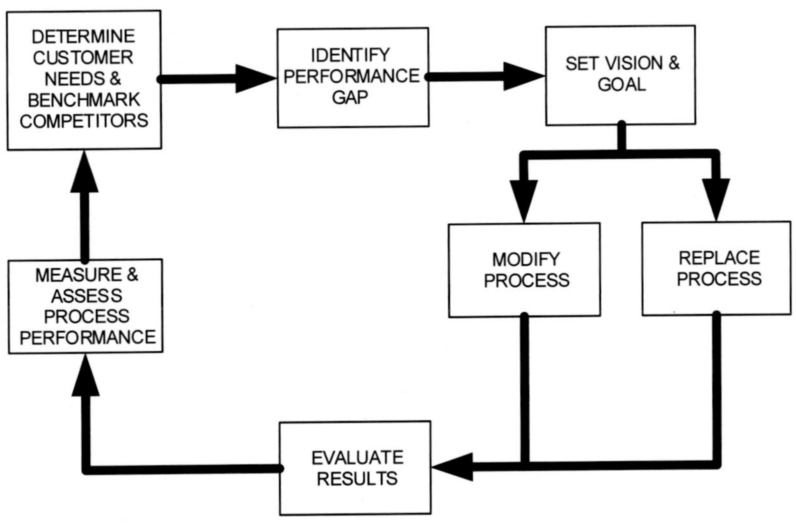

FIGURE 6.2

Process modification or replacement.

The outcome of Benchmarking entails more than just modifying or replacing a Process. An alternative outcome of benchmarking or auditing may be that a Process is not being applied but is still relevant. A Change Management strategy is required to encourage staff to apply the Process, whether it is to be computerized or not.

Full assessment of performance and risks is required. Chapter 30, Strategies and Actions, discusses the direction to be taken.

6.4 EVOLUTION OF BUSINESS PROCESS MANAGEMENT

Past practice has been as follows:

1. A Statement of User Requirements is developed, sometimes based on user "wants" as opposed to their "needs."
2. Information Technology Department develops a computerized process.
3. The process is "beta" tested.
4. The process is corrected.
5. Final application is approved by the user.

The process is sometimes not what the customer needs. This is generally depicted as in Fig. 6.3.

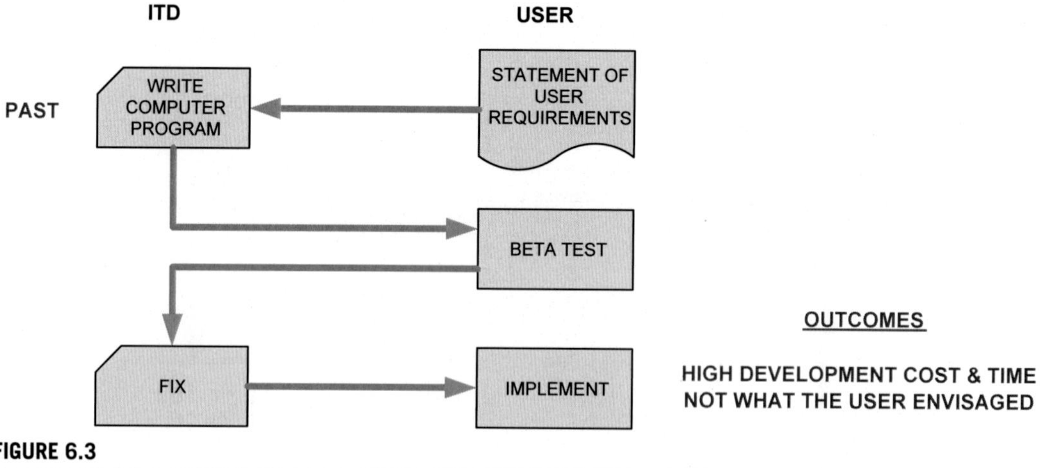

FIGURE 6.3

Manual approach.

With changing technology, the computerization of Processes is becoming cheaper and easier.

Box 6.2 gives examples of the move from a centralized mainframe approach to a cheaper and easier intranet approach.

BOX 6.2 BUSINESS PROCESS MANAGEMENT (BPM) APPLICATION USING THE INTRANET

Example 1: Project Management System

In the 1990s, a National Oil Company (NOC) decided that it needed a computerized project management system for the 200 projects it managed. The system was to have been mainframe-based.

The IT Department developed a "statement of user requirements" from interviews with all potential users of the new project management system. As it turned out, this was a wish list from all potential users, but was not necessarily what the project engineers needed.

To compound matters further, the development was carried out by inexperienced IT staff who had little knowledge of either the processes or Business Process Mapping.

After extensive cost and time overruns, the CEO decided to cancel the project, retaining only the cost control module, which was linked to the company accounting system.

Meanwhile, the intranet was introduced using Novell software. Two project planning engineers who knew the requirements of the system decided to write the remaining processes for the project management system in Dbase and run the software on the intranet. It cost less than $50,000 and was still in use 5 years later.

Example 2: Turnaround Critical Path Modeling

The same NOC used Artemis software on a mainframe computer for the Turnaround Critical Path Modeling (CPM) system. This entailed having IT specialist support and a service contract with Artemis.

With the introduction of the intranet, the potential for using intranet-based software that did not require a service contract or IT specialist support was explored. To this end, two turnaround planning engineers spent a year downloading the CPM models from the mainframe to the new Open Plan software for use on the intranet. This resulted in hundreds of thousands of dollars per year in service charges being saved. The processes for CPM were also effectively amended by the trained turnaround planning engineers without the services of an IT specialist.

Comment

Prior to this, Artemis was successfully used for the English Channel Tunnel project, but subsequently lost significant market share to intranet-based Open-Plan and Prima-Vera.

Current practice entails the following:

1. Flowcharting is carried out by the user and a BPM expert using specialist flowcharting software.
2. Flowcharting software is loaded into the Enterprise Resource Management (ERM) software.
3. ERM application is implemented.
4. Effectiveness is monitored.
5. ERM application is refined.

This can only be done with standard SAP/Oracle ERM computer processes using BPM tools, such as ARIS[1,2] or Bizagi[3] (see Fig. 6.4).

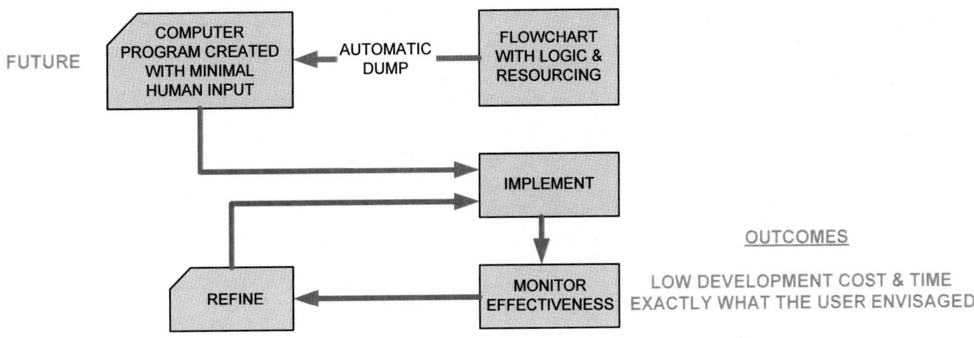

FIGURE 6.4

Automated approach.

Manual processes in written procedures still need basic flowcharting.

Note: Typically only about 50% of an end-to-end process is SAP-based (as per the outcome of the Unilever transformation program carried out in 2006).

Although the BPM tool is just part of the implementation process, it has been used successfully for all processes in an Oil and Gas company. Section 4.13, Integrated Management Systems: Specific and Generic Examples, includes discussion on the application of ARIS to development of "The Statoil Book" (see Box 4.8: The Statoil Book Processes).

BPM Implementation frameworks utilize the above software within a structured project to ensure buy-in and sustainability. A full project management approach with a strong emphasis on "change management" is required to ensure success.

Example:

The "7FE Framework" as described in "Business Process Management" by Jeston and Nelis.[4]

Box 6.3 gives an example where a company did not use a BPM Implementation Framework or a tool such as ARIS, and landed up with a dual system.

BOX 6.3 NATIONAL OIL COMPANY SAP IMPLEMENTATION

A National Oil Company (NOC) decided to streamline their business processes by implementing SAP modules for finance, materials, maintenance, human resources, and training.

However, the implementation entailed automating their existing ineffective manual systems without using a Business Process Management (BPM) Project Framework or any BPM software.

The Outcome: Parallel Systems

As a result, trust in the automated system was such that some people still kept manual files, and the envisaged efficiency improvements did not materialize.

Box 6.4 gives an example where a company did not use a tool such as ARIS, but utilized many consultants and SAP specialists for a successful implementation of Enterprise Resource Planning (ERP) in the company.

BOX 6.4 BHARAT PETROLEUM SAP IMPLEMENTATION

Bharat Petroleum embarked upon enterprise integration through the implementation of an SAP R/3 Enterprise Resource Planning (ERP) system.

The implementation of the ERP system was not conceived simply as an information systems project, but built upon a previous de-layering and restructuring of the company around six new strategic business units.

Implementation was carried out with assistance from PricewaterhouseCoopers, 24 SAP consultants, a team of 70 in-house SAP qualified consultants, and 6 full-time change coaches. All users were involved in training, focused on improving "organizational learning," and Visionary Leadership and Planning Programs.

In the year after completion of the implementation, Bharat Petroleum achieved 24% sales growth. SAP itself rated Bharat Petroleum in the top quartile of SAP ERP implementations.

6.5 PROCESS INDUSTRY FOCUS

In the process industry, the primary focus is on **asset integrity,** whereas in many commercial operations, the focus is on "processing invoices."

Thus plant process management systems and not commercial business systems dominate. These could be highly complex and involve a high level of risk where human interaction is desirable to question and challenge proposals. Box 6.5 outlines examples of the automation of high risk processes.

BOX 6.5 HIGH-RISK PROCESSES

Automated Plant Change Request Process

This was implemented in a Gas Plant. The process was that the Request be sent by automated sequenced emails through the intranet for each participant's approval. However, the mentality was that "the previous person has checked it so it must be OK."

An audit revealed that the automated approach did away with the requirement of face-to-face interaction, which encouraged questioning and challenging each other on the issues involved in approving the change. **This resulted in a high-risk situation with respect to changes to the process units** (see also and Box 20.2: Group Think: Challenger Space Shuttle O-Ring Failure).

Automated Permit To Work Process

An automated Permit To Work (PTW) Process was piloted in the Refining Business Unit of an Oil and Gas company. Various issues were ironed out, including the requirement to have hard copies of the PTW at the worksite.

The process was then rolled out to other Business Units of the company with success.

6.6 SUMMARY

Complex processes need mapping using structured BPM techniques to ensure an accurate depiction of what is required.

Flowcharting of processes is the starting point of BPM, and software packages are available that can help with charting complex Processes. These computerized Processes can be uploaded to ERM systems, such as SAP or Oracle, with minimal corrective action.

Changing existing Processes needs a structured Change Management process (see Chapter 30: Strategies and Actions).

However, there is a limit to business process automation, and a balance between automated and manual Processes has to be sought for maximum value addition. High-risk Processes related to process plant safety require particular attention.

REFERENCES

1. *ARIS intelligent guide to BPM*. Software AG; 2012.
2. Software AG, https://www.softwareag.com/corporate/rc/rc_perma.asp?id=tcm:16-134695.
3. Business Process Modelling Notation (BPMN). BizAgi Process Modeller. www.bizagi.com.
4. Jeston J, Nelis J. *Business process management*. Routledge; 2014.

FURTHER READING

1. Harmon P. *Business process change: a business process management guide for managers and process professionals.* Morgan Kaufman; 2014.
2. *Process classification framework.* APQC. https://www.apqc.org/; 2009.
3. *Business process management: a consortium benchmarking study best practice report.* APQC Publications; 2005.

PART

GOVERNANCE AND PERFORMANCE

2

OVERVIEW

WHY IS GOVERNANCE SO IMPORTANT?

Governance sets the basis on how we do business. It is how we ensure compliance with compulsory and voluntary external and internal restraints - laws, codes of conduct, standards etc.

THE OBJECTIVE

The objectives of Part 2 are the following:

- Identify good governance by reviewing selected international and national codes
- Identify a suitable framework, containing all the elements of good governance, to ensure performance enhancement
- Demonstrate how conflict of interest and degradation of ethics detrimentally impacts performance
- Show that vertical and horizontal alignment of the business is paramount for success

PICTORIAL VIEW

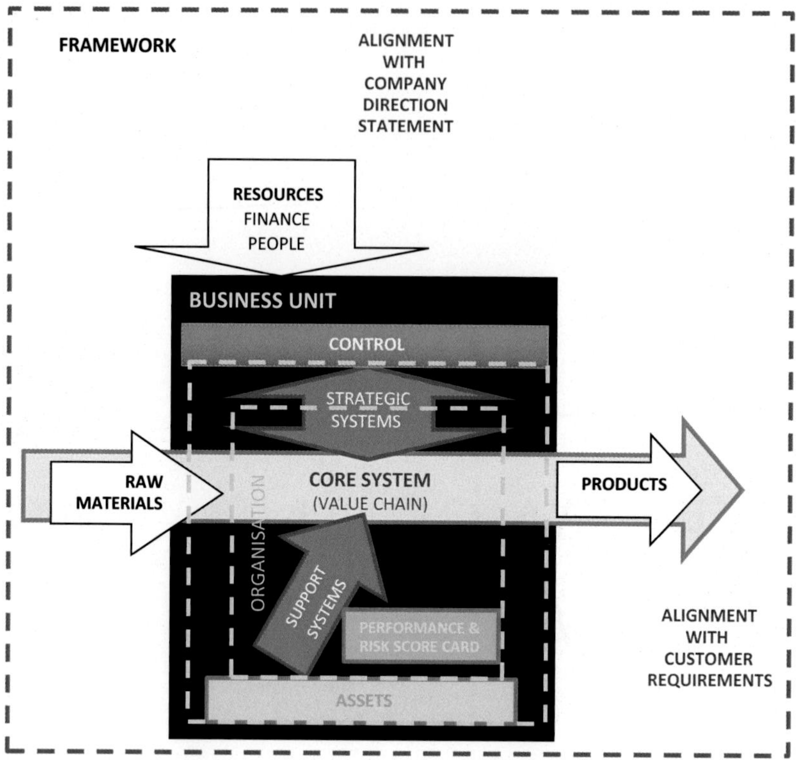

OUTLINE OF PART 2

- The company, as a legal entity, is defined.
- Principles of good governance are identified.
- A model framework for governance is presented, giving key elements that comply with the principles of good governance.
- The impact of poor performance, which is related to the lack of good governance, is demonstrated using diagrammatic representation of a decision-maker's attitude to risk (moral hazard) and adverse selection (asymmetric information).
- The role of the Board of Directors as the "fulcrum" of business performance is emphasized.
- The essential steps of a structured and transparent tendering and contract award process are detailed.
- The inter-relationship between governance, risk, integrity (of assets) and performance is demonstrated.
- The requirement of vertical and horizontal alignment of all aspects of the Business Unit for the survival of the company is explained by using the company direction statement (policy) as the guiding document.

CHAPTER 7

WHAT ARE THE PRINCIPLES OF GOOD GOVERNANCE?

7.1 INTRODUCTION

Core principles of governance, as applied to various codes/standards, are reviewed, and principles of good governance are identified. An appreciation of the importance of principles of good governance to the business for its survival is demonstrated.

Simple comparisons of Boards of Directors, Oil and Gas Producing Countries, and National Oil Companies (NOCs) are outlined, giving support to the principles. The approach of international rating agencies is also an essential factor for the financing of operations and investments.

7.2 TWO QUESTIONS

Two questions need to be answered:

1. What is the Company?

   and

2. How do we control the Company?

1. WHAT IS THE COMPANY?

Core Documents that define the Company are as follows.

Memorandum and Articles of Association: For Companies That Are Separate Legal Entities in Which Shareholders Have a Financial Interest

- Memorandum of Association is the constitution of the Company.
- Articles of Association are the rules adopted by the Company for the regulation of its internal affairs.

Mandate: For Directly Managed Business Units

- Similar to Articles of Association
- Defines boundaries and responsibilities

2. HOW DO WE CONTROL THE COMPANY?

The Company is controlled through a **framework** and set of **principles**.

Corporate Governance Framework

The Corporate Governance **Framework** is the documented control mechanism through which the Company is directed and controlled (see Chapter 8: Framework).

Performance Management for the Oil, Gas, and Process Industries. http://dx.doi.org/10.1016/B978-0-12-810446-0.00007-4

General Principles of Business Control

General Business **Principles** as discussed in this chapter are based on the following:

1. Cadbury Report (UK, 1992) and subsequent UK Corporate Governance Code (2010)[1]
2. Principles of Corporate Governance (OECD, 1998; 2004)[2]
3. Sarbanes-Oxley Act of 2002 (US, 2002)[3]
4. King III (2009) Corporate Governance Code for South Africa[4]

7.3 DEFINITIONS OF CORPORATE GOVERNANCE

Primary definitions are as follows.

INSTITUTE OF INTERNAL AUDITORS[5]

International Standards for the Professional Practice of Internal Auditing

> Corporate governance is the combination of processes and structures implemented by the board in order to inform, direct, manage, and monitor the activities of the organization toward the achievement of its objectives.

ORGANIZATION FOR ECONOMIC COOPERATION AND DEVELOPMENT

Principles of Corporate Governance

> Corporate governance involves a set of relationships between a company's management, its Board, its shareholders, and other stakeholders. Corporate governance also provides the structure through which the objectives of the company are set, and the means of attaining those objectives and monitoring performance are determined.

GENERIC DEFINITION 1[6]

> Corporate governance is concerned with the structures and systems of control by which managers are held accountable to those who have a legitimate stake in an organization.

GENERIC DEFINITION 2

> A control framework to ensure more informed risk-taking and decision-making to support the achievement of business objectives.

7.4 DISCUSSION OF CORPORATE GOVERNANCE CODES

Contemporary discussions on corporate governance tend to refer to **principles** raised in four documents listed in Section 7.2, released since 1990.

The Cadbury and OECD reports present general principles around which businesses are expected to operate to assure proper governance.

The Sarbanes-Oxley Act, informally referred to as Sarbox or Sox, was an attempt by the Federal Government of the US to **legislate** several of the principles recommended in the Cadbury and OECD reports. The US Sox Act of 2002 was intended to prevent reoccurrences of the Enron and WorldCom collapses, but nevertheless failed to prevent the world financial system collapse in 2008.

In summary, these principles are as follows:

- **Rights and equitable treatment of shareholders**: Organizations should respect the rights of shareholders and help shareholders to exercise those rights by openly and effectively **communicating** information and by encouraging shareholders to participate in general meetings.
- **Interests of other stakeholders**: Organizations should recognize that they have legal, contractual, social, and market-driven **obligations** to nonshareholder stakeholders, including employees, investors, creditors, suppliers, local communities, customers, and policy makers.
- **Roles and responsibilities of the board**: The board needs sufficient relevant skills and understanding to review and challenge management performance. It also needs adequate size and appropriate levels of **independence** and commitment to fulfill its responsibilities and duties.
- **Integrity and ethical behavior**: **Integrity** should be a fundamental requirement in choosing corporate officers and board members. Organizations should develop **codes of conduct** for their directors and executives that promote **ethical** and **responsible** decision-making.
- **Disclosure and transparency**: Organizations should clarify, and make publicly known, the roles and **responsibilities** of the board and management in order to provide stakeholders with a level of **accountability**. They should also implement procedures to independently verify and safeguard the integrity of the company's financial reporting. Disclosure of material matters concerning the organization should be timely and balanced to ensure that all investors have access to clear, factual information.

QUOTE FROM SIR ADRIAN CADBURY ("FATHER" OF CORPORATE GOVERNANCE)

> Governance yesterday focused on raising standards of board effectiveness; governance today [focuses] on the role of business in society; and the course of governance tomorrow is set by King III.

This is explained further as follows:

- The basis of US SOX (2002) is **Comply or Else**.
- The basis of UK Corporate Code (2010) is **Comply or Explain**.
- The basis of South Africa's King III (2009) is **Apply or Explain**.

When there are rules, there are those who skirt around the rules, which can result in a major failure of ethical standards, as occurred in the World Financial Collapse in 2008.

US and UK regulatory systems focus primarily on the shareholder, while others, such as the German, Japanese, and South African systems, focus on a greater balance of interests between shareholders and other external stakeholders.

King III (2009) is the latest Corporate Governance Code for South Africa with a primary emphasis on **leadership, sustainability, and corporate citizenship**. The code is designed on an "**apply or explain**" basis. This provides boards with the freedom to apply the recommendation differently, or apply another practice, if they consider that to be in the best interest of the organization, but they are then required to justify the departure from the recommendation. The principles in the code are drafted in such a way that they can be applied by any entity (public, private, or nonprofit).

The focus of governance systems could be reflected as shown in Table 7.1.

Table 7.1 Focus of Governance Systems

	Shareholder Focus	Stakeholder Focus
Benefits	For Investors: • high Rate of Return (RoR) • reduced risk For the Economy: • encourages entrepreneurship • encourages inward investment For Management: • independence	For Investors: • closer monitoring of management • longer-term decision horizon For Stakeholders: Deterrent to high-risk decisions
Disadvantages	For the Economy: • risk of short-termism • top management greed	For the Economy: • reduced financing opportunities for growth For Management: • potential interference • slower decision-making • reduced independence

Adapted from Johnson, Scholes, Whittington. Exploring corporate strategy. *Pearson; 2008.*

Shareholder focus in British Petroleum, under the Chairmanship of Sir John Browne, led to some major incidents. Box 9.5, Accountability, Responsibility, and Discipline gives details.

Box 7.1 discusses the issues of Shareholder versus Stakeholder focus.[7]

BOX 7.1 SHAREHOLDER VALUE AND/OR STAKEHOLDER VALUE?

"First, on a merely descriptive level, if one examines the relationship between the firm and the various groups to which it is related by all sorts of contracts, it is simply not true to say that the only group with a legitimate interest in the corporation are shareholders. From a legal perspective there are far more groups apart from shareholders that appear to hold a legitimate 'stake' in the corporation since their interests are already protected in some way. There are not only legally binding contracts to suppliers, employees, or customers, but also an increasingly dense network of laws and regulations enforced by society, which make it simply a matter of fact that a large spectrum of different stakeholders have certain rights and claims on the corporation…

A second group of arguments comes from an economic perspective. In the light of new institutional economics, there are further objections to the traditional stockholder view. For example, there is the problem of externalities: if a firm closes a plant in a small community and lays off the workers, it is not only the relation with the employees that is directly affected – shop owners will lose their business, tax payments to fund schools and other public services will also suffer – but since the company has no contractual relation to these groups, the traditional model suggests that these obligations do not exist…

Another, even more important aspect is the agency problem: one of the key arguments for the traditional model is that shareholders are seen as the owners of the corporation, and consequently the corporation has its dominant obligation to them. This view, however, only reflects the reality of shareholder's interests in a very limited number of cases. The majority of shareholders do not invest in shares predominantly to 'own' a company (or parts of it), nor do they necessarily seek for the firm to maximize its long-term profitability. In the first place, shareholders often buy shares for speculative reasons, and it is the development of the share price that is their predominant interest – and not 'ownership' in a physical corporation."

From Crane, Matten. Business Ethics: Managing Corporate Citizenship and Sustainability in the Age of Globalization. *Oxford University Press; 2010.*

KING IN BRIEF

Subjects covered are as follows:

- Structure
  - Composition
  - Executive committee
  - Nonexecutive directors
  - Independent nonexecutive directors
- Minimum number of directors
- Frequency of meetings
- Rotation
- Removal of CEO
- Chairman
- Board Committees
- Key Risk and Reporting Implications
  - Integrated Reporting
  - Combined Assurance
  - Annual Review of Internal Financial Controls
  - Risk-Based Auditing
  - IT Governance
  - Governing Stakeholder Relationships

King III is summarized in Fig. 7.1.

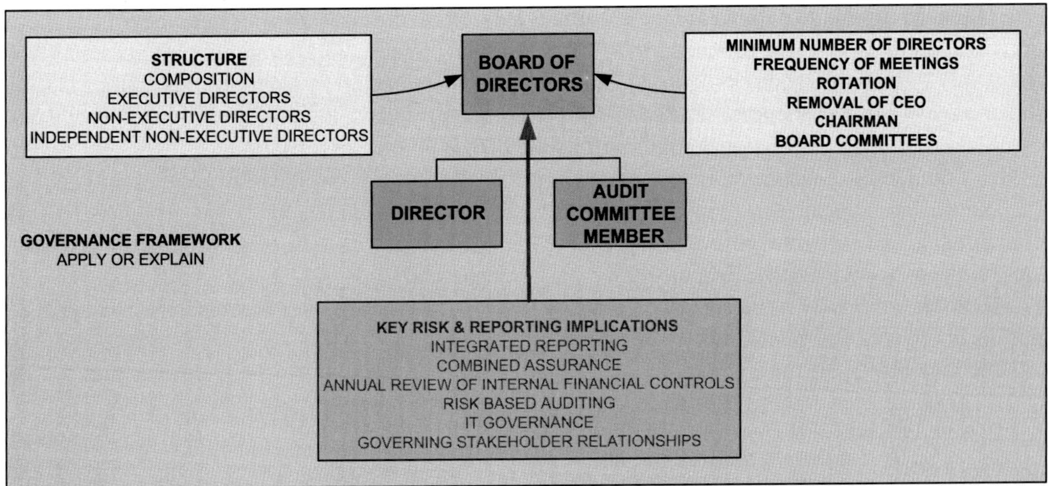

FIGURE 7.1

King III Governance Code.

Box 7.2 provides a good example of the contents expected in a simple Board (Governance) Manual. This is a primary part of the Governance Framework discussed in Chapter 8, Framework.

BOX 7.2 CORPORATE BOARD (GOVERNANCE) MANUAL EXAMPLE

Table of Contents

Corporate Governance Overview and Guidelines

1. Introduction
2. Director responsibilities
3. Director qualification standards
4. Board meetings
5. Board committees
6. Director compensation
7. Director orientation and continuing education
8. Director's access to management and independent advisors
9. Management evaluation, succession, and executive compensation
10. Code of ethics and conflict of interest
11. Annual performance evaluation of the board
12. Board interaction with shareholders, investors, press, customers, etc.
13. Periodic review of the corporate governance guidelines.

Appendix A: Audit and risk committee charter
Appendix B: Compensation committee charter
Appendix C: Nominating and governance committee charter
Appendix D: Social responsibility committee charter
Appendix E: Matters requiring board approval
Appendix F: Director independence standards
Appendix G: Commitments

7.5 CORE PRINCIPLES

The Organization for Economic Cooperation and Development (OECD) is an international economic organization of 34 countries founded in 1961 with a view to stimulate economic progress and world trade. It is a forum of countries committed to democracy and the market economy, providing a platform to compare policy experiences, seek answers to common problems, identify good practices, and coordinate domestic and international policies of its members.

In 2005, OECD published Guidelines on Corporate Governance of State-Owned Enterprises.

The Business Sector Advisory Group on Corporate Governance to OECD has articulated a set of core principles of corporate governance practices that are relevant across a range of jurisdictions. These are **F**airness, **A**ccountability, **R**esponsibility, and **T**ransparency.

A certain level of **I**ndependence of processes, decision-making, and mechanisms is required to minimize or avoid potential **conflicts of interest**.

The core principles must guide the interpretation of different governance **structures and systems** at the corporate level and the **D**iscipline to adhere these.

These core principles are summarized as follows.

FAIRNESS

All decisions taken, processes used, and their implementation will not be allowed to create unfair advantage to any one particular party.

ACCOUNTABILITY

Identifiable groups within the organization, for example, governance boards who take actions or make decisions, are authorized and accountable for their actions.

RESPONSIBILITY

Each contracted party is required to act responsibly to the organization and its stakeholders.

TRANSPARENCY

All actions implemented and their decision support will be available for inspection by authorized organization and provider parties.

INDEPENDENCE

All processes, decision-making, and mechanisms used will be established so as to minimize or avoid potential **conflicts of interest**.

DISCIPLINE

All involved parties will have a commitment to adhere to procedures, processes, and authority structures established by the organization.

Box 7.3 lists the OECD Principles of Corporate Governance (2004) contents headings.

BOX 7.3 THE OECD PRINCIPLES OF CORPORATE GOVERNANCE

1. Ensuring the Basis for an Effective Corporate Governance Framework
2. The Rights of Shareholders and Key Ownership Functions
3. The Equitable Treatment of Shareholders
4. The Role of Stakeholders in Corporate Governance
5. Disclosure and Transparency
6. The Responsibilities of the Board

From OECD Principles of Corporate Governance. *2004.*

7.6 THE EXTRACTIVE INDUSTRIES TRANSPARENCY INITIATIVE[8]

The Extractive Industries Transparency Initiative (EITI) is a global standard to promote open and accountable management of natural resources. It seeks to strengthen government and company systems, inform public debate, and enhance trust. In each implementing country, it is supported by a coalition of governments, companies, and civil society working together.

The 12 Principles of EITI, which were agreed by all stakeholders in 2003, lay out the general aims and commitments by all stakeholders. These are listed in Box 7.4.

BOX 7.4 THE EXTRACTIVE INDUSTRIES TRANSPARENCY INITIATIVE PRINCIPLES

The Extractive Industries Transparency Initiative (EITI) Principles provide the cornerstone of EITI. They are as follows

1. We share a belief that the prudent use of natural resource wealth should be an important engine for sustainable economic growth that contributes to sustainable development and poverty reduction, but if not managed properly, can create negative economic and social impacts.
2. We affirm that management of natural resource wealth for the benefit of a country's citizens is in the domain of sovereign governments to be exercised in the interests of their national development.
3. We recognize that the benefits of resource extraction occur as revenue streams over many years and can be highly price-dependent.
4. We recognize that a public understanding of government revenues and expenditure over time could help public debate and inform choice of appropriate and realistic options for sustainable development.
5. We underline the importance of transparency by governments and companies in the extractive industries and the need to enhance public financial management and accountability.
6. We recognize that achievement of greater transparency must be set in the context of respect for contracts and laws.
7. We recognize the enhanced environment for domestic and foreign direct investment that financial transparency may bring.
8. We believe in the principle and practice of accountability by government to all citizens for the stewardship of revenue streams and public expenditure.
9. We are committed to encouraging high standards of transparency and accountability in public life, government operations, and in business.
10. We believe that a broadly consistent and workable approach to the disclosure of payments and revenues is required, which is simple to undertake and to use.
11. We believe that payments' disclosure in a given country should involve all extractive industry companies operating in that country.
12. In seeking solutions, we believe that all stakeholders have important and relevant contributions to make.

From Extractive Industries Transparency Initiative (EITI). www.eiti.org/eiti/principles.

7.7 GOVERNANCE SURVEYS

Governance surveys provide support to the development of good governance.

Comparison of Boards, Evolution of Governance in Oil and Gas producing countries, Benchmarking of NOCs, the Role of International Rating Agencies, and various International Surveys are some exercises identifying shortcomings that are needed to be overcome in achieving the principles described in Section 7.5. These are described as follows.

BOARD OF DIRECTORS COMPARISON

Comparison of Boards of Directors (BoDs), using generic dimensions, help to assess compliance with codes and other elements of good governance. The following are some generic dimensions:

- Main tasks and roles
- Size of board, number of independent versus nonindependent members, professional background of board members
- Meeting's decision-making process and meeting frequency
- Board committee's roles and responsibilities, size, composition, and meeting frequency

Box 7.5 gives a sample of comparison of the Boards of two International Oil Companies (IOCs) and one NOC, using the previously mentioned generic dimensions.

BOX 7.5 BOARD OF DIRECTORS (BOD) COMPARISON STUDY

Generic Dimensions	BP	Shell	Saudi Aramco (SA)
Main tasks and roles	Purpose of BoD is to maximize long-term shareholder value through the allocation of its resources to activities in the oil, natural gas, petrochemicals, and energy businesses Main tasks: Pursue BP goal and account to shareholders for all actions of BP Govern BP by discharging its unique responsibilities Focus primarily on strategic issues and discuss economic, political, and social issues and any other relevant external matters that may influence or affect the development of BP's business Review and, where appropriate, determine the long-term strategy and annual plan for BP Monitor the decisions, actions, and performance of BP, including the implementation of, and performance against, the long-term strategy and the annual plan Ensure that systems and processes are in place for succession, evaluation, and compensation of the Executive Directors and other key members of senior management	BoD discusses, reviews, and takes necessary actions on reports submitted from the CEO, CFO, and various functions, including corporate. The BoD steers the company's overall strategy and management, corporate and capital structures, financial reporting and controls, and internal controls Main tasks: Strategy and management Oversee groups operations and management Review performance in the light of the group's strategy, objectives, business plans, and budgets Financial reporting and controls Approve preliminary announcements of interim and final results Declare dividends Contracts Approve major capital projects, investments, or contracts in excess of the amount delegated to the CEO Communication Approve resolutions and related documentation to put forward to shareholders at a general meeting Other Consider additional items that could be added based on meetings, items such as Internal Controls, Board membership and other appointments, remuneration, and governance matters	BoD members jointly oversee activities of SA and govern the organization by establishing broad policies and objectives Main tasks: Select and appoint chief executive to whom responsibility for administration of the organization is delegated Govern the organization by broad policies and objectives, formulated and agreed upon by the chief executive and employees, including to assign priorities and ensure the organization's capacity to carry out programs by continually reviewing its work Acquire sufficient resources for the organization's operations and to finance the products and services adequately Account to the public for products and services of the organization and expenditures of its funds Provide for fiscal accountability, approve the budget, and formulate policies related to contracts from public or private sources Accept responsibility for all conditions and policies attached to new, innovative, or experimental programs Provide vision for the future, develop and implement the long-term plan, and ensure that the organizational mission remains responsive to changes in the organizational realities
Size of board, no. of independent versus non-independent members, professional background of board members	15, Chairman, deputy chairman, CEO, CFO, 3 executive directors, 8 nonexecutive directors	13, Chairman, Deputy Chairman, CEO, CFO, Exploration & Production International, 8 nonexecutive directors	12. Chairman, President/CEO, 10 board members (3 nonexecutive and 7 executive)

Continued

BOX 7.5 BOARD OF DIRECTORS (BOD) COMPARISON STUDY—cont'd

Generic Dimensions	BP	Shell	Saudi Aramco (SA)
Meetings' decision-making process and meeting frequency	Meet 12 times per year	Meet 8 times per year	Not available
Board committee's roles and responsibilities, size, composition, and meeting frequency	Group Management Team (see management team benchmarking) Chairman Committee Safety, Ethic, and Environment Assurance Committee Audit Committee (see audit committee benchmarking) Remuneration Committee Nomination Committee	Group Management Team I(see management team benchmarking) Audit Committee (see audit committee benchmarking) Remuneration Committee Corporate Social Responsibility Committee Nomination and Succession Committee	Management Team (see management team benchmarking)

OIL AND GAS PRODUCING COUNTRY EVOLUTION COMPARISON

An Oil and Gas Producing Country may want to know what initiatives are needed to propel it to the next level of development.

Typically, Oil and Gas Producing Countries are categorized into the following development phases:

1. Entrepreneurship
2. Nationalization
3. Golden Age
4. Diversification

Firstly, countries are filtered. Key drivers for assessment are relative sector size, knowledge gap, and sector complexity. At the same time, the NOCs and IOCs are filtered to identify suitable comparative companies.

Box 7.6 gives an example showing Oil and Gas Sector Evolution Phases and selected countries.

BOX 7.6 GOVERNANCE BENCHMARKING OIL AND GAS PRODUCING COUNTRIES (C. 2010)

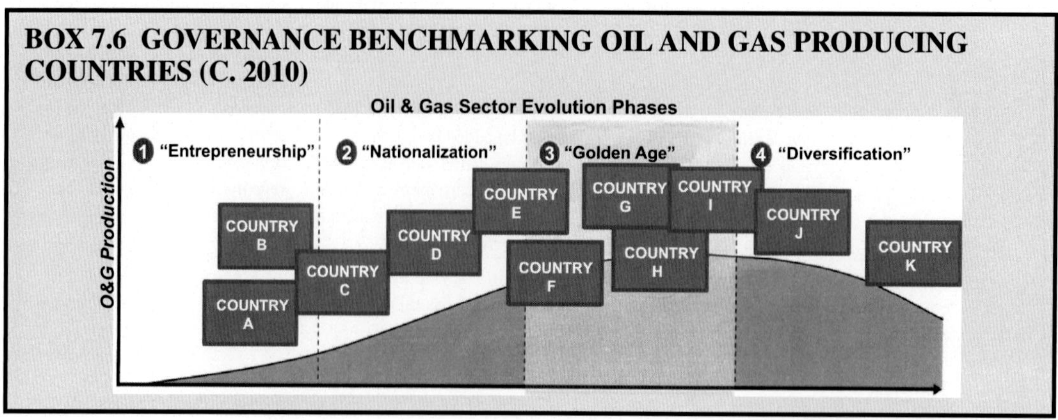

NATIONAL OIL COMPANY BENCHMARKING

Chatham House[9] has undertaken extensive benchmarking of NOCs. The benchmarking is based on five principles of good governance, which are as follows:

1. clarity of goals, roles, and responsibilities;
2. sustainable development for the benefit of future generations;
3. enablement to carry out the role assigned;
4. accountability of decision-making and performance; and
5. transparency and accuracy of information.

Box 7.7 shows the outline of the benchmarking survey with related questions.

BOX 7.7 CHATHAM HOUSE NATIONAL OIL COMPANY BENCHMARKING SURVEY

Five Principles of Good Governance

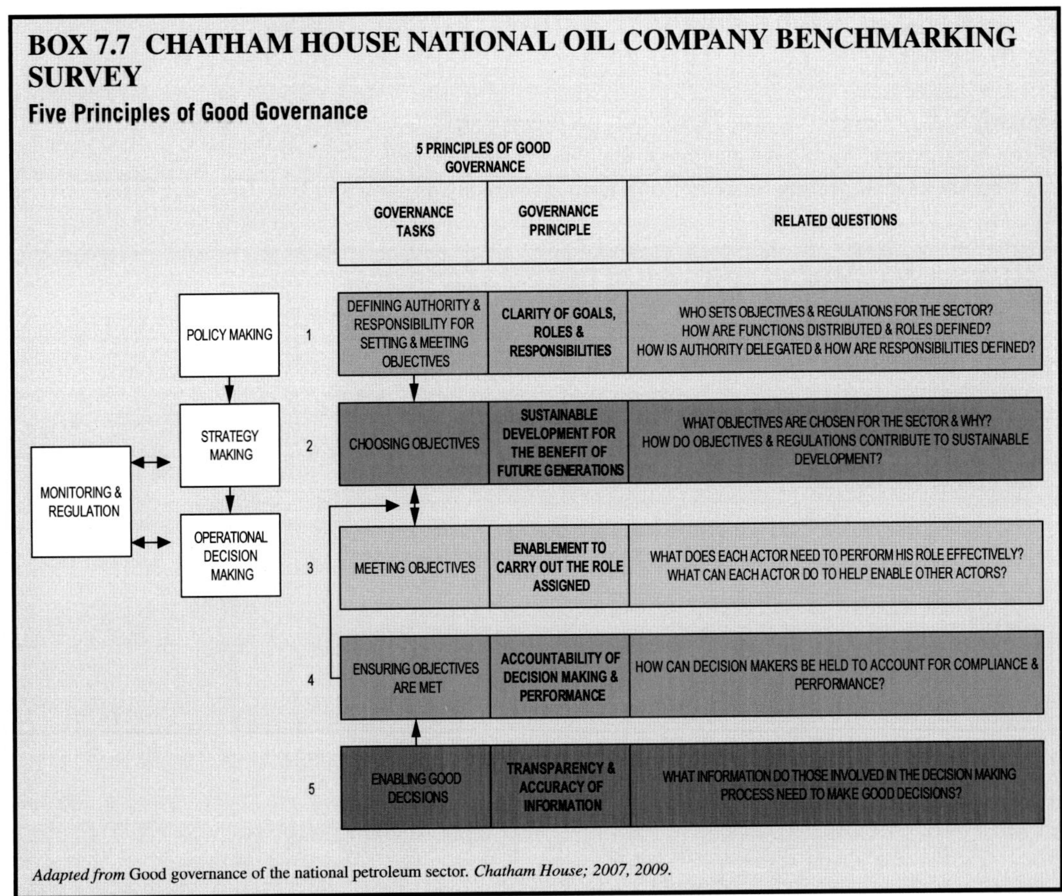

Adapted from Good governance of the national petroleum sector. *Chatham House; 2007, 2009.*

INTERNATIONAL RATING AGENCIES

Various international rating agencies, such as Standard & Poor's (S&P), Moody's, Fitch, and Dagong, carry out ratings of countries and large organizations. These ratings influence the ability to borrow funds for running a country or business and for specific major investments.

The acts leading to downgrading of South Africa's Electricity Supply Commission's once-prized bonds to "junk bonds" by international rating agencies are described in Box 9.6, The Jewel in the Crown.

Standard & Poor's Surveys[10]

S&P conducts Governance, Accountability, Management, Metrics, and Analysis (GAMMA) surveys to evaluate corporate governance practices. The process of their analysis involves the assessment of individual corporate and country practices and structures against the following broad principles:

1. **F**airness
2. **A**ccountability
3. **R**esponsibility
4. **T**ransparency

Moody's[11]

Each International Rating Agency has a scale that is used to assess ratings.

Moody's Credit Rating Scale for long-term investments is shown in Box 7.8.

BOX 7.8 MOODY'S CREDIT RATING SCALE

Range is from "Minimum credit risk" at the top to "In default, little prospect of recovery" at the bottom of the scale.

	Long-Term Rating	
Investment grade	Aaa	Smallest degree of risk
	Aa1	Very low credit risk
	Aa2	
	Aa3	
	A1	Low credit risk
	A2	
	A3	
	Baa1	Moderate credit risk
	Baa2	
	Baa3	
Noninvestment grade	Ba1	Questionable credit quality
	Ba2	
	Ba3	
	B1	High credit risk
	B2	
	B3	
	Caa1	Very high credit risk
	Caa2	
	Caa3	

INTERNATIONAL SURVEYS

Various international surveys are carried out regularly. Some examples are as follows.

Transparency International Corruption Perceptions Index[12]

The Corruption Perceptions Index was created in 1995 by Transparency International. Each year, about 200 countries are ranked according to the perceived level of corruption in each country. Transparency International also has a Bribe Payers Index.

In 2015, Denmark was placed first in the global anticorruption perceptions index.

Singapore, under the leadership of Prime Minister Lee, advocated a total intolerance to corruption, enforced by a Corrupt Practices Investigation Bureau.

Global Fraud Survey[13]

The annual Global Fraud Report, commissioned by Kroll and carried out by the Economist Intelligence Unit, is a very informative overview of worldwide fraud.

Reputation: Fortune's 50 Most Admired Companies[14]

Fortune magazine publishes "lists of the world's most admired companies." For their "50 most admired companies overall" list, Fortune's survey asks businesspeople to vote for the companies that they admire the most, from any industry.

7.8 LEADERS IN CORPORATE GOVERNANCE

Some Oil and Gas companies have built a reputation for being leaders in Corporate Governance, only to be let down by members of their BoD. Cleanups sometimes ensue, but it takes a long time to rebuild a reputation and regain stakeholder trust.

Box 7.9 shows two examples: an NOC and an IOC.

BOX 7.9 THE LEADERS IN CORPORATE GOVERNANCE!

Until recently, Petrobras and Shell were considered leaders in corporate governance.

Petrobras[15]

The Petrobras website shows how far they have progressed, especially with respect to **sustainability and corporate citizenship**, as emphasized by King. However, recent political interference into the affairs of Petrobras has occurred with allegations of corruption.

> *According to trial testimony, top executives at Petrobras accepted huge bribes from a cartel of companies, enriching themselves while also channeling funds to political figures and to the leftist Workers' Party. Although no testimony has emerged suggesting that Ms. Rousseff personally profited from the scheme, she was the chairwoman of the oil giant from 2003 to 2010, roughly corresponding to the period that the system of collusion, kickbacks and payoffs took shape. – The New York Times.*
>
> Ms. Rousseff was subsequently impeached and removed from the office of President of Brazil.

Shell[16,17]

The Oil and Gas reserves reported in Shell's Annual Report of 2002 was found to be overstated.

> *Ultimately, the alleged overstatement would prove to be 4.47 billion barrels of oil equivalent, or about 23% of the company's total. A joint investigation was conducted by the Securities and Exchange Commission and Federal Reserve, and Shell settled claims with the regulators for USD 120 million and GBP 17 million (or USD 28 million), respectively, without admitting to, or denying the findings of the Commission.*
>
> The Group Chairman, CFO, and E&P CEO left the company shortly after the reserves' revelations.

The ASEAN (Association of South East Asian Nations) 'flu' of 1997 triggered improvements in the regulation of the region's capital markets. Outcomes include the formation of the ASEAN Corporate Governance Scorecard and subsequently the ASEAN Corporate Governance Awards. Box 7.10 outlines the ASEAN Corporate Governance Scorecard.[18]

BOX 7.10 ASEAN (ASSOCIATION OF SOUTH EAST ASIAN NATIONS) CORPORATE GOVERNANCE SCORECARD

The Asian Development Bank (ADB) and the ASEAN Capital Market Forum (ACMF) together developed the ASEAN Corporate Governance Scorecard.

The ASEAN Corporate Governance Scorecard provides a rigorous methodology benchmarked against international best practice—including the OECD's principles of corporate governance—to assess the corporate governance performance of publicly listed companies (PLCs) in the six participating ASEAN member countries.

This common methodology provides foreign investors and external fund managers comparable information to form part of their investment decision-making process. The scorecard also provides assurance to foreign investors that corporate governance is a priority agenda in the region.

7.9 SUMMARY

Corporate governance is concerned with the structures and systems of control by which managers are held accountable to those who have a legitimate stake in an organization.

The company is defined by its Memorandum and Article of Association and controlled through a **governance framework** and set of **principles**.

Different Corporate Governance Codes have differing emphases, varying from strict enforcement to reliance on ethical management. The South African model (King III) appears to be the leader in legislation.

Definitions of corporate governance vary, but can be distilled into core principles, being Fairness, Accountability, Responsibility, Transparency, Independence, and Discipline. These principles will be demonstrated in Chapter 9, How Does Governance Affect Performance?

Various international rating agencies carry out ratings of countries and large organizations based on principles of good governance. These ratings influence the ability to borrow funds for running a country or business and for specific major investments.

Various international surveys are carried out regularly. These include the Corruption Perceptions Index, Global Fraud Survey, and Fortune's 50 most admired companies.

REFERENCES

1. *Cadbury report (UK, 1992) subsequent FRC UK corporate code (CC)*. June 2010.
2. *OECD principles of corporate governance*. 2004.
3. Sarbanes-Oxley (2002). *Section 404: a guide for management by internal controls practitioners*. 2nd ed. The Institute of Internal Auditors; 2002. January 2008.
4. *King code of governance for South Africa*. Institute of Directors Southern Africa; 2009.
5. The Institute of Internal Auditors (IIA) International Standards for the Professional Practice of Internal Auditing. https://na.theiia.org/standards-guidance/Public%20Documents/IPPF%202013%20English.pdf.

6. Johnson, Scholes, Whittington. *Exploring corporate strategy*. Pearson; 2008.
7. Crane, Matten. *Business ethics: managing corporate citizenship and sustainability in the age of globalization*. Oxford University Press; 2010.
8. Extractive Industries Transparency Initiative (EITI). www.eiti.org/eiti/principles.
9. *Good governance of the national petroleum sector*. Chatham House; 2007, 2009. www.chathamhouse.org.uk/research/eedp/current_projects/good_governance.
10. Standard & Poor's. *Criteria: GAMMA Scores*. April 2008.
11. Moody's credit rating scale. http://chartsbin.com/view/1175.
12. Transparency International. The corruption perceptions index. www.transparency.org.
13. Kroll. *Global fraud report 2015–16*. http://www.kroll.com/global-fraud-report.
14. Fortune's most admired companies. http://fortune.com/worlds-most-admired-companies.
15. New York Times, Sreeharsha V. *Brazilian senator and banker are arrested as Petrobras scandal widens*. November 25, 2015.
16. Tran M. *Shell fined over reserve scandal*. The Guardian; July 29, 2004.
17. CNN/Money. http://money.cnn.com/2004/08/24/news/international/royaldutchshell_sec.
18. ASEAN corporate governance scorecard: country reports and assessments 2013–2014. Asian Development Bank.

FURTHER READING

1. Naidoo R. *Corporate governance*. Lexis Nexis Butterworths; 2009.

CHAPTER

GOVERNANCE FRAMEWORK

8

8.1 INTRODUCTION

A structured Governance Framework (GF) model based on four elements (Control, Organization, Assets, and Systems) is described in this chapter.

An overview of this model and its associated GF document is outlined. This could be used to develop a governance system or upgrade a governance system to comply with the various corporate codes and associated principles. The model applies the Business Unit concept and the Systems approach described in Part 1, Systems.

The following is an example of a directive from a Chief Executive to ensure responsible parties carry out adequate business controls:

> No company can exist without controls of some form. For a business to function effectively and efficiently, there must be a clear and well-thought-out framework of controls that are appropriate to the ***Group of Companies*** and the particular ***Business Unit***.
>
> It is the responsibility of the Chief Executive Officers/Managing Directors (MDs)/General Managers and Corporate Managers to establish, maintain, operate, and demonstrate an appropriate framework of business controls so as to cover all activities of the ***Business Unit***.

8.2 BEFORE AND AFTER

If a Systems approach is not adopted from the start, companies tend to build silos resulting in duplication of effort, wasted resources, poor integration, and higher operating costs.

The objectives of a Systems approach are to have the following:

- Streamlined Processes within Systems (Part 1, Systems)
- Common integrated computer platform (Part 1, Systems)
- Integration of Systems (Part 1, Systems and Part 2, Governance and Performance)
- Reduced overall Business Risk (Part 3, Risk and Performance)
- Clear indication of high-level performance relative to targets (Part 4, Performance Indicator Selection)
- Efficient and effective use of resources: raw materials, human capital, and financial (Part 5, Asset Performance Management)
- Sustainability of physical assets (Part 5, Asset Performance Management)
- Long-term integrity of physical assets (Part 5, Asset Performance Management)
- Security of information (Part 5, Asset Performance Management)
- Lower costs (Part 6, Benchmarking)
- High-quality knowledge base available for decision-making (Part 7, Assessment, Strategies, and Reporting)

Performance Management for the Oil, Gas, and Process Industries. http://dx.doi.org/10.1016/B978-0-12-810446-0.00008-6

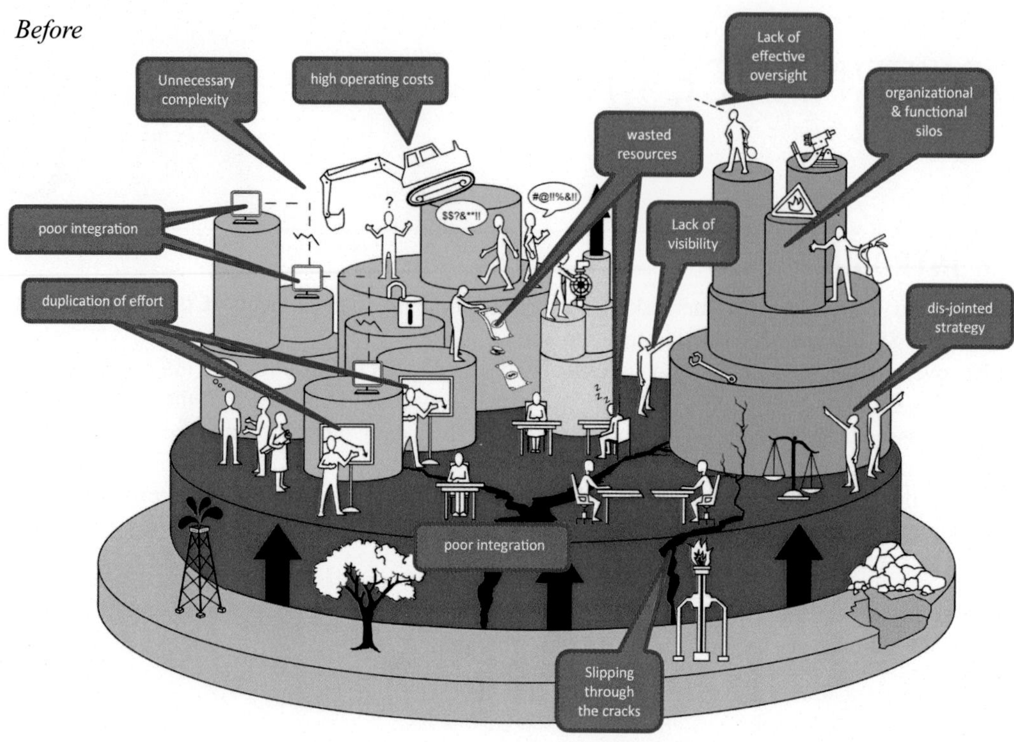

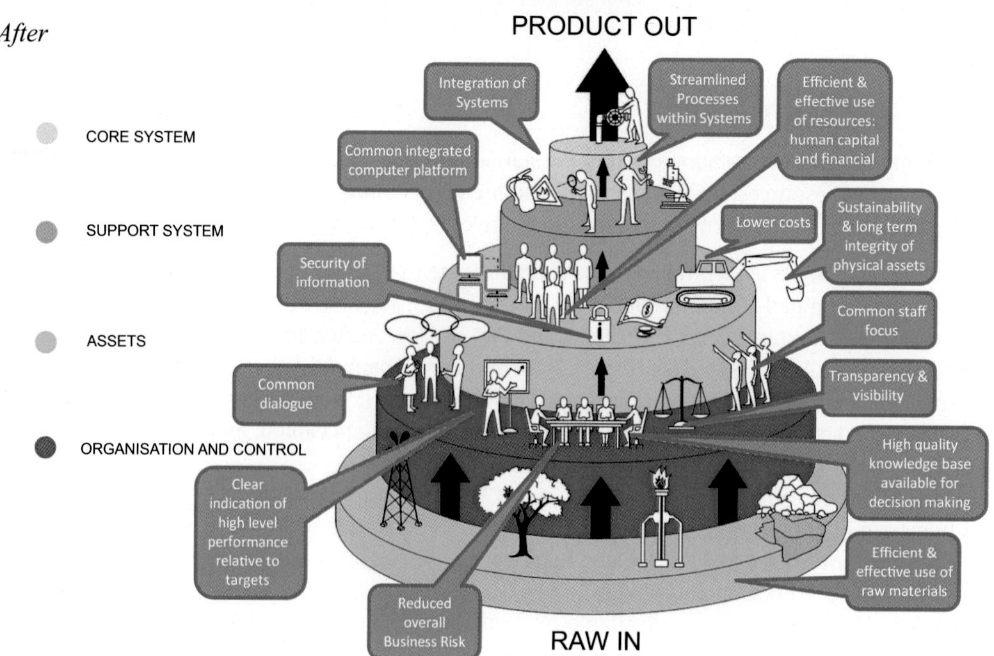

FIGURE 8.1

Before and after applying a governance framework and Systems approach.

- Common dialog and staff focus (Part 7, Assessment, Strategies, and Reporting)
- Transparency and visibility (Part 8, Business Oversight)

Fig. 8.1 illustrates possible "before" and "after" scenarios related to applying the GF and Systems approach to managing the business. This depiction also shows the extremes in maturity of a business.

Maturity will be discussed further in Chapter 29, Identification, Analysis, and Evaluation of Gaps and Chapter 39, Alignment to Achieve Recognition for Excellence.

8.3 THE MODEL GOVERNANCE FRAMEWORK (GF)

The principles of good governance described in the previous chapter are covered in the Model GF, which encompasses the complete Company as described in its **Memorandum and Articles of Association**. This could include a number of Business Units or just one Business Unit.

OUTLINE

- The **Corporate Objectives**, contained in the Company **Direction Statement**, are the source of an integrated GF.
- The **Corporate Objectives** should give guidance to the Business Management Systems in which the Company is engaged.
- The **Management Systems** provide a basis for considering which business controls are required.
- Individual controls for each **Business Unit** falls within one or another of three interlinking types termed "control mechanisms." These are People Management (Organizational Effectiveness), Process Management (Systems), and Performance Management.
- The business controls must be applied in the business operations in an effective and efficient manner.
- Managers should review and appraise:
  - Performance of the business activities against quantitative and qualitative yardsticks
  - The continued appropriateness of the GF
- It is an important responsibility of managers to review the appropriateness of their business controls whenever there are changes to:
  - The business environment
  - Company or departmental organizations
- Occasionally, a change in the business environment is such that the Corporate Objectives themselves may change, which would in turn trigger review and possible amendment of related parts of the GF.

ALIGNMENT: BUSINESS OBJECTIVES AND VALUE CHAIN

Corporate Objectives should flow from the **Vision** (Where we want to be) and **Mission** (Why we exist) of the company (Vertical Alignment). Horizontal Alignment should coincide with one or more **Core Management Systems** being the primary Value Chain. This is discussed further in Chapter 10, Alignment.

The Model GF is divided into four elements as follows:

1. Control: "How we control the company"
2. Organization: "How we are structured and behave"

3. Assets: "What we have"
4. Systems: "How we do things"

Fig. 8.2 depicts the Model GF.

1. Control

Performance is controlled by external and internal restraints generally as follows:

1. Industry Best Practices
2. International Standards (ISO, etc.)
3. Country or State Laws and Regulations
4. Company Policies and Regulations

Compliance with ISO Management Standards ensures implementation of the continual improvement cycle.

A Parent Company's Business Unit approach to Governance ensures responsibility for Control and Performance within each Business Unit in line with the Parent Company's Mission and Corporate Objectives, ensuring sustainable optimum value addition for the Business Unit.

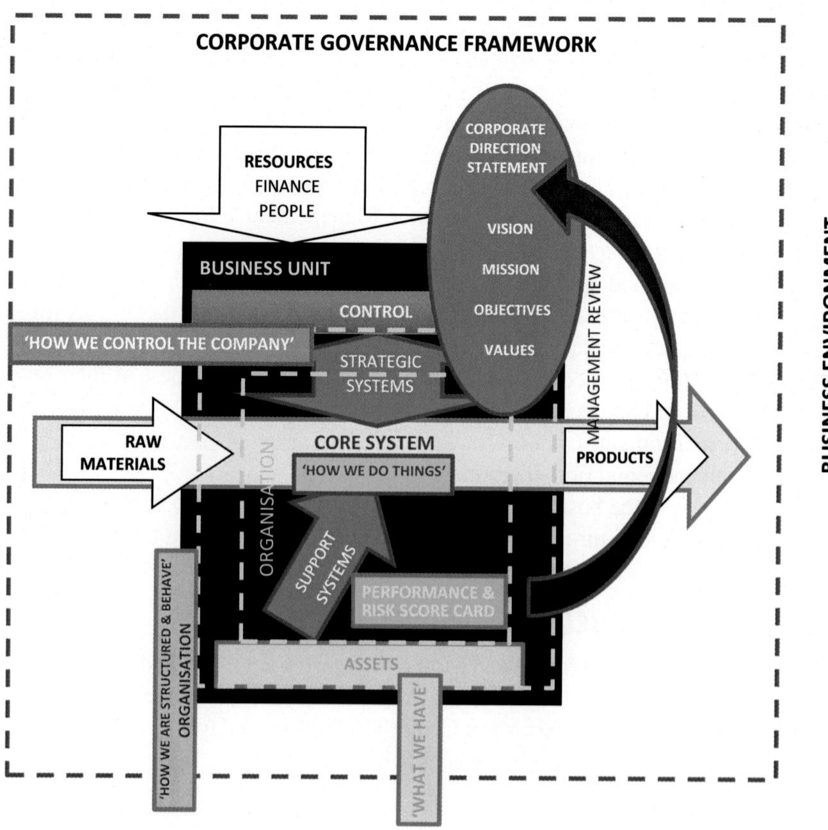

FIGURE 8.2

Model governance framework (GF).

The Business Unit is discussed in more detail in Chapter 2, The Business Unit.

Characteristics of Control

The overall control system must be geared to changes in risks and opportunities in the business environment. In relation to any particular risk, there are **four strategies** to choose from. These are shown in Table 8.1.

Table 8.1 Risk Strategies

Tolerate	Taking the risk where the consequences are small or the practicality of the other responses is unacceptable
Treat	Treating the risk by establishing effective and economical controls to reduce risks to an acceptable level
Transfer	Transferring the risk to other parties through, for example, insurance, hedging, or contracting out
Terminate	Terminating the risk by ceasing the activity or divesting from that particular business area

The previous is discussed further in Part 3, Risk and Performance.

Controls must contribute to the fulfillment of the Corporate Objectives. They need to be "fit for purpose."

People are the heart of a sound control system. The effectiveness of any system of control depends on the following:

- The "tone" set by senior management. This is the most important factor contributing to its success.
- Adequate time and resources being made available for operation, maintenance, and review of business controls.
- The quality of managers and staff generally, and the degree to which the system is understood, supported, and promoted by them.
- The continued appropriateness of the method of communication between them.
- The understanding of what information is necessary for them to perform their responsibilities. The corresponding data should be reliable, timely, and suitably presented.

2. Organization

Appropriate Roles, Responsibilities, Accountabilities, and Organizational Structures are critical for efficient implementation of Plans and Strategies so as to achieve Performance Targets. These have a direct bearing on the motivation of individuals and groups.

Organization is discussed further in the following chapters:

- Chapter 10, Alignment
- Chapter 20, Human Capital
- Chapter 34, Roles and Responsibilities

3. Assets

The following assets are primary contributors to Performance Management.

Physical

Most performance relates to the Production **Asset** Business Units.

Human Capital

Performance of individuals and groups is critical to the success of the business. Directly employed staff should undergo an annual appraisal cycle based on "SMART" (**S**imple, **M**easurable, **A**ligned, **R**epeatable, **T**imeliness) objectives cascaded from the Corporate Objectives.

Information

Timely and appropriate use of information is critical for decision-making related to Performance Management.

Finance

The timing and utilization of Financial Assets to ensure optimum value addition is essential.

Assets are discussed in detail in Part 5, Asset Performance Management.

4. Systems

Systems are as follows:

1. **Strategic**: These require cross company synergy/trade-off.
2. **Core**: These are the Value Chains of the **Business Unit**.
3. **Support**: These support the **Core Systems** to ensure optimum value addition.

Systems are discussed in detail in Part 1, Systems.

8.4 PHYSICAL PICTURE OF THE MODEL GOVERNANCE FRAMEWORK

The GF needs to be all-encompassing. The establishment of this on the company intranet with electronic Management System Manuals and printable PDF documents is the trend.

The Governance Framework must include the Board Manual, Corporate Policies and Standards, Committee Charters and Terms of References (ToRs), Levels of Authority (LoA), and the Internal Audit Process, possibly bundled into a Governance System.

The Governance Framework and an Integrated Management System (IMS) need to come together physically. The ***Statoil Book*** shown in Box 4.8, The Statoil Book Processes is an excellent example of how an IMS is embedded in the GF.

Typical contents of the Governance Framework are shown in Box 8.1.

THE GOVERNANCE FRAMEWORK DOCUMENT AND THE INTEGRATED MANAGEMENT SYSTEMS MANUAL

There sometimes appears to be a conflict between the GF document and an IMS Manual. The Governance Framework relates to a legal entity, and an IMS for ISO certification relates to an asset or group of assets.

BOX 8.1 GOVERNANCE FRAMEWORK CONTENTS

Strategic Systems (Examples)

Governance Management System

- **Governance Framework (GF) document**: Elements discussed in this chapter
- **Board (Governance) Manual**: Elements discussed in Chapter 7, What Are the Principles of Good Governance?
- Corporate Standards and Policies
- Committee Charters/ Terms of References (ToRs)
- Levels of Authority (LoA)
- Internal Audit Process (in line with International Standards for the Professional Practice of Internal Auditing)

Integrated Management System (Includes How All Documentation for Governance Framework is Managed Within Management Systems)

- **Integrated Management System (IMS) Manual:** Elements discussed in Chapter 4, Management Systems Determination and Requirements
- Procedures to comply with the requirements of certification to various codes, standards, and regulations (ISO, government, etc.)
- Matrix of compliance relating documents in GF to paragraphs of codes, standards, and regulations

  Strategic Planning System: Elements discussed in Chapter 5, The Business Cycle.

  Annual Planning and Budgeting System: Elements discussed in Chapter 5, The Business Cycle.

  Performance Management System: Elements discussed in **all chapters**

  Risk Management System: Elements discussed in Chapter 11, Risk Management, Chapter 12, Risk Control Mechanisms, and Chapter 13, Merging Performance and Risk.

  HSE/Asset Integrity Management System: Elements discussed in Chapter 22, Physical Assets.

  Human Capital Management System: Elements discussed in Chapter 20, Human Capital.

  Financial Management System: Elements discussed in Chapter 21, Finance.

  Investment Management System: Elements discussed in Chapter 33, The Opportunity Lifecycle.

  Affiliate Oversight Management System: Elements discussed in Chapter 34, Roles and Responsibilities and Chapter 35, Management Review.

  Hydrocarbons Management System: Elements discussed in Chapter 36, Hydrocarbon Accounting.

  ***Core Systems**: Elements discussed in* Chapter 4, Management Systems Determination and Requirements.

  ***Support Systems**: Elements discussed in* Chapter 4, Management Systems Determination and Requirements.

The complementary relationship between a **GF document** and a **Corporate IMS Manual** is shown in Table 8.2.

Table 8.2 Document Comparisons

Governance Framework (GF)	Corporate Integrated Management System Manual (IMS)

• **Scope** encompasses the complete Legal Entity
• **Focus** on framework and principles for managing the business based on:
  ◦ Corporate Policies
  ◦ Strategic Systems
  ◦ Business Risk
• **Audits** carried out based on risk to the business
• **Applied** to a Legal Entity or Group of Companies
• **Shareholder** focus (vertical alignment)
• **External restraints**
Country Laws, IFRS, OECD, IMO, UNFCCC, UN Global Compact, etc.
	• **Scope** encompasses the Business Unit or group of Business Units
	• **Focus** on certification based on:
	  ◦ HSEQ Policies
	  ◦ System Procedures
	  ◦ Management Review
	• **Audits** carried out against requirements for certification
	• **Applied** to Assets: Operations and Services
	• **Customer** focus (horizontal alignment)
	• **External restraints**
	ISO standards, Consent To Operate (CTO), etc.

HSEQ, Health Safety Environment and Quality; *IFRS*, International Financial Reporting Standards; *IMO*, International Maritime Organization; *UNFCCC*, United Nations Framework Convention on Climate Change.

8.5 THE MODEL GOVERNANCE FRAMEWORK DOCUMENT: DISCUSSION OF CONTENTS

Based on the Model Governance Framework, a "top" document is required to draw together all aspects of the company. This is often referred to as the "Corporate Governance Framework."

The outline of the document is as depicted in Fig. 8.3.

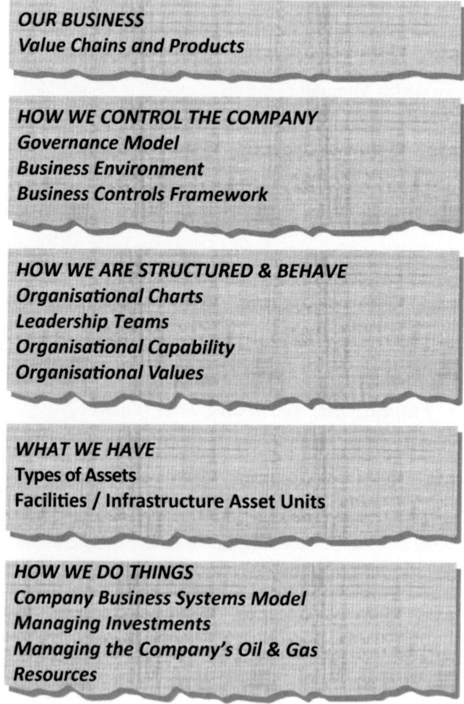

OUR BUSINESS
Value Chains and Products
HOW WE CONTROL THE COMPANY
Governance Model
Business Environment
Business Controls Framework
HOW WE ARE STRUCTURED & BEHAVE
Organisational Charts
Leadership Teams
Organisational Capability
Organisational Values
WHAT WE HAVE
Types of Assets
Facilities / Infrastructure Asset Units
HOW WE DO THINGS
Company Business Systems Model
Managing Investments
Managing the Company's Oil & Gas
Resources

FIGURE 8.3

Governance framework document.

This document could be intranet based, as shown in the **Statoil Book** example in Box 4.8, The Statoil Book Processes.

The contents of a Model Governance Framework are shown in Box 8.2.

Explanation of suitable content is outlined as follows:

SECTION 1.2: SCOPE

The scope is all-encompassing and based on the Memorandum and Articles of Association of the Company.

BOX 8.2 MODEL GOVERNANCE FRAMEWORK CONTENTS

P *denotes pointers on suitable content for inclusion as outlined in Section 8.5.*

1. Introduction	
1.1 Forward	
1.2 Scope	**P**
1.3. Our Business	
1.3.1. Local	
1.3.2. International	
1.4 Our Mission and Vision	
1.4.1. Responsibility	
1.5 Terminology	
2. Business Control Framework: "How We Control the Company"	**P**
2.1 Overview	
2.2. Foundations	**P**
2.2.1 Company General Business Principles	
2.2.2 Code of Conduct	
2.2.3 Statement of Risk	
2.2.4 Group Policies, Standards, Regulations, Frameworks, and Manuals	
2.3 Organizational Relationships	**P**
2.3.1 Corporate Entity Structure	
2.3.2 Legal Entities	
2.3.3 Applicability to Affiliates (Joint Ventures and Other Relationships)	
2.4 Activities	**P**
2.4.1 Delegation of Authority	
2.4.2 Strategy, Planning, and Appraisal	
2.4.3 Assurance and Compliance	
2.5 Business Environment	**P**
3. Managing Our Organization: "How We Are Structured and Behave"	**P**
3.1 Organizational Design	
a. Asset Units	
b. Service Units	
3.2 Organizational Structure	
3.3 Board of Directors (BOD)	
3.4. Managing Director (MD)	
3.5. Business Units	
3.5.1. Business Unit Heads	

Continued

BOX 8.2 MODEL GOVERNANCE FRAMEWORK CONTENTS—cont'd

3.5.2. Functions

3.6. Organization Charts

3.7. Leadership Teams

3.8 Organizational Capability

3.9 Interface Management

3.9.1. Responsibilities

3.9.2. Affiliates and Other Financial Interests

3.9.3. Service Level Agreements (SLAs)

3.9.4. Other Stakeholders

4. Managing Our Assets: "What We Have" P

4.1 Definition

4.2 Types of Assets

4.2.1. External

4.2.2. Natural

4.2.3. Human

4.2.4. Physical

4.2.5. Information

4.2.6. Financial

4.2.7. Stock

4.2.8. Intangible

4.2.9. Investment

4.3. Asset Integrity

5. Managing Our Systems: "How We Do Things" P

5.1 Company System Model

5.2. Strategic Systems Descriptions and Ownership

5.2.1. Governance

5.2.2. Strategic Planning and Budgeting

5.2.3. Allocation of Resources

5.2.4. Risk Management

5.2.5. HSEQ/Asset Integrity

5.2.6. Affiliate Management

Appendix A: Products
Appendix B: Company Group Structure
Appendix C: Company Asset and Service Units
Appendix D: Performance Report (PR) and Business Plan (BP) Formats
Appendix E: Related Documents
Appendix F: Definitions
Appendix G: Generic Roles and Responsibilities

SECTION 2: BUSINESS CONTROL FRAMEWORK: "HOW WE CONTROL THE COMPANY"

The Business Control Framework should cover, and link together, all Systems, Organizations, and Assets within the Company.

The following are the main elements:

1. **Foundations**
2. **Organizational Relationships**
3. **Activities**

This is depicted in Fig. 8.4, and further explained in the following sections.

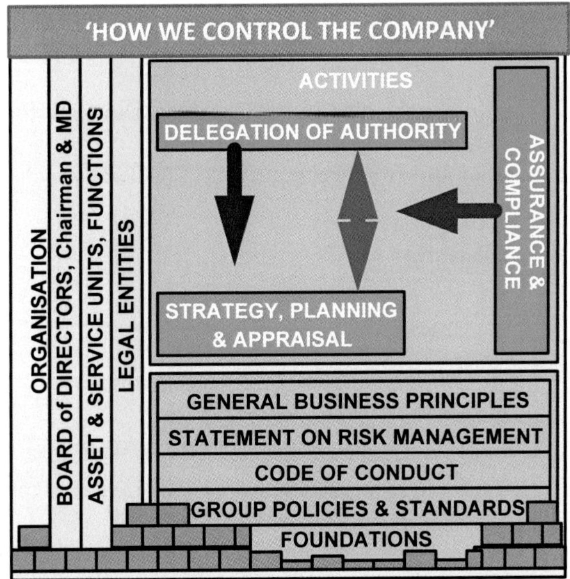

FIGURE 8.4

Three elements of business control framework.

Section 2.2: Foundations

Foundations consist of the following:

1. General Business Principles
2. Statement of Risk
3. Code of Conduct
4. Group Documentation: Policies, etc.

1. Company's General Business Principles

General Business Principles describe the following:

1. **Objectives of the Company**
2. **Core values**
3. **Responsibilities**
4. **Principles and behaviors** by which the Company does business
5. Commitment to contribute to **sustainable development**
6. Disclosure and **transparency**

These are as follows:

1. Objectives of the Company The Company Direction Statement generally consists of at least the **Corporate Vision** ("Where we want to be") and **Mission** ("Why we exist"). The **Corporate Objectives** describe the "Direction we need to take."

2. Core Values Core Values clearly set out how people are expected to perform. They define what is important to us and include values such things as professionalism, integrity, cooperation, continuous improvement, etc.

Note: Items 1 and 2 should be embedded in the Company Direction Statement. See Chapter 10, Alignment, for details.

3. Responsibilities The Board of Director's **responsibilities** generally include the following:

1. Establishing and reviewing Corporate Policy
2. Strategic Guidance and Management
3. Corporate Governance
4. Performance Monitoring
5. Risk and Audit Review

4. Principles and Behaviors by Which the Company Does Business

Text example:

> *The company employees shall:*
>
> a. *have an obligation to exercise utmost fairness, honesty, objectivity, and diligence, and maintain an exemplary level of ethical behavior in the performance of their duties for the* ***Company****, and shall reject bribery in all its forms*
> b. *avoid Conflicts of Interest whenever possible, and disclose them to the* ***Company****.*

Refer to **Code of Conduct** section for details.

5. Sustainable Development This is normally embedded in the decision processes for investment and operation. However, a generic statement on the subject should be provided.

6. Disclosure and Transparency Compliance with the requirements of a **Government** anticorruption watchdog is essential. **"Transparency"** and **"communication"** should be key words in the "values" part of the Company **Direction Statement**.

Statement of Risk Management

An example Corporate Risk Statement is shown in Box 11.2, Statement of Risk.

This statement is supported by further separate documentation cascading throughout the organization. See Part 2, Risk and Performance, for details.

Code of Conduct

Text example:

> *The **Code of Conduct** provides guidance to staff on how to comply with the law and regulations, and operate in accordance with the **General Business Principles**. The **Code of Conduct** should be clearly documented in Regulations covering Code of Ethics and Conflict of Interest.*

See Section 9.5, Conflict of Interest and Ethics Corporate Statements, for further details.

Group Documentation: Policies, Standards, Regulations, and Frameworks

Text examples:

> *The implementation of **Group Policies, Standards, Regulations, Frameworks, Manuals, etc.,** (Group documentation) is mandatory in all **controlled** Affiliates.*
>
> *A risk-based approach is adopted for their implementation in **Affiliates** and other companies that are **not controlled**. Group documentation is established for essential matters: matters that present significant **Group**-level risks and opportunities or matters that are subject to external stakeholder expectations and external disclosures. **Business** and **Functional** executives are required to understand these documents and ensure that they comply with them in their day-to-day operations.*

Section 32.3, Types of Companies, discusses control of companies.

Section 2.3: Organizational Relationships

Corporate Entity Structure

The **organizational structure** should be designed, aligning with Systems where possible, to achieve the Company's overall business objectives while respecting the separate legal identity of the individual companies through which it is implemented.

Text example:

> ***Organization** as documented in the **Governance Framework** relates to how the Company and the various legal entities relate to each other and how their business activities are organized and managed. The **Organization** consists of the **Corporate Entity Structure, Legal Entities**, relationships with Affiliates and other entities, the **Board of Directors (BoD)** structure, and generic relationships.*
>
> *The Company is governed via the **BoD** through the **MD** to the various **Asset** and **Service** Business Units. The Company internally organizes its activities along **Business Unit** and **Systems** lines with additional provisions to take into account country perspectives.*

Authorities are delegated to individual staff separately in their capacity as employees of a particular Affiliate company (corporate authorities). These separate corporate and organizational "delegation of authority" processes are described in Section 2.4.1 of the document Delegation of Authority.

This is discussed further in Chapter 34, Roles and Responsibilities.

Legal Entities

A description of the legal entities of the company is required.

Affiliate companies are the legal entities in which the parent company holds a controlling interest, either directly or indirectly.

This is discussed further in Section 32.3, Types of Companies.

Applicability to Affiliates (Joint Ventures and Other Relationships)

The Corporate Governance Framework applies to all companies in which a Parent Company, either directly or indirectly, has a **controlling interest**.

The concept of control and the application of standards are not always straightforward. An **Affiliate Oversight Manual** would provide more detailed instructions and guidelines on the Parent Company's expectations on the Governance and Management of both incorporated and unincorporated **Affiliates** (Joint Ventures and other business arrangements). Heads of Functional Business Support Units also provide additional discipline-specific guidance such as Finance; HR; IT; Health, Safety, and Environment (HSE); Integrity; Procurement; etc.

Section 2.4: Activities

Business Units should have extensive freedom in how they conduct their business and improve performance within the boundaries set by the **Group Governance Framework**, including the following activities:

Delegation of Authority

A Parent Company needs to have an integrated, consistent process to delegate authority from **BoD** to **Business Units**, individuals, and committees. This process must recognize the distinction between corporate authorities (related to legal entities) and business organizational authorities (related to the Parent Company's internal business and functional organizational structure).

Corporate Authorities Corporate Authorities need to be described. The principal corporate authorities are defined in:

a. **Memorandum and Articles of Association**
   - This sets boundaries for the Company's business activities and defines its operating principles.

b. **Matters Reserved for the BoD**
   - This contains the corporate authorities relating to the company specifically and organizational authorities on which the **BoD** has reserved the right to make decisions.

c. **Board Committee Terms of Reference**
   - This describes the roles of **Committees** set up by the **BoD.**

d. **Company Manual of Authority (MoA)**
   - This documents authorities delegated by the **MD** for the efficient and effective operation of the Parent Company itself. This covers Financial and Legal Authorities. Technical and other Authorities are generally decentralized.

e. **Other Corporate Authorities**
   - This documents all authorities delegated by legal entities to individuals and committees, in line with local legal requirements. Parent Company Shareholder Representatives should use their influence to ensure that Joint Venture Companies maintain similar manuals.

Business Unit Organizational Authorities Business Unit Authorities need to be described.

a. **Mandate for directly managed Business Units**
   - This provides limits within which these organizations are allowed to operate.

b. Business Unit (Affiliate) MoAs

- This defines organizational authorities within **Affiliates**, linking with relevant Parent Company documents.

Strategy Formulation, Planning, Appraisal (Annual Planning and Budgeting), and Evaluation and Control)

Text example:

> *The objective of* ***performance management*** *is to improve performance by:*
>
> 1. *having realistic plans that deliver the strategy;*
> 2. *implementing the plans within agreed boundaries; and*
> 3. *appraising performance to enable timely intervention.*

> ***Performance management*** *covers the strategy, planning, and appraisal processes as well as the processes that provide assurance that performance is delivered within the boundaries.*
>
> *The* ***MD****, supported by the* ***Corporate Leadership Team (CLT)****, is responsible for* ***performance management*** *at Company Group level.*

This expanded on in Chapter 5, Business Cycle.

Staff Appraisals

Text example:

> *Staff members should agree individual annual targets to help deliver their* ***Business Unit****'s objectives for the year. Performance appraisals against these targets should be undertaken generally on a common basis and performance reward processes should be executed consistently across the* ***Group****.*

Refer to Chapter 20, Human Capital, for complementary discussions.

Assurance and Compliance

Text example:

> The **Group** *should have Group-wide self and independent assessment processes for obtaining* ***assurance*** *on the adequacy of risk management and internal control.* ***Business Units*** *apply these processes and take action to remedy identified weaknesses. Each* ***Business Unit*** *has one or more* ***Management Review Committees,*** *which include on their agendas the monitoring of the* ***assurance program*** *and the actions required to remedy control weaknesses.*

Group Assurance Declaration Process

Text example:

> *At the end of each year, each* ***Asset Unit Head*** *submits a* ***Business Assurance Declaration*** *directly or indirectly to the* ***MD*** *on compliance with legal & ethical requirements and* ***Group Policies/Regulations/Standards/Manuals/Frameworks****. These declarations are based on structured self-assessment, which takes account of results of independent assessment, for example, by* ***Internal Audit Department (IAD)****.*
>
> *At the end of each year, each individual directly employed by the parent company submits a* ***Conflict of Interest Declaration*** *indirectly to the* ***MD*** *on* ***compliance*** *with legal and ethical requirements of the company.*

Internal Audit

Text example:

> ***Internal Audit Department (IAD)*** *provides the* ***MD, Audit Committee,*** *and ultimately the* ***BoD*** *with* ***independent assurance*** *on the design and operation of the system of internal controls.*
>
> *The* ***Group Internal Audit Manager*** *is accountable, in consultation with the* ***Business Units****, for the development and implementation of an* ***Annual Audit Plan*** *for approval by the* ***Audit Committee.***
>
> ***Business Units*** *are required to provide all the necessary assistance to enable* ***Internal Audit Department (IAD)*** *to carry out its duties.*
>
> *Internal Audit Department (The* ***IAD)*** *should also investigate fraudulent, compliance, and other control incidents. A fraud helpline and/or email address is useful for obtaining anonymous tips on fraudulent activities. These incidents should be reported to the* ***MD*** *and the* ***Audit Committee.***

Other Assurance-Related Activities There are a range of other assurance-related activities organized by ***Business Units***.

Business Units identify, investigate, share, and learn from incidents or compliance issues relating to other specific Corporate Policies/Standards.

The prevention and detection of fraud is built into the design and operation of financial and other controls.

Examples of other assurance programs are as follows:

1. Health Safety Environment and Quality (HSEQ) Management System (certification) Audits
2. Asset Integrity Reviews
3. Quantitative Risk Assessments
4. Risk-Based Inspection
5. Benchmarking
6. Project Value Assurance Reviews

Compliance Program A compliance team is required in countries where extensive State legislation controls the Oil and Gas Industry.**External Compliance interfaces** and responsibilities are described in Section 2.5, Business Environment.

Section 2.5: Business Environment: Discussion

External relationships are normally extensive with different focal points within the Parent Company.

Examples are interfaces with the external regulatory, legal, environment, health and safety, and certification authorities. These can be where:

- Company Registration with the appropriate authorities is required;
- a License to Operate may be issued by a Government Environmental Authority;
- ISO certification is issued by an Accredited Authority;
- the reporting of major incidents, as defined by the Government, is required.

SECTION 3: ORGANIZATION: "HOW WE ARE STRUCTURED AND BEHAVE"

A brief description is required including the following:

Section 3.1: Organizational Design

- Asset Business Units
- Service Business Units

Refer to Chapter 4, System Requirements, for complementary discussions.

Section 3.2: Organizational Structure

- BoD
- MD
- Asset Business Units
- Service Business Units
- Organization Charts
- Leadership Teams
- Organizational Capability (Organizational Effectiveness)
- Interface Management
  - Responsibilities
  - Affiliates and other financial interests
  - Other stakeholders

Refer to Chapter 20, Human Assets and Chapter 34, Roles and Responsibilities, for complementary discussions.

SECTION 4: ASSETS: "WHAT WE HAVE"

A description of the company assets is mandatory. A brief statement on how asset integrity is managed is also required.

Part 5, Asset Performance Management, gives guidance.

SECTION 5: SYSTEMS: "HOW WE DO THINGS"

A description of the company Systems is required with more detailed descriptions of each Strategic System.

The following are examples of Strategic Systems:

- Governance
- Strategic Planning and Budgeting
- Allocation of Resources
- Risk Management
- HSEQ/Asset Integrity
- Affiliate Management

Refer to Part 1, Systems, for a full discussion.

All documentation should be encapsulated into Systems.

Basic rules for any document are as follows:

1. Title, document type, and edition/revision date on front page;
2. Document identity (number), approving authority, and custodian on front page or second page;
3. Pages to be numbered x of y, y being the last page of the document; and
4. Each page to have page number and document identity.

The control of all documentation needs to be centralized within the IMS of the Business Unit. See Section 4.12, Integrating Systems in a Business Unit, for details.

8.6 EXAMPLES FROM INDUSTRY

INTERNATIONAL OIL COMPANIES

British Petroleum (Post-Gulf of Mexico Disaster)[1]

BP has a system of internal control, the contents of which are shown in Box 8.3.

BOX 8.3 BP'S GOVERNANCE FRAMEWORK CONTENTS

1. Control environment
- **1.1.** Board and executive governance of the group
  - **1.1.1.** Board governance principles including executive limitations
  - **1.1.2.** Board committees
  - **1.1.3.** Executive Committees
  - **1.1.4.** Group plan and planning processes
  - **1.1.5.** Financial framework
- **1.2.** The assignment of authority and responsibility
  - **1.2.1.** System of delegation
- **1.3.** Integrity and ethical values and legal compliance
  - **1.3.1.** Code of conduct
  - **1.3.2.** Certification
- **1.4.** Management philosophy and operating style
  - **1.4.1.** Group strategy
  - **1.4.2.** Organizational structure
- **1.5.** Competence framework
  - **1.5.1.** Leadership framework
  - **1.5.2.** Learning and development

2. Management of risk and operational performance
- **2.1.** Risk management
  - **2.1.1.** Risk management system
  - **2.1.2.** Group risk categories and group risks
  - **2.1.3.** **Operating Management System (OMS)**
  - **2.1.4.** Group standards
  - **2.1.5.** Processes and practices
- **2.2.** Monitoring performance and the management of risk
  - **2.2.1.** Operating performance reviews
  - **2.2.2.** Management information
  - **2.2.3.** Group financial risk committee
  - **2.2.4.** Group operations risk committee

3. Management of people and individual performance
- **3.1.** Clear lines of communication
  - **3.1.1.** Internal communications
  - **3.1.2.** External communications
- **3.2.** Management of people
  - **3.2.1.** Performance objectives
  - **3.2.2.** HR policies and procedures
- **3.3.** Employee concerns
  - **3.3.1.** Open Talk
  - **3.3.2.** Fraud and misconduct reporting standard

The framework is outlined in Fig. 8.5.

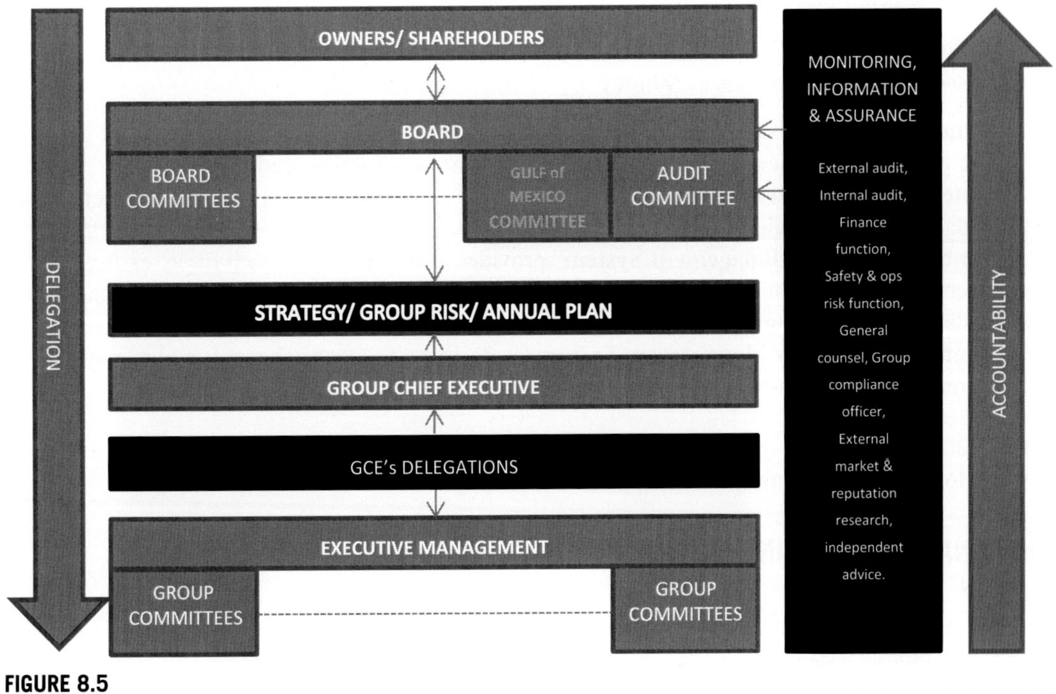

FIGURE 8.5

BP Governance Framework.

BP's **Operating Management System (OMS)** is outlined in Section 4.13, Integrated Management Systems: Specific and Generic Examples.

Comment

BP appeared to have restructured after the Gulf of Mexico disaster, but still has a complex control framework. After a major incident, controls tend to increase.

Shell (As of November 2007)[2]

Shell has a "**Shell Control Framework**" document, which documents a **single overall control framework** for Shell. Its components are Foundations, Organization, and Processes.

It refers to five enterprise-wide risk areas: Finance, Governance and Management, Health, Safety and Environment, Information Management, and People. Each of these is covered by one or more Group Standard. Additionally, there are Manuals containing instructions on how to apply the Standards and Guides containing good practices.

In support of the **"Shell Control Framework"** and relevant manuals, their management processes are divided into three primary categories: **Strategic management, Core business, and Resource management**.

Comment

Shell has a relatively simple Control Framework with a modular Business Unit approach and categorized management processes.

ExxonMobil[3]

ExxonMobil's documentation is as follows:

- "**Standards of Business Conduct**" are ExxonMobil'**s top documents.** These consist of guiding principles, 16 foundation policies, and open-door communication procedures.
- "**System of Management Control Basic Standards**" defines the basic principles, concepts, and standards that drive their business controls.
- "**Controls Integrity Management System**" provides a structured approach for assessing financial control risks, establishing procedures for mitigating concerns, monitoring conformance with standards, and reporting results to management.
- "**Operations Integrity Management System**" is designed to identify hazards and manage risks inherent to their operations and associated with the full life cycle of projects.

Comment

ExxonMobil's control documentation appears to be straightforward and simple.

NATIONAL OIL COMPANIES

Statoil[4]

Statoil's Governance Control framework is integrated into their **Statoil Book,** as shown in Box 4.8, The Statoil Book Processes.

Comment

Statoil appears to have a highly integrated Governance Control Framework with categorized management processes.

CONSULTANT'S RECOMMENDATION

Oil and Gas Process Levels

A four-level approach was recommended with the top two levels (0 and 1) described as follows:

Level 0 Management: Best Use of Finite Resources

Level 1

- Strategic Planning and Budgeting
- Risk Management
- Investment Management
- Talent Development
- Feedstock

Level 0 Business Support

Level 1

- Finance
- HR and Communication

- Procurement
- IT
- Control

Level 0 Technical
Level 1
- R & D
- Engineering and Projects

Level 0 Operational
Level 1
- Marketing and Sales
- Facility Operation and Maintenance

Elements of the Consultants process models include:

1. process cycle
2. process meetings
3. process flowchart
4. process authority matrix

8.7 BENEFITS OF A GOOD GOVERNANCE FRAMEWORK

Primary benefits would be:

1. effective decision-making based on a high-quality knowledge base, including a dynamic risk register and monitoring of top performance indicators;
2. single source database eliminating duplication of data (possibly conflicting), resulting in reduced cost of data storage and handling;
3. increased security for critical data;
4. streamlined **Processes** within clearly defined **Systems**, eliminating separate Processes in functional structures;
5. common staff focus on the **Company Vision and Customer Satisfaction** (alignment) and the **Strategies for Improvement** of performance, thus increasing productivity and eliminating departmental silos.

8.8 SUMMARY

A "before and after" depiction gives the two extremes of the maturity of a business, ranging from "no formal systems" to "fully integrated systems."

The model Governance Framework is based on a Systems approach and built on the following four elements:

1. Control: "How we control the company"
2. Organization: "How we are structured and behave"

3. Assets: "What we have"
4. Systems: "How we do things"

Use of the model governance framework enables the organization to migrate toward a common integrated knowledge base so that major risks and opportunities can be escalated to higher levels in the Business Unit and Group of Companies for balanced and effective decision-making. Resultant top performance indicators would also be common throughout the Company. Reduced duplication of effort and data would drive costs down. The **Company Vision** and **customer requirements** (alignment) would also be clear to all staff.

REFERENCES

1. BP, http://www.bp.com/en/global/corporate/investors/governance.html.
2. Shell, http://www.shell.com/investors/environmental-social-and-governance/corporate-governance.html.
3. Exxon-Mobil, http://corporate.exxonmobil.com/en/investors/corporate-governance/corporate-governance-guidelines/guidelines.
4. Statoil Book, http://www.statoil.com/en/About/TheStatoilBook/Pages/TheStatoilBook.aspx.

CHAPTER

HOW DOES GOVERNANCE AFFECT PERFORMANCE?

9

9.1 INTRODUCTION

Governance, Risk, Integrity, and Performance are interacting forces that optimize the overall performance of the business over its lifetime.

There is a direct correlation between good governance and excellent performance. Good governance sets the control parameters for performance to excel. If governance is too restrictive, it inhibits performance. If too lax, corruption sets in.

Generally performance is controlled by external and internal restraints as follows:

1. Industry Best Practices
2. International Standards (ISO, etc.)
3. Country or State Laws and Regulations
4. Company Governance Policies and Regulations

Business ethics are an integral part of both governance and performance.

Business ethics exists at two levels:

- At the macro level, there are issues related to the role of business in society. The broad ethical stance of an organization is a matter of Corporate Social Responsibility.
- At the individual level, business ethics is about the behavior and actions of the people within organizations.

Zero conflict of interest and high ethics are essential for good governance.

Fair tendering processes are essential to retain contractor interest and enable them to make a fair profit in order to stay in business.

9.2 GOVERNANCE PRINCIPLES

These principles, as discussed in Chapter 7, What are the Principles of Good Governance?, are now discussed from the performance perspective.

FAIRNESS, TRANSPARENCY, AND INDEPENDENCE

This is paramount for retaining a "competitive edge" when dealing with suppliers and customers. For suppliers, a transparent tendering process is required to ensure fair competition. This is discussed

Performance Management for the Oil, Gas, and Process Industries. http://dx.doi.org/10.1016/B978-0-12-810446-0.00009-8

further in Section 9.7. On the other hand, customers need to feel that they are being treated fairly and not being cheated.

ACCOUNTABILITY, RESPONSIBILITY, AND DISCIPLINE

This is a framework for **staff** to work within. Staff need to know their limits of authority and also need motivation. They must be meticulous in adhering to company policies, standards, and procedures and be held responsible for their actions. Performance management systems and competency assessment systems ensure optimum output of product. Chapter 20, Human Capital, discusses this in more detail.

9.3 ATTITUDE TO RISK AND ADVERSE SELECTION

The concepts of Attitude to Risk and Adverse Selection are discussed as follows.

ATTITUDE TO RISK (MORAL HAZARD)

Moral hazard is a situation where the behavior of one party may change to the detriment of another after the transaction has taken place.

A party makes a decision about how much risk to take, while another party bears the costs if things go badly, and the party insulated from risk behaves differently from how it would if it were fully exposed to the risk.

For example, a person with insurance against car theft may be less cautious about locking his car because the negative consequences of car theft are now (partially) the responsibility of the insurance company.

ADVERSE SELECTION (ASYMMETRIC INFORMATION)

Adverse selection is a term used in risk management. It refers to a market process in which "bad" results occur when buyers and sellers have unequal information (i.e., access to different information), resulting in the "bad" products or services being more likely to be selected.

For example, a seller of a second-hand car knows that the car has not been serviced regularly and withholds that information from the buyer. The buyer, thinking that the car has been routinely maintained, purchases the car and then experiences mechanical trouble.

Fig. 9.1 shows an example for contract tendering categorized into Governance, Risk, Compliance, and Assurance.

The balance of Governance, Risk, Compliance, and Assurance optimizes the outcome.

An example of the process is shown in Box 9.1.

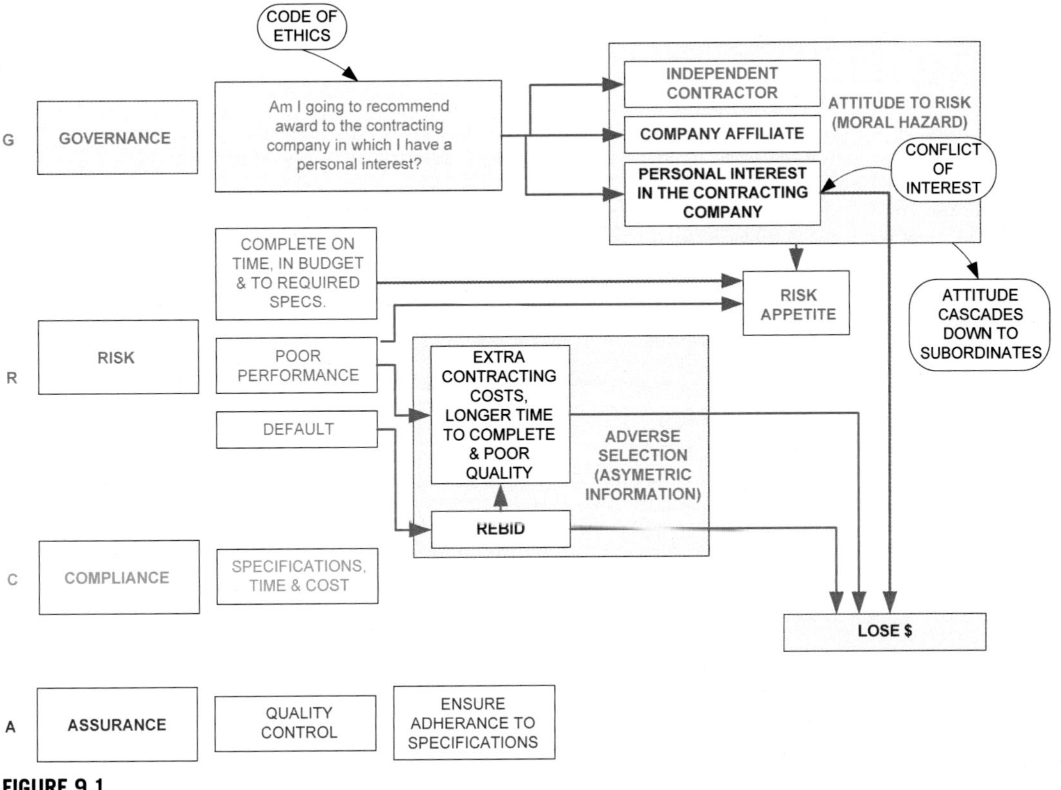

FIGURE 9.1

Contract Tendering: Governance, Risk, Compliance, and Assurance.

BOX 9.1 ATTITUDE TO RISK AND ADVERSE SELECTION: THE 2011 JAPANESE TSUNAMI

Extraction From Report[17]

May 15, 2011|By Tomoko A. Hosaka, Associated Press.

"FUDAI, Japan – In the rubble of Japan's northeast coast, one small village stands as tall as ever after the tsunami. No homes were swept away…

Fudai is the village that survived - thanks to a huge wall that once was deemed a mayor's expensive folly and that now has been vindicated as the community's salvation.

The 3,000 people living between mountains behind a cove owe their lives to a late leader who saw the devastation of an earlier tsunami and made it the priority of his four-decade tenure to defend his people from the next one.

His 51-foot-high floodgate, between mountainsides, took a dozen years to build and more than $30 million in today's dollars.

The floodgate project was criticized as wasteful in the 1970s. But the gate and an equally high seawall behind the community's adjacent fishing port protected Fudai from the waves that obliterated so many towns on March 11. Two months after the disaster, more than 25,000 Japanese are missing or dead.

Towns to the north and south also had seawalls, breakwaters, and other protective structures. But none were as high as Fudai's.

The town of Taro believed it had the ultimate fort – a double-layered 33-foot-high seawall spanning 1.6 miles across a bay. It proved no match for the tsunami.

In Fudai, the waves rose as high as 66 ft, so some ocean water did flow over the wall, but it caused minimal damage.

The gate broke the tsunami's main thrust. And the community is lucky to have two mountainsides flanking the gate, offering a natural barrier.

The man credited with saving Fudai is the late Kotaku Wamura, a 10-term mayor whose political reign began in the ashes of World War II and ended in 1987.

But Wamura never forgot how quickly the sea could turn. Massive earthquake-triggered tsunamis flattened Japan's northeast coast in 1933 and 1896. In Fudai, the disasters destroyed hundreds of homes and killed 439 people."

Analysis

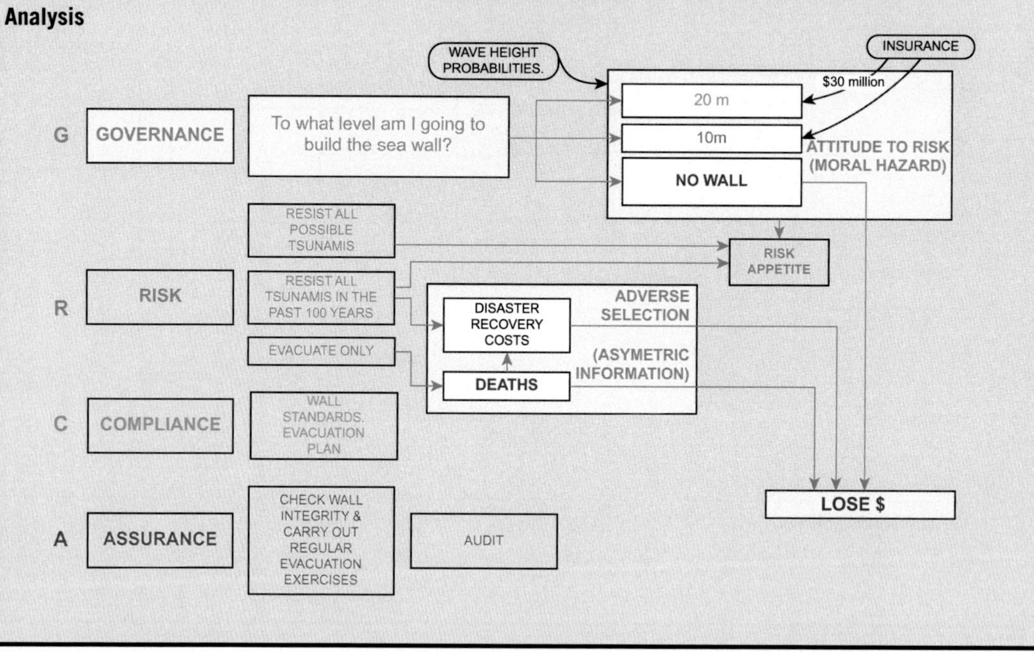

9.4 THE BOARD: THE "FULCRUM" OF BUSINESS PERFORMANCE[1–3]

OECD STATEMENT

"The board should be able to exercise objective judgment on corporate affairs independently, in particular, from management."

The Board of Directors is the "Fulcrum" of Business Performance. It is the central processor of the business brain. It draws data from the external environment and business operations and is responsible for making appropriate decisions to take the Company forward.

The Board must lead in the formulation of Policy and approve the Company Policy Statement (see Chapter 10: Alignment). It must also undertake strategic thinking to determine the focus within the Company's restrained resources (financial and human capital) and thus approve strategies for achieving the Company Policy, all within the **Company's Risk Appetite,** which has also been determined by the Board. This is recorded in the Corporate Risk Statement.

Basic requirements for Boards include the following:

- Boards must be seen to operate "independently" of the management of the Company, so the role of nonexecutive directors is increased.
- Boards must be competent to scrutinize the activities of managers. The collective experience of the board, its training, and the information at its disposal are crucially important.
- Directors must have the time to do their job properly. Limitations on the number of directorships that an individual can hold are also an important consideration.

It is the behavior of boards and their members that is likely to be most significant, whatever structural arrangements are put in place.

For example, respect, trust, "constructive friction" between board members, fluidity of roles, individual as well as collective responsibility, and the evaluation of individual director and collective board performance.[16]

Company laws generally impose fiduciary duties/duties of loyalty and good faith, and other duties on **Board Members,** including:

- the duty to act in good faith in the best interests of the Business Unit;
- to comply with confidentiality requirements;
- to ensure they have adequate skills;
- to avoid conflicts of interest;
- not to exploit their position for their own gain; and
- to make full and fair disclosure of all material matters to the board and/or the shareholders.

The Board must also approve Conflict of Interest and Ethics Corporate Statements (see Section 9.5).

Guidelines for effective corporate governance related to board appointments include the following:

- At least half of the Board should be nonexecutives, excluding the Chairman.
- The Chief Executive should not be Chairman, and the roles of each should be separate.
- Full-time executives should limit their work with other companies.
- A senior independent director should be appointed the champion of shareholders' concerns.
- The Chairman should be independent and not an executive of the Company.
- Nonexecutives should serve no more than two 3-year consecutive terms.
- Nonexecutives should receive no further remuneration, pensions, or share options other than their standard fees.
- A nominations committee should monitor time input of nonexecutives.
- Nonexecutives should not simultaneously sit on the audit, nomination, and remuneration committees.
- Nonexecutives and boards should be suitably trained and regularly appraised.
- Nonexecutives should be drawn more from nonboard members of other businesses and the public sector.
- There should be total transparency on appointments, training, and attendance of nonexecutives.

Box 9.2 discusses aspects of the above.

BOX 9.2 FOUR KEY PILLARS TO STRONG GOVERNANCE[15]

Article in New Straits Times, Tuesday, May 3, 2016, by Goh Thean Howe

Kuala Lampur: For an organization to be successful, it must have strong governance backed by four key pillars: board of directors, management team, internal auditors, and external auditors.

"If the four pillars are not working effectively, we will be seeing more high-profile corporate scandals such as FIFA, Olympus, Volkswagen and Toshiba, where 'toxic culture' can undermine good governance and ultimately destroy shareholder's value," said Institute of Internal Auditors Malaysia President Datuk Shabaruddin Ibrahim.

He said a far-sighted and effective nomination committee would keep an eye on the need for succession planning in the boardroom.

The board, meanwhile, should be strengthened by independent directors who will not have any connection with the company and, therefore, have no conflict of interest in discharging their duties.

"Any approval being sought at the board level needs to be approached with a watchful eye and inquiring mind," Shabaruddin said in a statement.

He said having independent directors was critical, as the interests of management, the company, and the shareholders differed.

9.5 CONFLICT OF INTEREST AND ETHICS CORPORATE STATEMENTS[4]

Ethics can be summarized as:

1. legality of the action;
2. fairness of the action; and
3. personal feeling of well-being with respect to the action taken.

CODE OF ETHICS STATEMENTS

Business Integrity and Code of Ethics examples are shown in Box 9.3.

BOX 9.3 BUSINESS INTEGRITY AND CODE OF ETHICS EXAMPLES

Petronas[18] Malaysia

Business Integrity Statement

"We will pursue our business with honesty, integrity and fairness respecting the different cultures in the countries in which we operate. Corporate funds and assets will only be used for lawful or proper purposes and all business transactions will be reflected accurately and fairly.

We recognize that transparency and open communication are essential. To this end, we will provide all relevant information as may be required by our stakeholders about our activities, subject only to overriding considerations of business confidentiality."

Petrobras[13] Brazil

Code of Ethics Statement: Purpose

"The **purpose** of the Code of Ethics is to clearly define the ethical principles which guide Petrobras' actions and conduct, both on its institutional and on its employees side, explaining the sense of ethics of the Mission and Vision of the Company and The Strategic Plan."

Tips on creating a code of ethics statement are described in Box 9.4.

BOX 9.4 KEY STEPS SUGGESTED AS BEST PRACTICE IN THE DEVELOPMENT AND IMPLEMENTATION OF EFFECTIVE CODES OF ETHICAL PRACTICE

1. Code of ethical practice must be directly **set in the context of the Mission and Values of the Company.** This is critical for presenting the code as a coherent, strategic document and serves to trigger mutual endorsement. This can sometimes be neglected when legal concerns take priority in the development of codes.
2. The code should **emanate from the board**, be embraced willingly and practically by all of the senior leadership team, and be introduced/endorsed by the Chairman and CEO.
3. In both development and implementation, it is vital that the code be **widely circulated,** both within the organization and externally. It must be presented in an appropriate form and simplified or translated as necessary. Recipients need to commit to the code in theory and in practice, but can only do so if they are able to fully engage with it.
4. The code must **contain sanctions for noncompliance and must be seen to be enforced.** Breaches should be reported to the Board and to the internal and external public as appropriate.
5. Implementation of the code must be **supported by training** to ensure that the code is understood and to effectively measure its impact on recipient values, attitudes, and behaviors.
6. To support the effective operation of the code, **senior management must put practical, fair, impartial, and, where necessary, confidential procedures in place for providing advice** to those with the task of making decisions in relation to the code.
7. Similar **supportive** provisions are required, in practice, with regard to the **process for raising an issue for investigation linked to the provisions of the code.**

Adapted from "Reputation and Responsibility" Module. Henley Business School.[14]

9.6 GOVERNANCE, RISK, INTEGRITY, AND PERFORMANCE

Poor **Governance** causes higher **Risks,** which compromises the **Integrity** of the physical assets and results in poor long-term **Performance.**

Example:

The award of a turnaround contract to a contractor, not experienced in turnarounds (governance), raises the risks of poor quality and late startup (risk), increasing the probability of leaks (integrity) at startup, thus reducing the long-term availability of the plant and making the plant unprofitable (Performance).

THE "TIPPING POINT"

This occurs when the assets deteriorate to such a point that a massive injection of funding is required to get the assets to a stable situation without major loss of containment. After a long period of neglect, the recovery process takes time and is difficult.

The "tipping point" could occur when the Board focuses primarily on the Shareholder, as discussed in Table 7.1, Focus of Governance Systems.

Box 9.5 discusses the BP situation.[5]

BOX 9.5 ACCOUNTABILITY, RESPONSIBILITY, AND DISCIPLINE: BP

The BP policy, under CEO John Browne (1995–2008), demonstrates the disastrous effect of cost reduction without ensuring the long-term integrity of the assets. BP managers had short assignments with incentives for cost reduction. The long-term integrity problem was passed on to the next manager, eventually with disastrous consequences in a number of cases.

The question is: Did BP reach the "tipping point" after a number of years of cost cutting?

Also see related discussion on BP in the following boxes:

1. Box 12.1, Swiss Cheese: BP Deepwater Horizon
2. Box 13.4, BP and the Oil Industry after the Texas City Refinery Disaster
3. Box 18.4, The Battering of BP's Reputation
4. Box 24.2, Integration Into the Business Example 1: Cost Reduction

The "tipping point" could also occur when political interference, incompetence of the Board, and the focus on "Affirmative Action" result in a long-term deterioration of the assets.

Box 9.6 discusses the long-term deterioration in the South African Electricity Supply Commission, resulting in Rating Agencies downgrading its credit to "junk bonds."[6,7]

BOX 9.6 THE JEWEL IN THE CROWN: THE RISE AND FALL OF SOUTH AFRICA'S ELECTRICITY SUPPLY COMMISSION

In the 1950s, the South African Government established the Electricity Supply Commission (ESCOM) as a Government Corporation to generate and distribute electrical power in South Africa. Sometime in the 1960s, the Government legislated that ESCOM become the sole generator of power from coal in the country.

ESCOM then built a number of the biggest coal-fired power stations in the world: paired 600 MW steam boilers and turbines. As a result, by the end of the 1960s, ESCOM reliably generated the cheapest power in the world.

In the early 1970s, ESCOM constructed its first nuclear power station: two sets of reactors and turbines with a generation capacity of 900 MW for each set. At the same time, they built two large pump storage schemes consisting of two 200 MW and four 250 MW pump-turbines. This complementary setup was decided upon because the nuclear power units operate optimally on a steady load and as such are used as a base load for the national electricity grid. In the middle of the night, when power demand was low, the pump-turbines would take excess power from the nuclear power units to pump water to an upper dam, and then generate electricity at peak demand times in the day.

In the 1980s, the UK led the privatization of power generation by breaking up the Central Electricity Generating Board, Britain's equivalent of ESCOM. The South African Government resisted the move to privatization of power generation, even after a 1998 White Paper suggested breaking up the ESCOM monopoly.

After the establishment of a democratic government in 1994, an "affirmative action" program was established by the government to balance the employment in government entities in line with the population distribution. ESCOM, as a State Company, was thus committed to fast-track training of previously disadvantaged ethnic groups. The ESCOM Board of Directors also became more influenced by politicians.

Inadequate experience and political interference caused things to unravel for ESCOM. Early indications were as follows:

Koeberg Nuclear Power Station

A 900-MW turbine rotor failed on start-up after a Turnaround and Inspection (T&I). The steam turbine rotor was stripped of its blades. Apparently, something was left inside the turbine when boxing up the rotor casing. It was believed to be sabotage or complete incompetence of the turnaround management, as the occurrence is so unlikely that spare rotors are not kept on site as critical spares. The South African Navy was requested to sail to France and return with a new rotor. The unit was thus out of service for many months. Lost generation capacity was equivalent to a significant part of the requirements of Cape Town City. Koeberg also requires major refurbishment by 2017 if it is to retain its nuclear licence from the International Atomic Energy Agency.

Majuba Power Station

This was completed in about 2000 as the last of a series of mega power stations. During the project, it was decided for some reason (probably cost or time) not to line the three concrete coal silos with polyurethane. Management was possibly not aware of the effects of high sulfur coal on concrete! One silo failed recently, and a second has been compromised, causing a major disruption in power generation.

The Cape Gas to Power Initiative

Shell discovered gas offshore near the Orange River mouth on the border with Namibia. This is referred to as the Kudu gas field.

In about 2002, a feasibility study was carried out by consultants for the development of the field, installation of a gas line to Cape Town, and the construction of two large gas turbines to generate 2000 MW of power. The project was viable, as it provided for the timely power needs of the country. It was predicted that if a government decision for establishing a new major power station of this order was not forthcoming in the near future, the country would have a power crisis (the decision to build large power generation units has to preempt the power demand by least 5 years).

The government had to decide to end ESCOM's generation monopoly and guarantee a long-term return on this project. The government did not make the decision, and a power crisis ensued 5+ years later.

Two New Power Stations: Medupi and Kusile

In April 2010 the World Bank approved loans to build two new coal-fired mega power stations on the condition that a percentage of the loans were used for renewable power generation. Wind turbines were built in various parts of the country, and a concentrated solar generator was built near Upington in the northwest of the country.

As I write, the two new mega coal-fired power stations are **years** behind schedule and **10 times** over budget.

Maintenance

With the new directors and top management appointees having little experience of long-term asset management, costs were cut and human capital experience levels deteriorated, resulting in poor maintenance of the assets over an extended period. The current maintenance backlog is estimated to be in excess of $7 billion.

Power Outages

In 2014, all South Africans started to experience routine power outages of varying degrees, inflicting major losses to businesses.

Board Games

In 2015, the President of South Africa appointed his Vice-President to investigate the management of ESCOM. The Minister of Energy suspended the Chief Executive, Finance Director, and Technology Director pending the investigation, and later dismissed the Chairman of the Board.

Junk Bonds

In early 2015, international rating agencies (Moody's and Standards & Poor's) downgraded ESCOM bonds to "Junk Bonds."

9.7 PROCUREMENT AND GOVERNANCE[8,9]

MODEL TENDERING (BIDDING) PROCESS

To optimize the Governance, Risk, Compliance, and Assurance elements of good governance, a structured transparent tendering system in line with international best practice is required. The basics are:

1. decide on a suitable contracting strategy for the particular Scope of Work;
2. Invitation to Bid package to be sent out with reasonable bid period so as to ensure a comprehensive and complete bid submission by all participants;
3. tenders to be sealed separately (technical and commercial) and submitted into a sealed tender box by a set time on a set day;

4. sealed tender box to be opened, bids registered, and technical bids opened by a multidisciplinary Tender Committee;
5. technical bids to be evaluated by experts and technical approvals submitted to multidisciplinary Tender Committee. Multidisciplinary Tender Committee only then opens technically approved commercial bids; and
6. commercial bids to be evaluated by experts, and commercial approvals and recommendations for award submitted to multidisciplinary Tender Committee before obtaining line management approval (Approving Authority).

Fig. 9.2 outlines the process.

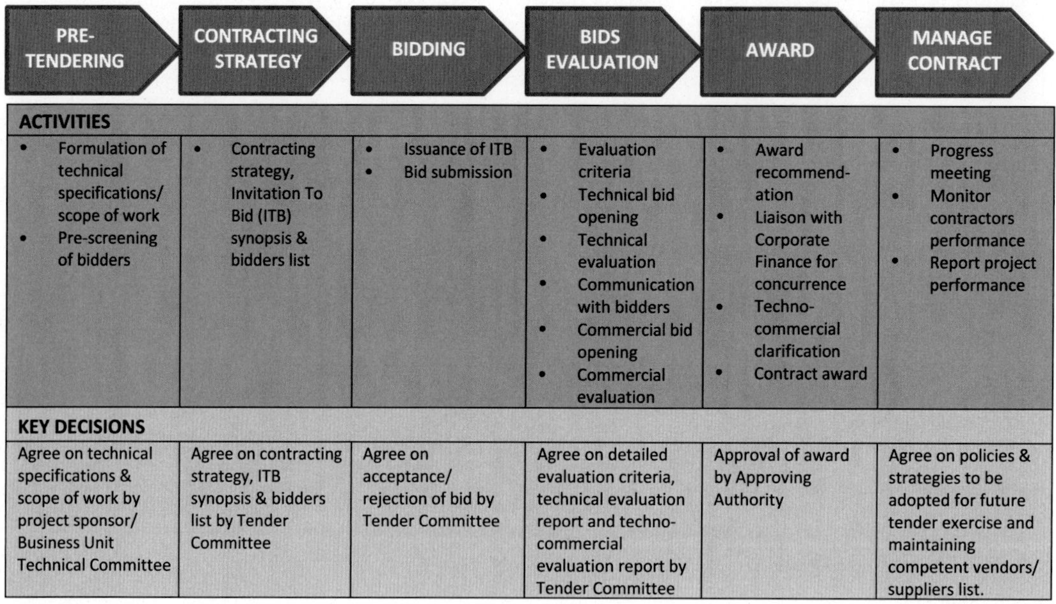

PRE-TENDERING	CONTRACTING STRATEGY	BIDDING	BIDS EVALUATION	AWARD	MANAGE CONTRACT
ACTIVITIES					
• Formulation of technical specifications/ scope of work • Pre-screening of bidders	• Contracting strategy, Invitation To Bid (ITB) synopsis & bidders list	• Issuance of ITB • Bid submission	• Evaluation criteria • Technical bid opening • Technical evaluation • Communication with bidders • Commercial bid opening • Commercial evaluation	• Award recommend-ation • Liaison with Corporate Finance for concurrence • Techno-commercial clarification • Contract award	• Progress meeting • Monitor contractors performance • Report project performance
KEY DECISIONS					
Agree on technical specifications & scope of work by project sponsor/ Business Unit Technical Committee	Agree on contracting strategy, ITB synopsis & bidders list by Tender Committee	Agree on acceptance/ rejection of bid by Tender Committee	Agree on detailed evaluation criteria, technical evaluation report and techno-commercial evaluation report by Tender Committee	Approval of award by Approving Authority	Agree on policies & strategies to be adopted for future tender exercise and maintaining competent vendors/ suppliers list.

FIGURE 9.2

Tendering process framework.

The type of contractual arrangement and its relative risk are discussed in Section 13.2, Procurement Risk and Performance.

EFFECT OF POOR TENDERING

Poor tendering practices could be highly detrimental to the performance of the Business Unit.

An example of procurement that is detrimental to the performance of the business is shown in Box 9.7. Auditing is essential (see Section 11.9: Assurance Services).

> **BOX 9.7 PROCUREMENT**
>
> **Raised Operational Risk Profile**
>
> A valve vendor was awarded a contract to supply hundreds of a particular process valve for a gas plant, based on the lowest price. Within a short time, some of the valves showed evidence of pinhole leaks, and a decision was made to replace all of these valves as and when sections of the plant were made available.
>
> Regrettably, the vendor had already been paid for the original valves, and the asset owner was thus not only landed with the bill for new values, but also with the cost of reinstallation. Fortunately, the valves did not cause much actual downtime, but nonetheless, the plant operated at a higher risk level until all the valves were replaced.
>
> Clearly, the main contractor's poor quality control was responsible, and the situation could possibly have been averted by the appointment of a third-party inspector who was tasked to ensure that the valves were manufactured according to the standards stipulated in the contract.

BYPASSING THE FORMAL TENDER PROCESS

Serious consequences can occur when a major project is poorly managed and formal tendering not adhered to. The contract management process is a critical part of the management of the project. Box 9.8 gives scenarios of the possible long-term consequences.

> **BOX 9.8 INVESTMENT PAYBACK**
>
> The following scenarios actually occurred, but the facts are difficult to verify. Obviously, names and dates have been withheld.
>
> **Scenario 1**
>
> A small, oil-rich country decided to invest in a new airport and create a hub for its airline to carry travelers east–west and north–south to compete with other airports and airlines in the region. The airport investment was to be recovered from airport taxes on each passenger, although the cost of the project was supposed to ensure that airfares would remain competitive.
>
> The prime minister had final say as to who was going to build the airport, and the project was awarded, without going through a formal bidding process, to a multinational private contractor (*the contractor allegedly paid the prime minister a percentage of the project value, but being a private company, was not required to disclose such payments*).
>
> The project ran years late and well over budget, and, as a result of high costs and delays in handover, the airport tax on each passenger had to be considerably increased. The airline lost business to other airlines operating from hubs in nearby countries and, naturally, lost market share.
>
> **Scenario 2**
>
> A major power station was built for a national power company with tenders for turbines and boilers awarded to companies with political connections. Cost and duration overruns ensued.
>
> As a result of high costs and delays in handover, the price of electricity had to be considerably increased to ensure payment of the investment loans. Consumers thus paid the price for this political interference. The knock-on effect was the resulting reduced competitiveness of companies in the country.

STANDARD CONDITIONS OF CONTRACT[10,11]

The use of Standard Conditions of Contract, such as those published by Fédération Internationale Des Ingénieurs-Conseils (FIDIC) or IChemE, reduces the risk to both the purchaser and the supplier and enhances contract performance if both parties are familiar with these conditions. This can contribute considerably in eliminating Asymmetric Information.

Example:

Familiarity with common standard conditions of contract reduces the probability of disputes arising.

9.8 THE CANCER OF CORRUPTION

DEFINITION OF CORRUPTION

Corruption is the abuse of entrusted power for private gain.

An old Chinese saying:

"The fish rots from the head."

To relate this to corruption:

If the head of a business is corrupt, the rest of the staff will ask, "Why can't we do what he is doing?"

The effect on the assets of the business (physical assets, reputation, financial, etc.) can have long-term implications.

An example:

If a company is known to favor certain bidders for procurement of services and materials, other suppliers will stop bidding and thereby leaving the company with too few uncompetitive bids. This would drive up the value of bids and possibly also have a detrimental effect on the quality of goods or services supplied, as well as increase the delivery time.

Box 9.9 gives two examples of corruption.

BOX 9.9 CORRUPTION

Example 1: Mr. Ten Percent

The Prime Minister of a Middle East country attempted to influence the award of a major automation contract for the National Oil Company (*he was known for getting 10% commission from each contract awarded to suppliers sponsored by him*). However, the company eventually got its preferred vendor due to influence of its International Oil Company (IOC) partner.

If the Prime Minister's choice of vendor had been accepted by the Company, it might have lived with a substandard automation system for the life of the asset.

Example 2: Lost Time

A turnaround contractor was awarded the contract for a turnaround, based on its political connections, without a track record for completion of turnarounds on time. Not surprisingly, the turnaround ran late.

Comment

Neither the lost production days nor the loss in revenue can ever be regained by the asset owner.

See Section 7.7, Governance Surveys, for other discussion on corruption.

9.9 WHISTLEBLOWING[12]

Whistleblowing is a powerful tool against illegal practices.

It is aimed at enabling employees, customers, suppliers, managers, or other stakeholders to raise concerns on a confidential basis in cases where conduct is deemed to be contrary to the Company's values. It may include:

- actions that may result in danger to the health and/or safety of people or damage to assets or environment;
- unethical practices in accounting, internal accounting controls, financial reporting, and auditing matters;
- criminal offenses, including money laundering, fraud, bribery, and corruption;
- failure to comply with any legal obligation;
- miscarriage of justice;
- any conduct contrary to the **ethical principles of the Company** embedded in Company Policies, frameworks, etc.; and
- any other legal or ethical concern and concealment of any of the above.

Note:

Ethical principles of the Company are discussed under **Values** in Section 10.5, The Development of a Policy or Direction Statement.

The reporting mechanism should be simple, using telephone, email/Internet, and surface mail communication channels from anywhere in the world. It needs to be directed to an independent service provider who removes all indications as to the identity of the informants before submission to designated persons in the Company. The program must also be monitored by the Company Audit Committee.

9.10 SUMMARY

External constraints of best practices, standards, laws, and Company Policy determine the limits of the business operation.

The focus on **F**airness, **T**ransparency, and **I**ndependence is more apparent for those interfacing with suppliers, customers, and other stakeholders.

The focus on **A**ccountability, **R**esponsibility, and **D**iscipline is more apparent for management of the business internally.

The Board of Directors is the "Fulcrum" of Business Performance. It draws data from the external environment and business operations, and is responsible for making appropriate decisions to take the Company forward.

Conflict of interest and ethics are demonstrated in the tendering context using concepts of "attitude to risk" and "adverse selection." A transparent and fair tendering process is critical for good governance.

A structured transparent tendering system, in line with international best practice, is required to assist in optimizing the Governance, Risk, Compliance, and Assurance elements of good governance.

Ethics is basically about the legality of the action, fairness of the action, and personal feeling of well-being with respect to the action taken.

Governance, risk, integrity (of the physical assets), and performance are interrelated, **and there could be a serious "knock-on effect."**

Whistleblowing is a powerful tool against illegal practices. Zero conflict of interest and high ethics are the goals for good governance, and these need to be embedded in the Company Direction Statement (see Chapter 10: Alignment).

REFERENCES

1. Garratt R. *The fish rots from the head*. Profile Books; 2010.
2. Statement of directors' responsibilities. May 2011. www.activerisk.com.
3. Schedule of matters reserved for the board. February 2011. www.activerisk.com.
4. Blanchard K, Peale NV. *The power of ethical management*. Cedar; 1991.
5. Steffy LC. *Drowning in oil: BP and the reckless pursuit of profit*. McGraw-Hill; 2011.
6. The Big Issue no 228. *The dark side of South Africa*. February–March 2015.
7. Noseweek no 186. *Eskom: it's downhill all the way*. April 2015.
8. Taylor RD. *Law of contract*. Blackstone Press; 1989.
9. Garrett GA. *World class contracting*. Chicago: CCH; 2001.
10. FIDIC – International Federation of National Associations of Independent Consulting Engineers.
    (a) *Conditions of contract for works of civil engineering construction (red book).*
    (b) *Conditions of contract for electrical and mechanical works including erection on site (yellow book).*
    (c) *Conditions of contract for design-build and turnkey (orange book).*
11. IChemE – Institution of Chemical Engineers UK.
    (a) *Lump sum the international red book.*
    (b) *Reimbursable the international green book.*
    (c) *Target cost the international burgundy book.*
    (d) *Subcontracts the international yellow book.*
12. *Petronas whistle blowing policy*. http://www.petronas.com.my/about-us/governance/Documents/PETRONAS-Whistleblowing-Policy.pdf.
13. Petrobras. *Code of ethics*. http://www.investidorpetrobras.com.br/en/corporate-governance/governance-instruments/code-ethics.
14. Hillenbrand C, et al. *Reputation and responsibility module*. Henley Business School; July 2011.
15. Howe GT. *New straits times*. May 3, 2016 (Tuesday).
16. Johnson, Scholes, Whittington. *Exploring corporate strategy*. Pearson; 2008.
17. Hosaka TA. Report on the Japanese tsunami. Associated Press; May 15, 2011.
18. Petronas. *Code of business ethics (CoBE)*. http://www.petronas.com.my/about-us/governance/Pages/governance/code-of-conduct-business-ethics.aspx.

FURTHER READING

1. Terms of reference: the audit committee. March 2011. www.activerisk.com.
2. *CEO Succession 2000–9: reference for trend in separation of chairman and CEO functions*. http://www.strategy-business.com/article/10208.
3. UK HSE INDG 417. *Leading health and safety at work*. 2007.

CHAPTER

ALIGNMENT

10

10.1 INTRODUCTION

Alignment is the key to the success of the Company and each Business Unit (BU). As discussed in Chapter 2, The Business Unit, there must be "value addition," and this comes from a focus on alignment.

Emphasis must be on the **Company Policy or Direction Statement** and **Core** "value addition" **Systems**. Otherwise, the function must be discarded or ring-fenced (see examples in Box 2.1: Focus on Core Business, and Box 2.2: Disposal of the Company School). Fig. 10.1 depicts vertical and horizontal alignment.

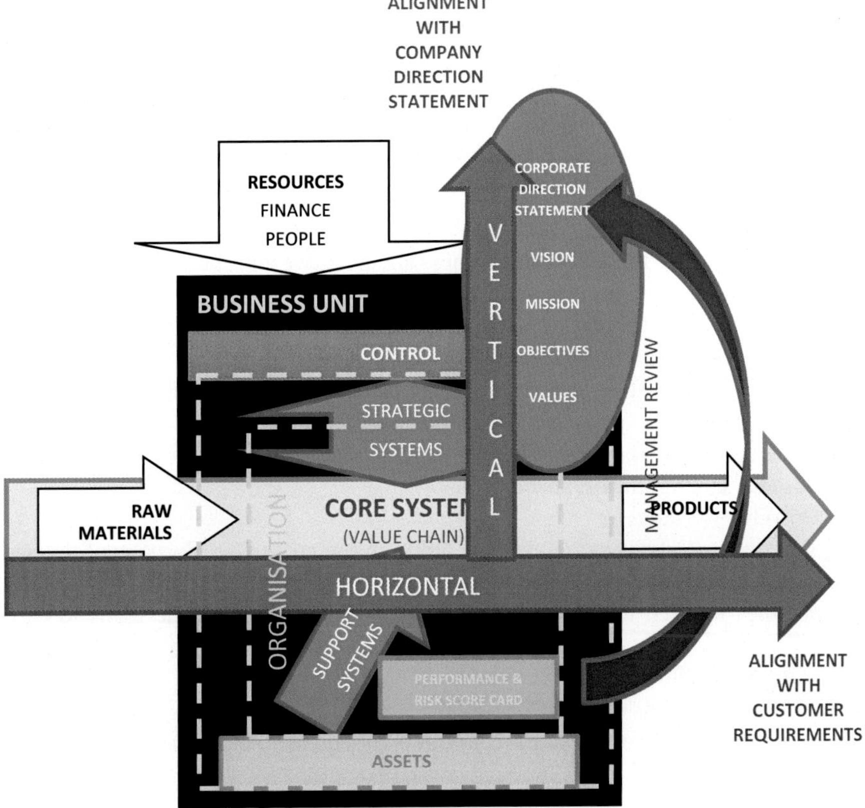

FIGURE 10.1

Vertical and horizontal alignment.

Performance Management for the Oil, Gas, and Process Industries. http://dx.doi.org/10.1016/B978-0-12-810446-0.00010-4

Vertical and horizontal alignment needs to be balanced. Too much focus on vertical alignment ("**Shareholder Focus**" as shown in Table 7.1 Focus of Governance Systems) could be detrimental to the long-term survival of the company (see Box 9.5: Accountability, Responsibility, and Discipline). Conversely, too great an emphasis on horizontal alignment could make the business uncompetitive and drive it out of business.

10.2 VERTICAL ALIGNMENT

The hierarchy is as follows:

- **Vision**: Where do we want to be?
- **Mission**: Why do we exist?
- **Core Values**: What we stand for.
- **Corporate Objectives/Strategy**: What direction do we need to take?
- **Value Chain**: Raw materials in to products out
- **Business Unit**: The legal entity
- **Core System:** An Element of the Value Chain.
- **Key Performance Indicators (KPIs)**: How do we measure progress?
- **Benchmark:** Where are we?
- **Targets**: What do we need to achieve?
- **Strategies and Plans**: Means to achieve targets. How do we get there?
- **Ownership:** Responsibility, Accountability, and Reward

Vision: Where do we want to be?

A **Vision** is the future state for which the business should strive and should be the focus for all staff. Everything anyone does in the Company should ultimately strive toward the Company Vision.

In other words, a vision statement is concerned with what the organization aspires to be.

Mission: Why do we exist?

This is a "value addition" statement. It describes briefly what the Company does to achieve the Vision.

A mission statement aims to provide employees and stakeholders with clarity about the overall purpose and "raison d"être" of the organization.

Core Values: What we stand for

Core values are the underlying principles that guide an organization's Corporate Strategy.

Corporate Objectives/Strategy: What direction do we need to take?

This is sometimes referred to as "Corporate Strategy" and identifies the focus to carry out the Mission.

Corporate Objectives/Strategy are statements of specific outcomes that are to be achieved.

Value Chain: Raw materials in to products out

This is the sum of all of the Core Systems connected, in sequence, as the product progresses through one or more BUs.

BU: The legal entity.

This is as defined in Chapter 2, The Business Unit.

Core System: An Element of the Value Chain

This is as defined in Chapter 4, System Determination and Requirements.

KPIs: How do we measure progress?

Obtaining the right KPIs for effective improvement is described in Chapter 16, Key Performance Indicator Selection Guidelines.

Benchmark: Where are we?

It is essential to compare what we are doing to the competition. This is discussed in Part 6, Benchmarking.

Targets: What do we need to achieve?

Short- and long-term targets are required so as to aspire to the Vision. These are discussed in Chapter 30, Strategies and Actions.

Strategies and Plans: Means to achieve targets. How do we get there?

Targets cannot be achieved without Strategies and Plans to achieve targets. These are discussed in Chapter 30, Strategies and Actions.

Ownership: Responsibility/Accountability/Reward

Nothing will happen without commitment from the Owners of the Strategies and Plans to achieve the targets. This is as discussed in Section 4.10, System and Process Ownership.

10.3 HORIZONTAL ALIGNMENT

Horizontal alignment must ensure:

1. customer orientation;
2. streamlined interfaces between departments/directorates/affiliates;
3. consistent approach;
4. focus on performance measurement; and
5. duplication/gaps are eliminated.

CUSTOMER ORIENTATION

All staff need to identify with the customers' needs and the fulfillment thereof. Well-designed customer surveys will give meaningful feedback on how well the company is achieving this.

STREAMLINED INTERFACES BETWEEN DEPARTMENTS/BUSINESS UNITS

Departmental silos must be eliminated by focusing on the Core System and tuning it to perfection. Organizational structures need to be designed in line with Systems as far as possible. Also, the "handover" between BUs needs to be perfected.

CONSISTENT APPROACH

Common Standard Processes and procedures need to be developed for all to use. Each Process must have a single custodian, even if they are partly owned by others. This ensures consistency and the elimination of duplication.

FOCUS ON PERFORMANCE MEASUREMENT

The measurement of simple KPIs that count toward continuous improvement is essential. Other KPIs need to be discarded. Chapter 16, Performance Indicator Selection Guidelines, helps select the right KPIs. Chapter 31, Reporting, gives suitable formats for performance reporting.

DUPLICATION/GAPS ARE ELIMINATED

As discussed previously, a custodian of Processes and/or procedures ensures elimination of duplication and may also identify gaps. Gaps are also identified through auditing of Systems.

10.4 ALIGNMENT AND SYSTEMS

Integration of **Systems** requires vertical and horizontal alignment. Fig. 10.2, depicts vertical and horizontal alignment for an Oil and Gas company.

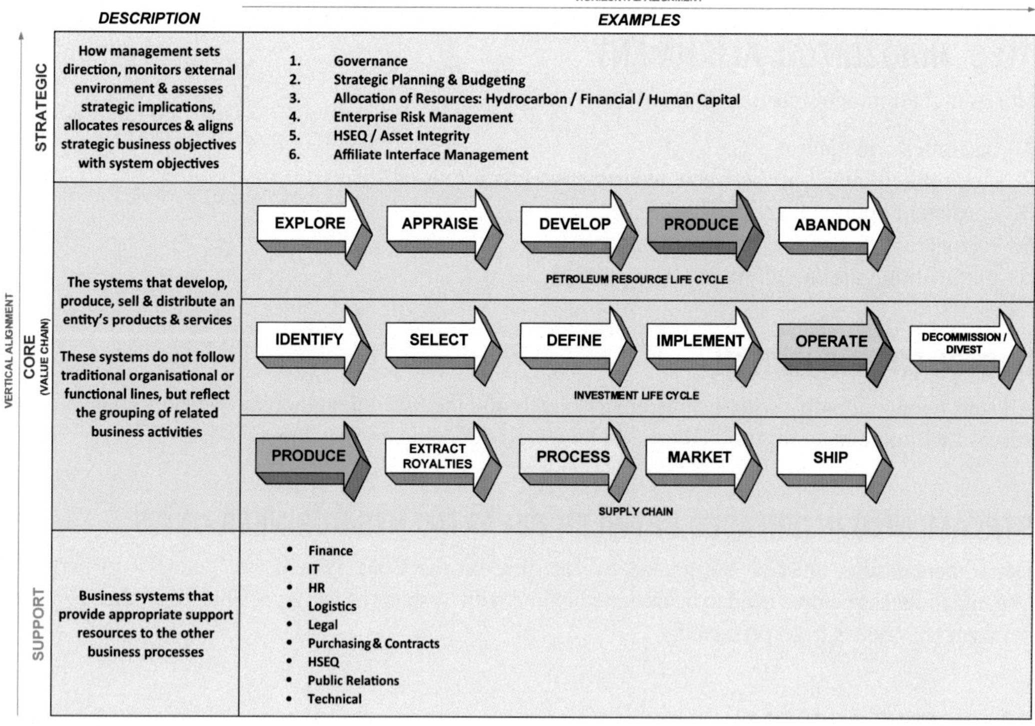

FIGURE 10.2

Oil and gas Systems alignment.

10.5 THE DEVELOPMENT OF A POLICY OR DIRECTION STATEMENT

This document must be in simple language that is easily understood by all staff.

THE STATEMENT

A **Policy or Direction Statement** generally includes the **Vision, Mission, Objectives, Commitment, and Values**.

In summary, the **Policy Statement** is the Company or BU's **top-level guiding document**. It tells us:

1. *where* we want to be (**Vision**);
2. *what* we have to do to get there (**Mission and Objectives**);
3. our *Commitment* to achieve our **Vision; and**
4. what we stand for, our **Core Values.**

THE VISION

A **Vision** is an ideal future state for which the business should strive. It is what we want to be. It needs to be easily understood and shared by all members of staff. In addition, it should, ideally, be quantitative, i.e., capable of being measured.

KEY REQUIREMENTS OF THE STATEMENT WORDING

- Future tense
- A simple, short sentence that is easily understood
- Each word carefully and meaningfully chosen

An example is shown in Box 10.1.[1]

> **BOX 10.1 VISION STATEMENT EXAMPLE: CHEVRON**
>
> *"to be the global energy company most admired for its people, partnership, and performance."*
>
> **Breakdown**
>
> - To be: future tense
> - Global energy company: involved with any part of the energy supply chain across most countries in the world.
> - Most admired: the energy industry and general population's perception that this is a good company to invest in and/or work for and/or partner with.
> - People: "human capital" is the most important part of a business.
> - Partnership: investment, technical, social, government.
> - Performance: investment, technical, operational, people.

THE MISSION

The Mission is the "purpose of business" or reason why the business exists. It should mention the main operating Objectives and how they are achieved. Ideally, it should be quantitative.

A mission statement must be:

1. real
2. unique to the company
3. concise
4. short and simple

Box 10.2 gives an example.

> **BOX 10.2 MISSION STATEMENT EXAMPLES**
>
> **State Oil Company Refining Business Unit Mission Statement**
>
> "Our mission is to ensure the State gets maximum benefit from its Petroleum Resources by engaging in all Refining activities that would add value to these resources"

Box 10.3 gives examples of the changing focus of National Oil Companies.

BOX 10.3 CHANGING FOCUS OF NATIONAL OIL COMPANY (NOC) MISSIONS

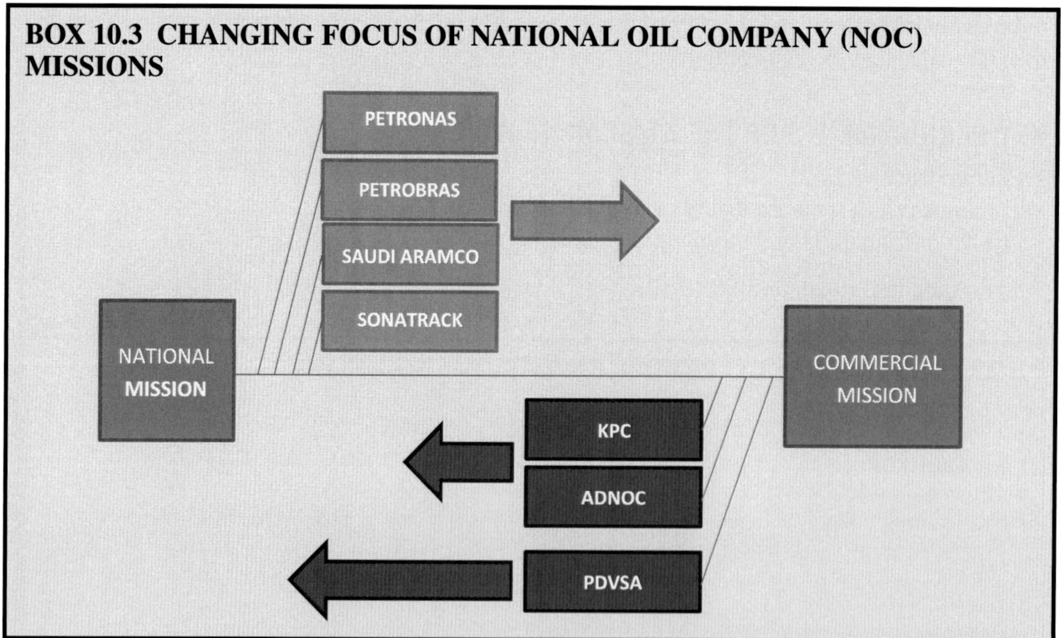

THE BUSINESS OBJECTIVES/STRATEGY

Objectives need to be quantitative and linked to clear **Vision and Mission** statements. Objectives are essential for organizational success as they state direction, aid in evaluation, create synergy, reveal priorities, focus coordination, and provide a basis for effective planning, organizing, motivating, and controlling activities. Objectives should be clear, measurable, consistent, reasonable, and challenging.

Notes:

1. **Annual Objectives** are short-term objectives linked directly to the Business **Objectives**.
2. **Goals** are sometimes stated as well. However, these are defined as open-ended statements where **Objectives** are defined as being desired end results.

Box 10.4 gives an example of a set of objectives for a Refinery.

BOX 10.4 REFINERY BUSINESS UNIT OBJECTIVES: EXAMPLE AND *EXPLANATION*

Our main objectives are as follows:

1. Achieve the highest possible production, maximizing product value, at minimum conversion cost.
   This is directly linked to the Refinery's top Solomon Index of Net Margin. The following shows linkage of the specific words used above with specific Key Performance Indices in the Solomon benchmarking studies:
   - *highest possible production: Utilization*
   - *maximizing product value: Volumetric Expansion Index*
   - *minimum conversion cost: all indices*
2. Operate the plants safely and in an environmentally responsible manner.
   Major safety/environmental incidents are likely to affect Net Margin. This is directly linked to a Health and Safety Index and an Environmental Index.
3. Meet customer's expectations with regard to Quality, Quantity, and Timely Delivery.
   Customer Satisfaction is the primary business driver.

COMMITMENT

Commitment statement shows willingness to achieve the Objectives. **This is a requirement for International Standards Organization (ISO) certification**. Box 10.5 gives an example of a commitment statement with an explanation of the words in the statement.

BOX 10.5 SAMPLE COMMITMENT STATEMENT AND *EXPLANATION*

We are committed through our Integrated Management System for the following:

- Communicate this Policy to our employees and make it available to interested Customers and Stakeholders.
  This is required to ensure that all employees are aware of this policy. Customers and stakeholders can also see that we are committed to the Continual Improvement Process.
- Implement Occupational Health and Safety, Environmental, and Quality Management Systems as an integral part of prime line responsibility at all levels of our organization.
  This statement is a requirement for compliance with International Standards Organization (ISO) 9001:2000, ISO 14001, and OHSAS 18001. This is to be reflected in Job Descriptions and Annual Appraisals.
- Ensure the implementation of a Continual Improvement Process.
  This statement is a requirement for compliance with ISO 9001:2000, ISO 14001, and OHSAS 18001.
- Comply with all relevant State Legislation and Regulations as well as International Standards and Agreements adopted by Refining Directorate.
  This statement is a requirement for compliance with ISO 9001:2000, ISO 14001, and OHSAS 18001.
- Ensure that Customer and Stakeholder requirements are complied with.
  This statement is a requirement for compliance with ISO 9001:2000, ISO 14001, and OHSAS 18001. A Stakeholder is any person or organization that is influenced by, or can influence, the activities of the Refining Business Unit (BU). The methods of measuring customer satisfaction will be detailed in the Integrated Management System (IMS) procedures.

VALUES

Values are the way people are expected to perform. It is what is important to us. These need to be agreed by all parties and include such things as professionalism, honesty, etc.

Management author Jim Collins defines core values as "the essential and enduring tenets of an organization – the ***very small set of guiding principles that have a profound impact on how everyone in the organization thinks and acts.****"*[4]

Core values:

- articulate what we stand for;
- are a set of beliefs that influence the way people and groups behave;
- are the "soul" of the organization;
- guide the business processes;
- guide the decision-making; and
- explain why we do business the way we do.

Core values are important because they:

1. influence behavior;
2. communicate what we really believe;
3. are sacred and do not change very often;
4. provide a moral compass;
5. provide continuity through change;
6. help people make tough decisions;
7. help to decentralize decision-making;
8. help people to be more proactive; and
9. are integrated into all levels and functions of the business.

Core values are supported by Company "ethics" and "conflict of interest" statements (see Section 9.5: Conflict of Interest and Ethics Corporate Statements).

Example[2]:

The Virgin Group's core values are paramount where emphasis is on the employee satisfying the customer's needs, even if it may cost the company a little. Thus management's focus is on looking after staff first, and the rest will follow.

An example of Core Values is shown in Box 10.6.

BOX 10.6 CORE VALUES

"While our key physical assets are the country's Oil and Gas reserves, our ability to maximize the economic returns to the State from these resources depends totally on the **quality and commitment of our staff**, whom we value as our greatest asset.

We also value our great **partners**, suppliers, and customers, and we shall endeavor to:

- *set an example in* ***personal and organizational integrity and transparency;***
- ***treat all people fairly and with mutual respect;***
- *have total participation of an excellent diverse workforce, rendering our* ***diversity a source of strength;***
- ***reward and promote based on merit, performance, and competence;***
- ***openly communicate*** *and work as a single integrated team."*

While holidaying in Sri Lanka, I came across a great Direction Statement that covered most of the requirements, **stated simply and effectively**. This is shown in Box 10.7.

BOX 10.7 KANDI HOTEL DIRECTION STATEMENT

Our Vision "අපගේ දැක්ම"

" To be World Class" "ලොවම දිනිය හැකි ආයතනයක් වීම "

Our Mission "අපගේ මෙහෙවර"

" To be a family of people and companies committed to provide legendary and innovative service with high stakeholder satisfaction "

" අපගේ සෑම පාරිභෝගිකයෙකුටම හා අත් කැපකරුවනට සදා මතකයේ රැඳෙන, නවතාවයකින් හා විශිෂ්ඨත්වයෙන් යුතු උතුම් ඝනයේ සේවා සැපයීමට පේට්වින්ගේ පවුලේ අප සැවොම ඇප කැප වෙමු ."

Our Core values – අපගේ මූලික හර හා වටිනාකම්

- ✓ At all times we will work with **Positive attitudes**.
  - සැමවිටම සුභවාදී ආකල්ප අනුව යමින් වැඩ කරමු .
- ✓ We will be **Committed to our service.**
  - සේවයට කැපවී බැඳී සිටිමු.
- ✓ Always we shall protect our **Integrity**.
  - අපගේ අවංක බව සැමවිටම රැකගනිමු.
- ✓ We shall work always with **Confidence**.
  - හැමවිටම ආත්ම විශ්වාසයෙන් කටයුතු කරමු.
- ✓ **Humility** will be an important value that we will protect.
  - නිහතමානිකම අගනා ගුණයක් ලෙස රකිමු.
- ✓ We shall extend **Sincerity** to all.
  - සැමට අව්‍යාජ සුහදතාව දක්වමු.
- ✓ All our services will be done with **Dedication**.
  - කරන සෑම කටයුත්තක්ම කැපවීමෙන් කරමු.
- ✓ Until a mission is completed we shall **Stay focused**.
  - අරමුණක පිහිටා එය මුදුන්පත් කර ගනිමු.
- ✓ We shall do our tasks with **Tenacity**.
  - ධෛර්යයෙන් සැමවිට ක්‍රියාකරමු.
- ✓ **Dependability** is a characteristic of our service.
  - විශ්වාසනීයත්වයෙන් යුතු සේවයක් සපයමු.
- ✓ **Pursuit of excellence** will be our clear objective.
  - විශිෂ්ඨත්වය සොයා යෑම අපගේ පරම අරමුණ කර ගනිමු.

The Balanced Scorecard[3] could be used as part of the vertical hierarchy. An example is shown in Box 10.8.

BOX 10.8 APPLICATION OF THE BALANCED SCORECARD

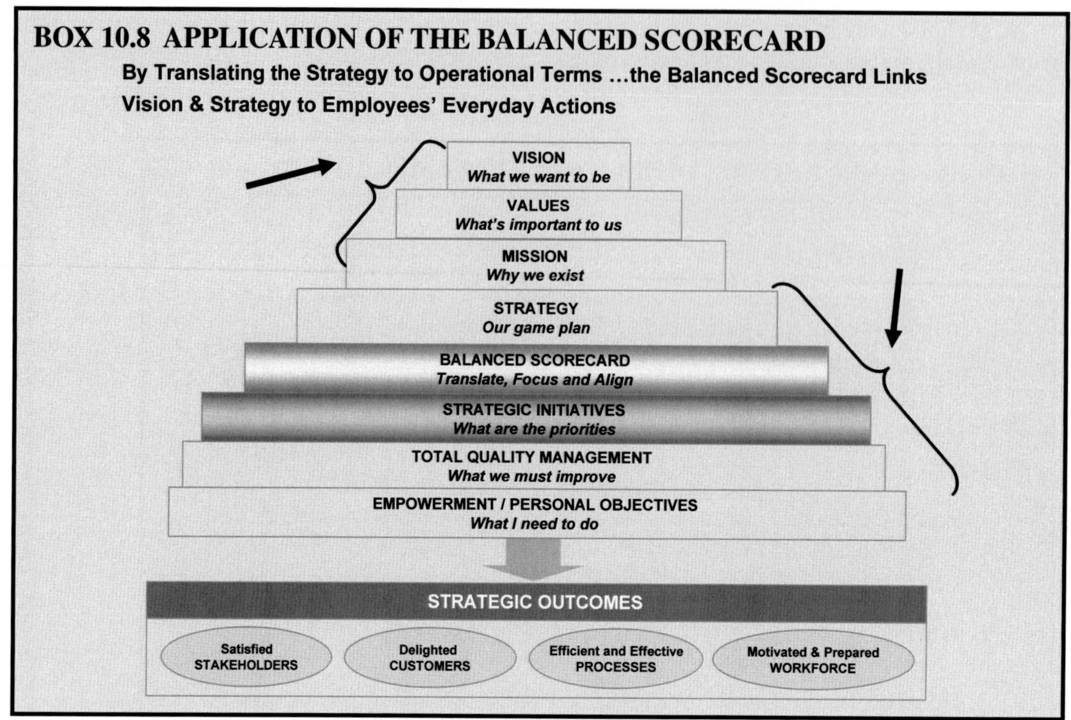

10.6 ALIGNMENT AND ORGANIZATION STRUCTURES

Hierarchical

These are the traditional structures found in an operating environment. The trick is to align these, as far as possible, with **Core and Support Management Systems**.

For example:

1. *All staff operating the process units and thus directly managing the Core Production System would be members of the Production or Operations Department.*
2. *All staff undertaking maintenance work, within the Support Maintenance System, would be members of the Maintenance Department.*

Matrix

These are applied in a project situation. The team is formed at the start of the project (including secondees from Operations and Maintenance Departments) and disbanded at the end of the project. The project could be the construction of a new process unit or it could be a Turnaround and Inspection of existing process units. It could also be a Management Improvement Project (see Section 30.4: Strategies, Plans, and Actions).

Mixed

These are generally applied when there is a major production target or operational improvement is to be achieved.

RESPONSIBILITIES

Chairman

The boss of the Board of Directors, **not** the company.

Board of Directors Member

Someone with the ultimate statutory legal accountability and liability for the performance and behaviors of the organization. In most countries, appointments must be registered.

Chief Executive Officer/Managing Director

The boss of the company.

If he/she and any other staff member becomes a Board Member, they require **two** employment contracts, one for the 80+% of time managing the operation of the company and the other for carrying out the duties of a director.

Senior Operations Staff

The management team.

Customer Interfacing Staff

The customer's perception of the organization is usually created by the customer-facing staff.

Other Staff

All other staff would have **personal performance targets** agreed by management, which align with the higher-level targets of the organization.

10.7 DOCUMENT HIERARCHY

Document hierarchy is important to prioritize compliance. This gives a clear indication that all higher documents have to be complied with. Fig. 10.3 gives an example.

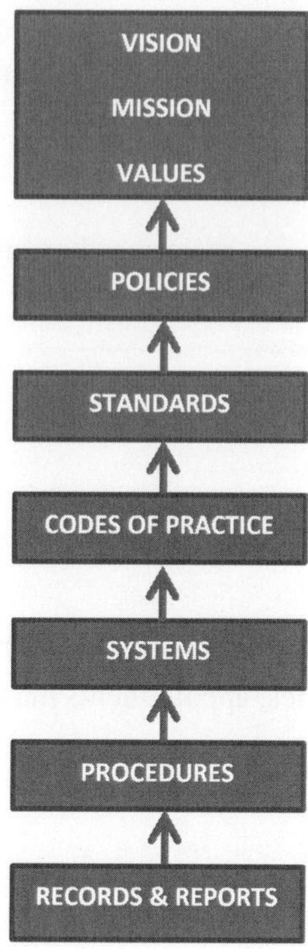

FIGURE 10.3

Document hierarchy example.

10.8 SUMMARY

Alignment is the key to the success of the Company and each BU. A requirement for integration of **Systems** is that vertical and horizontal alignment is balanced.

Vertical alignment hierarchy is as follows:

- **Vision**: Where do we want to be?
- **Mission**: Why do we exist?
- **Core Values**: What do we stand for?
- **Corporate Objectives/Strategy**: What direction do we need to take?
- **Value Chain**: Raw materials in to products out
- **BU**: The legal entity
- **Core System:** An Element of the Value Chain
- **KPIs**: How do we measure progress?
- **Benchmark**: Where are we?
- **Targets**: What do we need to achieve?
- **Strategies and Plans**: Means to achieve targets. How do we get there?
- **Ownership:** Responsibility, Accountability, and Reward

Core values are the underlying principles that guide an organization's Corporate Strategy and are supported by Company "ethics" and "conflict of interest" statements. They are "**the essential and enduring tenets of an organization—the very small set of guiding principles that have a profound impact on how everyone in the organization thinks and acts**."

Horizontal alignment must ensure:

1. customer orientation;
2. streamlined interfaces between departments/directorates/affiliates;
3. consistent approach;
4. focus on performance measurement; and
5. duplication/gaps are eliminated.

A Policy or Direction Statement, including **vision**, mission, commitment, and values, must be in simple language that is easily understood by all staff. They must, therefore, understand their part in achieving the **Vision of the Company** as well as satisfying the **Customers' requirements**.

REFERENCES

1. Chevron Vision. www.chevron.com/about/the-chevron-way.
2. Branson R. *The Virgin way*. Virgin; 2014.
3. Kaplan S, Norton DP. *The balanced scorecard*. HBS Press; 1996.
4. Collins J. *Good to great*. Random House; 2001.

PART

RISK AND PERFORMANCE

3

OVERVIEW

WHY IS RISK MANAGEMENT SO IMPORTANT?

Running a business involves taking certain "leaps of faith" or risks, regardless of the industry involved. The amount of risk, or risk appetite, is a level of risk that the company is prepared to tolerate and is determined by the Board of Directors. All high-level risks in the business need to be assessed and any exposure outside of the accepted limits must either be mitigated or eliminated since operating outside of the risk limits could detrimentally affect the performance of the business.

THE OBJECTIVE

The objectives of Part 3 are the following:

1. Give guidelines for establishing corporate-wide risk management
2. Identify some risk control mechanisms
3. Demonstrate that risk and performance management are intertwined and need to be managed jointly
4. Outline the requirement for incident preparedness and operational continuity management

PICTORIAL VIEW

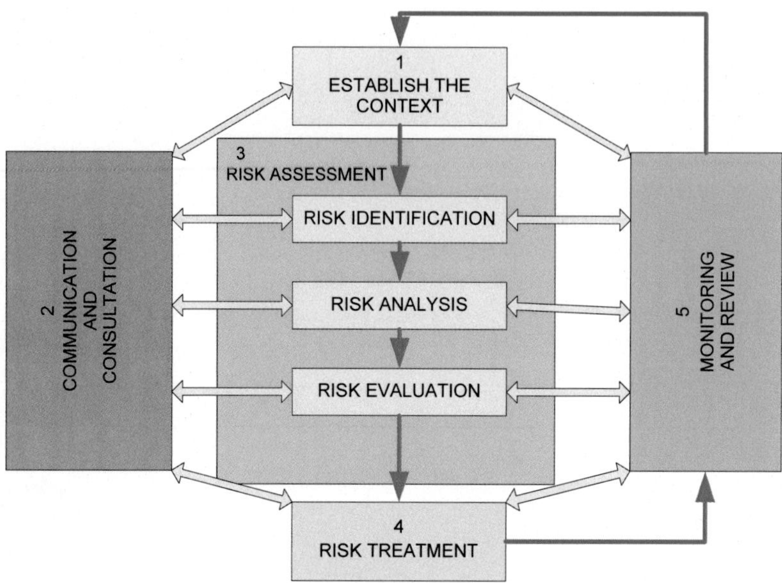

OUTLINE OF PART 3

- Risk is defined.
- The International Standards for Risk Management are outlined.
- Primary categories of risk are discussed.
- Corporate risk appetite is explained with a sample risk statement.
- Four strategies for risk control are defined.
- Roles and responsibilities, in line with three lines of defense, are presented.
- Types of reviews and audits are listed.
- Risk maturity assessment is outlined.
- Benefits of undertaking enterprise-wide risk management are listed.
- Interactive risk control mechanisms for application throughout the business are discussed.
- Scenario planning, as an integral part of strategic planning, is emphasized.
- The issue of "black swans" is demonstrated using examples.
- Root cause analysis tools are demonstrated using examples.
- The combined management of risk and performance is outlined, giving common prioritization.
- The relationship between ISO standards for health/safety and environment is discussed.
- Incident preparedness and operational continuity management requirements are discussed.

CHAPTER

RISK MANAGEMENT

11

11.1 INTRODUCTION

Risk is everybody's business.

This chapter discusses the embedding of risk management in all aspects of the business. Traditionally, the focus was solely on Occupational Health and Safety, but since recent major disasters, a more holistic approach has evolved.

The process generally involves the following steps:

1. **identifying and ranking** the risks inherent in the company's strategy (including its overall goals and appetite for risk);
2. **selecting the appropriate risk management approaches** and transferring, or avoiding, those risks that the business is not competent or willing to manage;
3. **implementing controls** to manage the remaining risks;
4. **monitoring the effectiveness** of risk management approaches and controls; and
5. **learning from experience** and making improvements.

Risk assurance is an independent check on risk control processes.

11.2 RISK DEFINITION

RISK: GENERIC DEFINITION

Risk is often expressed in terms of a combination of the impact of an event and the associated likelihood of occurrence of the event.

Risk = Probability × Consequence.

Risk/threat: condition/event that may prevent us from meeting desired objectives.

Probability: chance/probability that a risk/threat will occur.

Consequence: represents the impact that will be incurred if the risk/threat actually occurs.

ISO 31000 STANDARD: RISK MANAGEMENT PRINCIPLES: DEFINITION[1]

A risk is defined as an uncertain event or condition that, if it occurs, will have an impact on the business objectives.

RISK MANAGEMENT: GENERIC DEFINITION

Risk Management is the systematic application of management policies, procedures, and practices to the activities of communicating, consulting, establishing context, and identifying, analyzing, evaluating, treating, monitoring, and reviewing risks.

Performance Management for the Oil, Gas, and Process Industries. http://dx.doi.org/10.1016/B978-0-12-810446-0.00011-6

11.3 ISO 31000, RISK MANAGEMENT[1,2]

ISO 31000 is the international standard for risk management.

The key requirements of the process are to:

1. establish the context
   a. identify responsibility and reporting lines for risk
2. communication and consultation
   a. determine your risk appetite
3. risk assessment
   a. list risks within categories (suggested: strategic, operational, financial, compliance, environmental)
   b. rank risks by, for example, using a matrix diagram
   c. define appropriate action for risks according to the four Strategies of Risk Control (4 Ts)
4. risk treatment
   a. take action
5. monitoring and review
   a. monitor regularly through improved risk reporting
   b. learn the lessons and feed back into the process.

The process is depicted as shown in Fig. 11.1.

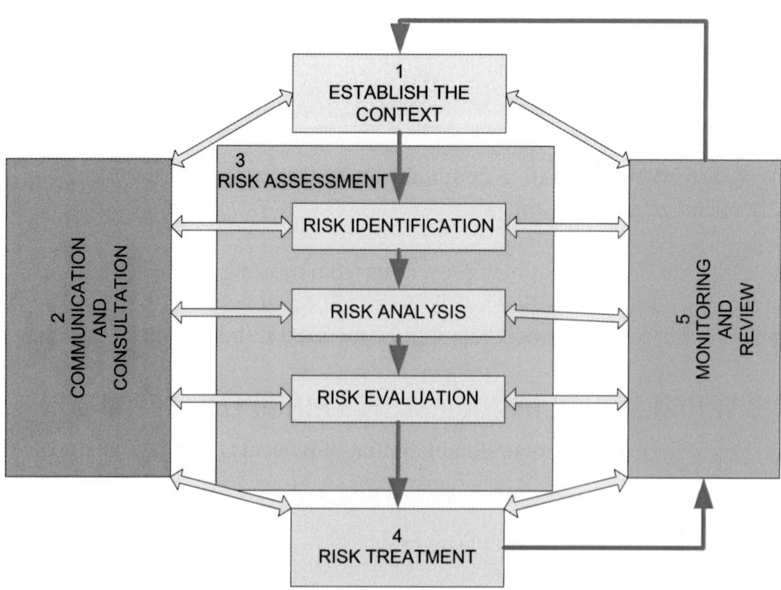

FIGURE 11.1

ISO 31000: Risk Management process.

11.4 CATEGORIES OF RISK

The Committee of Sponsoring Organizations (COSO) of the Treadway Commission (US)[3] gives four categories of risk:

1. strategic
2. financial/reporting
3. operational
4. compliance

1. **Strategic** risks are associated with macroeconomics and politics that may affect long-term planning and competitiveness. It affects strategy selection and deployment, for example, investment in alternative industries and technologies.
2. **Financing** involves risks such as inadequate financial appraisal and monitoring of performance, which may result in inefficient allocation of available financial resources and, consequently, poor financial returns.
3. **Operational** risks emerge during the execution of business functions to achieve business objectives. They can originate from People, Process, Performance (3 Ps) and/or technology involved in an operation.
4. **Compliance/Regulatory/legal risks** may arise from the ineffective or inefficient means of evidencing compliance with legal rules and regulations, for example, Health and Safety, Environment, Employment, International Trade, etc.

The four categories of risk are depicted as in Fig. 11.2.

STRATEGIC RISKS
Strategic risks relate to macro-economics and politics that may affect long term planning and competitiveness.
FINANCIAL RISKS
Financial risks relate to financial resources and financial returns.
Unexpected Loss
OPERATIONAL RISKS
Operational risks relate to operation of the physical assets that produces the product.
Approved By
REGULATORY/LEGAL RISKS
Regulatory / Legal risks relate to compliance with rules and regulations.

FIGURE 11.2

Committee of Sponsoring Organizations (COSO) risk categories.

COSO Enterprise Risk Management—Integrated Framework defines eight components with regard to the management of enterprise risk. These interrelated components are derived from the way management runs an enterprise and are integrated with the management process. The components are as follows:

1. internal environment
2. objective setting
3. event identification
4. risk assessment
5. risk response
6. control activities
7. information and communication
8. monitoring

These components are similar to the key requirements of ISO 31000.

Box 11.1 gives an example of a Corporate Risk Management System established by a National Oil Company (NOC). As seen in this example, the risk categories differ from the COSO categories, but this company finds these more suitable for their particular application.

BOX 11.1 THE ESTABLISHMENT OF ENTERPRISE RISK MANAGEMENT (ERM) IN A NATIONAL OIL COMPANY (NOC)

A dedicated **Corporate Risk Management Unit (CRMU)** was formed.

A **Risk Management Framework** was established, encompassing the following:

1. **Risk Strategy and Policy**
   A key requirement is that all managers are responsible for managing risk. The CRMU develops the Risk Framework, Policies, and Guidelines.
2. **Risk Measurement**
   Categories:
   a. Credit Risk
   b. Country Risk
   c. Trading Risk
   d. Finance Risk
   e. Insurance Risk
   f. HSE Risk
   g. Security Risk
3. **Risk Operation and System**
   A Risk Management Dashboard was created as a communication tool to monitor risk exposure and to assist senior management in making informed decisions on risk-related issues in their respective Business Unit.
4. **Risk Organization and Culture**
   Each Business Unit has a Risk Management Section. The Business Unit Heads report on risk to the Risk Management Committee, which in turn reports to the Company President and Management Committee.

11.5 RISK APPETITE

Risk appetite is the level of risk that one is willing to tolerate. If a risk is intolerable, it has to be removed or mitigated. This is the response. The risk before the response is referred to as the **inherent risk**, while the risk after the response is referred to as the **residual risk**.

The fundamentals of risk management are depicted in Fig. 11.3.

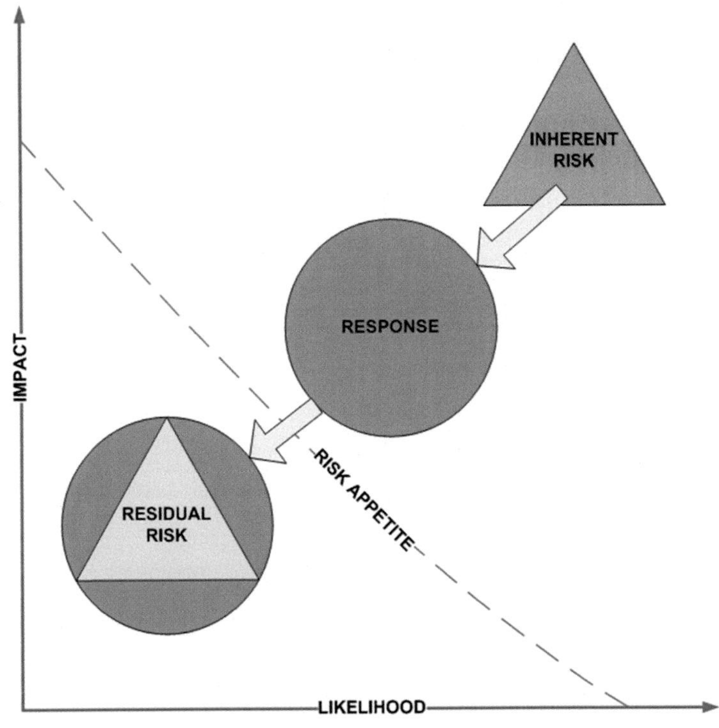

FIGURE 11.3

Risk appetite.

The Board of Directors sets the risk appetite of the company (see Section 9.4: The Board: The 'Fulcrum' of Business Performance).

Anything outside of the risk appetite line has to be brought within the line.

The common term for acceptance used in the industry is **ALARP, As Low As Reasonably Practical**. ALARP has three levels of risk: Intolerable, ALARP Region, and Broadly Acceptable Region. Within the ALARP Region, the risk can vary from a low-risk level, "tolerable if cost of reduction exceeds improvement gained," to an upper risk limit, "tolerable only if risk reduction is impracticable or cost/improvement ratio is grossly disproportionate." ALARP can be depicted as shown in Fig. 11.4.

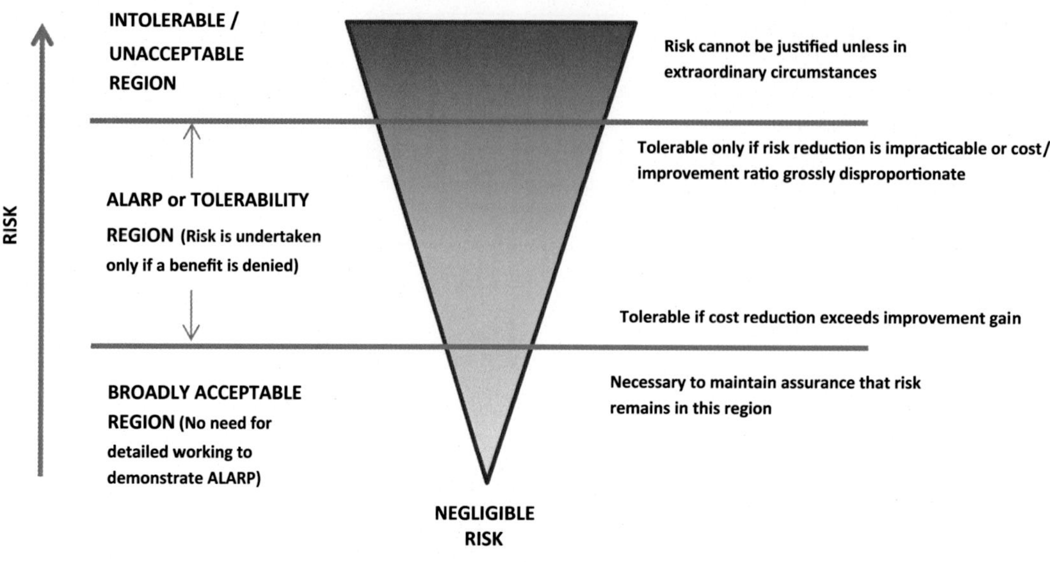

FIGURE 11.4

Risk tolerance.

11.6 STATEMENT OF RISK

This is the **Top Statement on Risk** in the Company and is normally contained in the **Governance Framework** document (see Section 8.5: The Model Governance Framework Document: Discussion on the Contents).

Box 11.2 shows a sample "Statement of Risk."

BOX 11.2 STATEMENT OF RISK

Sample for the Company Governance Framework

The following was a draft statement for inclusion in a company's Governance Framework:

> ***The company*** *operates in an environment of uncertainty and recognizes that risk management enables informed resource allocation.* ***The company****'s risk-based approach to internal control is described in this* ***Statement on Risk Management*** *and enterprise-wide instructions for its implementation. This risk-based approach applies to the overall design of the* ***company Governance Framework*** *and all of its components. It requires* ***Business Units*** *and* ***Functions*** *to understand the relations between the business environment, objectives, risk, and performance, and to establish appropriate risk responses and strategies to deal with both routine risks and the need to maintain business continuity in a range of reasonably foreseeable emergency scenarios and events.*

11.7 THE FOUR STRATEGIES OF RISK CONTROL

Typically there are four actions that can be taken to mitigate or remove a risk. These are described in Table 11.1.

Table 11.1 The Four Strategies (4 Ts) of Risk Control

Tolerate	Taking the risk where the consequences are small or the practicality of the other responses is unacceptable
Treat	Treating the risk by establishing effective and economical controls to reduce risks to an acceptable level
Transfer	Transferring the risk to other parties through, for example, insurance, hedging, or contracting out
Terminate	Terminating the risk by ceasing the activity or withdrawing from that particular business area

11.8 THREE LINES OF DEFENSE[4]

Common practice is to establish three lines of defense for Risk Management. The roles in Risk Management utilizing the three lines of defense approach would be split as follows:

1. Line Management
2. Corporate Risk Team
3. Internal Audit (IA) department

These are explained as follows:

1. Line Management: The **first line of defense** is the operational management and staff who have the day-to-day responsibility to manage and control risk. The first line implements the Policies, Standards, Frameworks, Processes, etc., which have been established by the second line of defense.
2. Corporate Risk Team: The **second line of defense** is the Risk Manager and his staff. They coordinate, facilitate, and oversee the management of risk and compliance within the company's **Risk Appetite**. (see Box 11.1, where a dedicated Corporate Risk Management Unit (CRMU) was established).
3. Internal Audit (IA) department: The **third line of defense** is the company's IA department supported by the **Board Audit (and Risk) Committee**. IA provides independent assurance with respect to the effectiveness of governance, risk management, and control.

The roles recommended by the Institute of Internal Auditors are described in Fig. 11.5. A sample audit recommendation for an ERM structure is depicted in Box 11.3.

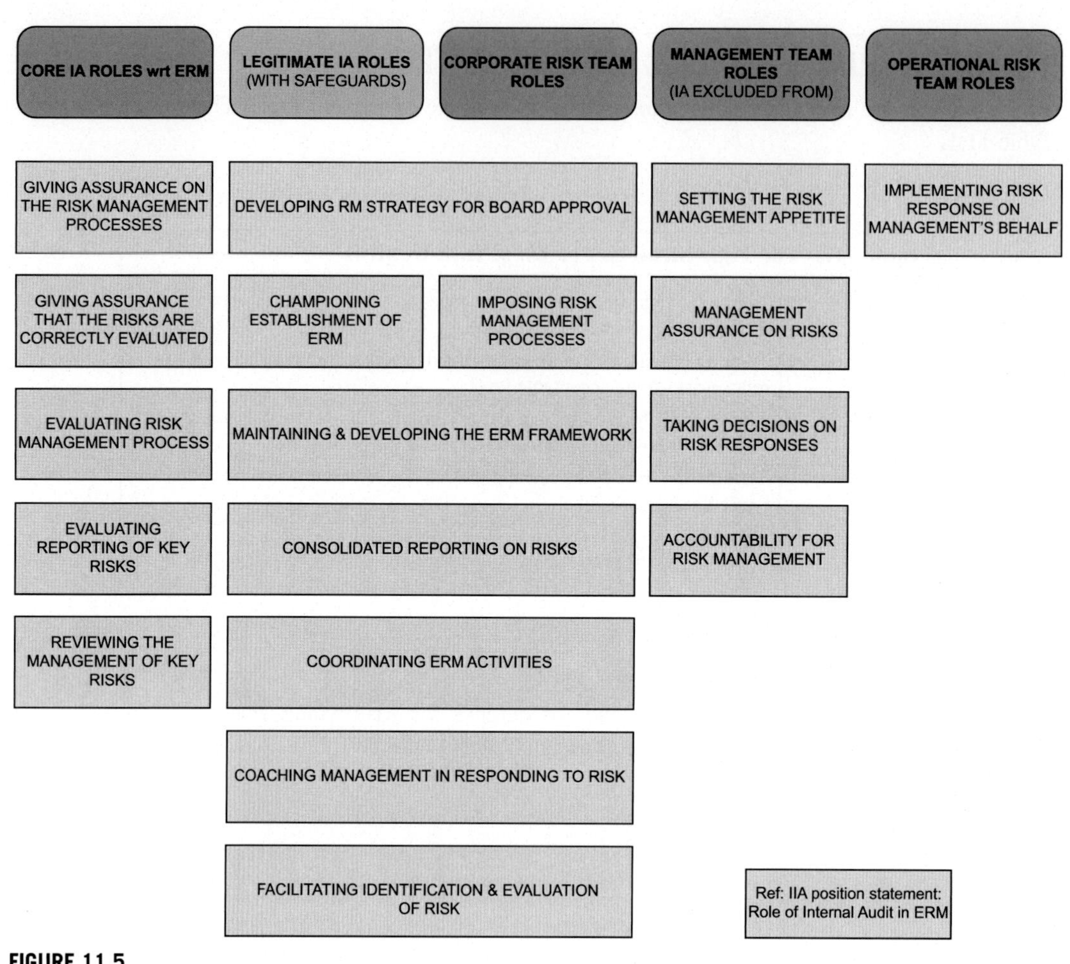

FIGURE 11.5

Three lines of defense roles. *IA*, internal audit.

11.9 ASSURANCE SERVICES[5]

Assurance Services are defined as:

> An objective examination of evidence for the purpose of providing an independent assessment on governance, risk management, and control processes for the organization.

Examples: financial, performance, compliance, system security, and "due diligence."

Some organizations are beginning to look more closely at their assurance needs and are asking questions like:

1. Who needs assurance both inside and outside the organization?

BOX 11.3 ENTERPRISE RISK MANAGEMENT STRUCTURE: AUDIT RECOMMENDATION

In an Audit of the NOC, the State Auditor recommended the introduction of Enterprise Risk Management (ERM). Implementation included the following proposed structure using three lines of defense.

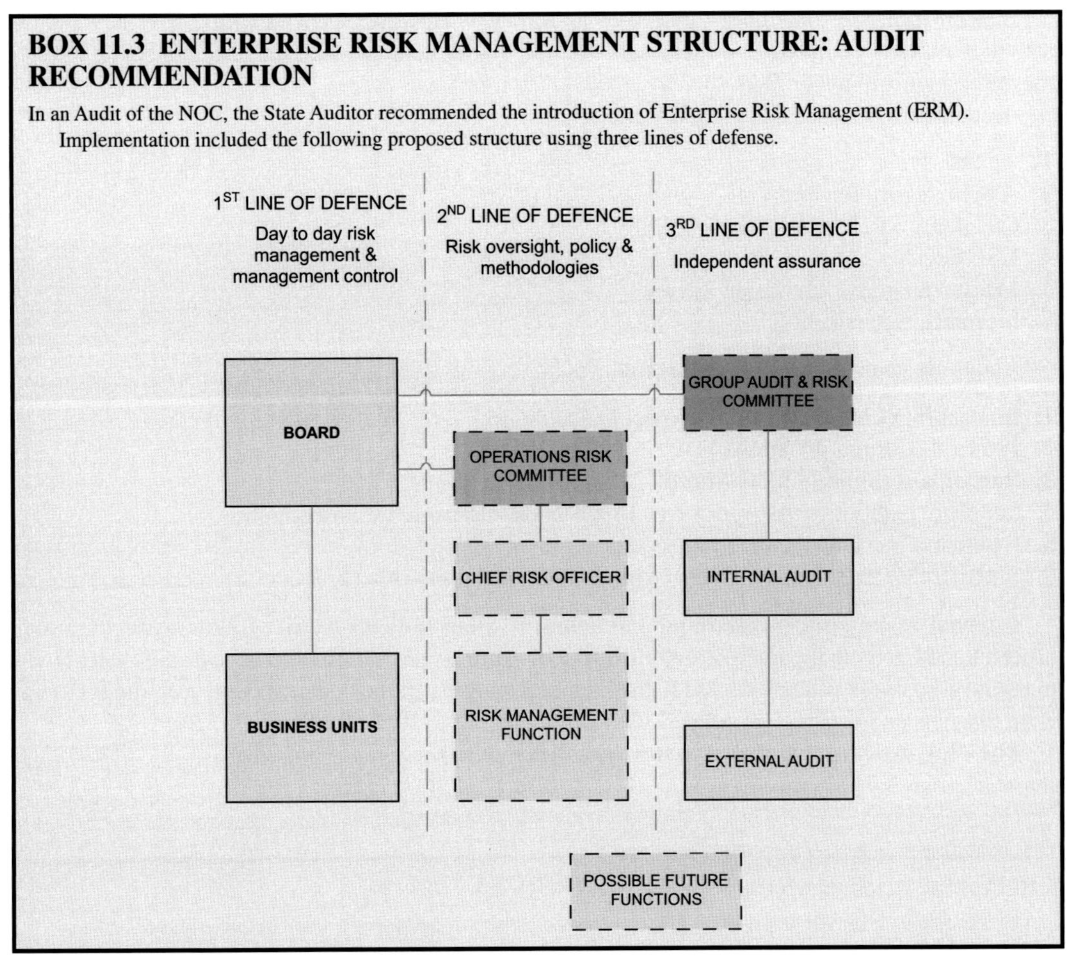

2. What do we require assurance on?
3. Who provides assurance?
4. How does assurance fit together?
5. Where are the gaps and overlaps?
6. Can assurance costs be controlled?

Fundamentally, it is about rationalizing the assurance provision so that both those governing the organization and the stakeholders know that the stated objectives are being achieved through the management of risk.

This is explicitly addressed in the Code of Governance Principles for South Africa, effective July 1, 2010, known as “King III,” whereby a combined assurance model is advocated to provide a coordinated approach to all assurance activity (see Chapter 7: What Are the Principles of Good Governance? for further discussion on “King III”).

Benchmarking is sometimes referred to as a review. However, benchmarking is a very focused activity with clear boundaries, which might not necessarily coincide with the business boundaries and may not capture all major risks.

Assurance Services include the following:

1. Audits
2. Asset Integrity Reviews
3. Quantitative Risk Assessment (QRA)
4. HSE Reviews
5. Measurement and Allocation Reviews
6. Insurance Assessment

Audits are generally categorized as follows:

1. Internal ISO Certification Audits (peer QA audits)
2. **Internal (Company) Audits**
3. **Participant (Shareholder) Audits**
4. Supplier Quality Control Audits (see Box 9.7: Procurement, for an example)
5. External (Financial) Audits
6. External ISO Certification Audits (Accreditation Authority)

Internal (Company) Audits and Participant (Shareholder) Audits follow a standard audit procedure in line with the International Institute of Auditors procedures and have traditionally tended to focus on business systems related to the financial integrity of the business. However, the focus now is on Systems and the risks associated with not achieving System Objectives.

Box 11.4 gives the outcomes of a survey of different risk assurance providers.

BOX 11.4 ASSURANCE SERVICES SURVEY

"Obtaining a full picture of internal and external providers of assurance can be time-consuming and complex, and this often means that no one assumes overall responsibility for this task. Our survey among heads of IA revealed that half of the participants felt that the people responsible for governance in their organization, such as the board of directors or governors, did not have a complete picture of assurance....

The table below lists the assurance providers you might find in an organization and confirms that many organizations have several sources of assurance, both internal and external":

Who Provides Assurance That Risks Are Being Managed and Controlled in Your Organization?	Percentage of Survey Respondents Who Say This Occurs in Their Organization.
1. Internal audit	90
2. Management: self-assessments	71
3. External audit	63
4. Management: Key Performance Indicator and performance reports	60
5. Risk management function	55

BOX 11.4 ASSURANCE SERVICES SURVEY—cont'd

Who Provides Assurance That Risks Are Being Managed and Controlled in Your Organization?	Percentage of Survey Respondents Who Say This Occurs in Their Organization.
6. Health and safety auditors	**45**
7. Compliance function	37
8. Quality auditors	**36**
9. Regulatory bodies	29
10. Information security auditors	26
11. External inspection agencies	19
12. Environmental auditors	**16**
13. Counter fraud team	11
14. Government agencies	10
15. Corporate social responsibility auditors	6
16. Complaints team	5
17. Funding and investment auditors	3

From: Coordination of Assurance Services.[5]

Items in **bold** are common in the oil and gas industry.

Comment

It appears from this survey that consolidation of audit services is necessary in many organizations.

11.10 INDIVIDUAL RISKS AND THE RISK MANAGEMENT FRAMEWORK

INDIVIDUAL RISKS

Each risk, which has been identified for entry into a risk register, needs to be clearly stated.

The structure of the statement must be in three parts, with clear distinctions between "**cause**," "**event**," and "**consequence**."

Examples:

As a result of ***the contracted rig being released late from a previous operation****, a* ***potential clash between marine installation and rig schedule*** *may occur, which will lead to* ***delays in project schedule****.*

Single mechanical shaft seals *on gas compressors have a* ***higher potential for leakage*** *than double seals and could result in an* ***explosion*** (see Box 18.1: Requirement for a New Standard: Potential Seal Failure).

It is clear from this that the **cause** must be understood in order to mitigate the **consequence**.

Each identified risk has a lifespan. The life of each risk is flowcharted from identification and registration through assessment, monitoring, and mitigating actions until it is no longer regarded as a risk, at which time, it is removed from the register. This is generally depicted as in Fig. 11.6.

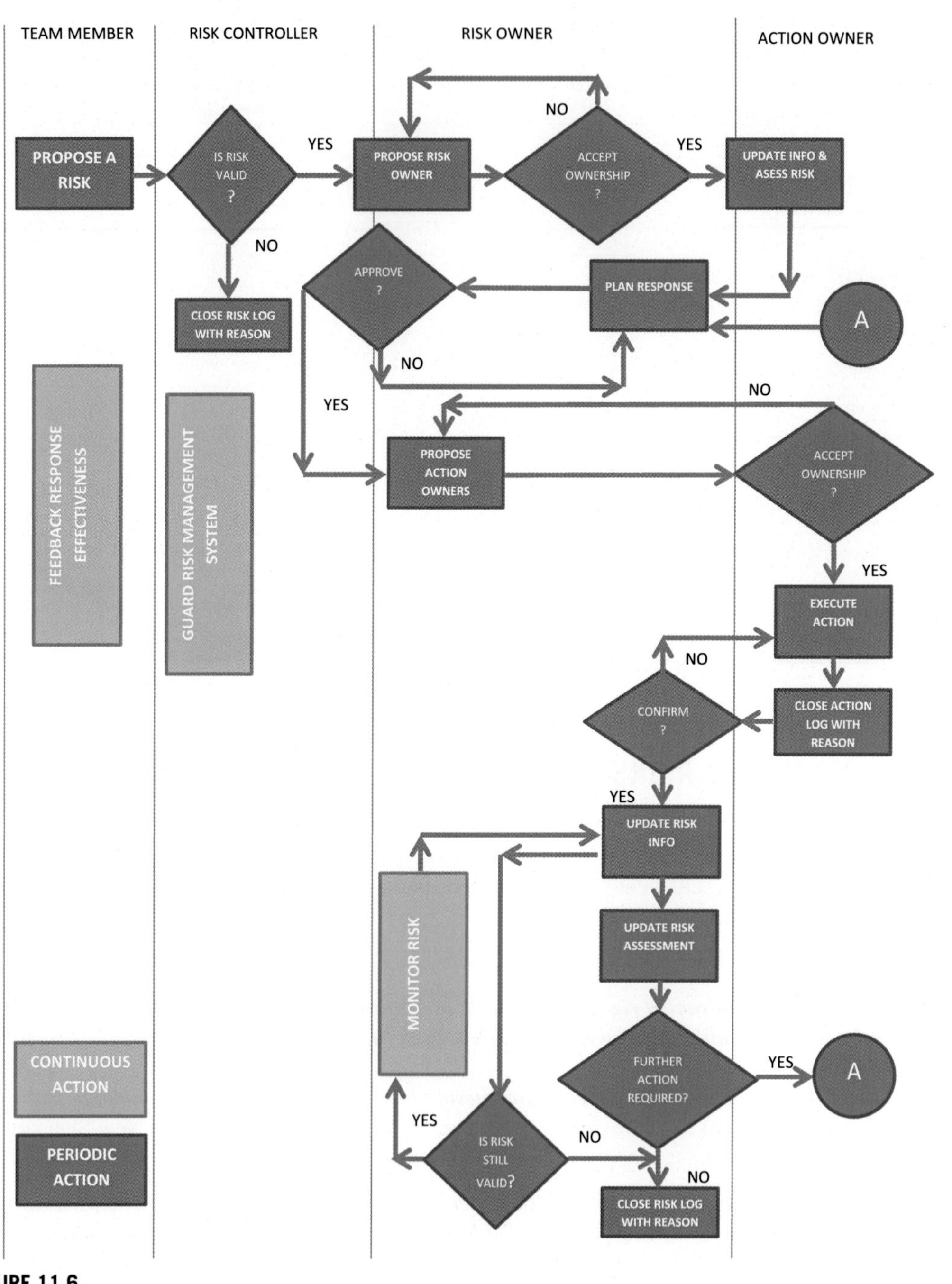

FIGURE 11.6

The life of a risk.

RISK MANAGEMENT FRAMEWORK

It is essential to establish a risk management framework to work within. A sample risk management framework is depicted in Fig. 11.7.

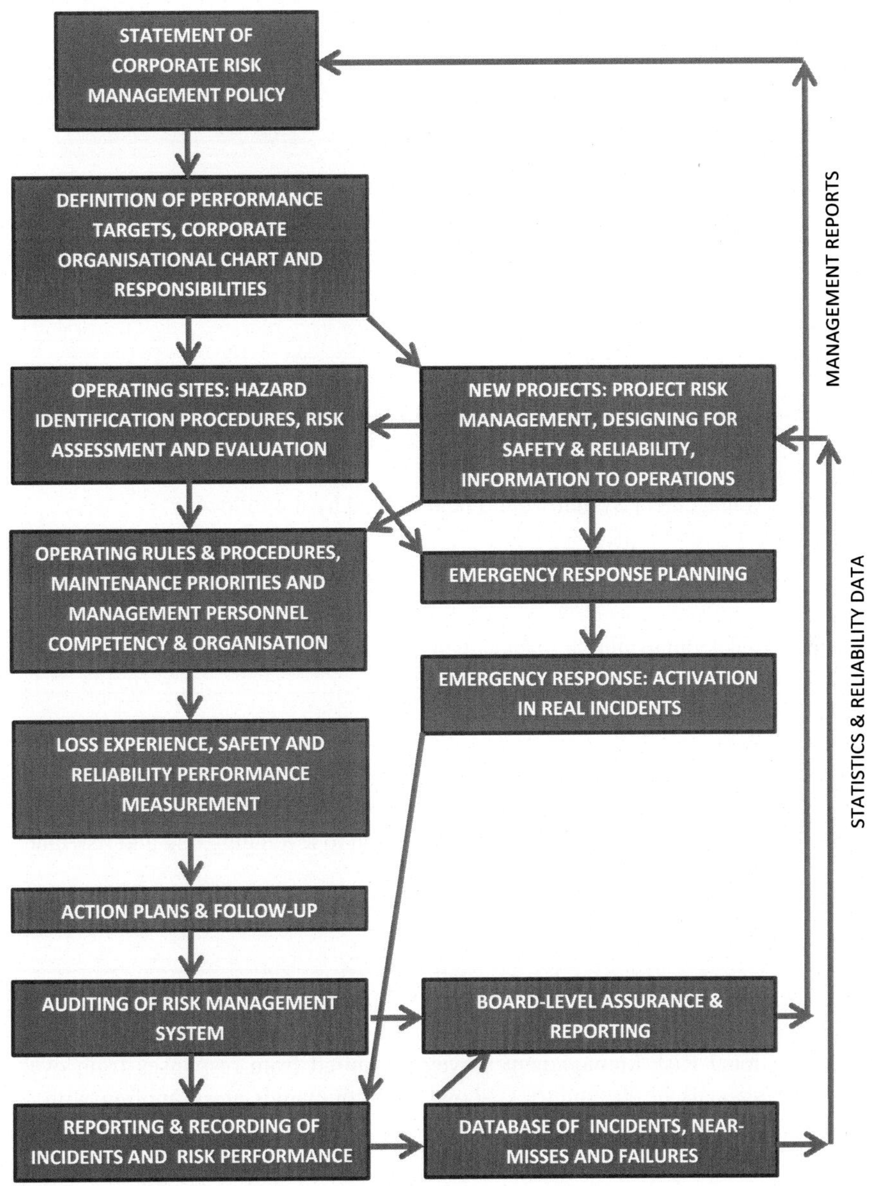

FIGURE 11.7

Risk Management Framework example.

11.11 RISK MATURITY

Risk Management Maturity needs to be assessed to show potential System improvements.

RMRP–2002-02, "Risk Management Maturity Level Development" (April 2002),[6] describes a simple framework with the following four levels of assessment for Risk Management Maturity:

1. ad hoc
2. initial
3. repeatable
4. managed

The Level of Maturity is assessed in the following categories:

1. definition
2. culture
3. process
4. experience
5. application

This tool could be used to assess an organization prior to embarking on establishing an Enterprise Risk Management System or Project Risk Management System, and determining the eventual required outcome after implementation.

A mature Risk Management System would be indicated by the following:

1. board-level commitment to ERM;
2. a dedicated risk executive in a senior-level position, driving and facilitating the ERM process;
3. an ERM culture that encourages full engagement and accountability at all levels of the company;
4. engagement of stakeholders in risk management strategy development and policy setting;
5. risk communication transparency;
6. integration of risk information for decision-making;
7. risk-based processes;
8. use of sophisticated quantification methods to understand risk and demonstrate "added value" through risk management; and
9. moving from focusing on risk avoidance and mitigation to leveraging risk and risk management options that extract value.

11.12 RISK MANAGEMENT SURVEY[7]

Aon, a leading global provider of risk management insurance, performs an annual risk management survey. Its 2015 Global Risk Management survey, compiled from responses from over 1400 risk management professionals in 60 countries, shows that companies are grappling with similar new risks, but differ on how they rank them and on how best to address them.

Threats to companies' **reputations** and **cyber risk** rank high among business leaders' top risk concerns, while economic and regulatory risks, along with increased competition and the inability to retain or recruit needed talent, continue to rank high.

The survey results offer answers to many key risk management questions including:

- How do your risk concerns compare to those of your peers worldwide?
- What's driving increased or decreased concern over certain exposures?
- How do some risk concerns vary by region or industry?

11.13 POTENTIAL BENEFITS OF ENTERPRISE RISK MANAGEMENT

Some potential benefits are listed as follows:

- consolidated reporting and visibility of disparate risks at board level;
- greater likelihood of achieving business objectives efficiently;
- improved understanding of the key risks and their implications;
- more informed risk-taking and decision-making;
- ERM facilitates Risk-Based resource allocation corporate-wide;
- identification and sharing of cross-business risks;
- greater management focus on the issues that matter;
- fewer surprises or crises;
- more focus internally on doing the right things in the right way and at the right time;
- capability to take on greater risk for greater reward;
- improved stakeholder confidence and trust; and
- *"it is a powerful tool for enhancing overall business performance."*

11.14 SUMMARY

A risk is defined as an uncertain event or condition that, if it occurs, will have an impact on the business objectives. This impact is normally negative.

The Management of Risk entails having the right information to make the right decision at the right time. Thus, knowledge of the company's biggest risks and the optimum mitigation actions are critical for Risk Management.

Risk is everybody's business, as risks are evident throughout the company.

ISO 31000, Risk Management, describes a generic process for managing risk as follows:

1. establish the context
2. communication and consultation
3. risk assessment
4. risk treatment
5. monitoring and review

At corporate level, there are four basic risk categories: Strategic, Financial/Reporting, Operational, and Compliance.

The Company's risk appetite differentiates between acceptable and unacceptable risk and is defined by the Company Board of Directors.

A "Statement of Risk" is required in the company's top governance framework document.

The 4 Ts are tolerate, treat, transfer, or terminate.

Three lines of defense (line management, corporate risk team, and audit) are advocated for full control of corporate risk.

Assurance Services are defined as "an objective examination of evidence for the purpose of providing an independent assessment on governance, risk management, and control processes for the organisation." Assurance Services include Audits, Asset Integrity Reviews, QRA, HSE Reviews, Measurement and Allocation Reviews, and Insurance Assessment.

A risk management framework needs to be established in which the Risk Management System can operate.

Risk Management Systems need to be assessed for maturity so as to initiate improvement of the System.

Surveys, such as those undertaken annually by Aon, help companies with their strategic risk planning.

Enterprise Risk Management has significant advantages for performance improvement of the business. These include bringing together all the top risks into a single risk register.

The interaction for risk mitigation and control takes place at all levels in the business. These mechanisms are discussed in Chapter 12, Risk Control Mechanisms.

REFERENCES

1. ISO 31000:2009. *Risk management – principles & guidelines.*
2. ISO 31010:2009. *Risk management – risk assessment techniques.*
3. Committee of Sponsoring Organizations (COSO) of the Treadway Commission (USA). *Enterprise Risk Management – Integrated Framework*; September 2004. http://www.coso.org/publications/erm/coso_erm_executivesummary.pdf.
4. *The role of internal audit in enterprise risk management position paper.* Institute of Internal Auditors – UK and Ireland; 2004.
5. *Coordination of assurance services. Professional guidance for internal auditors.* Institute of Internal Auditors – UK and Ireland; 2010.
6. *RMRP – 2002 – 02 Version 1.0. Risk management maturity level development.* INCOSE risk management working group: PMI RMSIG, UKAPM RSIG. April 2002.
7. *Aon risk management survey.* http://www.aon.com/2015GlobalRisk/.

FURTHER READING

1. Gemes A, Golder P. *What is your risk appetite?* Booz. http://www.strategy-business.com/article/00010?gko=cb35c.
2. Phillips S. *The rise of the CRO.* The Peninsula; 21 September, 2011.
3. *Risk management fundamentals.* The Institute of Internal Auditors – UK and Ireland; 2006.
4. *An approach to implementing risk based internal auditing – professional guidance for internal auditors.* Institute of Internal Auditors – UK and Ireland (IIA); 2003.
5. The Institute of Internal Auditors (IIA). *International standards for the professional practice of internal auditing.* https://na.theiia.org/standards-guidance/Public%20Documents/IPPF%202013%20English.pdf.

6. *The risk IT practitioner guide*. 2009. www.isaca.org. Good comparisons with COSO framework and ISO guide 73.
7. Allen F, Bloodworth P. *Operational risk and resilience*. PWC; 2001.
8. *Applying COSO's enterprise risk management integrated framework*. www.theiia.org.
9. *Enterprise risk management readiness guide*. www.activerisk.com.
10. Price E. *Embedding a risk management culture from the top down*. www.activerisk.com.
11. Active Risk Manager software. http://www.sword-activerisk.com/products/active-risk-manager-arm/.
12. *High level framework for process safety management*. Energy Institute. https://www.energyinst.org/technical/psm/PSM-framework.
13. D'Aquino R, Berger S. Baker report: analyzing BP's process safety program. *Chem Eng Prog* February 2007.
14. Saling P. Implementing enterprise risk management at MOL Group – a case study. *Palisade user Conference London 23–24th April 2007*.
15. @RISK software from Palisade. http://www.palisade.com/risk/?gclid=CjwKEAjwxce4BRDE2dG4ueLArHMSJADStCqMPeCd4YPzvhHnF_MjoarKYWj_Eab5jyu4-a_JOioF1BoCnlrw_wcB.
16. *EasyRisk Manager-functionality incorporated into Synergi Life risk management module*. DNV-GL https://www.dnvgl.com/news/dnv-gl-software-launches-a-complete-risk-management-solution-28589. EasyRisk was used by Shell for the Pearl GTL project.
17. *The 9 hallmarks of top-performing ERM programs*. www.information-management.com/news/erm_risk_management.

CHAPTER

RISK CONTROL MECHANISMS 12

12.1 INTRODUCTION

One of the four strategies for risk mitigation and control is to **treat** the risk by establishing effective and economical controls to reduce risks to an acceptable level. This chapter discusses some risk control mechanisms for understanding control of risk so that mitigation is effectively focused.

Risk control mechanisms supplement, and do not replace, common sense, experience, and thinking outside the box.

12.2 ACCIDENT/INCIDENT CAUSATION THEORY

BASIC INCIDENT CAUSATION THEORY[1]

Incidents occur when inadequate barriers fail to prevent the things that can cause harm to escalate to undesirable consequences.

The barriers can be of different types. These are discussed in Section 12.4. The barriers are put and kept in place by people with the competence to do so, in line with required standards. Incidents happen when people make errors and fail to keep barriers functional or in place. The following are the primary areas of influence to reduce the probability of an incident occurring:

1. equipment design and controls layout
2. display and alarms
3. work practices and procedures
4. work management and authorization
5. task design and individual or team workload
6. process safety culture

Taproot Root Cause Analysis (RCA) software[3] defines the root cause of an incident more specifically in the following categories:

1. Procedures
2. Training
3. Quality Control
4. Communications
5. Management System
6. Human Engineering
7. Immediate Supervisor

RCA is discussed in Section 12.8.

Performance Management for the Oil, Gas, and Process Industries. http://dx.doi.org/10.1016/B978-0-12-810446-0.00012-8

12.3 THREE Ps

A company's "genetic" advantage is determined by the technology that is applied during the design of the process units. During this stage, there should be a strong emphasis on "fail-safe" operation.

> Trevor Kletz: "What you don't have can't leak."[4]

Operations, Maintenance, and Engineering staff are trained to operate the plant at the sustainable design throughput for the life of the investment. The systems and processes ensure orderly and predictable operation and maintenance of the plant. However, performance often has to be "stretched" to retain a competitive edge. All of the above is in the context of risk management. Thus everything we do relates to our approach to risk management. Higher performance without attention to other risk control mechanisms tends to raise the risk.

The interactions for similar risk levels are explained using three Interactive Risk Control Mechanisms and the underlying technology.

These are as follows:

1. **Process**: Processes within the Integrated Management System (IMS)
2. **People**: Organizational Effectiveness (OE)
3. **Performance**: Performance Monitoring and Management
4. **Technology**: Underlying design of the plant

PROCESSES

This entails the formal IMS consisting of Strategic, Core, and Support Systems and their relevant Processes, whether documented or embedded in computer systems. These are discussed in detail in Part 1, Systems.

PEOPLE

OE covers the behavioral aspects of performance enhancement, which generally involve the following subjects:

1. leadership skills
2. teamwork and cooperation
3. problem-solving skills
4. fact-based decision-making
5. results focus
6. change management
7. mentoring/coaching
8. skills to do the work
9. open and honest communication
10. behavior-based performance management
11. rewards for results

PERFORMANCE

This entails the measurement of performance and the required improvement process. It includes the use of **leading performance indicators that mitigate risk**.

TECHNOLOGY

The 3 Ps (People, Processes, and Performance) sit on the technology that has been established during the design and construction phases of the investment cycle.

Fig. 12.1 depicts the basic three risk control mechanisms (People, Processes, and Performance) within a Technology triangle.

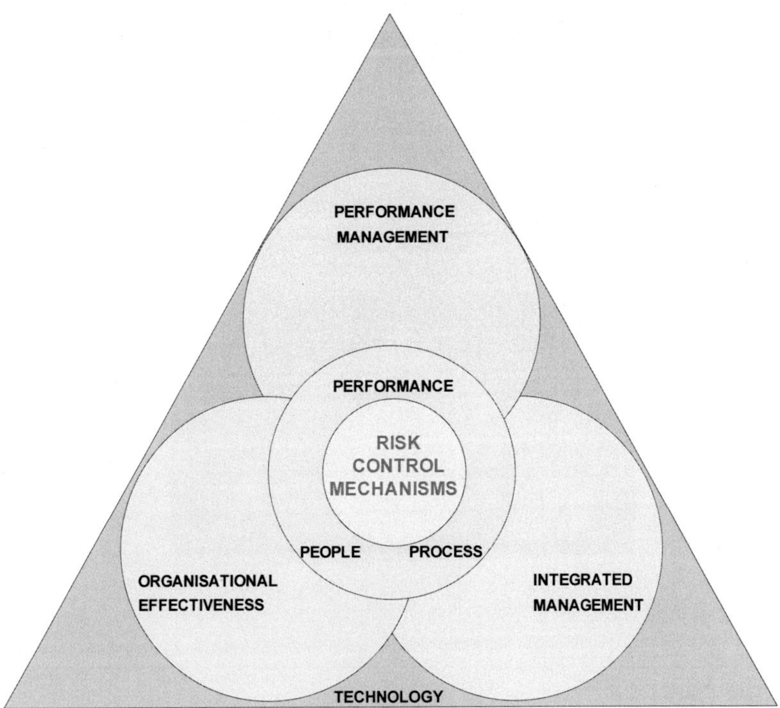

FIGURE 12.1

The risk control triangle.

Examples of interaction between the 3 Ps:

1. *A higher level of competence (**people**) could accommodate a reduction in necessary documentation (**process**).*
2. *Reduced **performance** monitoring could necessitate increased documentation (**process**) and visa versa.*
3. *Proactive (leading) Indicators (**performance**) could replace predictive reactive (lagging) indicators with improved **process** management.*

HOW DO THE 3 PS FIT INTO THE BUSINESS UNIT?

As mentioned in Chapter 11, Risk Management, risk is everybody's business. Risks crop up all over the business. However, **any risk that is outside the company's risk appetite has to be registered and addressed**.

Fig. 12.2 is a depiction of the 3 Ps being applied anywhere in the Business Unit.

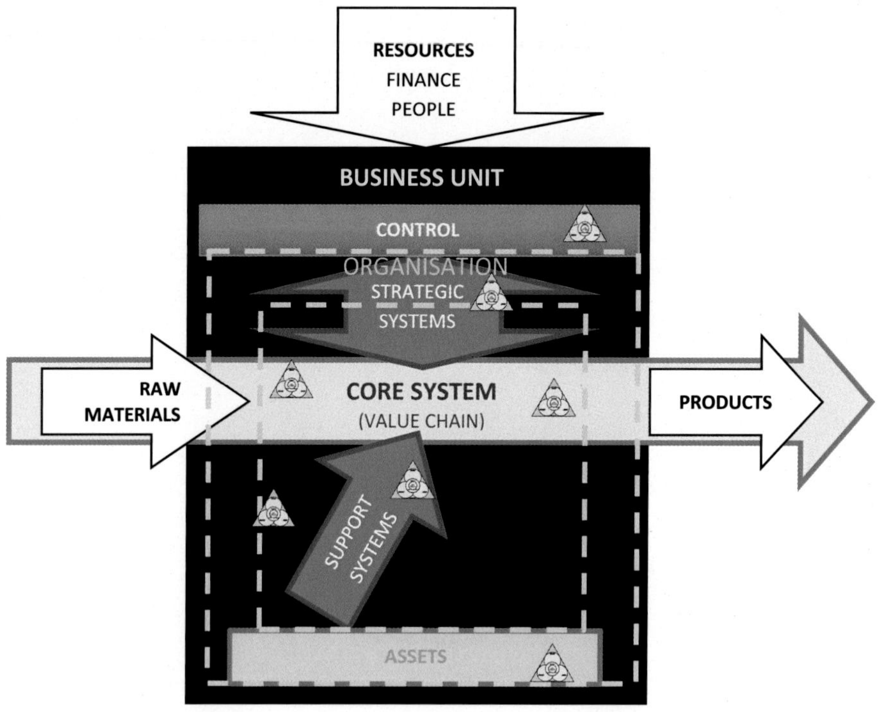

FIGURE 12.2

The business unit with 3 Ps.

12.4 SWISS CHEESE

The Swiss cheese model depicts each barrier as a slice of Swiss cheese stacked together side by side. The holes in the slice represent weaknesses in parts of that barrier. Incidents occur when holes in the slices momentarily align, permitting "a trajectory of accident opportunity" so that a hazard passes unimpeded through several barriers, leading to an incident. The severity of the incident depends on how many barriers (cheese slices) have holes that line up at the same time.

The potential holes in each slice have to be minimized by active management such as follows:

PREVENT

Loss of Primary Containment has to be minimized by enforcement of **processes**, training of competent **people**, and monitoring with the use of suitable **performance** indicators (3 Ps), as well as auditing, to ensure compliance.

DETECT

At the design stage, careful consideration is needed to install suitable gas and flame detection systems that will alarm and/or shut the process down. These have to be maintained and tested as per **recommended practice**.

CONTROL

Independent control systems are designed for critical processes that may incur serious damage to the process units if left to manual intervention or intervention by the primary Distributed Control System. These are referred to as Emergency Shutdown Systems (ESDs). These have to be maintained and tested as per recommended practice.

MITIGATE

All of the elements of Business Continuity Management have to be developed and adhered to with regular Emergency Response Plan exercises based on simulation of a potential real event [this is discussed in Chapter 14: Incident Preparedness and Operational (Business) Continuity Management].

Part 5, Assets, Chapter 22, Physical Asset Performance Management, has a list of recommended Performance Indicators to monitor management of the previous.

Fig. 12.3 depicts the process as applied in the process industry when the holes in each layer of Swiss cheese line up and result in an event.

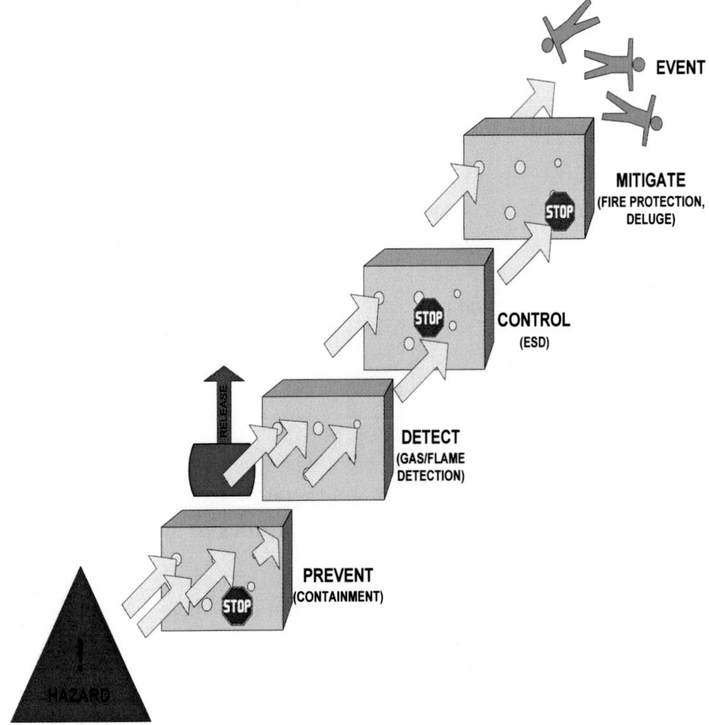

FIGURE 12.3

The Swiss Cheese model applied to the process industry.

Box 12.1 shows an example of the Swiss Cheese model as applied the BP Deepwater Horizon well blowout in the Gulf of Mexico in April 2010.

BOX 12.1 SWISS CHEESE: BP DEEPWATER HORIZON[8]

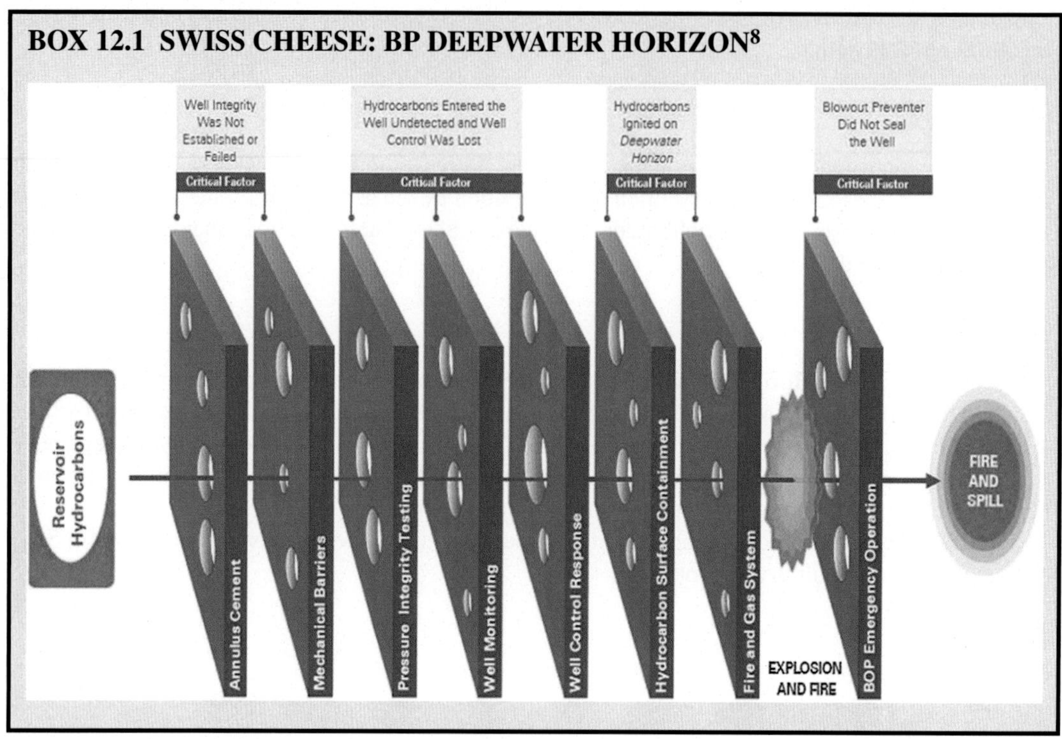

The Swiss cheese model covers both **active** and **latent** failures.
Active failures are unsafe acts or equipment failures directly linked to an initial hazard event.

Example:

The ESD of a Fluidic Catalytic Cracking Unit (FCCU) was taken out of service for repairs while the unit was operating.
Although the ESD had a battery backup in case of power supply failure, the modules needed to be replaced and were therefore removed.
During the delay in getting replacement spares, power to the FCCU was lost and the unit shut down rapidly ("crashed"), causing thermal shock damage.
Comment: The function of ESD systems is for emergencies and thus should never be without power. Spare battery backup units must be available in stock as spares for criticality A equipment (see Section 22.7: Maintenance).

Latent failures are additional factors in the system that may have been present for some time, but not corrected until they finally contribute to the incident.

Example:

Referring to Box 12.1 the Blowout Preventer was the last slice of cheese and it failed to operate.

A variation on the "3 Ps" theme discussed in Section 12.3 is depicted as part of a Swiss Cheese event shown in Box 12.2. This is based on BP's Operating Management System framework (see Section 4.13: Integrated Management Systems: Specific and Generic Examples).

BOX 12.2 BPS "4PS" DEPICTED IN A SWISS CHEESE MODEL

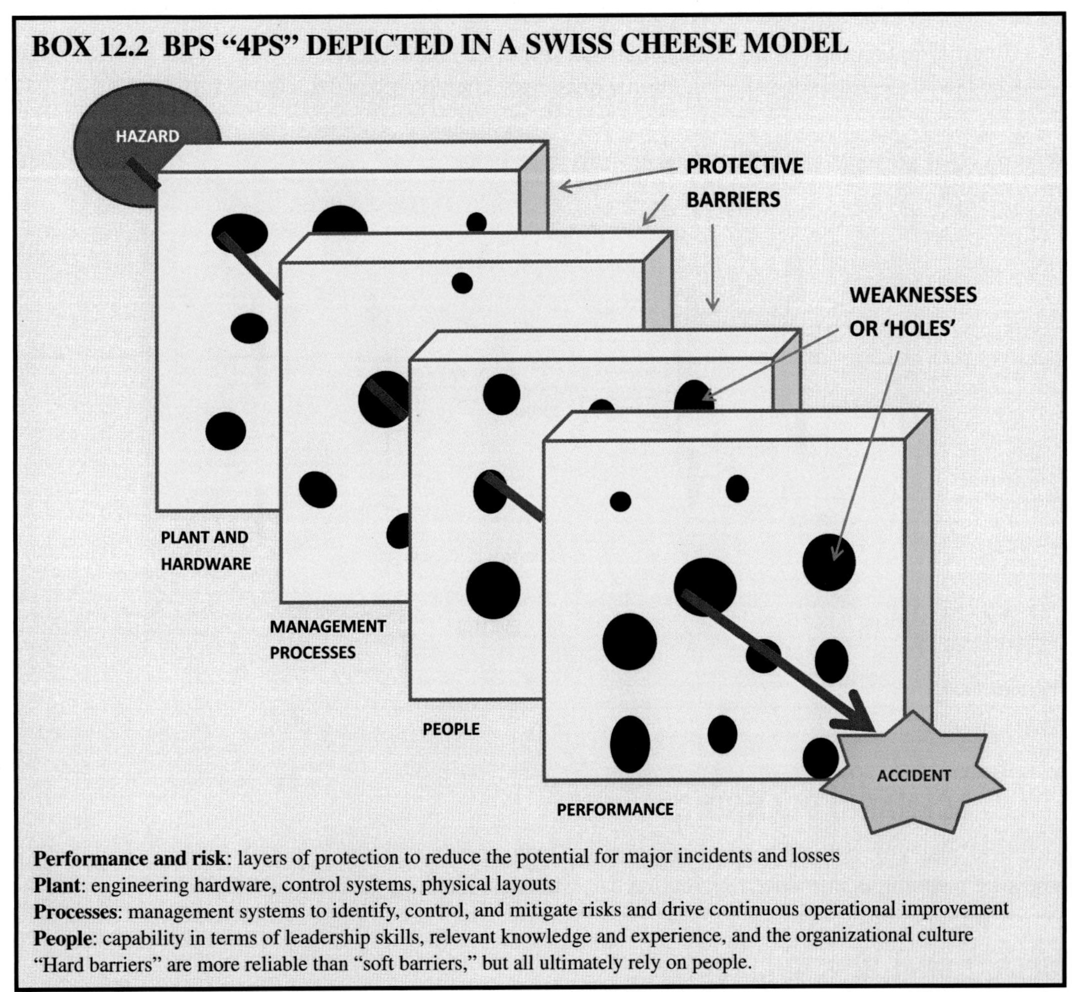

Performance and risk: layers of protection to reduce the potential for major incidents and losses
Plant: engineering hardware, control systems, physical layouts
Processes: management systems to identify, control, and mitigate risks and drive continuous operational improvement
People: capability in terms of leadership skills, relevant knowledge and experience, and the organizational culture
"Hard barriers" are more reliable than "soft barriers," but all ultimately rely on people.

12.5 BOW TIE

The bow tie representation, which is used as a risk mitigation tool, serves to document threats and consequences for a situation with controls/barriers for each real or potential event.

The main elements are as follows:

- **Hazards** form the major ways in which damage or injury can occur.
- **Threats** are the ways hazards can be released.
- **Hazardous event** is the event one wishes to avoid.
- **Consequences** are the outcomes that have to be avoided (see Event Potential Matrix in Section 12.8).
- **Barriers and controls** are ways in which threats and consequences are countered to ensure that the **hazard event** does not result in an **unwanted outcome or consequence.**

Controls and barriers could be of a management or technical nature.

The bow tie is used for risk mitigation workshops. The bow tie is depicted in Fig. 12.4.

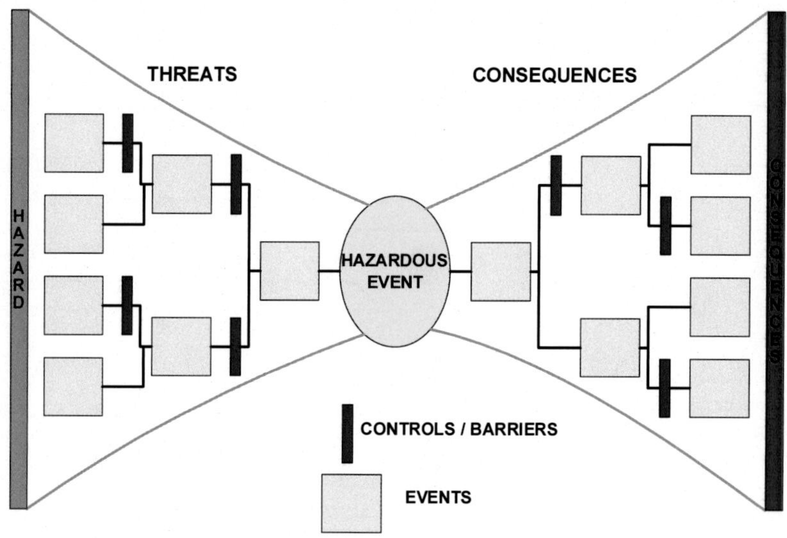

FIGURE 12.4

Bow tie.

12.6 SCENARIO PLANNING

Scenario planning risk analysis takes place on both operational and strategic levels. Strategic level scenario planning is introduced in Section 5.2, Strategic Planning (Strategy Formulation).

OPERATIONAL RISK ANALYSIS

Oil and gas companies do extensive scenario planning using case studies in order to practice approaches to managing events (see Section 14.10: Scenarios for Emergency Response Planning).

STRATEGIC RISK ANALYSIS

Strategic risk analysis entails identification and mitigation of external risks.

Strategic risks are not always foreseen. Box 12.3 shows some major world events that were not foreseen.

BOX 12.3 MAJOR WORLD EVENTS SINCE THE 1970S

The following lists some recent major world events that were not foreseen:

- 1973: World Oil Crisis as a result of OPEC shutting off their supply of oil to the world
- 1977: Tenerife Airport Disaster: the deadliest aircraft crash in history
- 1990: Saddam Hussain's invasion of Kuwait, resulting in a major oil crisis
- 1998: Asian Financial Crisis, resulting in financial hardship in these countries for many years
- 2000: Technology Bubble Crash on the New York Stock Exchange, resulting in financial loss to many people
- 2001: September 11 terrorist attack on the New York Word Trade Center and the Pentagon, and the subsequent economic downturn
- 2002: Enron, WorldCom, and other accounting frauds that resulted in the SOX legislation in the US
- 2008: Derivative-based financial meltdown, leading to some of the biggest banks in the world going bankrupt
- 2011: Japanese Tsunami, resulting in multiple nuclear power station meltdowns
- **2010 US Gulf Coast Oil Spill: the biggest ever**
- 2014: Three Malaysian aircraft losses in less than 1 year in unrelated incidents

The reoccurrence of some of the types of events, as shown in Box 12.3, could be mitigated with scenario planning. Nevertheless, Sarbanes-Oxley (SOX) legislation, which was introduced after the accounting frauds of 2002, still did not prevent the 2008 financial meltdown. Even ratings given by Rating Agencies, such as Moody's, Standard and Poor's, etc., which were based mainly on historical performance, did not give any forewarning of the 2008 financial crisis.

Thus legislation, performance ratings, and historical performance are not enough. Technical innovation, at the design and construction stages, as well as solid processes and **competent and honest people** are essential to reduce risk. Performance analysis based on **predictive or leading** indicators is also critical. The risk triangle mentioned earlier attempts to describe the interactions.

Three **scenarios** for a study case are usually described: worst, best, and most likely case. The most likely case is normally carried forward for developing the strategic plan.

12.7 BLACK SWAN[6]

In risk terminology, a Black Swan is an "unknown unknown," which is an unforeseen event. Some of the events in Box 12.3 could be regarded as Black Swans. It goes without saying that there are events that cannot be predicted, although scenario planning may include for events such as tsunamis, but might not cater for the magnitude of the tsunami.

CAN ONE MITIGATE THE IMPACT OF A BLACK SWAN?

With reference to Box 9.1, Attitude to Risk and Adverse Selection: The 2011 Japanese Tsunami, Fundai obviously took serious mitigation steps. Conversely, in hindsight, Tokyo Electricity Power Company appears to have taken insufficient mitigation steps since a number of its nuclear power stations suffered reactor meltdowns as a result of the tsunami.

Chapter 14, Emergency Planning and Operational Continuity Management, discusses getting the business back on track after a Black Swan has occurred.

12.8 TOOLS FOR OPERATIONAL RISK AND PERFORMANCE ANALYSIS AND EVALUATION

Risk analysis and evaluation makes use of various tools. It should be noted that each tool has specific applications and various limitations.

COMMON RISK TOOLS

Table 12.1 gives a list of common tools used primarily in the oil and gas industry.

Table 12.1 Common risk tools

Acronym	Description	Use
FIREPRAN	Fire Protection Analysis	Project
FMEA	Failure Mode Effects Analysis	Project
FEA	Finite Element Analysis	Project
EERA	Evacuation, Escape, and Rescue Analysis	Project (offshore)
ESSA	Emergency System Survivability Analysis	Project
EIA	Environmental Impact Assessment	Project
SIA	Safety Impact Assessment	Project
FMECA	Fault modes, Effects, and Criticality Analysis	Project
HSECES	Health, Safety, and Environment Critical Equipment and Systems	Project and Change Management
HAZID	Hazard Identification Study	Project and Change Management
LOCPA	Loss of Containment Protection Analysis	Project and Change Management
HAZOP	Hazard and Operability Study	Project and Change Management
LOPA/IPF/SIL [a]	Layers of Protection Analysis/Instrument Protective Functions/Safety Integrity Levels	Project and Change Management
PHA	Process Hazard Analysis	Project and Change Management
QRA	Quantitative Risk Analysis/Assessment	Project and Change Management
HEMP[a]	Hazards and Effects Management Process (a Shell process)	Project and Change Management
RCA[a]	Root Cause Analysis (Tri-beta and Taproot are common products used in the oil and gas industry)	Incidents
RCM[b]	Reliability-Centered Maintenance	Operations
RBI[b]	Risk-Based Inspection	Operations
SAFOP	Safe Operations	Operations

[a]See the following for examples.
[b]See Chapter 22, Physical Asset Management.

Layers of Protection Analysis/Instrument Protective Functions/Safety Integrity Levels

The key value of Layers of Protection Analysis (LOPA)/Safety Integrity Levels (SIL) is that it enables one to allocate safeguards that are proportional for risk reduction. Careful consideration of ALL safeguards is required.

The layers of protection approach to Safety are depicted in Fig. 12.5.[5]

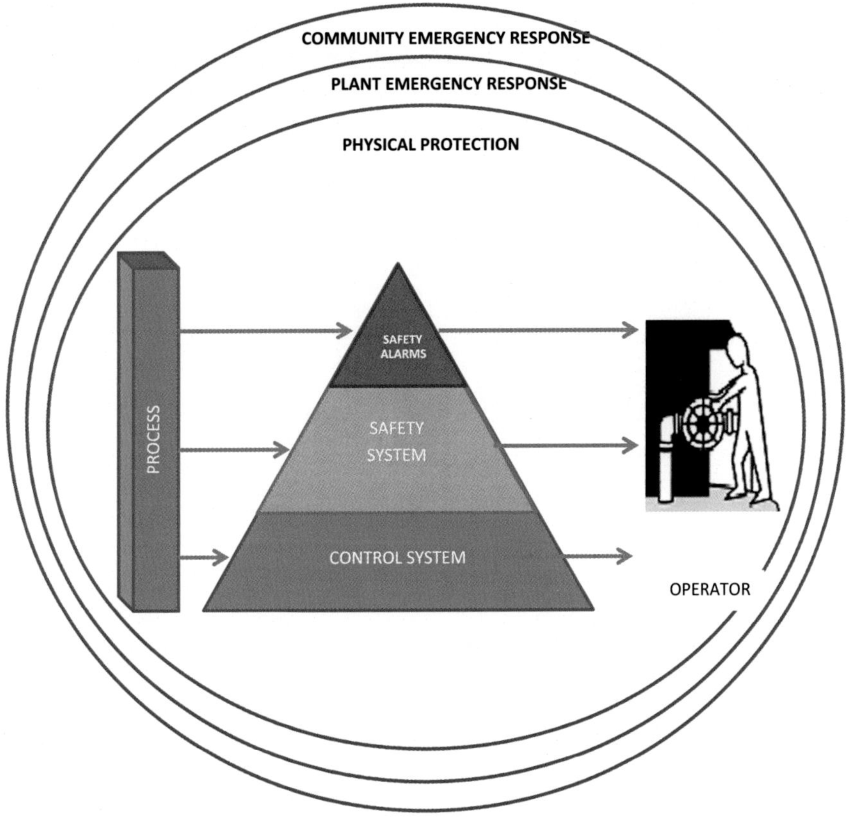

FIGURE 12.5

Layers of protection.

Emergency response is discussed in Chapter 14, Emergency Planning and Operational Continuity Management.

Physical Protection is discussed in Chapter 20, Human Capital and Chapter 22, Physical Asset Management.

Hazard Evaluation and Management Process

Box 12.4 identifies the Shell Hazard Evaluation and Management Process (HEMP) cycle, which is commonly used in the Oil and Gas industry.

BOX 12.4 RISK MANAGEMENT PROCESS: HAZARD EVALUATION AND MANAGEMENT PROCESS

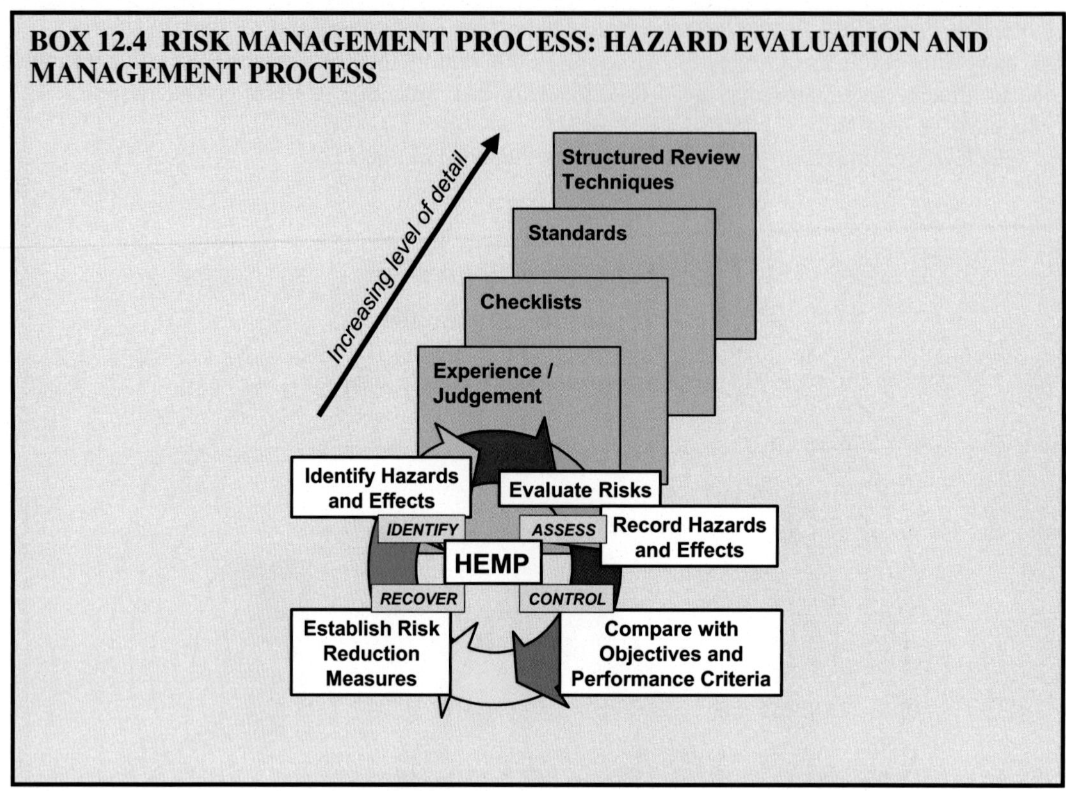

Root Cause Analysis (RCA)

When an incident does occur, a thorough investigation into the cause of the incident is required. This is referred to as RCA. Depending on the severity of the incident, different levels of management are involved in the investigation to ensure objectivity.

The aim of RCA is to establish:

1. What was the sequence of events?
2. How did it happen and what barriers failed?
3. Why did the barriers fail?

RCA tools could be either generic or proprietary.

Generic tools include the use of cause and effect diagrams.

Box 12.5 shows an example of the analysis process using a cause and effect diagram.

Use of "Tap Root" and "Tripod" **proprietary tools** are common in the Oil and Gas industry.

BOX 12.5 CAUSE AND EFFECT DIAGRAM: AMINE SYSTEM CORROSION

An amine plant was shut down due to a reboiler failure. The Root Cause Analysis (RCA) determined that there were three root causes of failure resulting in **corrosion in three systems**. This is shown in the following cause and effect diagram:

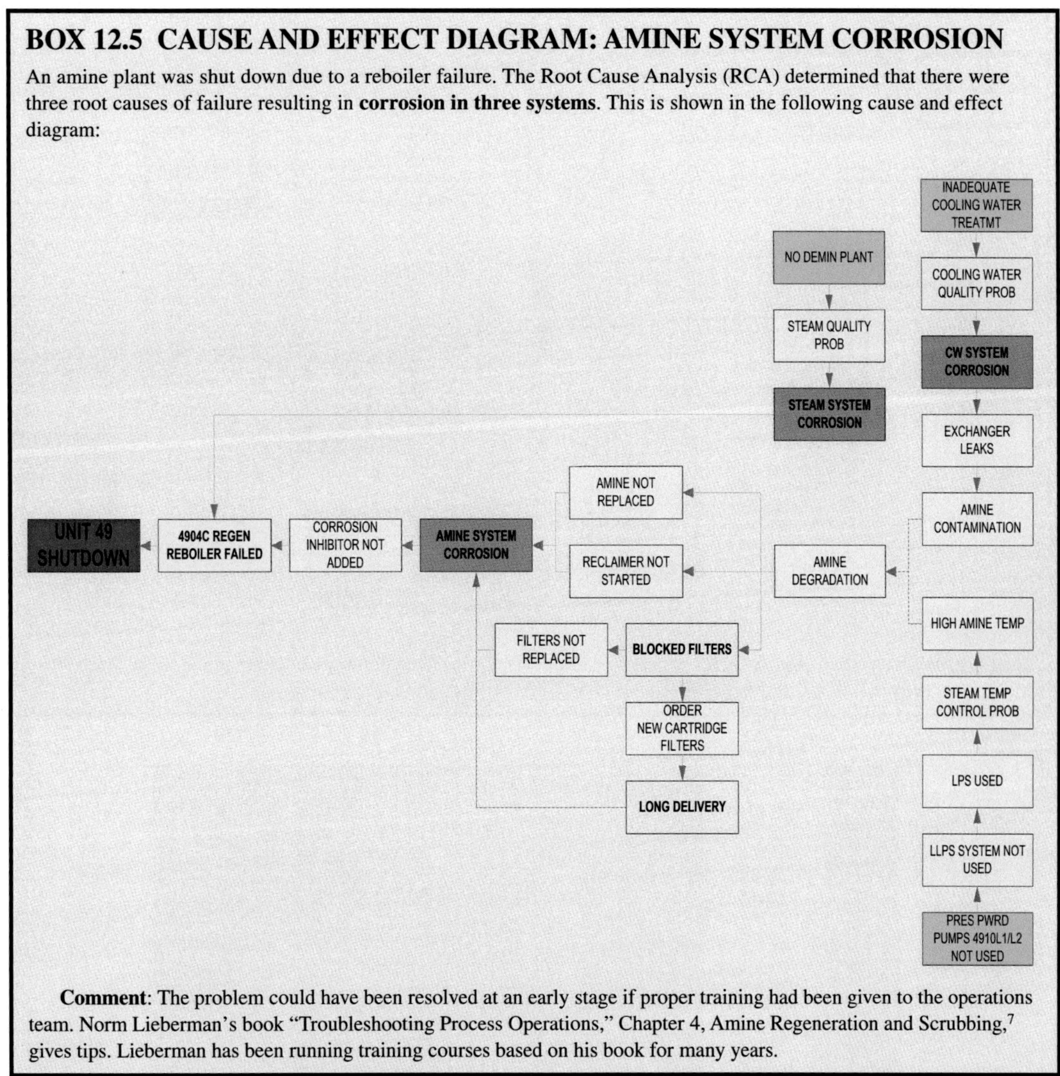

Comment: The problem could have been resolved at an early stage if proper training had been given to the operations team. Norm Lieberman's book "Troubleshooting Process Operations," Chapter 4, Amine Regeneration and Scrubbing,[7] gives tips. Lieberman has been running training courses based on his book for many years.

Box 12.6 shows an example of the analysis process using "Tripod."[2]

RISK MATRICES

Variations of two basic matrices are used as shown in the following sections.

Qualitative Assessment

An example of a qualitative assessment applied to risk is shown in Fig. 12.6.

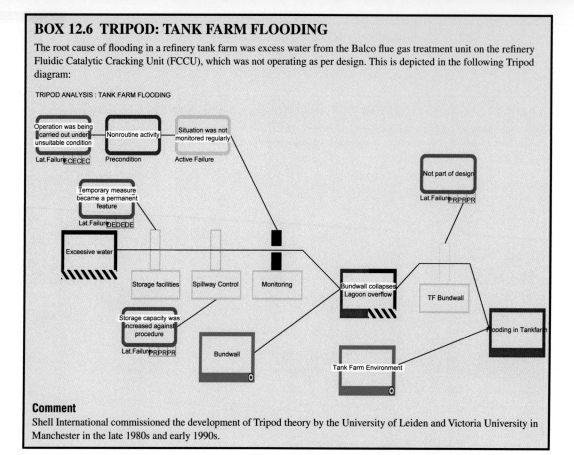

BOX 12.6 TRIPOD: TANK FARM FLOODING

The root cause of flooding in a refinery tank farm was excess water from the Balco flue gas treatment unit on the refinery Fluidic Catalytic Cracking Unit (FCCU), which was not operating as per design. This is depicted in the following Tripod diagram:

TRIPOD ANALYSIS : TANK FARM FLOODING

Operation was being carried out under unsuitable condition
Lat.Failure ECECEC
Nonroutine activity
Precondition
Situation was not monitored regularly
Active Failure
Not part of design
Lat.Failure PRPRPR
Temporary measure became a permanent feature
Lat.Failure DEDEDE
Exceesive water
Storage facilities
Spillway Control
Monitoring
Bundwall collapses Lagoon overflow
TF Bundwall
Storage capacity was increased against procedure
Lat.Failure PRPRPR
Bundwall
Flooding in Tankfarm
Tank Farm Environment

Comment

Shell International commissioned the development of Tripod theory by the University of Leiden and Victoria University in Manchester in the late 1980s and early 1990s.

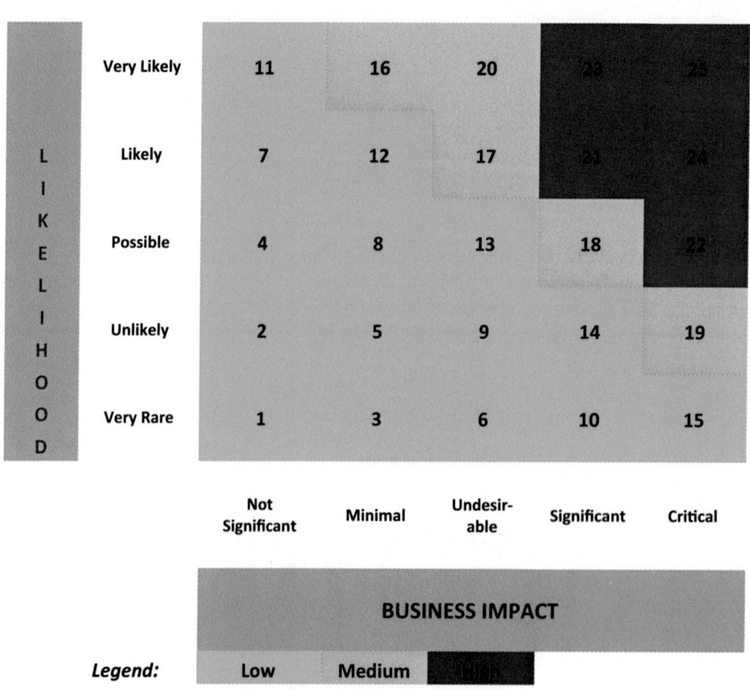

FIGURE 12.6

Qualitative risk assessment.

This tool has been used to establish an Enterprise Risk Management System from scratch (see Box 12.7).

BOX 12.7 ENTERPRISE RISK MANAGEMENT QUALITATIVE MATRIX EXAMPLE

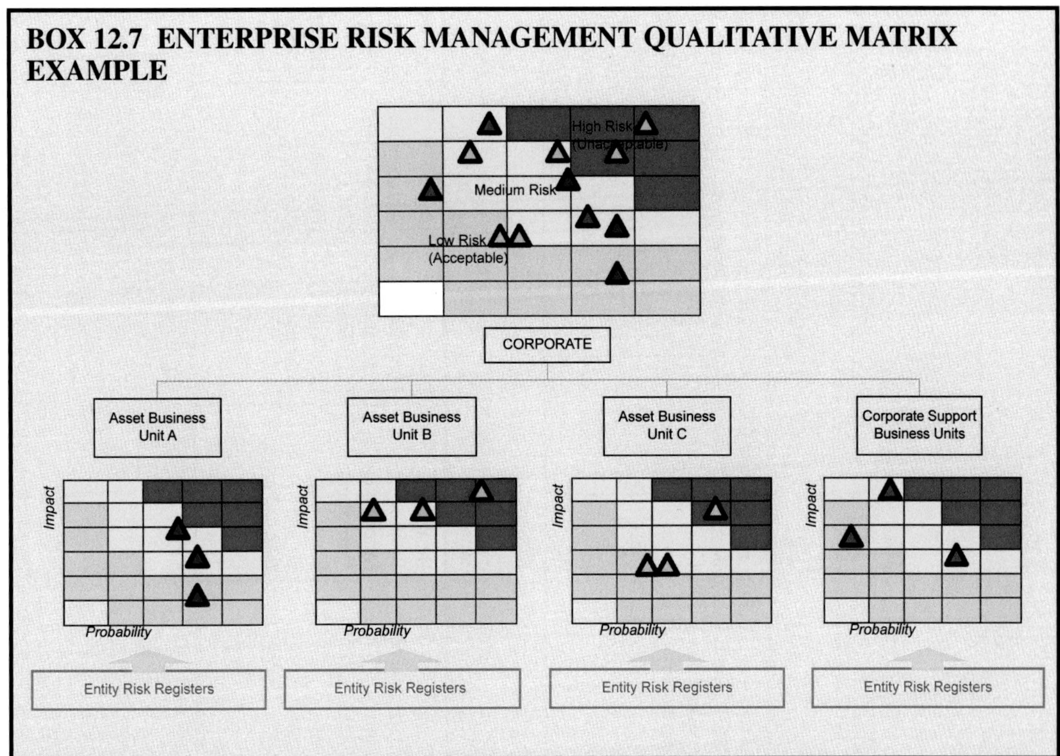

The inputs into the Risk Matrices, as shown in Box 12.7, were developed in a spreadsheet as shown in Table 12.2.

This spreadsheet contains subjective elements and it is best, were possible, to use an objective tool such as an Event Potential Matrix (EPM), as discussed in the following section.

Quantitative Risk Assessment: Event Potential Matrix (EPM)

The EPM, which consists of a Risk Assessment Matrix (RAM) and a Risk Value Matrix (RVM) is a very useful tool, most commonly used to assess how important corrective action is.

An EPM is normally used to determine both the level of involvement as well as the priority/urgency of remedial action.

- **Risk Value = Probability × Rating**
- **Probability** = Likelihood of potential consequence happening
- **Rating (Severity)** = Cost or effect of the **consequence** (actual or potential)

Table 12.2 Steps for Developing a Risk Matrix

	Step 1	Step 2	Step 3	Step 4			Step 5		
Risk reference:	Business Objective description	Key Deliverables/ necessary prerequisites to achievement of Business Objective	Related Process(es)	Description of Risk Event Impacting Key Deliverables	Root cause of risk event	Risk Owner (Position Identifier)	Likelihood of Risk Event Occurring **1.** Very rare **2.** Unlikely **3.** Possible **4.** Likely **5.** Very likely	Business Impact Assessment of Risk Event **1.** Not significant **2.** Minimal **3.** Undesirable **4.** Significant **5.** Critical	Inherent Risk Management rating High/medium/ low

Step 6	Step 7			Step 8	Step 9			Step 10
Description of Existing Control Procedures	Likelihood of Risk Event occurring **1.** Very rare **2.** Unlikely **3.** Possible **4.** Likely **5.** Very likely	Business Impact Assessment of Risk Event **1.** Not significant **2.** Minimal **3.** Undesirable **4.** Significant **5.** Critical	Residual Risk Management rating High/medium/ low	Results of Routine Independent Compliance Checking Support Residual Risk rating Yes, No, None	Risk Management Remediation Plan Name/ Location/ Reference	Risk Management Remediation Plan Owner (Position Identifier)	Risk Management Remediation Plan anticipated completion date	Date Risk-Based Business Plan last reviewed/ updated

Consequences are classified as follows:

1. People: Risk to safety or health
2. Asset/Production: Risk to assets or production
3. Environment: Risk to the environment
4. Reputation/Customer: Risk to the company's reputation or customer satisfaction

Consequence values are assigned within five levels of severity, from low to high.

Probability (likelihood) is categorized from A (never heard of in industry) to E (occurs several times a year in this plant).

A rating or class is given from 1 to 5.

Risk tolerance (low, medium, high, or intolerable) is determined from both how often the incident has occurred as well as the rating.

The EPM is depicted in Fig. 12.7.

EVENT POTENTIAL MATRIX

EXPANDED VERSION FOR REFINERY

SEVERITY	RISK ASSESSMENT MATRIX (RAM) — CONSEQUENCES				RISK VALUE MATRIX (RVM) — PROBABILITY: LOW <= LIKELIHOOD => HIGH (OF POTENTIAL CONSEQUENCE OCCURRING)					RATING / CLASS
	PEOPLE (RISK TO SAFETY / HEALTH)	ASSET / PRODUCTION (FINANCIAL RISK)	ENVIRONMENT (RISK TO ENVIRONMENT)	REPUTATION / CUSTOMER (RISK TO CUSTOMER OR REPUTATION)	NEVER HEARD OF IN INDUSTRY (IMPROBABLE)	HAS OCCURRED IN INDUSTRY (POSSIBLE)	HAS OCCURRED IN COMPANY	OCCURS SEVERAL TIMES A YEAR IN COMPANY	OCCURS SEVERAL TIMES A YEAR IN REFINERY	
	Pe	As	En	Cu	A	B	C	D	E	
=> LOW	SLIGHT HEALTH EFFECT / INJURY (FIRST AID)	SLIGHT DAMAGE (<MR 25 000), NO DISRUPTION TO OPERATION	SLIGHT EFFECT	INTERNAL QUALITY ISSUE, WAIVER, SLIGHT IMPACT	LOW					1
ACTUAL OR POTENTIAL CONSEQUENCE	MINOR HEALTH EFFECT / INJURY (LOST TIME INJURY)	MINOR DAMAGE (MR 25 000 to MR 250 000)	MINOR EFFECT	FAILURE TO SUPPLY, DELIVERY OFF SPEC, LIMITED IMPACT		MEDIUM				2
	MAJOR HEALTH EFFECT / INJURY	LOCAL DAMAGE (MR 250 000 to MR 1.25 million)	LOCALIZED EFFECT (IN REFINERY)	MULTIPLE COMPLAINTS, NATIONAL IMPACT			HIGH			3
	SINGLE FATALITY OR PERMANENT TOTAL DISABILITY	**MAJOR DAMAGE (MR 1.25 million to MR 25 million)**	**MAJOR EFFECT (IN INDUSTRIAL CITY)**	**LOSS OF CUSTOMER/S, REGIONAL IMPACT**				INTOLERABLE		4
HIGH<=	MULTIPLE FATALITIES	EXTENSIVE DAMAGE (>MR 25 million)	MASSIVE EFFECT (IN AREA / COUNTRY)	LOSS OF MARKET SHARE, INTERNATIONAL IMPACT						5

Risk Assessment used for Incidents that have occurred in Refinery **Risk Value & Risk Assessment** used for determining the Risk Value for potential risks

FIGURE 12.7

Event Potential Matrix.

Using the previously mentioned EPM, Box 12.8 describes an assessment that was not escalated after the second incident.

BOX 12.8 GAS PLANT RISK ASSESSMENT

A feed exchanger for a gas plant failed. Since the plant had only this single feed exchanger, the entire plant had to be shut down when the exchanger failed.

A subsequent investigation used the Event Potential Matrix (EPM) to assess the damage, which was rated with a potential severity of 3 and a probability of C, giving it a "medium risk." Thus the investigation into the failure was at departmental level.

However, the exchanger failed for the second time within a year, again shutting down the entire plant. The staff ignored the previous occurrence and so gave the incident the same "medium risk" rating.

A technical audit later discovered this and recommended that thc investigation be elevated to E, "occurs several times a year at this site," giving it a "high risk," thereby elevating the investigation to interdepartmental level with top management oversight.

As a result of the interdepartmental investigation, the exchanger was rerated as a criticality A item of equipment, and a spare exchanger bundle with an improved material specification was purchased.

(See Section 22.7: Maintenance.)

Box 12.9 gives an assessment of damage to refinery assets using the previously mentioned EPM.

BOX 12.9 EVENT POTENTIAL MATRIX REFINERY EXAMPLES

Asset

Tank collapse

- Vent blocked, causing collapse under vacuum
- Repair cost MR 400,000 to MR 750,000
- In White range of Risk Assessment Matrix (RAM) with Rating 3, but similar incident has happened previously in the company: Likelihood C on Risk Value Matrix (RVM) (Yellow range), therefore declare Major Incident
- In-depth incident analysis and discussion at management level

Spiral Exchanger Over Pressure

- Contractor over pressured spiral on hydro-test.
- Spiral replacement cost in excess of MR 100,000
- In White range of RAM with Rating 2 and no previous occurrence in company, therefore low risk
- Manage within Maintenance Business System

Fluidic Catalytic Cracking Cyclone Refractory Failure

- Refractory failure inside cyclones
- Loss of production ± 6 weeks with lost revenue of ±MR 200,000/day
- Repair costs >MR 100,000
- Estimate total cost to exceed MR 25 million for replacement of cyclones
- In Red range on RAM with Rating 5, therefore declare Major Incident
- In-depth incident analysis with management involvement

Cat Poly-Reactors Shock Loading

- Water carry over caused mechanical damage to support grid assemblies
- Repair costs estimated to be MR 10,000 to MR 100,000
- Reduced operating capacity resulting in operating losses in excess of MR 100,000
- In White range of RA and no previous known occurrence in the company, therefore low risk
- Manage within Inspection Business System

Currency: *MR*, Malaysian Ringits

12.9 ENVIRONMENTAL ASPECTS AND RISK REGISTERS AS REQUIRED BY ISO 14001 AND OHSAS 18001

The development of Environmental Aspects and Risk Registers requires an overall assessment of all "risks" and "environmental aspects" in the Business Unit.

Environmental aspects also need to relate to the Company's Consent to Operate, which stipulates specific values to be maintained or to be achieved in the future.

Health and Safety performance has always been based on hard historical data, such as "Lost Time Incidents."

The increased emphasis on leading performance indicators emphasizes the need for objective data. When developing an aspects register or a risk register, it is best to use objective data with a quantitative approach. However, bear in mind that some subjective (qualitative) data is unavoidable.

12.10 SUMMARY

Incident Causation Theory identifies people as the cause of most incidents. Incidents happen when people make errors and fail to keep barriers functional or in place.

The slices of Swiss cheese are depicted as Prevent, Detect, Control, and Mitigate. Potential holes in each of these slices have to be minimized to prevent an unwanted event.

Three interactive risk control mechanisms (processes, people, and performance) applied together but in different proportions could result in the same risk value (probability × consequence). Risks occur throughout the business at all levels, and so the three interactive risk control mechanisms could be applied anywhere in the Business Unit.

The Bow Tie is used for identification of threats and consequences.

Scenario planning and enactment are important tools for preparing for "known unknowns"—events that are known to happen in the process industry.

"Black Swans," which cannot be foreseen, are regarded as "unknown unknowns." These could be events that occurred in a related industry, but not within the process industry. Nonetheless, these events need to be explored to assess their possible occurrence in the process industry.

Various Risk Analysis and Assessment Tools can be applied at different stages of the life of the investment. Tools include the following:

1. RCA using generic tools or proprietary tools, such as Taproot or Tripod
2. HEMP
3. LOPA/Instrument Protective Functions/SIL

Qualitative Risk Assessment and Quantitative Risk Assessment, using matrices, are common. The EPM consisting of an RAM and an RVM, gives a quantitative assessment.

REFERENCES

1. Tripod Beta Incident Analysis Primer. October 2006.
2. Tripod Beta RCA software. http://publishing.energyinst.org/tripod.

3. TapRoot RCA software. http://www.taproot.com/.
4. Kletz T. *Process plants: a handbook for inherently safer design*. Taylor & Francis; 1998.
5. Shah GC. What every manager should know about layers of protection analysis. *Hydrocarb Process* April 2010.
6. Taleb N. *The black swan*. Penguin; 2010.
7. Lieberman N. *Troubleshooting process operations*. PennWell; 1991.
8. BP deepwater horizon accident investigation report. http://www.bp.com/content/dam/bp/pdf/sustainability/issue-reports/Deepwater_Horizon_Accident_Investigation_Report.pdf.

FURTHER READING

1. *Guidelines for risk based process safety*. CCPS Wiley; 2007.
2. ISO 14001:2004. *Environmental management.*
3. ISO 45001:2016. *Occupational health & safety management systems – requirements.*
4. API 580. *Risk based inspection*. 2nd ed. November 2009.
5. API RP 581: Risk based inspection technology – table 4.3 inspection effectiveness categories.
6. ISO 17776:2002, *Petroleum & natural gas industries – offshore production installations – guidelines on tools & techniques for hazard identification & risk assessment.*
7. Guidance on simultaneous operations (SIMOPS) International Marine Contractors Association. http://www.imca-int.com.
8. *Health safety and environment case guidelines for mobile offshore drilling units*. IADC; 2010.
9. *Health, safety and environment case guidelines for land drilling contractors*. IADC; 2007.
10. ISO 19011:2002, *Guidelines for quality and/or environmental management systems auditing.*
11. *Guidelines for risk based process safety*. AIChE Center for Chemical Process Safety (CCPS) Wiley – Interscience; 2007.
12. IEC 61882:2001, *Hazards & operability application guideline.*
13. *Guidance on risk assessment for offshore installations*. Health & safety executive (HSE). Offshore information sheet No. 3/2006.
14. Offshore Installations (Safety Case) Regulations. *Regulation 12. Demonstrating compliance with the relevant statutory provisions*. Health & Safety Executive (HSE); 2005. Offshore information sheet No. 2/2006.
15. *The key to managing major incident risks*. OGP; December 2008. Asset integrity. Report no 4.
16. *Managing major incident risks*. OGP; April 2008. Workshop Report No. 403.
17. Canaway R.T. Management of risk. *5th annual Russian/CIS petroleum refining conference in Milan, Italy 20th - 22nd October 2003.*
18. Brafman O, Brafman R. *Sway: the irresistible pull of irrational behaviour*. NY: Doubleday; 2008. http://www.pennwellbooks.com/.
19. Schrage M. *Boards of prevention - corporate directors can, and should, play a much more active role in overseeing risk and avoiding major crises*. Strategy & Business; June 14, 2010.
20. *Joint venture risk – and how to manage it.* The Joint Venture Exchange Water Street Partners. February 2013. Issue 55.
21. Thomson JR. *High integrity systems and safety management in hazardous industries*. Butterworth-Heinemann; Jan 2015.

CHAPTER

MERGING PERFORMANCE AND RISK

13

13.1 INTRODUCTION

Risks are managed by monitoring both performance indicators **and** progress on mitigation strategies. Essentially, the process for Risk Management and Performance Management is the same.

Firstly, the risks or performance gaps are identified and then followed by an assessment process, which identifies and registers those risks that are outside the company's risk appetite, as well as any significant performance gaps. Next, strategies, responsibilities for implementation, and timing are agreed, and lastly, progress on implementation is monitored and reviewed. Both significant risks and performance gaps might require some investment, and these are then also placed in an investment register (see Chapter 33: Opportunity Life Cycle).

13.2 PROCUREMENT RISK AND PERFORMANCE

The tendering process and the use of a "Standard Conditions of Contract" are discussed in Section 9.7, Procurement and Governance.

The procurement of materials (feedstock, process, maintenance, etc.) and services (maintenance, turnaround, logistics, etc.) are major contributors to risk control and performance improvement.

The risk mitigation related to the type of contract is very important. Table 13.1 shows the risk spectrum for selection of the most appropriate contract for a specific requirement. Risk is transferred to the contractor, depending on the type of contract.

Example 1:

A Lump Sum or Fixed Price contract will ensure no change in price if no variation orders are issued by the client. The scope needs to be very clearly defined before going to tender. The ***advantage*** *to the client is that the risk is passed to the contractor. However, a potential* ***disadvantage*** *is that if variation orders are issued, part of the risk moves back to the client, and the cost normally goes up.*

Example 2:

A Cost Reimbursable contract price is determined by the final measurement of work. The contract is determined by a Bill of Quantities (BoQ) with estimates of work to be done and the associated rates for each piece of work. The ***advantage*** *to the client is that work can be fast-tracked, as the design does not have to be complete when awarding the contract. The* ***disadvantage*** *to the client is that there is a high risk of the cost being greater than initially estimated.*

Performance Management for the Oil, Gas, and Process Industries. http://dx.doi.org/10.1016/B978-0-12-810446-0.00013-X

Table 13.1 Contracting Risk Spectrum

No.	Issue	Lump Sum	Unit price (Bill of Quantities; BoQ)	Target Cost	Cost Reimbursable	Target Man Hours (For Turnarounds)
1	**Financial objectives of client and contractor**	Different but reasonably independent	Different and in potential conflict	Considerable harmony. Reduction of actual cost is a common objective, provided cost remains within the incentive region.	Both, based on actual cost, but in potential conflict.	Considerable harmony. Reduction of actual man hours is a common objective.
2	**Contractor's involvement in design**	Excluded if competitive price based on full design and specification	Usually excluded	Contractor encouraged to contribute ideas for reducing cost.	Contractor may be appointed for design input prior to execution.	Contractor intimately involved with Critical Path Modeling (CPM) to determine estimated man hours for the project.
3	**Client involvement in management of execution**	Excluded	Virtually excluded	Possible through joint planning.	Should be active involvement.	Jointly involved with contractor.
4	**Claims resolution**	Very difficult, no basis for $ evaluation	Difficult, only limited basis for $ evaluation	Potentially easy, based on actual costs. Contract needs careful drafting.	Unnecessary except for fee adjustment. Usually relatively easy.	Unnecessary as actual man hours plus bonus (based on a formula) basis for payment.
5	**Forecast final cost at time of bid**	Known, except for unknown claims and changes	Uncertain, depending on quantity variations and unknown claims and changes	Uncertain. Target cost usually increased by changes, but effective joint management and efficient working can reduce final cost below an original realistic budget.	Unknown.	Estimated based on previous CPM model for turnaround of the same plant.
6	**Payment for cost of risk events**	Depending on contract terms, undisclosed contingency, if any, in contractor's bid; otherwise by claim and negotiation	Depending on contract terms, undisclosed contingency, if any, in contractor's bid; otherwise by claim and negotiation	Payment of actual cost of dealing with risks as they occur and target adjusted accordingly.	Payment of actual costs.	Payment for actual man hours.

Adapted from Wideman RM. Project and program risk management PMI; 1992.

APPROACHES TO IMPROVE PROCUREMENT PERFORMANCE

The following are some examples on how to improve procurement performance with the As Low As Reasonably Practical risk approach.

Improved Contracts System Key Performance Indicators

Key Performance Indicators (KPIs) need to be linked to real improvement in the overall management of the business.

> *Example:*
>
> *Critical long-term deliveries of equipment and materials for arrival on a specific critical date should be monitored individually. This is very important for a scheduled turnaround.*

Scope Definition

Clear scope definition is required for a tender to be priced. If this is not possible, then a BoQ contract may be more suitable. The changing risk profile is shown in Table 13.1.

Competence of Contractors

Major National Oil Companies (NOCs) and International Oil Companies (IOCs) classify the competence of their contractors prior to bidding. This is based on the contractors' experience in certain types of work, the complexity of the work, and the size of contract. They then invite only those in a certain classification to bid for the work.

Prequalification

In addition to competence classification, prequalification is required for activities that are of higher risk, and thus the experience of the contractor in the specific required field is crucial (see Box 13.3, Shell Pearl GTL Maintenance Contracts).

Eliminate Postbid Negotiations

International practice for formal bidding does not allow for postbid price negotiation. Regrettably, this practice removes the transparency and fairness entailed in a formal sealed bid process. With this practice, contractors are often pushed to reduce their profit margins to a high-risk situation so as to get the job.

Strategies

Strategies need to be tailor-made for the situation.

> *Example: Process Turnarounds (planned shutdowns)*
> *Process Turnarounds are planned in detail by the client using critical path modeling often with input from the turnaround contractor/s. Resources and time are refined after each turnaround* (see Section 23.4: Basic Requirements for Optimization of the Turnaround,item 5: Critical Path Modeling).
> *Most contracts in the industry are now incentive-based to reduce the turnaround duration. These are generally based on target hours with a shared incentive for beating the target. The saving to the client is generally very significant* (see Section 23.4: Basic Requirements for Optimization of the Turnaround,Item 6: Contractor Strategy and Table 13.1).

A Catalytic Cracker (FCCU) shutdown could have a lost opportunity cost (lost profit) well in excess of $100,000 per *day.*

13.3 OUTSOURCING: A RISK REDUCTION AND PERFORMANCE IMPROVEMENT TOOL

Outsourcing is defined as the procurement of products or services from sources that are external to the company. Outsourcing is a long-term relationship between the service provider and the company, with a high degree of risk sharing.

Outsourcing is a tool that allows management to focus on core competencies. Core competencies are skill and knowledge sets that are unique to maintain competitive advantage. Some inspection and maintenance services tend not to be core competencies for oil, gas, and process companies.

Primary benefits of outsourcing the following:

1. cost saving;
2. time saving;
3. removal or exposure of hidden costs;
4. focus on core activities;
5. freeing up of investment in support assets;
6. access to talent that is not available in-house; and
7. access to tools and technology that may not be available in-house.

Disadvantages of outsourcing include the following:

1. loss of control of the process, thus quality of service must be closely monitored;
2. reversibility: once a process has been handed over to an outsider, it is difficult and costly to bring it back in-house;
3. relationships: normally associated with retrenchments of staff and can damage morale of existing staff;
4. multiple clients: service providers normally have multiple clients and may not be able to give dedicated attention;
5. loss of intellectual property;
6. loss of critical skills; and
7. loss of flexibility.

Reasons for the failure of outsourcing include the following:

1. expected cost savings not realized;
2. service providers did not meet expectations;
3. failure to implement a rigorous contract management procedure;
4. poor supplier selection; and
5. failure to handle the human aspect efficiently.

Aspects of outsourcing to avoid failures are as follows:

1. Outsourcing must be linked to the business strategy.
2. Proper management of interfaces between outsource and service provider is to be ensured.

3. Common objectives and a scorecard are to be established in advance and used to monitor performance.
4. Appropriate risk sharing is to be ensured.

Box 13.1 lists inspection and maintenance activities that could be outsourced.

BOX 13.1 OUTSOURCING INSPECTION AND MAINTENANCE ACTIVITIES

The following is a list of possible inspection and maintenance activities:

1. leak sealing
2. on-site machining
3. refractory repair
4. tube bundle extraction
5. sandblasting and painting
6. heat treatment
7. catalyst handling
8. machining/milling/turning
9. fitting, including precision fitting
10. hot tapping
11. electrical installation
12. instrumentation
13. electric motor overhaul
14. pump overhaul
15. mechanical valve overhaul, including relief valve and control valve reconditioning
16. high-pressure cleaning
17. chemical cleaning
18. inspection (nondestructive examination/testing, radiography, etc.)
19. scaffolding
20. lagging, including cladding
21. boiler making
22. welding
23. rigging
24. mobile cranes

A structured approach is essential to streamline the change towards outsourcing without raising the corporate risk profile. An example of a bumpy transition is given in Box 13.2.

BOX 13.2 OUTSOURCING INFORMATION TECHNOLOGY (IT)

Chevron South Africa refines, markets, and distributes petrol, diesel, and other related products.

Chevron decided to outsource its IT function in the 1990s. This was actioned overnight with a contract having been established for the outsourcing, and all Chevron IT staff were offered jobs with the outsource contractor. Many IT staff, however, chose to look elsewhere for work, and thus the service from the IT contractor became precarious for a time.

At this time, the Chevron South Africa Refinery in Cape Town was planning a major turnaround. The transition time was unfortunate for those planning the turnaround, since not only was the new IT contractor in place who could not offer a stable service, but it also included the use of new Windows-based software, which was hosted on the company intranet. In order to ensure greater stability, Refinery management chose to delink the planning of the turnaround from the company intranet by creating its own mini intranet, thus being able to successfully complete planning of the turnaround [however, progress of the turnaround was another story; see Box 23.2, Critical Path Modeling (CPM)].

Longer-term call-off contracts for outsourcing noncore business services can be established. Contractors can establish their infrastructure and core management team with a long-term commitment to staff, as well as building technical competence.

Examples of successful outsourcing:

1. *Sasol has outsourced maintenance and turnaround activities to Fluor South Africa, a company in which Sasol has shares.*
2. *Under the name of Al Shaheen GE Services Company, Qatar Petroleum is in partnership with GE for maintenance of turbine, compressor, and related auxiliary equipment.*
3. *Shell Pearl GTL has outsourced maintenance and turnaround activities to two companies* (see Box 13.3).

BOX 13.3 SHELL PEARL GTL MAINTENANCE CONTRACTS

After startup of the new GTL complex in Qatar, Shell decided to outsource part of its maintenance and turnaround activities to contractors who have extensive experience in these fields.

An advantage was to boost the maintenance experience level in the complex quickly and thus mitigate the risk of low levels of experience, as often happens in a new plant with staff going through a rapid learning curve.

The bidding process went through a prequalification stage to ensure the required capabilities were obtained and resulted in two long-term maintenance contracts being successfully established for the complex.

13.4 HOW DO PERFORMANCE MANAGEMENT SYSTEMS AND ENTERPRISE RISK MANAGEMENT COME TOGETHER?

Risk and performance are complementary. However, the interrelationship is complex.

Example:

Health and Safety statistics and Cost indicators could look really good. However, if adequate leading performance indicators are not present and/or management does not understand the fundamentals of asset integrity, the company could have a series of nasty surprises that will be difficult to control.

Box 13.4 discusses the situation in BP just after the Texas City disaster.

Risk reduction and the closing of performance gaps are joint strategies for achieving KPI targets.

As seen in Fig. 13.1, the risk register and performance scorecard are side by side.

The Balanced Scorecard (BSC), as introduced in Section 15.11, Scorecards and Numbers does not consider risk. Box 13.5 discusses this topic.[5]

BOX 13.4 BP AND THE OIL INDUSTRY AFTER THE TEXAS CITY REFINERY DISASTER

Background

- 1997: "Getting Health, Safety and Environment right" established as BP's HSE framework, benchmarked with industry peers
- 1999–2005: Declining injury rates **among the best in the industry**
- March 23, 2005: Texas City incident caused BP to look deeply and urgently into what can be learned
- BP and US Chemical Safety and Hazard Investigation Board (CSB)[2] separately investigated the incident
- Baker Panel established at recommendation of CSB in October 2005
- Panel comprises a diverse team of distinguished experts led by former US Secretary of State James A. Baker III
- Baker Report published January 16, 2007[3]

Discussion

Prior to the Texas City disaster in mid-2000, BP Grangemouth had three separate accidents within 2 weeks, resulting in a record criminal fine. **These did not appear to show in the overall safety statistics**. Nonetheless, after the Texas City disaster, BP restructured and increased focus on HSE and Integrity in line with the recommendations of the Baker Report.

Attitude to risk control in the industry changed after publication of the CSB and Baker Reports. UK Health and Safety Executive published HSE guide 254, "Developing Process Safety Indicators," in 2006, giving a special focus to leading indicators. American National Standards Institute (ANSI)/American Petroleum Institute (API) followed in 2010, with the publication of ANSI/API RP 754, "Recommended Practice Process Safety Indicators for the Refining and Petrochemical Industries."

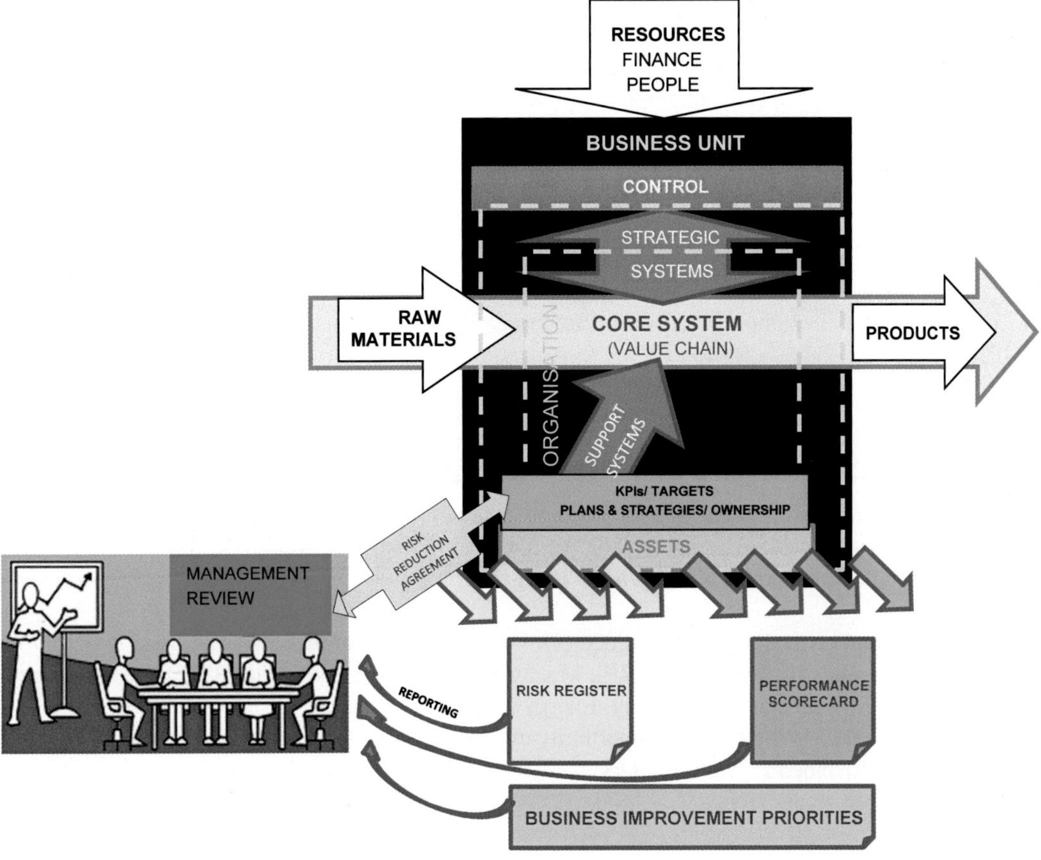

FIGURE 13.1

Risk and performance review.

BOX 13.5 PERFORMANCE MANAGEMENT AND RISK

Balanced Scorecard Quote:

"The Balanced Scorecard (BSC) has remained an enduring tool used by thousands of organizations to align business activities with strategy."

Kaplan, of BSC fame, has been quoted as saying:

If I had to say there was one thing missing that has been revealed in the last few years, it's that there's nothing about risk assessment and risk management. My current thinking on that is that I think companies need a parallel scorecard to their strategy scorecard — a risk scorecard.

Discussion:

Why have a parallel scorecard? The Oil and Gas industry has always emphasized risk as the most important aspect of strategy. The BSC should be replaced with a common sense approach that is flexible and lasting. Furthermore, a continuous improvement process has to be embedded in all aspects of the business.

13.5 PRIORITIZATION OF RISKS AND OPPORTUNITIES

The **value of a risk** is normally calculated from probability multiplied by value if the event had to occur.

The **value of an opportunity** is normally determined by a Return On Investment or other calculation. This is based on a specific expenditure, date of start of generation of revenue, and revenue to be generated. However, these are based on a project risk management process. This is discussed further in Section 28.7, Evaluation and Prioritization.

The format of a risk register is generally well-established. The format of a performance gap register is not.

Performance Systems, where certification is required, have Nonconformance Reports, which are logged in a register. Also, after an assessment is carried out, a report is generated with plans/strategies for improvement. An example is shown in Box 30.2, Example Refinery Business Improvement Plans (BIPs).

13.6 THE RISK AND OPPORTUNITY CYCLE

The top risk and opportunity categories could be based on Committee of Sponsoring Organizations (see Chapter 11: Risk Management). The oil and gas industry, however, would have a different slant on compliance, being **safety and integrity** compliance as opposed to general legal compliance.

The fundamental concept is an embedded continual improvement process as required by ISO 9001, Quality Management, and ISO 31000, Risk Management, as well as the other ISO Systems based on this concept: Environmental Management, Energy Management, etc.

The linkage between the risk and opportunity cycles is depicted in Fig. 13.2.

This will be expanded in ***later chapters***.

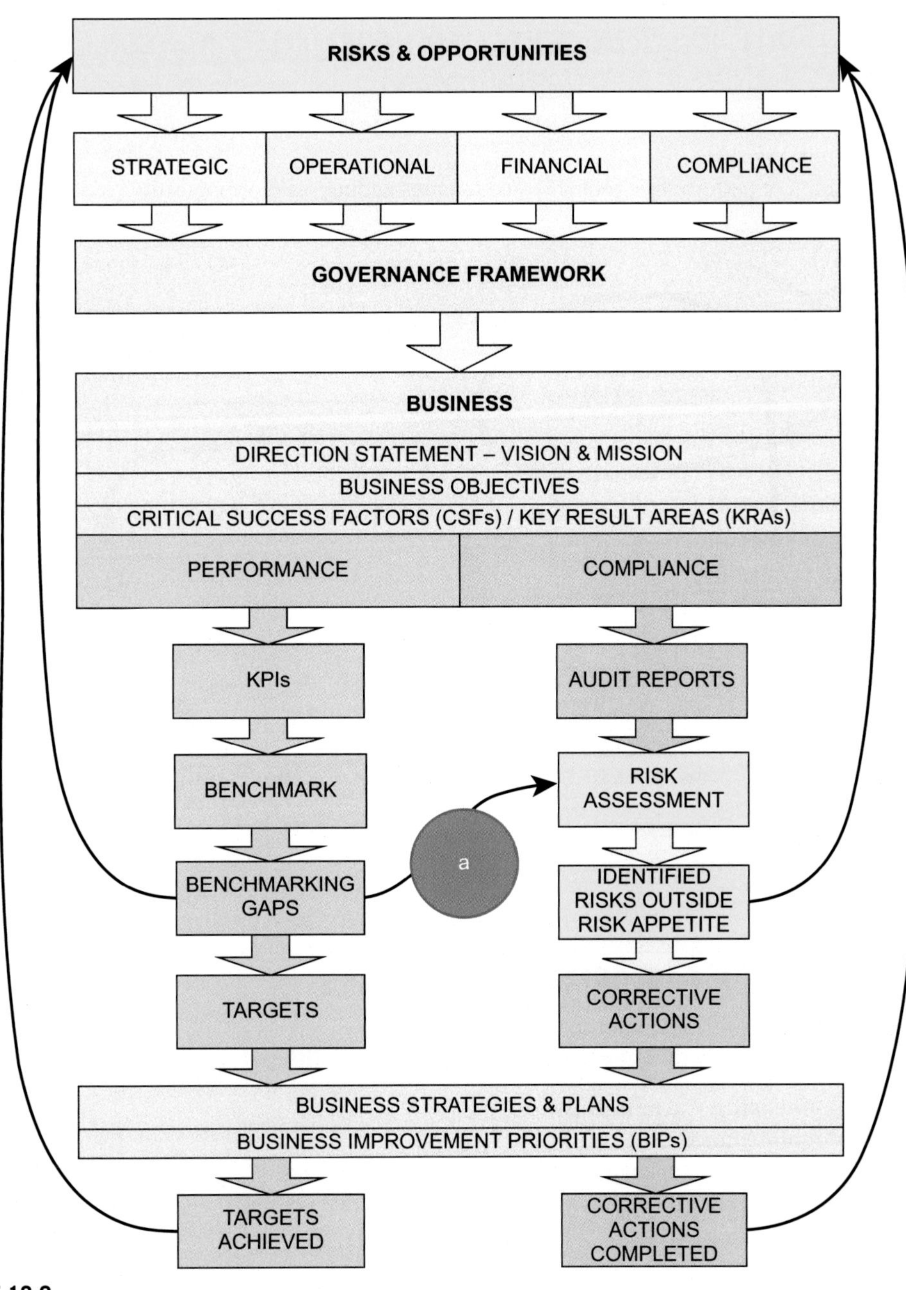

FIGURE 13.2

Risk and opportunity cycles.

13.7 ISO 14001, ENVIRONMENTAL MANAGEMENT, AND ISO 45001, HEALTH AND SAFETY MANAGEMENT

The Approach of the previously mentioned two standards is the same. This is depicted in the two standards as shown in Fig. 13.3.

The integration of performance probably just requires adding "**and opportunity**" to the word "**risk**."

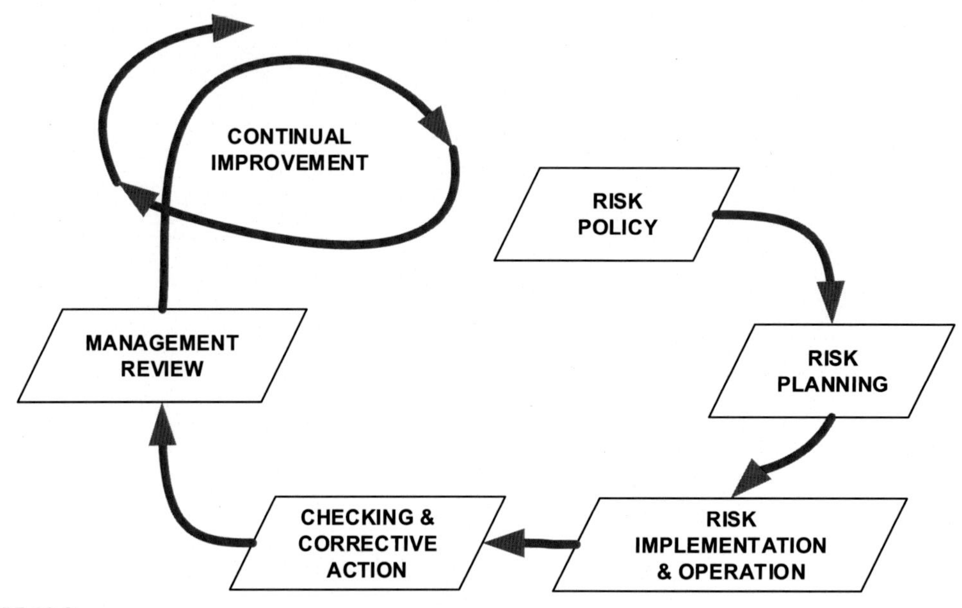

FIGURE 13.3

Continual improvement loop.

13.8 PRODUCTION PERFORMANCE AND RISKS

Lost production is a significant factor in lost profits.

Lost production is typically categorized in terms of **Availability** as follows:

1. **planned maintenance (turnarounds)**;
2. **unplanned maintenance**;
3. consequential downtime (upstream or downstream restrictions);
4. economic or contractual (for example, a feed gas contractual ceiling);
5. slowdowns (normally less than 75% for more than a day); and
6. other shutdown time.

PLANNED MAINTENANCE (TURNAROUNDS)

This is referred to as a shutdown or turnaround, which is planned from more than a year in advance to ensure a run length in line with industry norms with minimal unplanned down days.

Chapter 23, Turnaround, discusses the optimization of the operating cycle to maximize production over the long term. Box 23.2, Critical Path Modeling, gives an example of major lost profits due to a turnaround running 2 weeks late.

UNPLANNED MAINTENANCE

Unplanned down days for maintenance are unpredicted. This is also referred to as **Reliability**. This may or may not be the result of an incident, such as Loss Of Primary Containment.

All **major incidents** involving loss of production as a result of unplanned down days or consequential downtime are managed in accordance with processes described in Chapter 14, Incident Preparedness and Operational Continuity Management.

Section 30.5, Mechanical Availability Drill-Down, discusses **Availability** in more detail.

13.9 ERNST & YOUNG "TOP RISKS AND OPPORTUNITIES"[4]

On a regular basis, Ernst & Young publishes Oil and Gas Industry "top 10 risks and top 10 opportunities." These should be used for input into the annual company strategic review.

The risks and opportunities don't tend to change much over the years; however, the ranking changes. Some of these are listed in Box 13.6.

BOX 13.6 ERNST & YOUNG RISK AND OPPORTUNITIES EXAMPLES FOR THE OIL AND GAS INDUSTRY

Risks

1. Access to reserves: political constraints and competition for proven reserves
2. Uncertain energy policy
3. Cost containment
4. Worsening fiscal terms
5. Health, safety, and environmental risks
6. Human capital deficit
7. New operational challenges, including unfamiliar environments
8. Climate change concerns
9. Price volatility
10. Competition from new technologies

Opportunities

1. Frontier acreage
2. Unconventional sources
3. Conventional reserves in challenging areas
4. Rising emerging market demand
5. National Oil Company (NOC)–International Oil Company (IOC) partnerships
6. Investing in innovation and R&D
7. Alternative fuels, including second-generation biofuels
8. Cross-sector strategic partnerships
9. Building regulatory confidence
10. Acquisitions or alliances to gain new capabilities

13.10 SUMMARY

Risk Management and Performance Management need to be drawn together for effective decision-making. They go hand in hand throughout the life of the investment.

The process for Risk Management and Performance Management is basically the same:

> The risks or performance gaps are identified and followed by an assessment process, which identifies and registers those risks that are outside the company's risk appetite, as well as any significant performance gaps. Strategies, responsibilities for implementation, and timing are agreed, and progress on implementation is monitored and reviewed. Both significant risks and performance gaps might require some investment, and these are then also placed in an investment register (see Chapter 33: Opportunity Life Cycle).

Procurement can affect risk and performance. A well-thought-out procurement strategy can reduce risks considerably. Risk mitigation, related to the type of contract, is very important. Risk reduction initiatives include prequalification and contractor competency assessments. Outsourcing could reduce the risk to the company by transference, allowing the company to focus on the Core System.

ISO Standards for Environment (14001) and Health and Safety (45001) require registers for each of their applications. Significant risks/issues in these registers need to be rolled up into common registers with other risks and opportunities for each level of the business, while significant risks and performance issues affecting company strategy should be promoted to a corporate-wide reporting system.

Production performance is directly related to downtime, resulting in lost profits. As a result, downtime has to be optimized for attainment of long-term profits.

Ernst & Young's "Top Risks and Opportunities" provide a useful starting point for strategic risk and opportunity assessment in the Oil and Gas industry.

REFERENCES

1. Wideman RM. *Project and program risk management.* PMI; 1992.
2. BP Texas City. *US Chemical Safety and Hazard Investigation Board (CSB) investigation report: refinery explosion and fire*. March 23, 2005. Report 2005-01-TX. https://www.propublica.org/documents/item/csb-final-investigation-report-on-the-bp-texas-city-refinery-explosion.
3. *The report of the BP US refineries independent safety review panel.* January 2007. http://www.documentcloud.org/documents/70604-baker-panel-report.
4. *Top ten risks & opportunities oil & gas sector.* Ernst & Young; 2011.
5. Silverthorne S. *Risk management: the balanced scorecard's missing piece.* http://www.cbsnews.com/news/risk-management-the-balanced-scorecards-missing-piece/.

FURTHER READING

1. Achenbach J. *Riches and disasters on exploration's far frontier.* WP-Bloomberg; September 30, 2010.
2. Turk MA, Mishra A. *Process safety management: going beyond functional safety.* Hydrocarbon Processing. www.hydrocarbonprocessing.com/Article/3161534/Process-safety-management.
3. Oracle Primavera Pertmaster. https://www.oracle.com/applications/primavera/products/risk-analysis.html.
4. Why move from Pertmaster to Safran Risk? https://www.youtube.com/watch?v=wvfs9wVEYiM.

CHAPTER

INCIDENT PREPAREDNESS AND OPERATIONAL (BUSINESS) CONTINUITY MANAGEMENT 14

14.1 INTRODUCTION

A major incident affects the performance of the whole business. Consequently, Incident Preparedness and Operational Continuity Management (IPOCM) is an important part of managing the business [this is often referred to as Business Continuity Management (BCM), but note that IPOCM covers a wider scope].

IPOCM overlaps with Information Security Management and Information Technology Management and is an integral part of Risk Management. The relationships are illustrated in Fig. 14.1.

Information Security Management and IT Management, as shown in Fig. 14.1, are discussed in Chapter 19, Information.

FIGURE 14.1

Incident Preparedness and Operational Continuity Management relationship with Risk, Information Security, and IT Management.

This chapter outlines the basics of how to prepare for incidents and, should they occur, how to get the business back to normal as quickly as possible.

Performance Management for the Oil, Gas, and Process Industries. http://dx.doi.org/10.1016/B978-0-12-810446-0.00014-1

The key objectives are to:

a. minimize the probability of an incident arising;
b. reduce the impact of any potential incident; and
c. shorten the period of disruption.

Item (a) is covered using risk mitigation processes described in Chapter 12, Risk Control Mechanisms. Items (b) and (c) are minimized using rehearsed case studies of probable scenarios.

Fig. 14.2 depicts the loss of business operation due to an incident and the reduction of the period of disruption when applying a structured IPOCM.

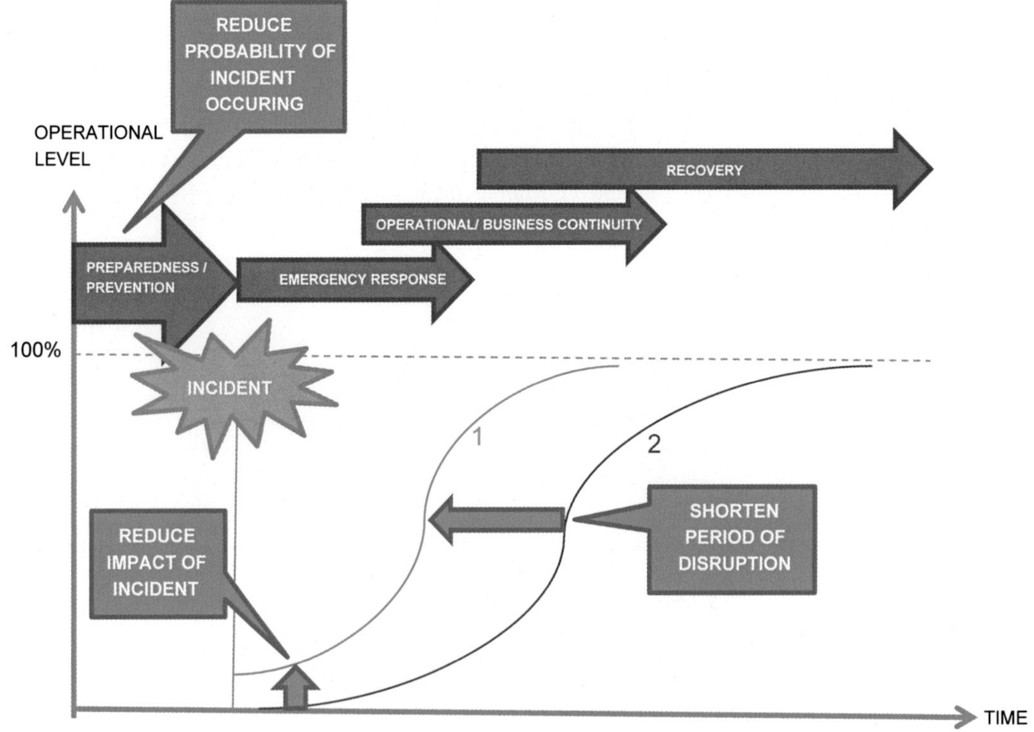

FIGURE 14.2

Illustrative benefits of Incident Preparedness and Operational Continuity Management.

IPOCM is applied primarily to **people, information, and physical assets.**

Interested parties and stakeholders require organizations to proactively prepare for potential incidents and disruptions in order to avoid suspension of critical operations and services. However, should operations and services be disrupted, then there should be preparations in place so that they resume as soon as possible.

14.2 DEFINITIONS

IPOCM is a holistic management process that identifies potential impacts that threaten the business, and it also provides a framework for minimizing their effect.

As per International Standards Organization (ISO) 22301: 2012,[1] **BCM** is a:

> holistic management process that identifies potential threats to an organization and the impacts to business operations that those threats, if realized, may cause, and which provides a framework for building organizational resilience with the capability for an effective response that safeguards the interests of its key stakeholders, reputation, brand and value-creating activities.

Business Continuity is the capability to plan for, and respond to, incidents and business disruptions in order to continue operating at an acceptable level.

Business Continuity Planning (BCP):

As per The Business Continuity Institute Dictionary of Business Continuity Management Terms,[2] BCP is:

> "The process of developing prior arrangements and procedures that enable an organization to respond to an event in such a manner that critical business functions can continue within planned levels of disruption. The end result of the planning process is the BCP."

14.3 APPLICABILITY OF STANDARDS

ISO 22399:2007, Guideline for Incident Preparedness and Operational Continuity Management,[3] is applicable to the whole IPOCM life cycle, including emergency planning. The key elements are Policy, Planning, Implementation and Operation, Performance Assessment, and Management Review. **The application of this guideline is totally relevant to the process industry.**

ISO 22301:2012, Business continuity management,[1] specifies the requirements for a management system to protect against, reduce the likelihood of, and ensure your business recovers from disruptive incidents. ISO 22301 follows the standard Plan Do Check Act cycle, but **excludes the activities of emergency planning**.

HSG 191, Emergency Planning for Major Accidents,[4] published in 1999, has been an excellent basis for establishing emergency planning in the process industry. It provides guidance for emergency planning under the UK Control Of Major Accident Hazards Regulations 1999 (COMAH). The guidance is aimed at those with responsibilities for emergency planning, on-site and off-site, at sites containing major hazards, and includes operators, local authorities, emergency services, and health authorities/boards.

14.4 ESTABLISHING INCIDENT PREPAREDNESS AND OPERATIONAL CONTINUITY MANAGEMENT

For each IPOCM **plan**, the life cycle of IPOCM implementation is as follows:

BUSINESS IMPACT ASSESSMENT

Identify the impact of disruption to critical processes from the loss or unavailability of critical assets due to any of the credible threats occurring.

RESPONSE STRATEGY

Identify appropriate actions and resources to ensure continuity of critical processes.

DOCUMENT AND IMPLEMENT PLAN

Document and implement continuity response facilities, resources, and procedures.

TRAINING, TESTING, AND MAINTENANCE

Ongoing training, testing, and maintenance to ensure continued viability of the plan and the organization's capability.

Fig. 14.3 outlines the previously mentioned IPOCM cycle.

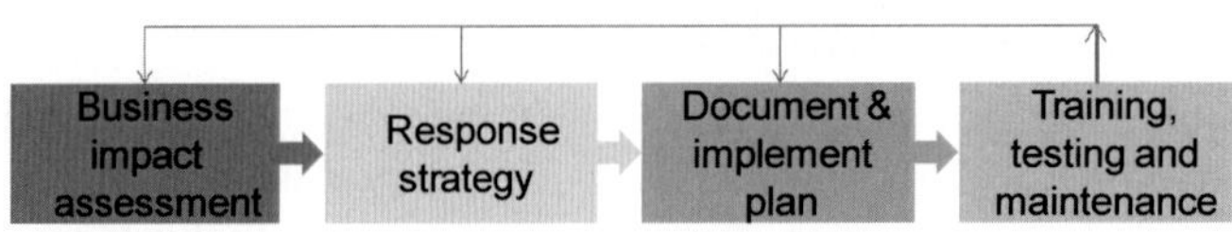

FIGURE 14.3

Incident Preparedness and Operational Continuity Management life cycle.

14.5 INCIDENT PREPAREDNESS AND OPERATIONAL CONTINUITY MANAGEMENT PLANS

There are four primary plans.

EMERGENCY RESPONSE PLAN

This is the main focus for the process industry.

It is the planned immediate response to an operational incident to secure the site and to minimize casualties, operational disruption, and damage to the environment.

Crisis Management Plan

The planned quick response to a crisis to limit damage (to reputation, market position, and financial bottom line) and to preserve or restore stakeholders' confidence.

Business Continuity Plan

The planned response to a prolonged disruption to minimize impact on critical processes and delivery to customers, and return to normal business operations.

IT Disaster Plan

The planned response to a prolonged outage to restore critical IT infrastructure and/or applications, and minimize disruption to the business.

Fig. 14.4 summarizes these plans.

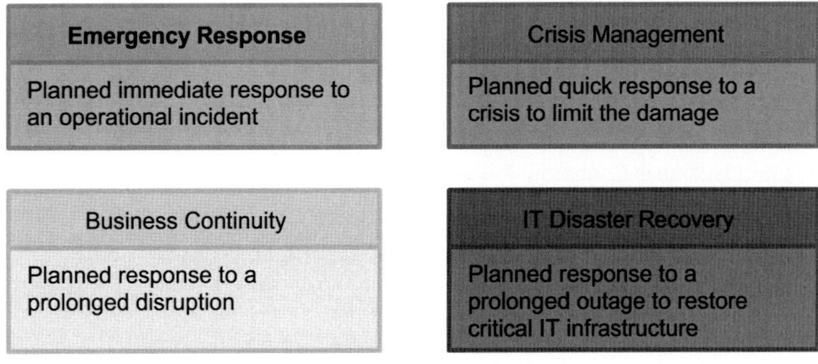

FIGURE 14.4

Incident Preparedness and Operational Continuity Management plans.

14.6 THE RELATIONSHIP OF INCIDENT PREPAREDNESS AND OPERATIONAL CONTINUITY MANAGEMENT PLANS TO ISO 31000, RISK MANAGEMENT PROCESS

The risk is effectively managed within the Enterprise Risk Management (ERM) framework and the Risk Management Process (RMP). However, should the risk materialize, the various plans then come into action: **Emergency Response, Crisis Management, Business Continuity, and Disaster Recovery**. This is depicted in Fig. 14.5.

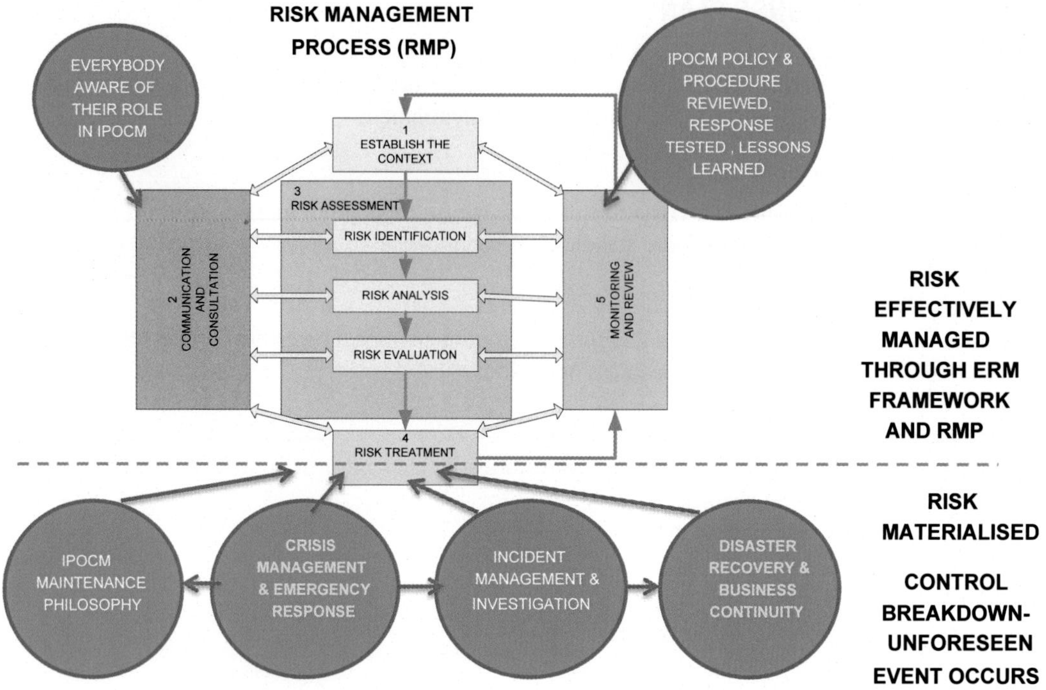

FIGURE 14.5

ISO 31001 process and Business Continuity Plans.

14.7 INCIDENT PREPAREDNESS AND OPERATIONAL CONTINUITY MANAGEMENT SEQUENCE

Fig. 14.6 shows the decision tree and when to apply each plan.

14.8 EMERGENCY RESPONSE PLAN DEVELOPMENT

HSG 191, Emergency Planning for Major Incidents,[4] is a guide for establishing an Emergency Response Plan (ERP) in compliance with the UK COMAH Regulations. These regulations were promulgated in line with the European Community (Seveso II) Directive. This has generally been accepted in the process industry worldwide.

The guide is aimed at those with responsibilities for emergency planning, on-site and off-site, at major hazards establishments, including operators, local authorities, emergency services, and health authorities/boards.

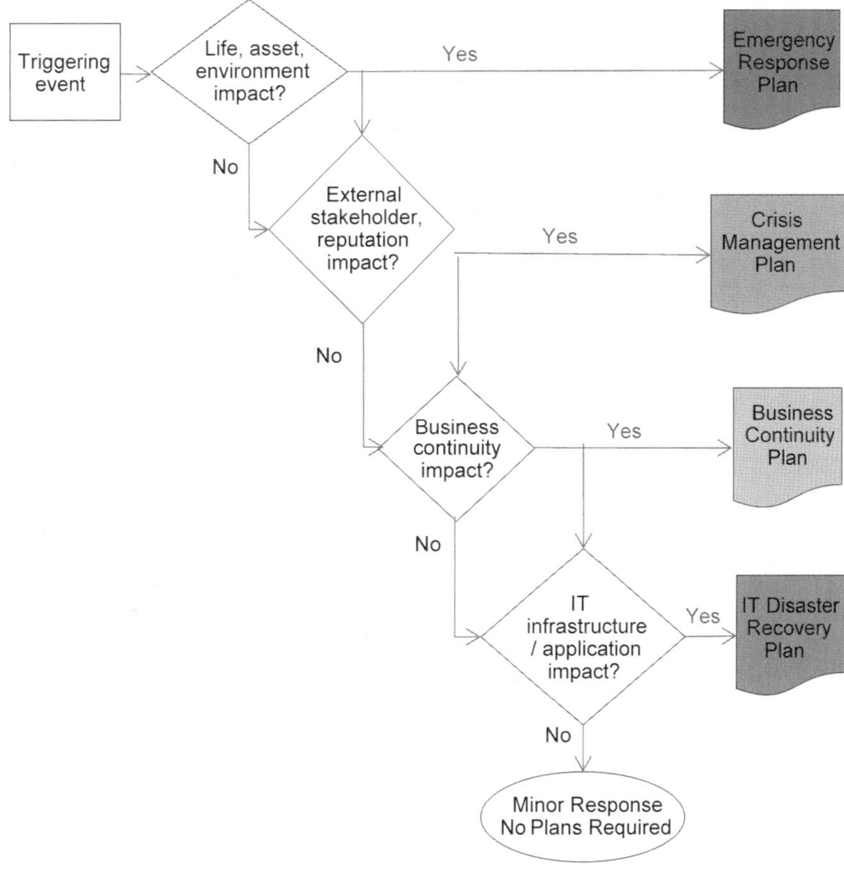

FIGURE 14.6

Incident Preparedness and Operational Continuity Management decision tree.

Box 14.1 shows a mind map for the development of an ERP for a refinery, generally in line with HSG 191.

14.9 SAFETY CASES

DEFINITION

> "A documented body of evidence that provides a convincing and valid argument that a system is adequately safe for a given application in a given environment."

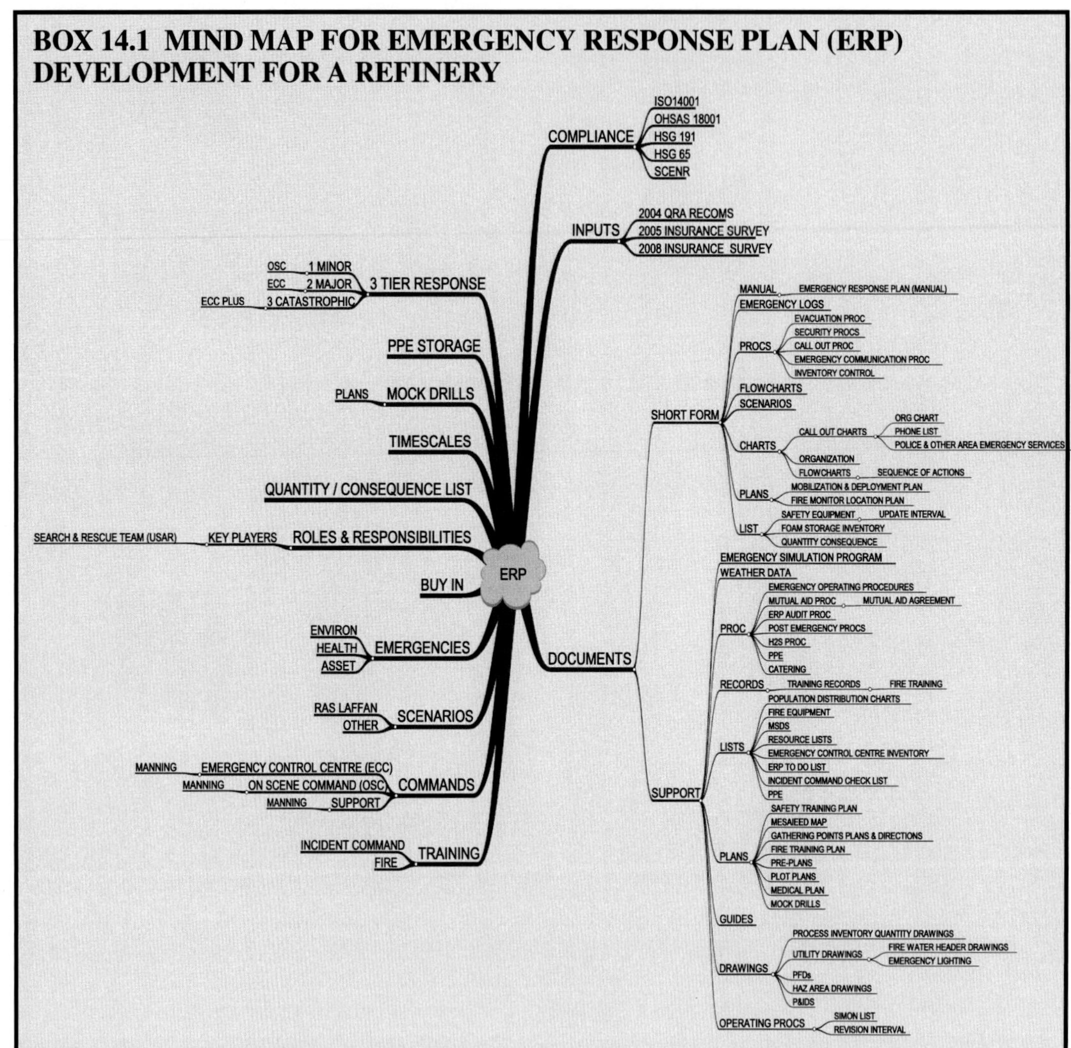

Typical contents of a safety case:

1. Facility description
2. Health Safety and Environment Management System
3. Formal Safety Assessment (FSA)
4. Summary
5. Safety critical elements and performance
6. Standards
7. As low as reasonably practical demonstration
8. Fitness to operate

Safety Cases were introduced after the Piper Alpha disaster in 1988 to minimize the probability of a major incident. Box 14.2 outlines the event.

BOX 14.2 PIPER ALPHA EXPLOSION IN THE NORTH SEA[5]

On July 6, 1988, the Piper Alpha oil platform experienced a series of catastrophic explosions and fires. This platform, located in the North Sea approximately 110 miles from Aberdeen, Scotland, had 226 people on board at the time of the event, 165 of whom perished (in addition, two emergency response personnel died during a rescue attempt). The platform was totally destroyed.

The outcome of this incident was a major tightening of Safety Regulations for offshore operations in the UK and elsewhere around the world.

14.10 SCENARIOS FOR EMERGENCY RESPONSE PLANNING

Scenarios are an important part of Emergency Response Planning. They are used to simulate possible emergencies for a range of exercises from "tabletop" to "full-scale" response exercises.

Box 14.3 is an example of a set of scenarios for a large Industrial City and Port with many Oil, Gas, and Petrochemical complexes.

BOX 14.3 SCENARIO EXAMPLES

Industrial City and Oil and Gas Export Port

Land-based:

- aircraft crash (water/land) ditching
- building fires
- criminal act
- hazardous materials release
- major road traffic accident
- tank/pipeline hydrocarbon release

Marine-based:

- man overboard
- leak/fire/explosion on Liquefied Natural Gas (LNG) vessels/oil tankers
- leak/fire/explosion on LNG vessels/liquid product berths
- leak/fire/explosion/major failure at dry cargo, sulfur, and service berths
- oil spill
- vessel moves off berth
- vessel collision
- vessel grounding/collision with breakwater/berth
- fires onboard vessels

14.11 BUSINESS RESILIENCE

DEFINITIONS

Business resilience comprises far more than disaster recovery; it also helps companies recover from and adjust easily to unplanned events, and take advantage of new opportunities.

Cisco.

> Business resilience is the ability of the company not to incur long term damage, physically or otherwise.
>
> **Anonymous.**

Box 14.4 gives an example of resilience.

BOX 14.4 RESILIENCE EXAMPLE

The 2012 RasGas Cyber Attack[6]

In June 2012, RasGas came under cyber attack, which seriously infected all business systems. Swift steps were taken to minimize the damage. Firstly, all infected PCs were quarantined (locked up in a room) and replaced. Next, all email traffic was terminated and replaced with faxes for critical communication with the outside world. Further, ExxonMobil (a major shareholder) sent in support staff to address the problems and get the business systems up and running again.

Fortunately, there was no effect on actual production, just a nuisance disruption of business processes, since the company was not able to produce business reports for a short time.

A major outcome was that firewalling on the process computers was enhanced so that only one-way traffic from the process computers to the business computers was enabled, thereby further improving cyber security in the process plant.

IT DISASTER RECOVERY PLANNING

An Oil and Gas IQ quote from 2012 statistics:

> 40% of all cyber attacks on critical infrastructure assets targeted the energy sector.

Planning to minimize downtime of IT systems is crucial. A critical element of resilience is the establishment of a remote alternative site ("second site") that mirrors the IT data center of the company, with continual backing up of data to this site.

Box 14.5 gives an example of the establishment of an IT backup to improve business resilience and ensure Business Continuity.

BOX 14.5 IT SECURITY AND BUSINESS CONTINUITY

Chevron South Africa: IT "Second Site"

In the 1980s, Chevron South Africa established an IT "second site" far from the company IT Data Center. If the primary site goes down for any reason, the computers at the second site kick in instantly. The site has Uninterruptable Power Supply and a standby generator in case the power supply is cut to the second site as well.

Comment

Aramco was believed to have had a second site in line with industry norms. However, the value of such a second site is now questionable in the light of case of the virus attack on Aramco in 2012.[7]

CRITICAL DOCUMENTS

Secure fireproof storage is essential for important original and other documents which are essential for the business to survive.

Inclusions are:

- legal and commercial documents, such as Agreements and Contracts; and
- key technical process plant documentation such as Process Data Sheets, Process Flow Diagrams, and Piping and Instrument Diagrams.

Box 14.6 gives an example of the establishment of a critical document backup to improve business resilience.

BOX 14.6 INFORMATION SECURITY AND BUSINESS CONTINUITY

Chevron South Africa: Critical Document Store

Apart from an IT second site (Box 14.5), the company also has a remote secure and fireproof storage facility for original and crucial documents, which are required for the continued operation of the company.

Should the entire Chevron Refinery be destroyed, duplicate technical documents are available from the critical document store to rebuild the refinery from scratch.

14.12 ACCIDENTS AND KNOWLEDGE TRANSFER: DO WE LEARN FROM MISTAKES?[8]

Dr. Trevor Kletz (a world authority on process safety):

> "There is an old saying that if you think safety is expensive, try an accident."

Major accidents/incidents can destroy a company. Learning from mistakes made in emergency drills, as well as the misfortunes of others, is essential in order to remain in business.

LEARNING FROM THE EMERGENCY DRILLS

One must undertake a **comprehensive Performance Assessment** of the whole IPOCM as per **Section 8 of ISO 22399**, including analysis and assessment of emergency drills.

LEARNING FROM WHAT HAS HAPPENED AROUND THE WORLD

Losses in varying degrees occur in all countries, in all industries, in all companies, and on all sites.

History suggests that losses will continue to occur. These may be due to "Black Swans" or due to reduced vigilance.

POINTERS

How can we learn from mistakes? Some pointers:

1. Recognize barriers of time and distance.
2. Recognize and reduce cultural barriers.
3. Look at other industries.

4. Share positives and negatives.
5. Recognize limitations of national and international standards.
6. Extract value from your insurance relationship.

These are expanded as follows:

1. Recognize Barriers of Time and Distance.

- Distance: an incident on the other side of the world may be what could potentially happen on your plant.
  - Promote local awareness and preemptive regulations/standards.
- Company strategy, professionalism, and technical capabilities may differ between multinationals and National Oil Companies (NOCs).
  - Joint Ventures could reduce risks by analyzing internal incidents that have occurred elsewhere and distributing outcomes to affiliates as lessons learned.
- "Time makes us forget": Do you remember big incidents from the past that might relate to your plant/industry?
  - Promote refresher awareness training.
- Loss of experienced people is increasing.
  - Develop industry cooperative learning programs, such as the PetroSkills cooperative, and other skills transfer programs.
  - Increase the use of "simulator" training where it is appropriate.
- Loss of corporate memory is an ongoing challenge.
  - Systems and Processes need to be documented, applied, and updated.
- Young companies lack skills and experience.
  - Recommend partnering with experienced companies for "fast-track" technology transfer.

2. Recognize and Reduce Cultural Barriers.

- The "it has not happened here" syndrome.
  - The same conditions as on your plant may have caused a major incident in another plant.
- Focus on Personnel safety versus Process safety.
  - Personnel safety has historically been based on **lagging** performance indicators, where focus is now more on Process safety, utilizing **leading** performance indicators.
- Fear of blame and the difficulty in challenging upwards in some cultures.
  - Promote a culture of learning.
  - Implement a behavior-based safety program: "all are equal in safety."

3. Look at Other Industries.

- Tunnel vision syndrome: within a Business Unit, within a company, and within the industry.
  - We **can** learn from other industries.

4. Share Positives and Negatives.

- Negatives get into the news.
- Positives don't get into the news and are more difficult to analyze.
  - The question has to be asked: "Why has the competition 'company x' not had any major incidents?".

5. Recognize Limitations of National and International Standards.

- Standards do not cover everything. They normally only evolve after a series of bad experiences or a risk assessment.
- International standards emerge from national standards, industry standards, and company standards (see Box 18.1: Requirement for a new Standard).

6. Extract Value From Your Insurance Assessor.

- Insurance Assessors have large databases that may be exploited.
  - Use the potential knowledge transfer opportunity.

14.13 INCIDENT PREPAREDNESS AND OPERATIONAL CONTINUITY MANAGEMENT KEY PERFORMANCE INDICATORS

IPOCM Key Performance Indicators (KPIs) should cover those listed in Table 14.1.

Table 14.1 Incident Preparedness and Operational Continuity Management Key Performance Indicators

Key Performance Indicator (KPI)	Unit	Definition
Unplanned events	Number	All unplanned events affecting production
Unplanned events: corrective action implementation within prescribed time	%	Target should be 100%
Emergency drills, Tier 1	%	Actual versus planned: managed by the operating staff
Emergency drills, Tier 2	Number	Internal fire department called out
Emergency drills, Tier 3	Number	Support required from civil defense and other industries in the area
Safety case implementation for onshore	%	Safety cases
Safety case implementation for offshore	%	Safety cases
Mandatory Health Safety and Environment training	%	% Complete
Fire and explosions	Number	All
Incident Preparedness and Operational Continuity Management (IPOCM) risk actions beyond the agreed implementation date	Number	

14.14 SUMMARY

A major incident affects the performance of the whole business.

The key objectives are to:

a. minimize the probability of an incident arising;
b. reduce the impact of any potential incident; and
c. shorten the period of disruption.

ISO 22399:2007, Guideline for incident preparedness and operational continuity management, is applicable to the whole IPOCM life cycle, including emergency planning.

The life cycle of IPOCM implementation is:

a. business impact assessment,
b. response strategy,
c. document and implement plan, and
d. training, testing, and maintenance.

The risk is effectively managed within the ERM framework and the RMP. However, should a risk materialize, the various plans come into action: emergency response, crisis management, business continuity, and IT disaster recovery.

Safety cases are "a documented body of evidence that provides a convincing and valid argument that a system is adequately safe for a given application in a given environment."

Regular practice exercises with likely scenarios are necessary in anticipation of any likely eventuality.

Business resilience is the ability of the company not to incur long-term damage, physically or otherwise. Mitigation includes IT disaster recovery planning and secure fireproof storage of critical documents.

KPIs are required to assess the performance of the IPOCM processes.

Learning from mistakes (ours and others) requires an open mind and continual reinforcement.

In summary, a structured approach to IPOCM has to be embedded in the way we do business, with ISO 22399:2007 as the basis of the company IPOCM.

REFERENCES

1. ISO 22301:2012. *Societal security—business continuity management systems—requirements.*
2. *Dictionary of business continuity management terms.* The Business Continuity Institute.
3. ISO 22399:2007. *Societal security—guideline for incident preparedness and operational continuity management.*
4. *Emergency planning for major accidents. Control Of Major Accident Hazards Regulations 1999 (COMAH).* HSG191. HSE Books; 1999.
5. Piper Alpha disaster. http://www.aiche.org/ccps/search/piper%20alpha%20summary .
6. Rasgas cyber attack. http://www.scmagazine.com/natural-gas-giant-rasgas-targeted-in-cyber-attack/article/257050/.
7. Aramco cyber attack. http://money.cnn.com/2015/08/05/technology/aramco-hack/.
8. Clough I. *Accidents and knowledge transfer: do we learn from our mistakes?* Presentation to IChemENorthern Branch; January 2008.

FURTHER READING

1. ISO 22313:2012. *Societal security—business continuity management systems—guidance.*
2. *A guide to the Control of Major Accident Hazards Regulations 1999 (as amended). Guidance on Regulations.* HSE.

3. *Good practice guidelines*. Business Continuity Institute; 2010.
4. PAS 77:2006. *IT service continuity management code of practice*.
5. ISO 27001:2013. *Information security management* (Formerly ISO 17799).
6. NFPA 1600:2004. *Standard on disaster/emergency management and business continuity programs*. National Fire Protection Association (USA).
7. *BCM good practice guidelines*. 2007.
8. *Large property damage losses in the hydrocarbon industries. The 100 largest losses 1972–2009*. Marsh's Energy Practice.
9. Hiles A. *The implications of the Turnbull report for business continuity management*. Kingswell International. http://www.kingswellinternational.com/news/the-turnbull-report/.
10. *BP deepwater horizon accident investigation report*. http://www.bp.com/content/dam/bp/pdf/sustainability/issue-reports/Deepwater_Horizon_Accident_Investigation_Report.pdf .
11. Business Resilience Certification Consortium International. http://www.brcci.org/.

PART

PERFORMANCE INDICATOR SELECTION

4

OVERVIEW

WHY IS IT IMPORTANT TO SELECT THE RIGHT PERFORMANCE INDICATORS?

Every performance indicator that is selected must contribute to, and align with, the company direction statement and must ensure customer satisfaction. The value or amount of influence the indicator has on the overall performance of the business needs to be assessed so that improvement can be accelerated by focusing resources on those few indicators that have the most influence.

THE OBJECTIVE

The objectives of Part 4 are the following:

1. Identify the most influential indicators
2. Guarantee vertical and horizontal linkage of these indicators
3. Ensure balancing of short-term performance of cost, integrity, reliability, etc., to maximize positive long-term performance

PICTORIAL VIEW

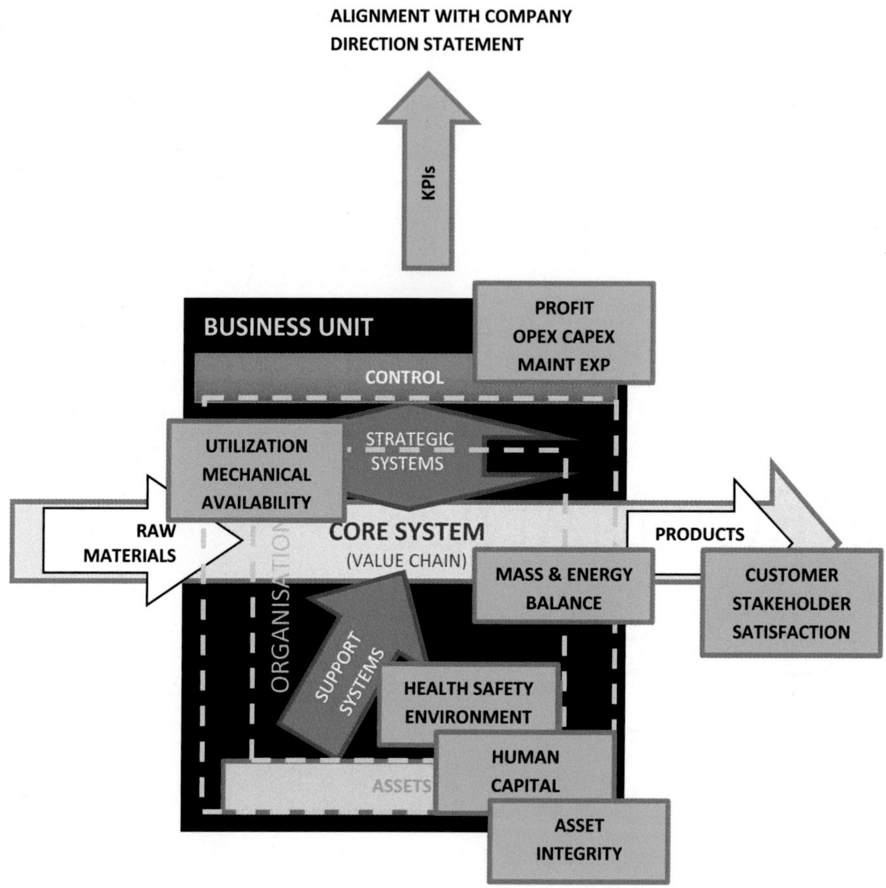

OUTLINE OF PART 4

- Perspectives of a performance indicator are demonstrated, depending on the users' requirements.
- Primary indicators are discussed.
- Competiveness and efficiency categories are defined using examples.
- Closing the gap between the current situation and the industry leader or "pacesetter" is elaborated.
- Elements of good performance indicators are listed and explained.
- Different approaches to selection are outlined—top down, benchmarking, and bottom up.
- Examples of the consequences of poor selection and definition are presented.
- Physical asset integrity key performance indicators (KPIs) are introduced.
- Examples of high-level performance indicators in the refining and gas processing industries are given.
- Relationships between indicators are discussed, including the positive and negative effects of one on another.

CHAPTER

PERFORMANCE FOCUS

15

15.1 INTRODUCTION

Performance Indicators (PIs) are management tools. Simply the process of defining PIs can in itself help managers and stakeholders in initial problem analysis and articulation of results expectations.

By verifying change, PIs help to demonstrate progress when things go right and provide early warning signals when things go wrong.

The purpose of having PIs is ultimately to support the effectiveness of managing the Company, Business Unit, System, or Process throughout the process of planning, management, and reporting.

The key to good PIs is **credibility**. Chapter 26, Data Verification, discusses how important credibility is for "buy-in" of benchmarking results, especially when they are not favorable.

The challenge is to meaningfully capture data that is substantively valid and is practical to monitor. PIs fundamentally **only** indicate; they do **not** explain.

Questions:

1. How do we know how well we are performing?
2. What should a Key Performance Indicator (KPI) be?
3. Are we on the right track? Are the selected KPIs in line with the business direction and customer requirements?
4. Are the set performance targets realistic?

In the process industry, the primary focus is on Health, Safety, Environment, and Asset Integrity. **Risk-related indicators are thus dominant**.

15.2 DEFINITIONS

KEY PERFORMANCE INDICATORS/PERFORMANCE INDICATORS

KPIs/PIs represent a set of measures focusing on those aspects of performance that are the most **critical** for the current and future success of the Company/Business Unit/System/Process.

Various authorities use different terminology for the same definition. Box 15.1 describes some of these.

The definitions used in this book are simple and generic. The terms **KPIs** and **PIs** are used interchangeably. For example, a PI at a high level in the company or Business Unit may be a Key Performance Indicator (KPI) in a Management System or Process.

Performance Management for the Oil, Gas, and Process Industries. http://dx.doi.org/10.1016/B978-0-12-810446-0.00015-3

> **BOX 15.1 INDICATOR DEFINITIONS**
>
> *A **leading** Process Safety Performance Indicator (PSPI), as defined in HSG 254,[1] or an "Activities Indicator," as defined by OCED,[2] would constitute a Performance Indicator (PI), according to Parmenter.[3]*
> *A **lagging** PSPI, as defined in HSG 254,[1] or an "Outcome Indicator," as defined by OCED,[2] would constitute a Results Indicator (RI), according to Parmenter.[3]*
>
> 1. "HSG 254" refers to definitions in **UK HSE guide 254, Developing Process Safety Indicators**.
> 2. "OCED" refers to **Organization for Economic Cooperation and Development** definitions.
> 3. "Parmenter" refers to definitions in **Introduction to Winning KPIs**, D. Parmenter, August 2009.

LAGGING VERSUS LEADING INDICATORS

Lagging indicators are KPIs that record actual integrity failures.
Leading indicators are KPIs that can be used to assess the health of the safeguards and controls that make up the barriers. They require a routine systematic check that key actions or activities are undertaken as intended.

HSG 254[1] defines leading and lagging indicators as follows:

"**Lagging indicators** reveal failings or holes in the barrier discovered following an incident or adverse event. The incident does not necessarily have to result in injury or environmental damage and can be a near miss, precursor event, or undesired outcome attributable to a failing in the Risk Control System."
"**Leading indicators** identify failings or holes in processes or inputs essential to maintain critical aspects of Risk Control Systems (i.e., to deliver the desired safety outcomes)."

The terminology for lagging and leading indicators adopted in this book are the same as the HSG 254 definitions.

STRATEGIC OBJECTIVES (VERTICAL ALIGNMENT)

A set of top Company/Business Unit Objectives that could be part of a Company/Business Unit Direction Statement, or have direct links to a Direction Statement.

This is one of the starting points for the development of KPIs. Box 10.4, Refinery Business Unit Objectives Example and Explanation, gives more details.

CRITICAL SUCCESS FACTORS

A list of issues or aspects of organizational performance that determine ongoing health, vitality, and well-being.

Critical Success Factors (CSFs) give focus in determining KPIs. Section 16.3, Approach to Selection, discusses Business Unit CSFs.
Success Factors may be critical for one but not for another, depending on the value to the addressee.

Example:

*Maintenance CSFs are at the top of the list for a Maintenance System, and **all** contribute to a set of Business Unit CSFs.*

CSFs are sometimes referred to as Key Result Areas.

CUSTOMER (HORIZONTAL ALIGNMENT)

The customer is "the receiver of goods or services." Quantity, quality, and timelines are key requirements that have to be embedded into KPIs.

Section 10.3, Horizontal Alignment, gives details.

15.3 DISCUSSION OF APPROACH

BACK TO BASICS

The process of converting data to a point where it can be used to make decisions has to be well structured, fault-free, and fact-based. The following is the progression for process plant data.

1. Data

Data is extracted from every instrument tag in the process plant through the Distributed Control System to the Real-Time Information System (RTIS). This can be selectively downloaded to Excel by stipulating tag number, date and time for start of record, date and time for end of record, and interval of extraction.

Example: Reading from a flow meter every 4 h for "feed to crude unit" in barrels (bbl) per day.

Time	00h00	04h00	08h00	12h00	16h00	20h00
Tag Reading	9500	10,050	9000	999	10,040	9980

2. Information

The data in Excel needs to be screened for spurious and out-of-range records, which need to be eliminated. The data can then be processed further into information.

Example: Average flow to crude unit for past 24 h in bbl per day.

Time	00h00	04h00	08h00	16h00	20h00	Average (excl. 12h00 reading)
Tag Reading	9500	10,050	9000	10,040	9980	9714

3. Knowledge

The information is then inputted into an equation to get valued knowledge.

Example: Average flow of crude per day divided by unit capacity (10,000 bbl per day) to get Utilization of the unit. $100 \times 9714/10000 = 97.1\%$.

4. Decision-Making

A decision can then be made based on the knowledge created.

Example: A question can be asked as to why the unit is not operating at full capacity, and thus corrective action can be taken.

The data progression could be depicted as in Fig. 15.1.

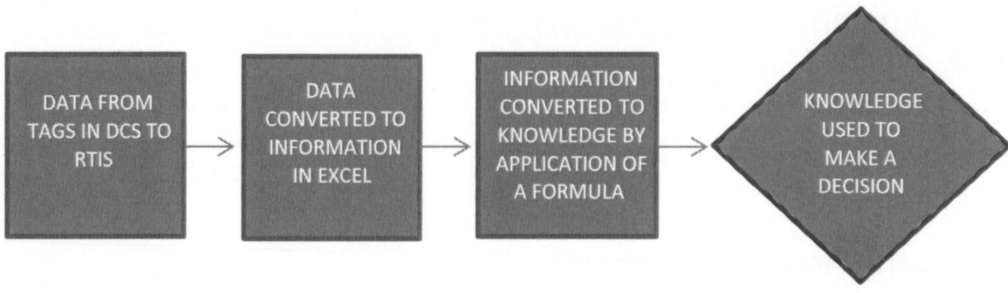

FIGURE 15.1

Data progression.

PRESENTATION FOR DECISION-MAKING

The question is: "How easy is it for a decision-maker to make a decision based on a KPI, or set of KPIs, and supporting information?"

Selection of the best means of presentation in order to elicit a decision is needed. Methods include graphics, table, traffic lights, numbers, words, etc. These will be discussed in Chapter 31, Reporting.

INDICES

A company could be locked in on a particular benchmarking consultant for some indices.

Example: Solomon Associates Energy Intensity Index

Simple generic indices need to be chosen, possibly linked to complex indices, for staff to understand and utilize so as to action improvement.

COMMON DENOMINATORS

Similarly, a company could be locked in on a particular benchmarking consultant for some common denominators.

Example: Solomon Associates cost per Equivalent Distillation Capacity

Simple common denominators, such as "replacement value" (RV), could be used but it is important to be aware of the definition of this.

Example: RV could be derived as the estimated value given by the insurance assessor, or it could be calculated based on current construction costs.

Definitions of the above indices and common denominators are given in Chapter 25, Common Denominators and Indices Used in Benchmarking.

QUANTITY OF INDICATORS

This is dependent on the Business Unit: 10 to 20 may be a manageable set of **top** PIs for any Business Unit. A balance needs to be struck between information overload, which should be avoided, yet maintaining sufficient detail to be able to analyze a top KPI that is not performing well.

USE FOR COMPARISON

Although it is convenient to use indicators that have common definitions in the industry, they might not always be readily understood or easily calculated. They then need to be interpreted, deconstructed, and cascaded in order to obtain ownership of the simpler contributing indicator. Box 16.2, Terminology: Refinery Utilization Cascade, gives an example.

ENABLEMENT: DISCUSSION AND ACTION

An indicator needs to be understood by everyone involved with agreeing to a target and actioning the strategies and plans to achieve it. It is crucial that the definition of the indicator be easy to grasp. Box 16.3, Refinery Mechanical Availability Relationships, gives an example.

ILLUSTRATIVE EXAMPLES OF SUITABLE INDICATORS

Qualitative Indicators

Existence (yes/no)

a. Risk As Low As Reasonably Practical? yes/no
b. Certified to ISO 9001? yes/no

Category

a. Level of risk policy awareness: " high," "medium," or " low"
b. Progress on implementation of a strategy/plan: red, amber, or green traffic lights

Quantifiable Indicators

Number

a. Of operators trained
b. Of nationals in substantive positions

Percentage

a. Share of operations budget allocated to maintenance
b. Process unit availability

Ratio

a. Reserve/production (R/P) ratio for a gas or oil well
b. Profit per bbl of oil

15.4 HOW TO SELECT INDICATORS

The essential challenge in selecting indicators is to find measures that can meaningfully capture key changes, combining what is substantively relevant to what is practically realistic.

CRITERIA

The following criteria and questions may be helpful in selecting indicators:

1. Valid: Does the indicator capture the essence of the desired result?
2. Practical: Are data actually available at reasonable cost and effort?
3. Precise meaning: Do stakeholders agree on exactly what to measure?
4. Clear direction: Are we sure if an increase is good or bad?
5. "Owned": Do stakeholders agree that this indicator makes sense to use?

DO'S AND DON'TS

In selecting indicators:

Do:

- Look for indicators that have clear meaning.
- Set targets that are realistic.
- Agree with beneficiaries and partners.
- Look for data that is easily available; avoid major data collection.
- Keep data sources and monitoring responsibilities in mind.

Don't:

- Lose sight of Company/Business Unit Objectives.
- Assume that data will be available.
- Set targets that cannot be achieved.
- Impose or insist on any one indicator.
- Overinvest in attempts to quantify.
- Use indicators that need expert analysis.
- Use more indicators than necessary.

BEST PRACTICES FOR ALL KEY PERFORMANCE INDICATORS

1. Ensure you have quality, consistent data. Bad data will undermine any attempt to deliver metrics.
2. Establish KPI targets and motivate employees to help achieve targets.
3. Monitor KPIs on an ongoing basis. However, not all KPIs will need to be measured and updated with the same frequency.
4. Adjust KPIs as necessary: As business or operations conditions change, KPIs may require updates to align with new Corporate or Business Unit Strategic Objectives and Initiatives.

15.5 TARGET AUDIENCE FOR A COMMON DATA SOURCE

Views of what is required are defined in three primary areas:

1. Strategic
2. Benchmarked
3. Operational

The view of a single slice of data can be different, depending on which of the above areas is the focus and who requires the data.

Example: a Strategic view looks at a dollar/ton, whereas an Operational view focuses on tons per day. Benchmarking looks at what is common among the competition, which might be the amount of energy consumed to produce a ton of product.

General Audience groups are as follows:

- Operations actioners
- Business unit management
- Business unit board of directors
- Corporate management
- Corporate board of directors
- Shareholders
- Other stakeholders

Fig. 15.2 gives an example of different KPIs, all with a common data source.

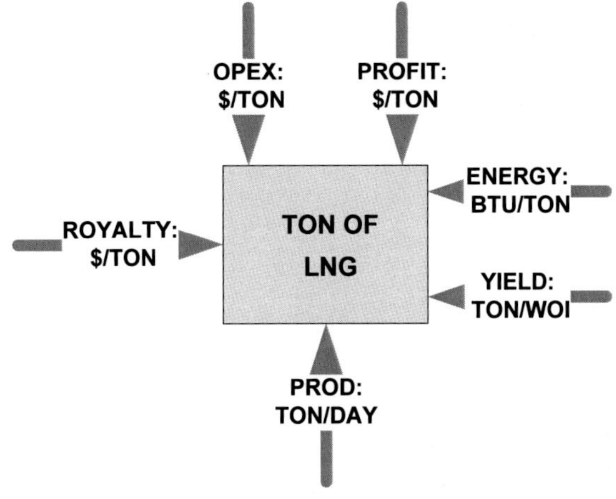

FIGURE 15.2

Views of a ton of liquefied natural gas.

Table 15.1 gives examples of different views from the same categories of data.

Table 15.1 Data Categories and Audience

Category	Business Unit Management	Board of Directors
Field development/production	Throughput	Reserve/production ratio
Finance	Opex	Margin
Technical	Availability	Utilization
Health, safety, security, and environment	Lost time incidents	Major incidents
Human capital	Vacancies	Head count
Assurance	Outstanding audits beyond the due date	Risks outside the company risk appetite

15.6 KEY ELEMENTS OF A KEY PERFORMANCE INDICATOR

KPIs require the following elements:

- Business System or Process
  - A KPI must be attached to a Business System or Process.
- Ownership
  - A KPI must be owned by the individual who has the most influence in improving the indicator and meeting the agreed targets.
- Quantitative (if possible)
  - A KPI must be an easily measurable value.
- Values
  - KPI values are as follows:
    - Planned or target value (short-term)
    - Planned or target value (long-term)
    - Actual value
    - Variance (planned/actual)
- Direct relation to business direction statement
  - KPIs should have a direct relationship with the business direction statement.
- Value addition
  - A positive change in a KPI must add significant value to the Business System or Process.

15.7 PRIMARY INDICATORS

Primary **short-term** focus relates to restrained resources: finance, human capital (people), and raw materials, where **long-term** focus relates to physical asset integrity.

General grouping is as follows:

- **Business** indicators relate to profit and return on capital employed.
- **Operating** indicators relate to how much one can squeeze out of a bbl of oil, million standard cubic feet (mmscf) of gas, or ton of petrochemical raw materials. Thus for the Oil & Gas industry, hydrocarbon balance must always be within prescribed limits.
- **Integrity** indicators relate to loss of primary containment.

Alignment is required for all PIs from the individual employee to shareholders aspirations. Retention and training of staff (human capital) is therefore critical for the success of the business.

15.8 PERFORMANCE GAP

Use of the correct primary indicators enables one to identify opportunities by comparing performance. By applying company and industry best practices, the performance gap is closed.

The steps are as follows:

1. Identify the gap by benchmarking
2. Apply best practices to close the gap
3. Remeasure the gap, by benchmarking, to record closure

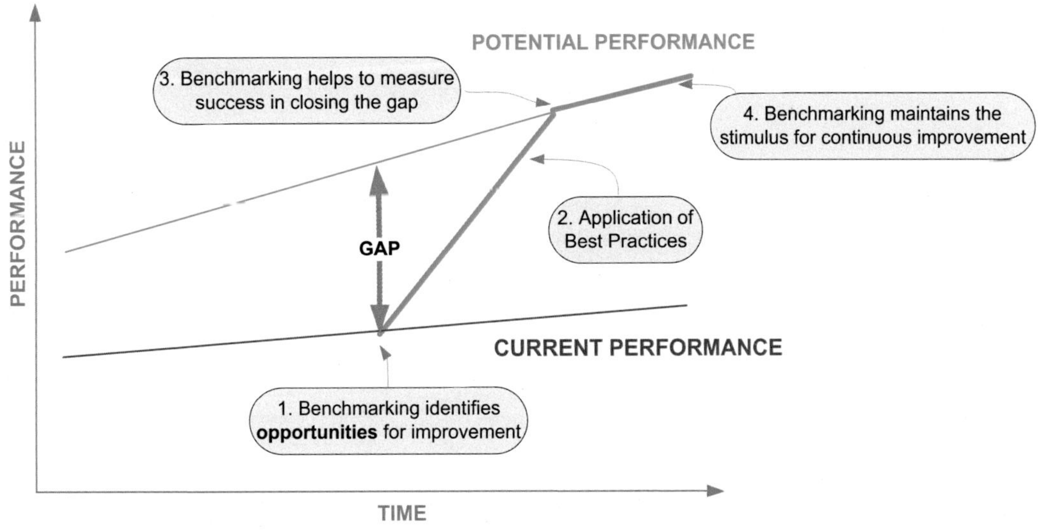

FIGURE 15.3

The performance gap.

This is depicted in Fig. 15.3 (note that the figure shows the rate of improvement for the top performers is faster than the rate of improvement for poor performers. Thus a constant level of performance over time is actually a deterioration of performance).

15.9 COMPETITIVENESS AND EFFICIENCY

Performance Management can be categorized into Competitiveness and Efficiency.

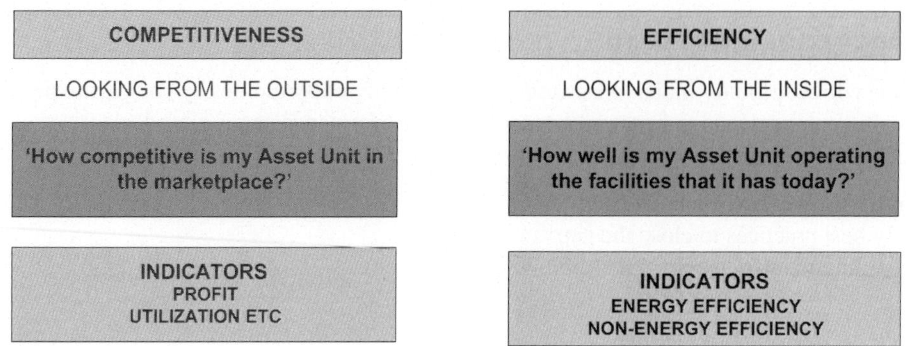

FIGURE 15.4

Competitiveness versus efficiency.

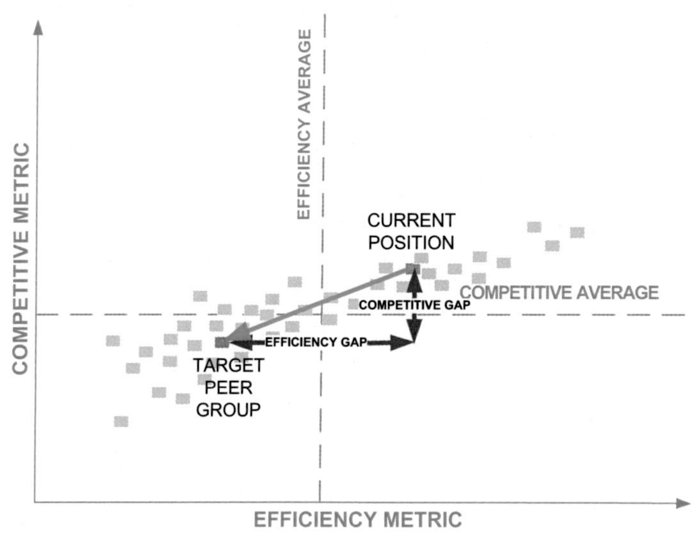

FIGURE 15.5

Competitive and efficiency gap relationship.

Competitiveness is focused on looking from the outside as to how well the company is doing compared to the competition.

Efficiency is focused on getting the maximum out of what is given to the operator.

The focus for private companies tends toward Competitiveness, while the focus for State-controlled companies tends toward Efficiency.

This is depicted as shown in Fig. 15.4.

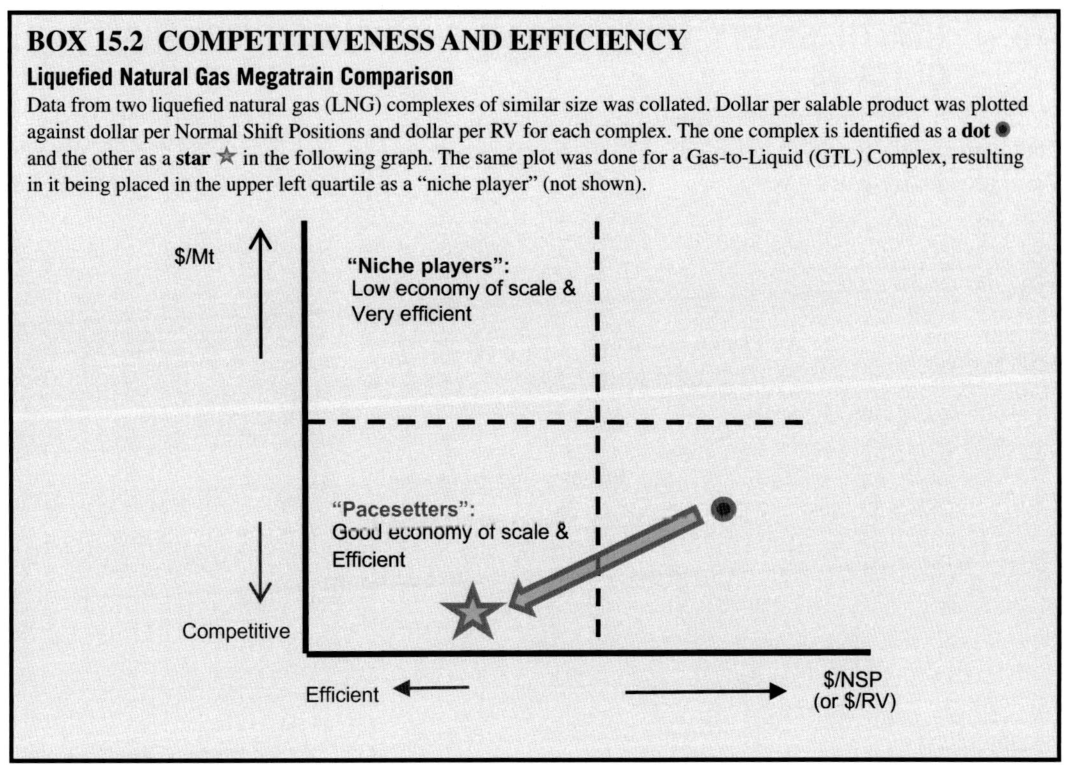

BOX 15.2 COMPETITIVENESS AND EFFICIENCY

Liquefied Natural Gas Megatrain Comparison

Data from two liquefied natural gas (LNG) complexes of similar size was collated. Dollar per salable product was plotted against dollar per Normal Shift Positions and dollar per RV for each complex. The one complex is identified as a **dot** ● and the other as a **star** ☆ in the following graph. The same plot was done for a Gas-to-Liquid (GTL) Complex, resulting in it being placed in the upper left quartile as a "niche player" (not shown).

The relationship between the Competitive Gap and Efficiency Gap in closing the Performance Gap is depicted as shown in Fig. 15.5.

Example: A simple hydro-skimming refinery produces a surplus of fuel oil with a relatively unattractive price and demand. These tend to be small refineries who have difficulty obtaining the low dollar/bbl operating expense that large complex integrated refineries, with economy of scale, are achieving (competitive metric), even though their efficiencies are good. See also Box 21.5, Reliance Refinery: Profits and Investment and Box 37.5, "Hydro-Skimming" Versus "Hydro-Cracking" Refineries.

Box 15.2 compares two similar-sized LNG complexes and one Gas-to-Liquid complex.

Gaps can be closed by continuous improvement or step improvement. Continuous improvement tends to be improvement of processes and enhancing competencies. Step improvement tends to be a structured change management process, which takes people through a paradigm shift. Chapter 29, Identification, Analysis, and Evaluation of Gaps and Chapter 30, Strategies and Actions, give details identifying gaps and actioning of improvements.

It is interesting to note that ExxonMobil use a similar indicator to dollar/RV for assessing performance of their Business Units. Box 21.6, ROCE Application: ExxonMobil, gives details.

15.10 HEALTH, SAFETY, ENVIRONMENT, AND INTEGRITY: CHANGE OF FOCUS

Due to well-structured Health, Safety, and Environment (HSE) systems, enforcement of legislation, and certification to ISO standards for Occupational Health, Safety, and Environment, incident rates have progressively improved.

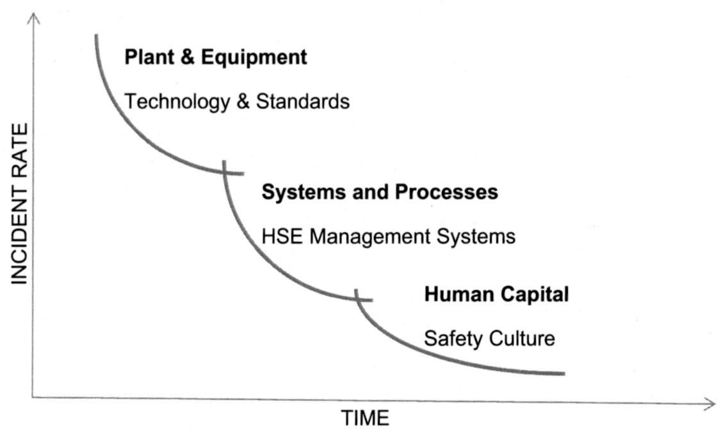

FIGURE 15.6

Falling incident rates over time.

Initially, engineering improvements to plant and equipment, with a focus on safety, brought incident rates down, but then the incident rates leveled off.

Following that, Integrated HSE Management Systems, with the requirements of reporting, assurance, competencies, and risk management, brought the incident rates down further, but then they leveled off again.

Now, the focus is on a **safety culture** with the associated requirements of visible leadership, personal accountability, shared purpose, and aligned performance commitment. This is serving to bring incident rates down even further.

Fig. 15.6 depicts the progress over time.

Note: Incident rates may give a false sense of security if other long-term performance issues, such as asset integrity, are not addressed. Box 13.4, BP and the Oil Industry after the Texas City Refinery Disaster, is a good example of this false sense of security.

BOX 15.3 THE BALANCED SCORECARD AND THE DANGERS OF APPLYING IT IN THE PROCESS INDUSTRY

The Balanced Scorecard (BSC) was introduced by Harvard Business School Professor **Robert Kaplan** and colleague **David Norton**. According to Kaplan and Norton, the BSC is a strategic management concept that enables companies to clarify their vision and strategy and translate these into action. The components are as follows:

- Vision and strategy
- Financial
- Internal business process
- Customer
- Learning and growth

Comment

This approach has been developed by so-called "bean counters" and does not take a holistic approach to performance. Professor Johnston, an opponent to Kaplan and Norton, is quoted as saying:

> *In time, this teaching contributed to the modern obsession in business with "looking good" by the numbers, no matter what damage [it] does to the underlying system of relationships that sustain any human organization.*

He says if companies simply focus on the "means" (for instance, designing a production system that makes errors visible and correctable the moment they occur), they wouldn't have to worry about enforcing targets and goals. Error counts would naturally get lower. The "ends" would take care of themselves.

Risk management is not even mentioned in the BSC approach. However, after the financial meltdown in 2008, Kaplan admitted it lacked a "crucial piece." See Box 13.5, Performance Management and Risk, for further discussion.

15.11 SCORECARDS AND NUMBERS

SCORECARDS

Scorecards are a collection of KPIs for use by various levels of management. However, they are only part of the reporting process and are simply PIs with actuals versus targets, since they do not give any supporting information.

There are many "all singing and dancing" tools on the market, such as Honeywell's "KPI Manager," which extracts data from the plant process RTIS, Enterprise Resource Planning Systems, and many other systems that produce a "display" of just about every PI in the plant. The crux is to configure the system so as to get credible "knowledge" for effective decision-making (see Section 15.3: Discussion of Approach). Remember the old adage: "garbage in, garbage out."

Box 15.3 discusses the pitfalls of one such tool.[4]

THE OBSESSION WITH NUMBERS

British Petroleum, under the reign of Sir John Brown, is a prime example of focusing on numbers and, as a result, destroying the assets. See the following boxes for further details:

1. Box 9.5, Accountability, Responsibility, and Discipline

2. Box 12.1, Swiss Cheese: Deep Water Horizon
3. Box 13.4, BP and the Oil Industry After the Texas City Refinery Disaster
4. Box 24.2, Integration into the Business, Example 1: Cost Reduction

ESCOM, the South African Electricity parastatal, has also experienced the long-term degeneration of its assets through poor governance and a short-term focus. Box 9.6, The Jewel in the Crown: The Rise and Fall of South Africa's Electricity Supply Commission, gives details.

SCORECARDS VERSUS DASHBOARDS

Various authorities argue that scorecards are different from dashboards. The view taken in this book is that a dashboard is an online summary of a few indicators and targets, and a scorecard is part of a more comprehensive report on performance taken at a certain moment in time. The audience in each case can be anything from Board level to System Owner.

Using scorecards and dashboards for reporting is discussed in Chapter 31, Reporting.

15.12 SUMMARY

By verifying change, PIs help to demonstrate progress when things go right and provide early warning signals when things go wrong.

The definitions used in this book are simple and generic. **KPIs** and **PIs** are interchangeable.

CSFs give focus in determining KPIs.

Leading indicators identify failings or holes in processes or inputs essential to maintain critical aspects of risk control systems and are thus useful in preempting an incident.

Alignment with the shareholders' aspirations is always the primary guide for selection of KPIs. Shareholder aspirations generally focus on Return on Investment (RoI). However, compliance with the law and maintenance of reputation are required to achieve RoI in the long term.

Targets are divided into short-term and long-term, with short-term focus on utilization of restrained resources (finance, human capital, and raw materials), and long-term focus on the integrity of the physical assets using leading indicators. The cascading of PIs is crucial to enable staff to take ownership of those indicators over which they have influence.

Presentation for decision-making, quantity of indicators at the top, and enablement (discussion and action) are some prime factors in the selection of indicators.

Criteria for selection include validity, practicality, having a precise meaning, indicating a clear direction, and ownership.

The audience for a common source of data sees the data from three perspectives: strategic, benchmarked, and operational.

Primary indicators relate to three areas of the company: business, operating, and integrity.

Closing a performance gap entails identifying the gap by benchmarking, applying best practices to close the gap, and remeasuring the gap, by benchmarking, to record closure.

Competitiveness is focused on looking from the outside as to how well the company is doing compared to the competition. Efficiency is focused on getting the maximum out of what is given to the operator.

Safety and integrity focus has changed toward leading indicators and instilling a safety culture among staff and contractors.

The dangers of just looking at numbers without context, as well as the effect on other aspects of the company, cannot be understated.

REFERENCES

1. *Developing process safety indicators. UK HSE guide 254*. HSE Books; 2006.
2. Organization for Economic Cooperation & Development (OCED). www.oecd.org.
3. Parmenter D. *Key performance indicators: developing, implementing & using winning*. KPIs John Wiley & Sons; 2007.
4. Kleiner A. *What are the measures that matter? A 10-year debate between two feuding gurus sheds some light on a vexing business question*. Organizations & People First Quarter 2002/Issue 26. January 9, 2002.

FURTHER READING

1. Parmenter D. *Pareto's 80/20 rule for corporate accountants*. John Wiley & Sons; 2007.
2. ANSI/API RP 754. Recommended practice. Process safety indicators for the refining and petrochemical industrics. April 2010.
3. *Process safety leading and lagging metrics*. CCPS; 2008.
4. Barr S. *Top 50 resources to improve your performance measures*. Performance measurement resource list. www.stacybarr.com.
5. *What is a key performance indicator (KPI)?* Balanced Scorecard Institute. https://balancedscorecard.org/Resources/Performance-Measures-KPIs.
6. Marr B. *Key performance indicators*. Pearson; 2012.
7. Huff D. *How to lie with statistics*. Penguin; 1993.
8. *Signposts of development: selecting key results indicators – in the context of the UNDP strategic results framework (SRF)*. UNDP; 13 May, 1999.

CHAPTER

KEY PERFORMANCE INDICATOR SELECTION GUIDELINES

16

16.1 INTRODUCTION

This chapter discusses Key Performance Indicator (KPI) selection guidelines for obtaining the most significant performance indicators: indicators that have a direct or indirect effect on the bottom line of the Business Unit.

16.2 ELEMENTS OF GOOD PERFORMANCE INDICATORS

It is essential to comply with the elements of good performance indicators to ensure appropriate selection.

Elements of good performance indicators are as follows:

1. specific and realistic
2. easy to calculate and monitor
3. understood and agreed upon by all interested parties
4. benchmarked (where possible)
5. quantitative preferred to qualitative
6. emphasis on leading indicators

The key is to measure the right things that stakeholders and staff find value in, and **not** try to measure everything.

1. SPECIFIC AND REALISTIC

The indicator must be easily identifiable with linkage to the corporate direction statement.

2. EASY TO CALCULATE AND MONITOR

Accuracy of data for calculating the indicator and the simplicity in calculating and monitoring the indicator reduces errors and questioning of the validity of the reported value.

3. UNDERSTOOD BY ALL INTERESTED PARTIES

The indicator has meaning if the interested parties are aware of the factors that affect the reported value.

Performance Management for the Oil, Gas, and Process Industries. http://dx.doi.org/10.1016/B978-0-12-810446-0.00016-5

4. BENCHMARKED

It is essential to have the same indicators as the rest of the particular industry to identify competitiveness and efficiency gaps.

5. QUANTITATIVE PREFERRED TO QUALITATIVE

Quantitative indicators are objective and based on facts, whereas qualitative indicators tend to be *subjective* unless clear standards for assessment are adhered to.

6. EMPHASIS ON LEADING INDICATORS

As a result of some major incidents in the Oil and Gas Industry, there is an increased focus on leading indicators, which might preempt an impending disaster.

Example of the use of leading indicators:

> An increase in the number of Emergency Work Orders may indicate a loss of control over the maintenance of a process unit and thus place the Hydrocarbon pressure containment at higher risk.

16.3 APPROACH TO SELECTION

Three approaches are required which need to be tied together:

1. top-down
2. industry comparisons (benchmarked)
3. bottom-up

1. TOP-DOWN

The Business Improvement Priorities are first determined.

These could include the following:

1. business control framework optimization
2. integrated business planning
3. operations and maintenance
4. project delivery
5. safety and integrity
6. human capital and capability development
7. raw materials availability

The Critical Success Factors (CSFs) or Key Result Areas (KRAs) of the company are then agreed. Typical Business Unit CSFs include the following:

1. market (demand)
2. supply
3. profit
4. Opex

5. Capex
6. Health, Safety, and Environment (HSE)/Integrity
7. Human Capital

CSFs/KRAs relate directly to Measurement Areas, which identify the requirement for KPIs. An example of the sequence is shown in Fig. 16.1.

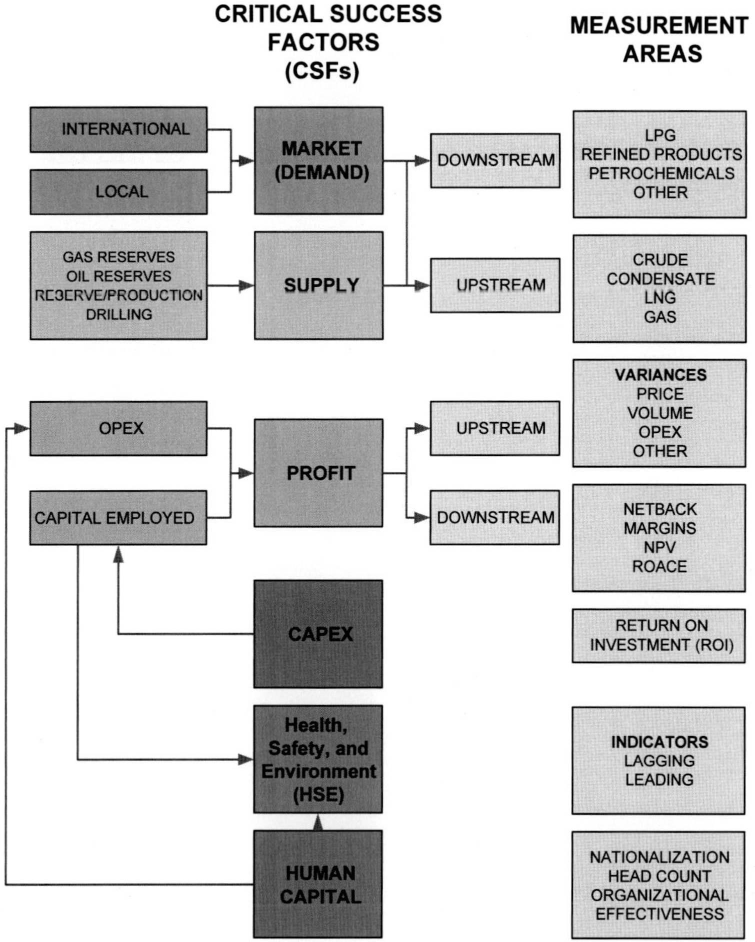

FIGURE 16.1

Critical Success Factors and measurement areas: Oil and Gas Industry.

The following is an example of CSF linkage to Objectives of a Refining Business Unit (see Box 10.4: Refinery Business Unit Objectives and Explanation and Box 10.6: Core Values, for details).

Example:

1. *market (demand):* ***highest possible production***
2. *supply: not covered, as this is a dedicated internal supply from the company's own oil wells*

3. *profit:* ***maximizing product value***
4. *Opex:* ***minimum conversion cost***
5. *Capex: not covered, as there was no major capital expenditure at the time*
6. *HSE/Integrity:* ***operate the plant safely and in a responsible manner***
7. *Human Capital: covered in company Core Values*

CSFs are not always required. CSFs were actually omitted from the development of the KPIs of the previously mentioned Refining Business Unit example since the identification of KPIs was based on Benchmarking values. Box 16.1 shows this example of the Business Unit Objectives linking directly to the Business Unit KPIs.

BOX 16.1 EXAMPLE OF LINKED OBJECTIVES, KEY PERFORMANCE INDICATORS, AND STRATEGIES

Refining Directorate

Integrated Management System: Objectives and Targets

No.	Objectives	Key Performance Indicators (KPIs)	KPI's Sponsor	2004 Measurements	2004 Best GCC Performer	Targets	Strategies
1.	Policy Mission Statement: add value	Net Cash Margin (NCM)		+$1.66/bbl	$5.72/bbl	>$3.60/bbl	All strategies below
2.	Improve the Net Cash Margin	Return on Investment (ROI)		+4.6%	19.6%	>15%	All strategies below
3.	Improve the Net Cash Margin	Cash Operating Expenses		30 US cents/ UEDC	18 US cents/ UEDC	<23 US cents/ UEDC	• SAP utilization with corrected cost centers • Quarterly Business Review • Energy savings
4.	Improve the Net Cash Margin	Personnel Index		223.4 wh/100 Equivalent Distillation Capacity (EDC)	79.9 wh/100 EDC	<150 wh/ 100 EDC	• Staff incentives • Minimum OT and absenteeism
5.	Policy objective: highest possible production	Utilization		75.9%	92.7%	>85%	• Integrated shutdown strategy • Downstream unit debottlenecking • Max mechanical availability
6.	Policy objective: highest possible production and maximum product value	Volumetric Expansion Index		+40	+98	>+69	• Volumetric measurement accuracy and materials accounting • Fuel consumption and losses quantification • Process Improvement Projects (PIPs)

GCC, Gulf Cooperative Council; *UEDC*, utilized equivalent distillation capacity (Solomon term).

Comment The previous data is a sample from a set of 12 KPIs giving actuals from the previous year, the best performer in the region, and the target for the next round of benchmarking. The agreed top strategies for achieving the targets are also listed. The complete document was approved by every KPI sponsor.

2. INDUSTRY COMPARISONS

These are based on generic measurement areas that are relevant across the industry. These could be grouped as follows:

1. **Financial**: Profit, Opex, Maintenance Expense, Capex
2. **Input–Output Balance**: Mass and Energy
3. **Operational**: Utilization, Mechanical Availability
4. **Health, Safety, Environment, and Asset Integrity**
5. **Stakeholder**: Human capital, customer, and other

These are depicted as shown in Fig. 16.2.

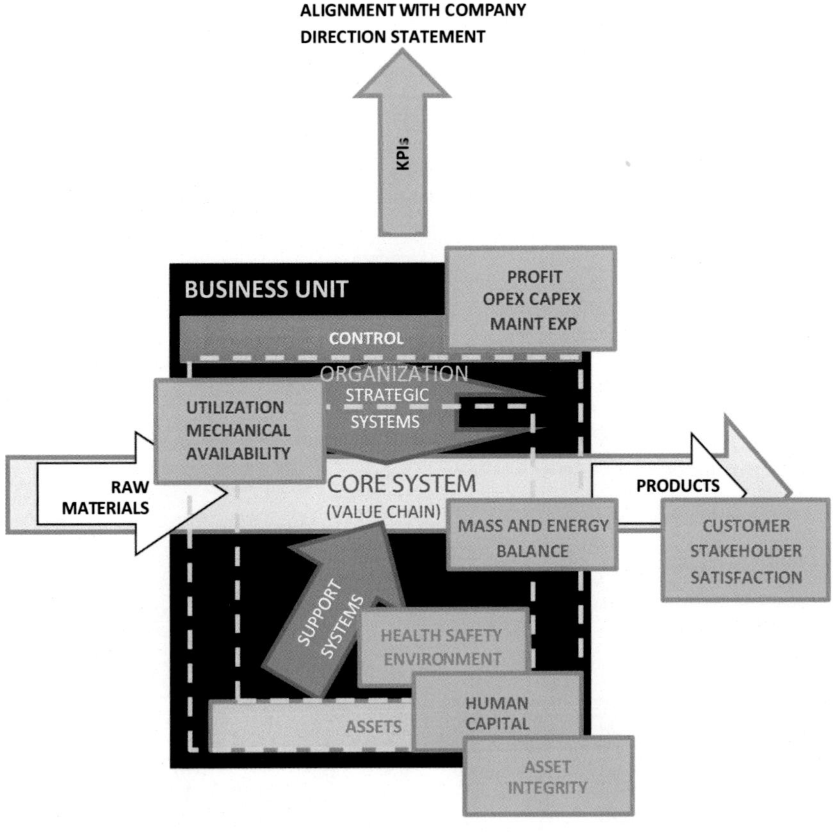

FIGURE 16.2

Generic industry measurement areas.

3. BOTTOM-UP

Specific System Indicators need to be developed relating to:

1. Core Systems: The Value Chains of the business
2. Support Systems: Systems supporting the previous

Indicators need to be defined in the user's terms and then linked to higher-level indicators. Box 16.2 discusses the use of terms that users can understand.

BOX 16.2 TERMINOLOGY: REFINERY UTILIZATION CASCADE

An Integrated Management System was developed for a refinery. Top KPIs were defined according to the terminology used by the benchmarking consultants.

Those staff members who were required to report on some of these KPIs were not always able to fully understand how they related to what they were doing and so were unable to effect improvement. To make it clear, cascade diagrams (organograms) were created on the intranet to show the linkage.

The following webpage shows each process unit's target utilization. A trend graph displays when a unit is selected.

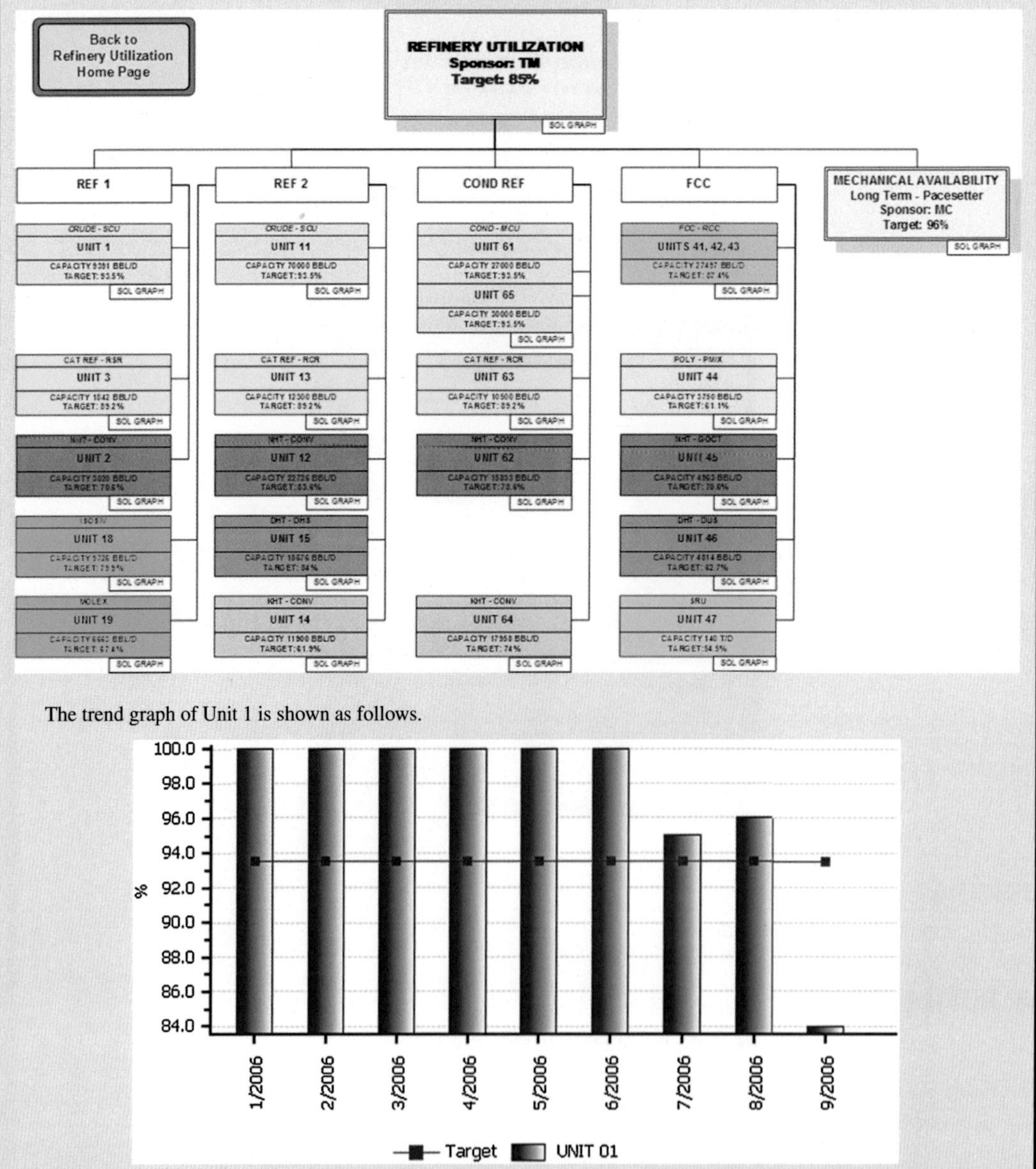

The trend graph of Unit 1 is shown as follows.

BOX 16.2 TERMINOLOGY: REFINERY UTILIZATION CASCADE—cont'd

Operators understand the Utilization of their process units, and the Performance Model calculates the weighted Refinery Utilization from inputs from each process unit. The cross-link to Mechanical Availability is also shown. Refinery Mechanical availability is calculated in a similar way (see Box 16.3).

Box 16.3 shows how various indicators within the Maintenance Management System have a direct impact on Mechanical Availability.

BOX 16.3 REFINERY MECHANICAL AVAILABILITY RELATIONSHIPS

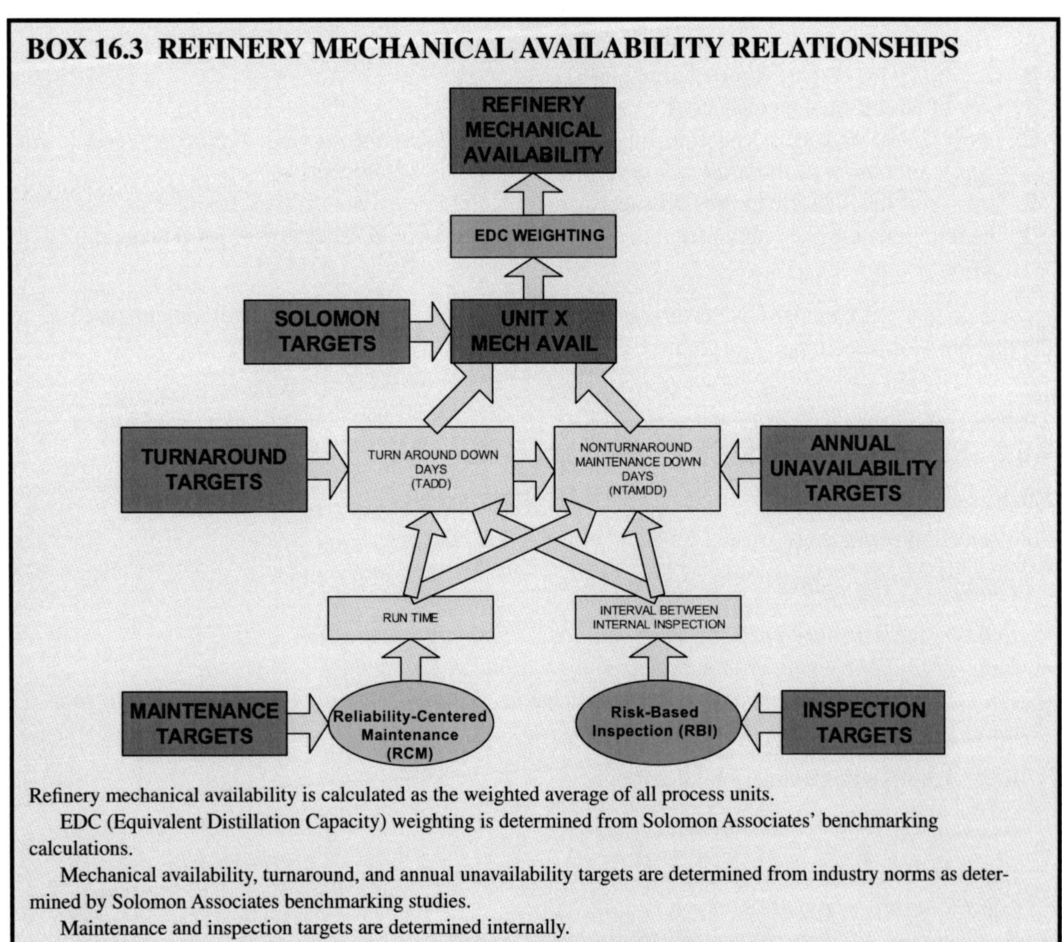

Refinery mechanical availability is calculated as the weighted average of all process units.

EDC (Equivalent Distillation Capacity) weighting is determined from Solomon Associates' benchmarking calculations.

Mechanical availability, turnaround, and annual unavailability targets are determined from industry norms as determined by Solomon Associates benchmarking studies.

Maintenance and inspection targets are determined internally.

16.4 DEVELOPMENT PROCESS

The development of KPIs normally takes place when a team of specialists meets to brainstorm possible alternatives. Prior to the meeting, research has to be carried out as to the likely KPIs for adoption. Most KPIs in the process industry are already well-established, and it only requires a decision on those few indicators that have the most influence on the performance of a System or Business Unit.

The team development process could entail the creation of a spreadsheet with the following columns:

1. description of the performance measure
2. explanation as to how the performance measure is calculated
3. person responsible for obtaining measurement
4. system where data is sourced from or to be gathered
5. refinements that may be required to produce the information
6. recommended display (type of graph, etc.)
7. how often it should be measured
8. likely "cause and effect" relationship (*for example: if unplanned down days are reduced, it will lead to increased availability and reduced costs*)
9. linkage of the measure to the CSFs or Business Unit Objectives (see Box 16.1, for an example)
10. the required delegated authority that staff will need to have in order to take immediate remedial action

A National Oil Company went through the previously mentioned team development process, and the outcome for selection of Corporate KPIs for is shown in Box 16.4.

16.5 CONSEQUENCES OF POOR SELECTION AND DEFINITION

POOR SELECTION

Poor selection could cause wasted time and effort.

Example 1: Throughput

- ***Refinery total throughput*** *not taking into account the percentage slop*
- *Refinery total throughput not maximizing high value added product*
- *Poor adherence to specification limits resulting in "giveaway" of high value components in a lower value product (e.g., excess Ethane in LNG)*

Box 16.5 discusses Refinery total throughput.

Example 2: Actual versus planned

- *Planned value not set at* ***maximum sustainable throughput***
- *Gas Plant throughput not seasonally adjusted!*
  - *Condensate to Gas ratio varies from summer to winter*

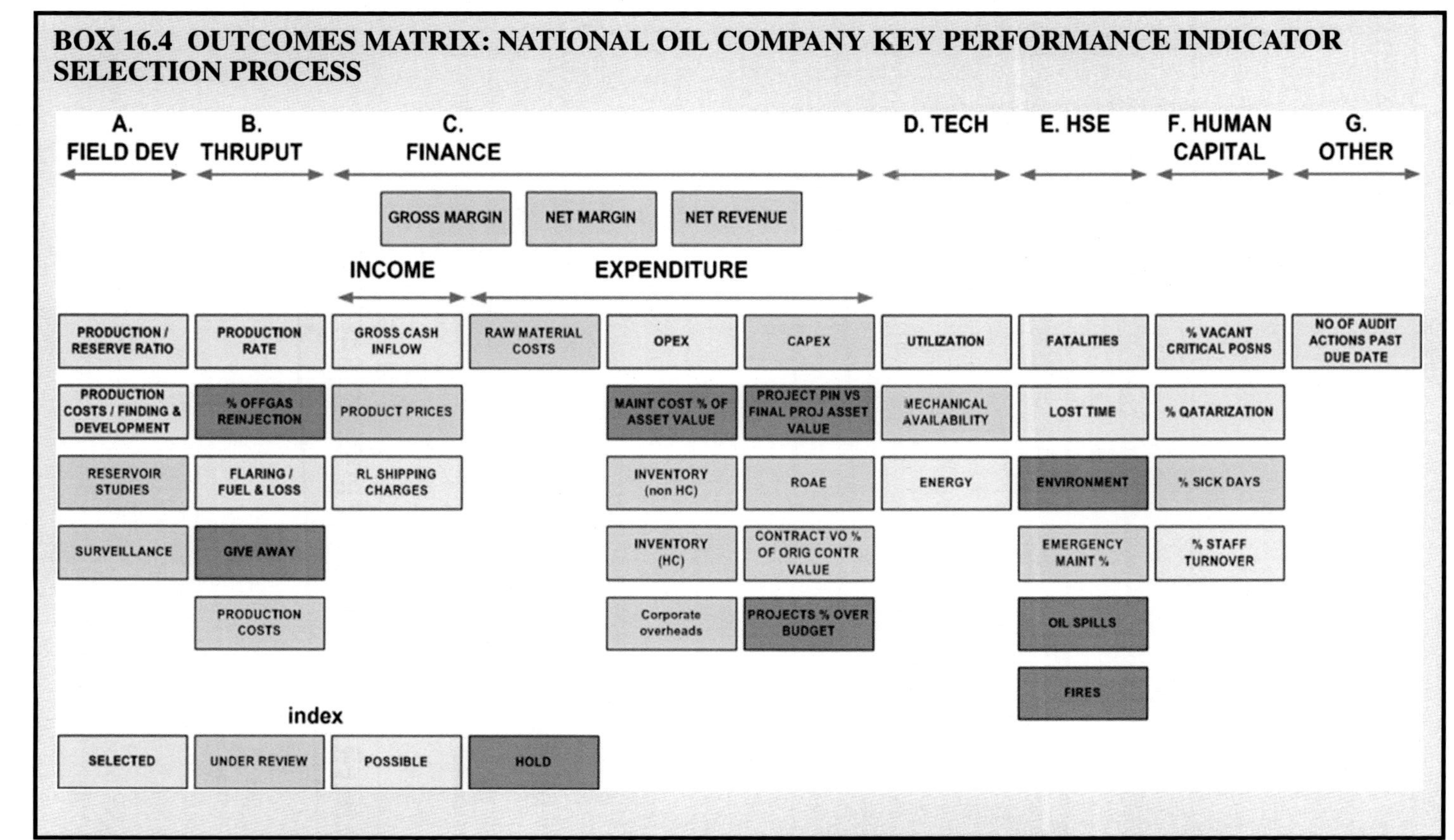

BOX 16.4 OUTCOMES MATRIX: NATIONAL OIL COMPANY KEY PERFORMANCE INDICATOR SELECTION PROCESS

BOX 16.5 CONSEQUENCES OF POOR SELECTION AND DEFINITION

Crude Throughput as a Key Performance Indicator

The Operations Manager of a National Oil Refinery wanted Crude throughput as a primary indicator. However, slops, which were fed back to the crude tanks, were an astonishing 10%. Slops were meant to be spread between three tanks, but instead were fed to just one tank. This threw out the crude assay for production simulation and product split.

It was decided that slops and "giveaway" (or yield) were better indicators, and, as a result, slops came down to less than 2%.

Box 16.6 discusses maximum sustainable throughput in a Gas Plant.

BOX 16.6 CONSEQUENCES OF POOR SELECTION AND DEFINITION

Maximum Sustainable Throughput

A gas plant operator would use the design capacity as the basis for utilization calculations. The operator would always achieve 100%, as the plant was actually capable of running at 110%. If there was a major upset in a month, it was easy to make up production the next month.

The shareholder then stipulated a maximum draw off from the wells per day and requested utilization to be based on maximum sustainable capacity based on extensive operating trials.

The operator utilization can now be compared with industry indicators showing the exact categories of losses in utilization for analysis and corrective action where required.

Example 3: Turnarounds

- *Actual versus planned time* ***without reference to industry norms***
- *Actual versus planned cost* ***without reference to industry norms***

POOR DEFINITION

Poor definition could give conflicting results between reporting periods as well as against benchmarked competitors. Small green "apples" would possibly be compared to juicy ripe "apples."

16.6 CENTER FOR CHEMICAL PROCESS SAFETY[2] FOCUS

Center for Chemical Process Safety (CCPS) focuses on **leading** metrics grouped as follows:

A. Maintenance of Mechanical Integrity
B. Action items follow-up
C. Management of Change (MOC)
D. Process Safety training and competency (and training competency assessment)

CCPS's "Process Safety Leading and Lagging Metrics" gives an excellent set of metrics for application in the above categories. These are outlined as follows:

MAINTENANCE OF MECHANICAL INTEGRITY

1. (Number of inspections of safety critical items of plant and equipment due during the measurement period and completed on time/Total number of inspections of safety critical items of plant and equipment due during the measurement period) × 100%
2. (Length of time plant is in production with items of safety critical plant or equipment in a failed state, as identified by inspection or as a result of breakdown/Length of time plant is in production) × 100%

ACTION ITEMS FOLLOW-UP

3. (Number of process safety action items past due date and/or having approved extension/Total number of active or open action items) × 100%

MANAGEMENT OF CHANGE

4. Percentage of audited MOCs that satisfied all aspects of the site's MOC procedure
5. Percentage of audited changes that used the site's MOC procedure prior to making the change
6. Percentage of startups following plant changes where no safety problems related to the changes were encountered during recommissioning or startup

PROCESS SAFETY TRAINING AND COMPETENCY

7. Training for Process Safety Management Critical Positions
8. Training Competency Assessment
9. Failure to follow procedures/safe working practices

Aspects of the previously mentioned categories are discussed in the following section.

16.7 PHYSICAL ASSET INTEGRITY KEY PERFORMANCE INDICATORS: AN INTRODUCTION

Quote (UK HSE[1]):

> 60% of major hazard loss of containment incidents are related to technical integrity and, of those, 50% have aging as a contributory factor.

KPIs are used to evaluate asset integrity performance against stated goals. Because major loss-of-integrity events are relatively rare, it is important to record and monitor even minor incidents.

The method of setting indicators requires the following questions to be answered:

1. What can go wrong?
2. What controls are in place to prevent major incidents?

3. What does each control deliver in terms of a "safety outcome"?
4. How do we know they continue to operate as intended?

KPIs selected should:

1. align with the risk management processes for the Business Unit
2. be used to aid the management of the five steps of asset integrity risk management (as per ISO 31001) as follows:
   a. establishing the context
   b. communication and consulting
   c. risk assessment
   d. risk treatment
   e. monitoring and review
3. cover all three aspects of incident prevention as follows:
   a. Plant: the Physical Assets and Technology
   b. Processes and Systems
   c. People: Organizational Effectiveness

For each risk control system:

1. The **leading indicator** identifies failings or "holes" in vital aspects of the risk control system discovered **during routine checks** on the operation of a critical activity within the risk control system.
2. The **lagging indicator** reveals failings or "holes" in that barrier discovered **following an incident or adverse event**. The incident does not necessarily have to result in injury or environmental damage and can be a near miss, precursor event, or undesired outcome attributable to a failing in the risk control system.

The method for determining suitable KPIs is to include for the following:

1. Immediate causes of significant releases to be identified.
2. Various risk control systems are identified for each hazard; typically each control system will contribute to risk reduction for more than one type of incident scenario.
3. Each risk control system is analyzed to define suitable site-specific **lagging and leading** KPIs.
4. The **minimum required risk control systems** are as follows:
   a. maintenance
   b. instruments and alarms
   c. inspection
   d. plant change
   e. permit to work
   f. emergency arrangements
   g. operating procedures
   h. staff competence
   i. plant design

Discussion on the **minimum required risk control systems** is referenced as follows:

a. Maintenance
   i. This is based on Reliability-Centered Maintenance as discussed in Section 22.7, Maintenance, and Section 22.8, Inspection and would be embedded in a Maintenance Management System. An example format is outlined in Section 4.9, System Examples.
b. Instruments and alarms
   i. This is based on Safety Instrument Systems, including Emergency Shutdown Systems and Fire and Gas Systems, as discussed in Section 22.7, Maintenance, and would be embedded in a **Maintenance Management System**. An example format is outlined in Section 4.9, System Examples.
c. Inspection
   i. This is based on Risk-Based Inspection as discussed in Section 22.8, Inspection, and would be embedded in an **Asset Integrity Management System**. An example format is outlined in Section 4.9, System Examples. This could also entail certification to ISO 55001, "Asset Management."
d. Plant changc
   i. This would be embedded in a **Safety Management System**. An example format is outlined in Section 4.9, System Examples.
e. Permit to work
   i. This would be embedded in a **Safety Management System**. An example format is outlined in Section 4.9, System Examples. It would link to Operating procedures and this could also entail certification to ISO 45001, "Occupational Health and Safety Management."
f. Emergency arrangements
   i. These are discussed in Chapter 14, Incident Preparedness and Operational Continuity Management. These would be as required by ISO 22399, "Guideline for Incident Preparedness and Operational Continuity Management."
g. Operating procedures
   i. This set of procedures support Processes as part of the **Core System** of the Business Unit as discussed in Chapter 4, System Requirements.
h. Staff competence
   i. This is discussed in Chapter 20, Human Capital.
i. Plant design
   i. This is based on Company, Industry, and International Standards. Aspects are discussed in Chapter 33, The Opportunity Life Cycle.

Section 20.5, Process Safety Management and Safety Culture, discusses the previous encapsulated in a Safety Management Framework.

Chapter 22, Physical Asset Performance Management, discusses the previously mentioned subject further and gives specific examples.

Examples of lagging and leading indicators for the previously mentioned control systems are given in Table 16.1.

Table 16.1 Risk Control Systems and Example Key Performance Indicators (KPIs)

Control System	Lagging Indicator	Leading Indicator
1. Maintenance (including onstream and turnaround, pressure relief systems, fire protection systems, etc.)	Number of loss-of-containment incidents/downtime (leaking exchanger, etc.)	% Safety critical plant/equipment that performs to specification when inspected or tested % Pressure containment items outstanding after scheduled inspection date % Pressure containment items excluded from risk-based inspection management system
2. Instrumentation and alarms [including Emergency Shutdown Systems (ESD), Fire and Gas (F&G), distributed control systems]	Number of incidents linked to failure of instrumentation or alarms	% Function tests of alarms/trips completed on schedule
3. Inspection (including onstream and turnaround, pressure relief systems, fire protection systems, etc.)	Number of loss-of-containment incidents/downtime (leaking exchanger, etc.)	% Maintenance work orders (priority A: emergency), which are completed to specified timescale
4. Plant change management (Management of Change; MOC)	Number of plant changes undertaken without complete MOC process	% Piping and instrument diagrams not matching actual plant configuration Average time taken to fully implement a change once approved
5. Permit to Work (PTW)	Number of incidents where errors in PTW process are identified as a contributory cause	% PTWs sampled where all hazards were identified and all suitable controls were specified % PTWs sampled where all controls listed were fully in place at worksite
6. Emergency arrangements	Number of emergency response elements that are NOT fully functional when activated in a real emergency	% ESD valves and process trips tested, as per schedule, defined in a relevant standard or facility safety case
7. Operational procedures	Number of operational errors due to incorrect/unclear procedures	% Operational procedures under review or development
8. Staff competence	% of weld failures	% Qualified first aiders
9. Plant design	Number of incidents where errors in plant design are identified as a contributory cause	Number of poststartup modifications required by operations Number of deviations from applicable codes and standards % Safety critical equipment/systems fully in compliance with current design codes

16.8 EXAMPLES OF HIGH-LEVEL KEY PERFORMANCE INDICATORS

Refineries and Gas Plants have been selected to give examples of high level KPIs. One of these could well be adapted to suit other process industries.

REFINERY

Refinery top KPIs are listed in Section 37.5, Refinery KPIs.

The example in Fig. 16.3 shows the relationships between KPIs as follows:

1. **Strategic**: indicated by **Business Plan KPIs**
2. **Benchmarking**: showing the linkage between strategic and operational KPIs, but defined differently
3. **Operational**: allocated to **departments** to give ownership and identity to a KPI that can be addressed at operational level

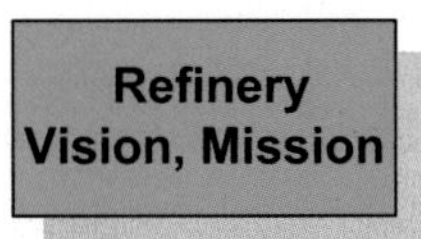

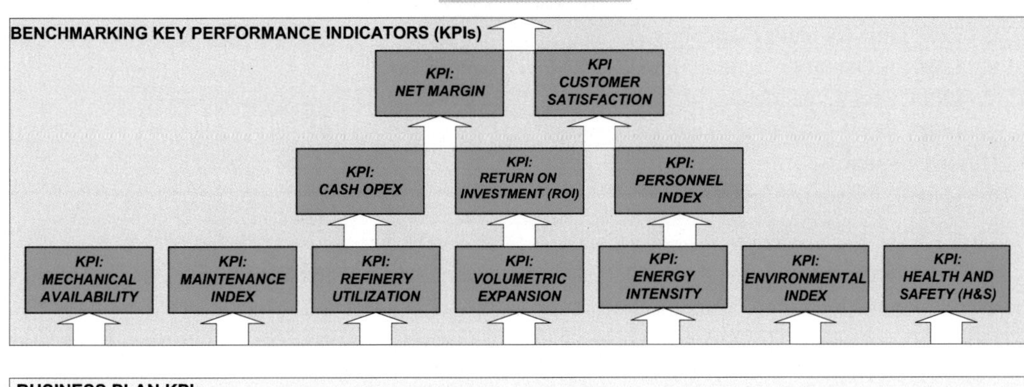

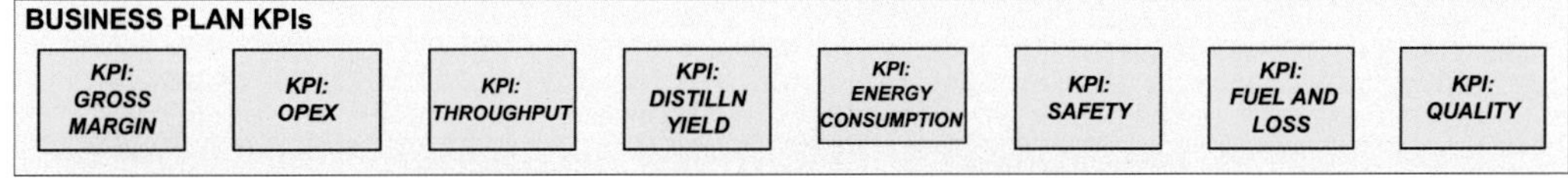

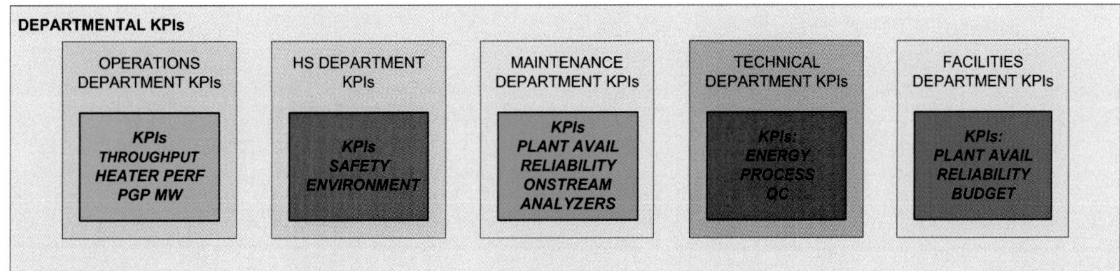

FIGURE 16.3

Refinery Key Performance Indicator categorization example. *LPG*, liquefied petroleum gas; *NPV*, net present value; *PGP*, power generation package.

LIQUEFIED NATURAL GAS PLANT KEY PERFORMANCE INDICATORS

Besides the top financial indicators of net profit after tax, return on capital employed, etc., the following are important:

1. Energy
   a. Energy use on intake: % woi
   b. Fuel and Loss: % woi

   c. Greenhouse Gas (GHG): % LNG
   d. GHG: % sales/flowing gas
   e. Flaring: % woi
   f. Total LNG loaded
   g. Total sales/flowing gas
2. Financial
   a. Fixed cost per ton of salable product
   b. Operating cost per ton of salable product
3. Physical Asset
   a. Reliability
   b. Availability
   c. Fatal Accident Rate
   d. Lost Time Incident Frequency
   e. Loss Of Primary Containment (LOPC) level 1 incidents
   f. LOPC level 2 incidents
   g. LOPC level 3 selected leading indicators
4. Human Capital
   a. Overall Personnel Index
   b. Employee turnover

Note the focus on energy, as a gas plant has to flare what it cannot consume or sell. Also be aware that there are no benchmarking consultant proprietary common denominators or indices except possibly the overall personnel index. This index can be deceptive as discussed in Box 26.1, Exclusions: Body Count.

Fig. 16.4 depicts the primary interests of an LNG plant shareholder.

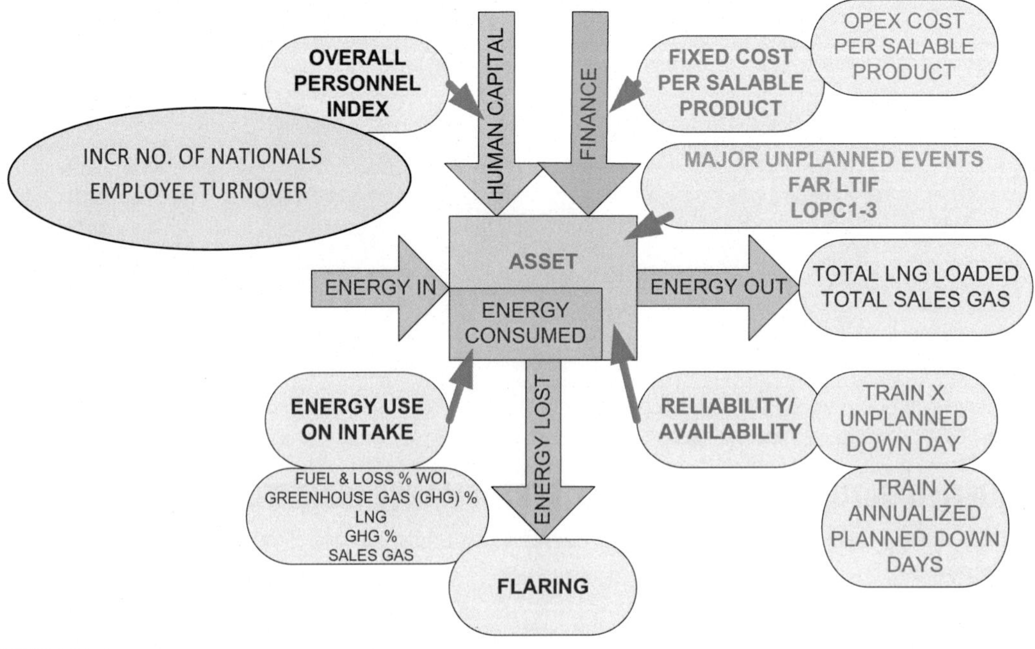

FIGURE 16.4

Gas plant Key Performance Indicators. *LTIF*, lost time injury frequency.

EXPLORATION

For Oil and Gas Companies undertaking exploration, the Reservoir/Production (R/P) ratio is the most important indicator for the future of the company.

> For example, an R/P ratio of 10 indicates a potential production at the current rate for the next 10 years.

Box 7.9, Leaders in Corporate Governance, discusses the importance of the R/P ratio.

16.9 SUMMARY

The elements of good performance indicators are: specific and realistic, easy to calculate and monitor, understood and agreed by all, benchmarked, quantitative, and with the emphasis on leading indicators.

Three approaches are adopted and consolidated: top-down, bottom-up, and benchmarked.

The awareness of poor KPI selection, and the resulting consequences, helps to define the KPIs more clearly.

Each KPI should, where possible, have a single primary owner who has the most influence over the performance of the KPI and who agrees on targets and strategies to achieve these targets.

Asset integrity KPIs determine the availability of the physical assets to produce at design capacity in the long term. The focus needs to be on **leading** indicators.

Asset integrity KPIs that are selected should align with the risk management processes for the Business Unit, be used to aid the management of the five steps of asset integrity risk management (as per ISO 31001), and cover all three aspects of incident prevention: Plant (Physical Assets and Technology), Processes and Systems, and People (Organizational Effectiveness).

A refinery example shows the KPIs from three perspectives: top (business plan), benchmarking, and operations (departmental).

The LNG shareholder perspective differs slightly from the refinery perspective, where the primary focus is on maximum utilization of gas from the wellhead, GHG emissions, and primary containment of the pressure envelope.

REFERENCES

1. *Plant ageing: management of equipment containing hazardous fluids or pressure: HSE Research Report RR509*. HSE Books.; 2006.
2. *Process safety leading and lagging metrics*. CCPS; 2008.

FURTHER READING

1. *Human factors performance indicators for the energy and related process industries*. Energy Institute; December 2010.
2. *Global maintenance and reliability indicators harmonised indicators*. 2nd ed. EFNMS Maintenance Benchmarking Committee SMRP Best Practice Committee; October 2009.
3. *Safety performance indicators: OGP Report No 2011s*. May 2012.

CHAPTER

RELATIONSHIPS

17

17.1 INTRODUCTION

This chapter discusses relationships between indicators. The links between indicators could be either hard or soft.

Different approaches are applied by the two primary benchmarking consultants in the petrochemical industry: Solomon Associates (SA)[1] and Philip Townsend Associates International (PTAI).[2] The reasons for selection of these consultants are discussed in Chapter 24, Benchmarking Process and Consultant Selection.

17.2 SIZE AND COMPLEXITY

This is required to compare small with large, simple with complex, and single units with multiple units. Chapter 25, Common Denominators and Indices Used in Benchmarking, covers this in detail.

17.3 MASS, VOLUME, AND HEATING VALUE

Refineries and gas plants are discussed in this section. However, some of these aspects could also be applied elsewhere in the process industry.

REFINERIES

Volume and Mass

Intake

The "crude slate," identifying components, viscosity, contaminants, etc., is used as input to a linear programming model to determine optimum production based on market demand and/or profit within the restraints of the design capability of the particular refinery.

Products

Products are normally measured for sale in:

1. barrels for transport fuels (at ambient temperature); and
2. tons for asphalt, bunker fuels, petrochemical naphtha, solids (at ambient temperature), and liquefied gases: Propane and Butane.

Plant balance must ensure mass loss and volume increase, as shown in Fig. 17.1.

Performance Management for the Oil, Gas, and Process Industries. http://dx.doi.org/10.1016/B978-0-12-810446-0.00017-7

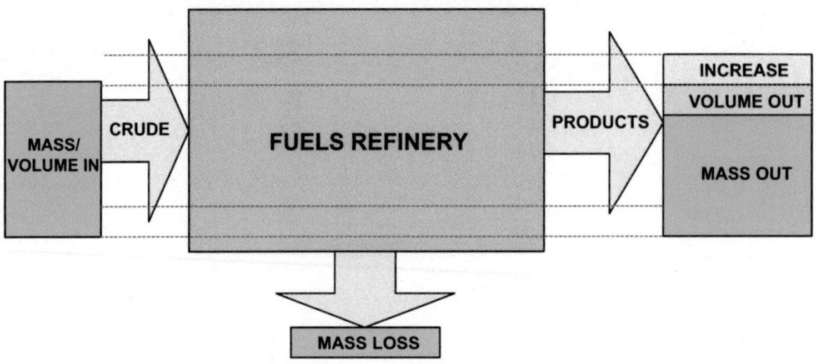

FIGURE 17.1

Refinery Mass and Volume Balance.

Using the **Profile II Model** described in Chapter 3, Models, the mass balance is cross-checked against the energy balance, thus giving a quick check on any anomalies. Solomon Volumetric Expansion Index (VEI) and Energy Intensity Index (EII) are required to be in specific ranges, respectively. Box 17.1 gives an example.

> **BOX 17.1 MASS, VOLUME, AND HEATING VALUE: REFINERY**
>
> **Profile II: Check on Mass Balance**
>
> During monthly analysis and reporting using the Profile II, it was found that the Volumetric Expansion Index (VEI) was too low. In other words, there was both a mass loss **and** a volume loss, which is not possible for a fuels refinery, so there had to be a miscalculation in mass and volume balance.
>
> Further investigation found that there had been a plant upset, at which time the slops flow was out of range of the flow meter, resulting in a low reading for an abnormally high flow of slops.
>
> **Comment**
>
> Cross-checks using different tools are useful to ensure a correct mass and energy balance.

VEI and EII are defined in Section 25.5, Proprietary Indices.

GAS PLANTS

Volume, Heating Value, and Mass

Intake

Intake is normally measured in Volume (MMSCF) and % hydrocarbon component composition (to determine heating value in BTUs).

Products

Products are normally measured for sale in:

1. barrels for liquids (at ambient temperature);
2. tons for solids (at ambient temperature), liquefied gases: LNG, Ethane, Propane, Butane; and
3. BTUs for sales/flowing gas.

Plant Balancing ensures mass and components are measured to identify specific yields and losses. Fig. 17.2 shows the process.

Mass and Volume balance is discussed in detail in Chapter 36, Hydrocarbon Accounting.

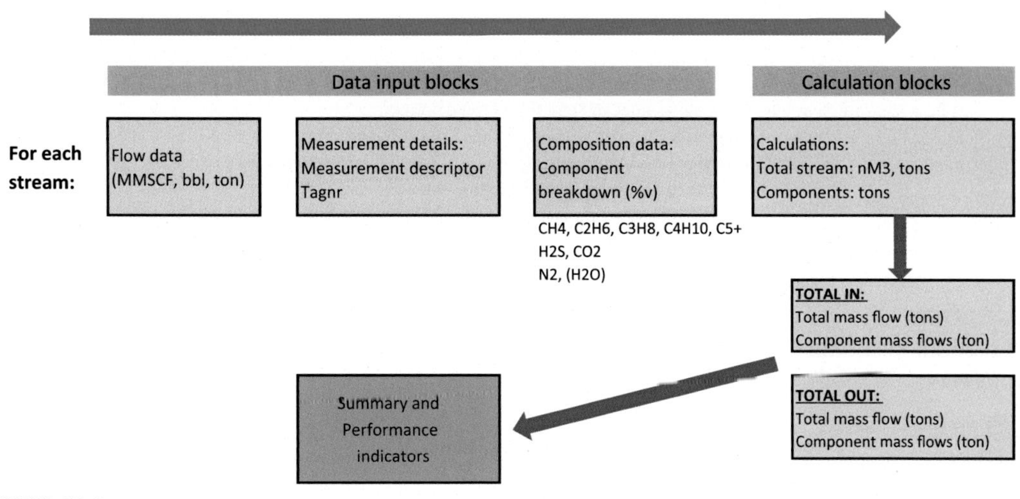

FIGURE 17.2

Gas Plant Mass Balance Process.

17.4 UTILIZATION AND AVAILABILITY

The ability of the plant to achieve **maximum sustainable capacity** relates to two key indicators: Utilization and Availability.

UTILIZATION

Utilization is the answer to the question:

"What is the actual throughput against design or maximum sustainable throughput?"

This is measured as a percentage of maximum sustainable throughput/capacity.

Maximum sustainable capacity can be determined as follows:

1. design capacity;
2. Engineer Procure and Construct (EPC) project commissioning tests; and
3. maximum sustainable production rate based on a time frame of at least 30 days without damage or premature aging of catalyst or equipment.

Refineries

Maximum Sustainable Capacity and actual throughput is based on feed flow for all refining units (barrels per day or tons per annum) except the Sulfur Recover Unit, where throughput is based on produced Sulfur (tons per day).

Box 17.2 discusses the correction of Maximum Sustainable Capacity for a refinery unit.

> **BOX 17.2 UNIT PERFORMANCE EXCEEDING 100% FOR AN EXTENDED TIME**
>
> During the monthly performance analysis in a refinery, it was found that a particular process unit exceeded 100% utilization based on design capacity.
>
> All flow measurement into the unit was checked and found to be correct.
>
> Readings for the unit were found to be at 110% of design capacity for a period in excess of 30 days. The unit maximum capacity was then amended so that maximum utilization was 100%. This is in line with the process for determining maximum capacity of a unit, according to Solomon Benchmarking practices.

Gas plants

Maximum Sustainable Capacity and actual throughput are based on production of primary products as follows:

A. LNG:
- LNG rundown capacity/production in Million Tons Per Annum

B. flowing/sales gas:
- flowing gas capacity/production in Billion Cubic Feet Per Annum

Petrochemical Plants

Maximum Sustainable Capacity and actual throughput are based on production of tons of primary products.

Power Plants

This is based on the design capacity of the turbo generators. At times this is exceeded, but sometimes never achieved. Box 17.3 gives an example of a steam turbine never achieving the required capacity.

> **BOX 17.3 LOST CAPACITY NEVER RECOVERED: ATHLONE POWER STATION**
>
> In the early 1960s, the City of Cape Town built a 180-MW coal-fired power station. The boiler contracts were given to two vendors: four boilers to each vendor, John Thompson and John Brown. Oerlikon of Switzerland was awarded five of the six 30-MW steam turbines, and one 30-MW steam turbine was awarded to Hitachi of Japan. All contracts had performance penalties. The boilers and Oerlikon turbines achieved their agreed capacities, but the Hitachi steam turbine could not and the performance penalty was consequently imposed on Hitachi.
>
> The power station was decommissioned in the 1990s. Over the life of the power station, the production losses from the Hitachi apparently far outweighed the gain from the performance penalty. However, the design of the Hitachi was simpler, and it was therefore considered to be more reliable than the Oerlikon turbines.

Comment

It would have been interesting to compare the records over the life of the power station to see which turbines were better with respect to life cycle costing, i.e., the total cost of investment and operation and maintenance versus revenue generated.

Athlone Power Station in its Heyday

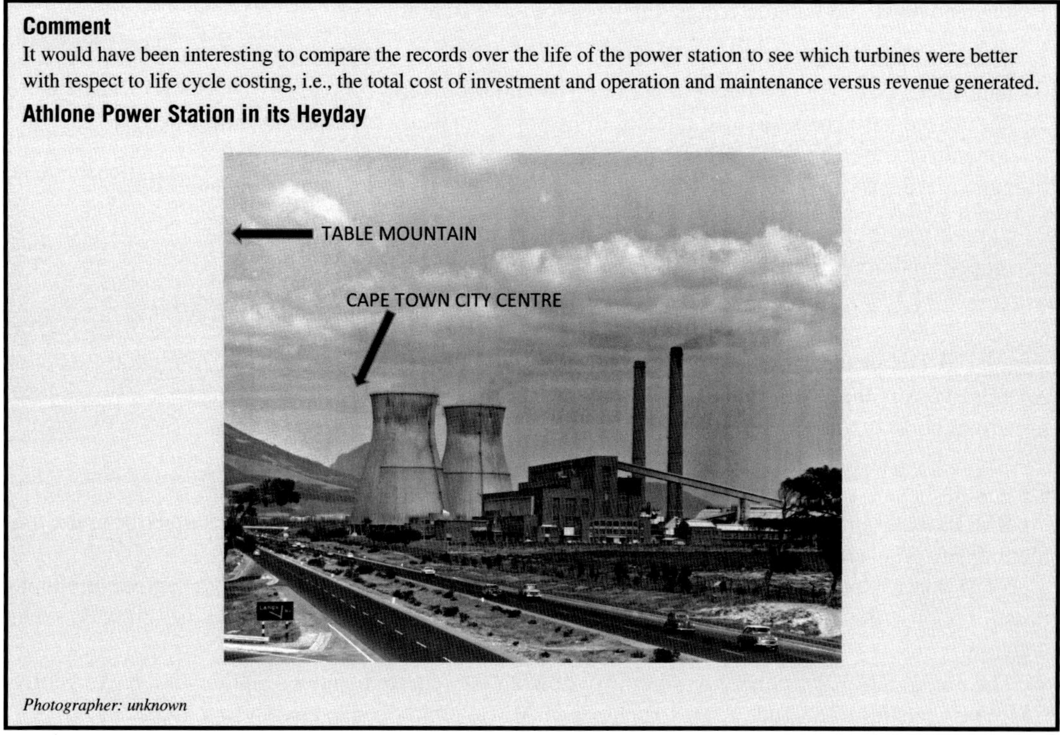

Photographer: unknown

AVAILABILITY

Availability is determined by the following:

1. unplanned maintenance
2. planned maintenance
3. consequential downtime (upstream or downstream restrictions)
4. economic or contractual (for example, a feed gas contractual ceiling)
5. slowdowns (normally less than 75% for more than a day)
6. other shutdown time

Availability is primarily dictated by maintenance downtime in days. This is split into unplanned and planned. This normally referred to as **Mechanical Availability (MA)**.

Unplanned down days for maintenance: These are, as the name indicates, unpredictable. These days can also be represented as **Reliability** when calculated as (total days in a year – unplanned maintenance days) *100/total days in a year.

Planned down days for maintenance: This is referred to as a Shutdown or **Turnaround,** which is usually planned from more than a year in advance to ensure a run length in line with industry norms with minimal unplanned down days.

Planned down days are determined in a number of different ways:

1. actual;
2. annualized based on last actual interval (PTAI and Solomon);
3. annualized based on estimated interval if the actual intervals have been erratic;
4. annualized based on fixed interval, such as 6 years; and
5. annualized based on future intervals as published in the annual Business Plan and Budget.

Simple MA is calculated from total lost maintenance days for a particular year.

As per Solomon, Annualized **MA** is generally calculated as follows:

$$MA = (\text{days in current year} - TADD - RMDD) \times 100/\text{days in current year}$$

where TADD = annualized down days taken from the last reported turnaround. Annualized days are calculated by dividing the reported days by the number of equivalent years between the last two reported turnarounds.

$$TADD = \text{days down} \times 365.25/\text{interval between turn arounds in days.}$$

RMDD = the annualized down days for unplanned down days that are not accounted for in the turnaround data. The actual days for the last 2 years are averaged for a year.

PTAI doesn't average the unplanned down days for the last 2 years. It takes the actual unplanned down days for the study year (PTAI benchmarks every year where Solomon benchmarks every 2 years).

The partial build of Refinery Availability showing the required planned shutdown days for Fluidic Catalytic Cracking and Crude Units to achieve 96% availability is depicted in Box 17.4.

BOX 17.4 REFINERY AVAILABILITY

Based on Benchmarking norms, the following was used to illustrate what a particular refinery needed to achieve.

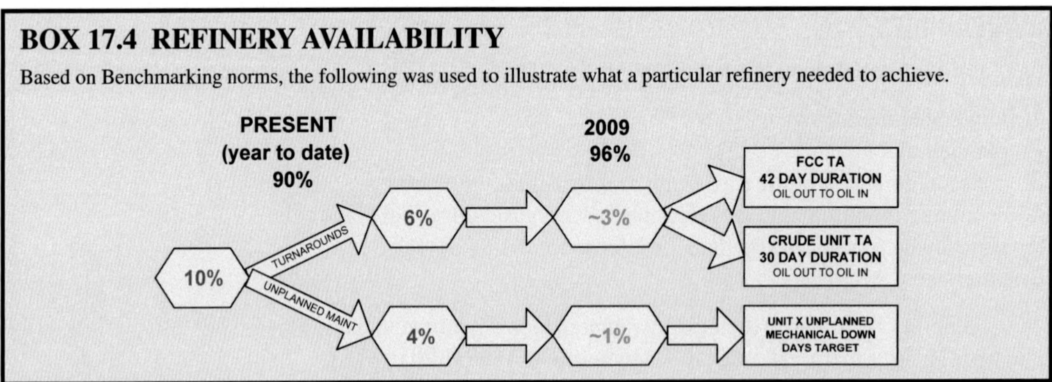

The importance of availability has a direct impact on revenues, as shown in Fig. 17.3. This is an example of "effectiveness."

Section 28.6, Performance Focus, discusses the previous in relation to competitiveness and efficiency.

The relationship between Utilization and Availability is commonly depicted as in Fig. 17.4.

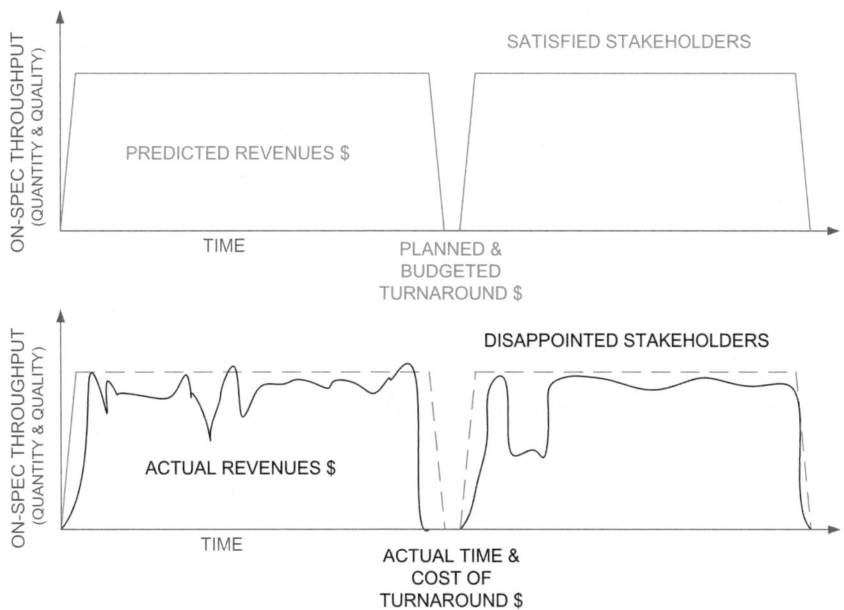

FIGURE 17.3

Availability comparison.

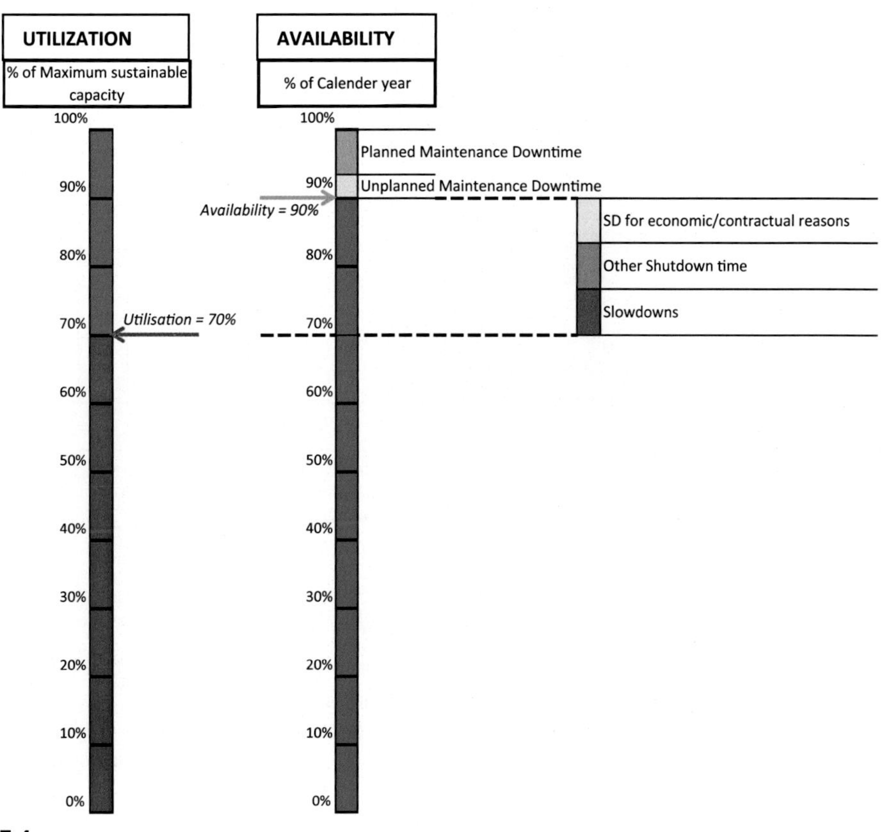

FIGURE 17.4

Relationship between Utilization and Availability.

However, PTAI defines both Utilization and Availability in capacity terms, as shown in Fig. 17.5.

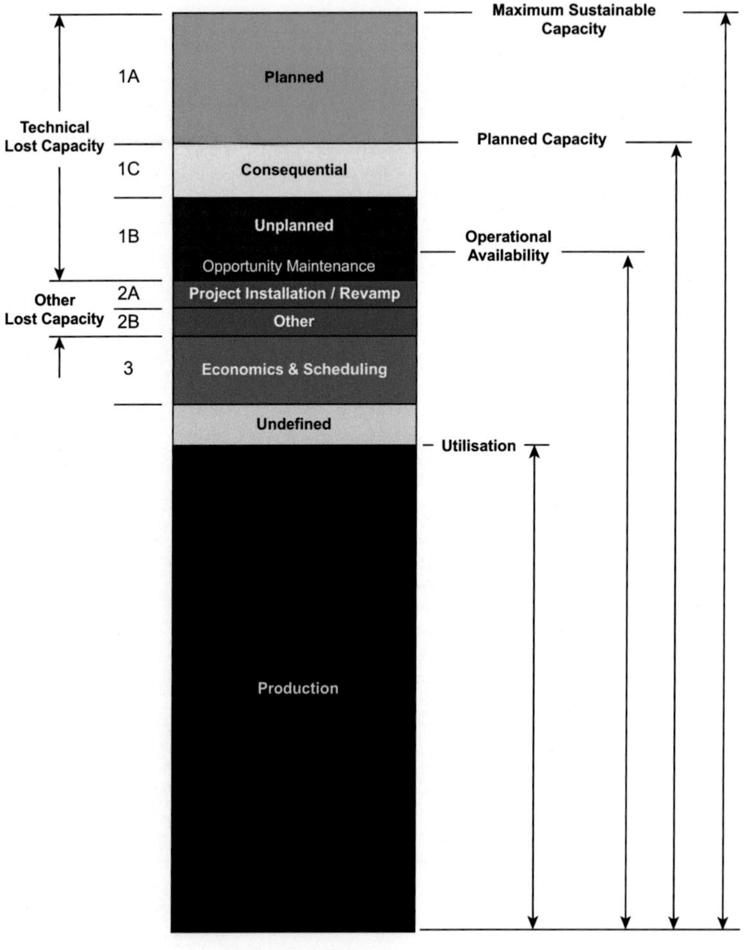

FIGURE 17.5

Relationship between Utilization and Availability as per Philip Townsend Associates International (PTAI) benchmarking consultants.

17.5 MONEY OF THE DAY AND REAL TERMS

It is important to understand the relationship between Money of the Day (MOD) and Real Terms (RT) for short- and long-term projections of expenditure and income.

MOD is the actual value of money at the time of the transaction.

RT is money corrected to the value of constant purchase power by taking inflation into account.

Box 17.5 gives an example of the relationship.

BOX 17.5 MONEY OF THE DAY/REAL TERMS RELATIONSHIP

Budget Target

The shareholder of two gas companies stipulated, in its planning guidelines, that the target for cost per ton of LNG be maintained at a certain value per ton for the next 5 years at an inflation rate of 4%. This is illustrated in the following table.

Year	Money of the Day (MOD)	Real Terms (RT) at 4% Inflation	Comment
Year 0	1	1	
Year 1	1	1.04	Maximum budget inflation
Year 2	1	1.082	Maximum budget inflation
Year 3	1	1.125	Maximum budget inflation
Year 4	1	1.17	Maximum budget inflation
Year 5	1	1.217	Maximum budget inflation

Return on Investment calculations for expenditure and income for major projects use the principle of RT for Discounted Cash Flow. This is calculated over a number of years, typically 5 to 10 years.

17.6 HUMAN CAPITAL

The two primary benchmarking consultants take different approaches.

SOLOMON ASSOCIATES

SA's calculation of personnel performance is as follows:

Personnel Index = work hours/ (refinery EDC + 100) = Number work hours per 100 EDC.

Equivalent Distillation Capacity (EDC) is defined in Section 25.3, Consultants Proprietary Common Denominators.

PHILIP TOWNSEND ASSOCIATES INTERNATIONAL

PTAI use the Site Personnel Index (SPI) methodology that was originally designed by Shell, more than 30 years ago, for European oil refineries. It is now routinely applied worldwide for refineries, LNG plants, gas plants, chemical plants, terminals, and similar facilities.

The SPI is a measure of the personnel effectiveness for a plant. The basic principle of the method is to relate net hours worked to the amount of equipment (**site complexity**) that is operated and maintained.

SPI = net hours worked per unit of **complexity**.

Normalized Shift Positions (NSP) is used as a common denominator and is calculated as follows:

NSP = (number of operators on shift) / (number of shifts) ×100/SPI

Complexity is explained in Section 25.3, Consultants Proprietary Common Denominators.

Net hours worked per year exclude vacation, public holidays, illness, etc.

Net hours worked are related to "Standard Activities," which only include normal operations, maintenance, administration, and management of the facility.

"Nonstandard Activities" are excluded from the calculation of NSP. These include new construction, services to third parties, maintaining housing or club facilities, product packaging, etc.

All maintenance hours are adjusted using the Maintenance Scaling Factor.

The shortcomings of benchmarking with the NSP approach, using "Standard Activities," is discussed in Box 26.1, Exclusions: Body Count.

Contractor hours for turnarounds are annualized.

17.7 ENERGY

The primary problem with measuring energy is to give a standard value to all types of energy for comparison. The two primary benchmarking consultants take slightly different approaches to a generic Reference Gas approach. These approaches are described as follows.

SOLOMON ASSOCIATES

Energy evaluation is based on the typical (Standard) energy consumption for each process unit or technology type.

Examples are as follows:

- *Standard Crude Unit*: Energy Standard = 3 + 1.23 × API gravity kbtu/bbl
- *Resid Catalytic Cracker (RCC):* Energy Standard = 70 + [40 × (Coke Yield, wt% of Fresh Feed)] kbtu/bbl.

  The American Petroleum Institute gravity, or **API gravity**, is a measure of how heavy or light a petroleum liquid is compared to water: if its API gravity is greater than 10, it is lighter and floats on water; if less than 10, it is heavier and sinks.

PHILIP TOWNSEND ASSOCIATES INTERNATIONAL

PTAI sets the Standard Energy Cost as $1/mmbtu, which is similar to the Reference Gas approach.

REFERENCE GAS

In Qatar, National Reference Gas (NRG) is used and is generally defined as natural gas with Higher Heating Value of 1000 btu/scf.

The basis is derived as per Table 17.1.

Table 17.1 Reference Gas Heating Values

Higher Heating Value (HHV)		Lower Heating Value (LHV)	
BTU/scf	**kcal/kg**	**BTU/scf**	**kcal/kg**
1000.0	12,330	902.3	11,125

All energy production inside and outside a complex is converted to NRG.

All energy carriers are normalized to NRG using conversion factors, e.g.:

1. Steam at 5 bar/177 C: 18.1 t/tNRG
2. Steam at 42 bar/385 C: 15.5 t/tNRG
3. Electrical Power: 4.44 MWh/tNRG

Conversion factors allow comparisons between Companies with different utilities schemes. Conversion factors are based on the following:

- common swing producers (e.g., boiler for steam)
- standardized efficiencies
- Intake = Dry wellhead gas + imported liquefied petroleum gas

Box 17.6 gives an example of comparing equivalent energy generation of two gas plants.

BOX 17.6 GAS PLANT ENERGY BALANCE

Imported Versus Produced Power

Two equivalent companies consume 1200 MWh of electrical power.

Company A imports 1200 MWh of electricity. Company B needs 3871 MWh of fuel gas to generate 1200 MWh of electricity (condensing steam turbine at 31% efficiency).

Company B appears to consume more energy than company A (3871 versus 1200 MWh), but both companies have the same consumption efficiency (1200 MWh for similar output).

It is therefore required to introduce the equivalent energy concept. The basis for equivalent energy is a standard fuel gas @ 1000 BTU/SCF (= National Reference Gas; NRG)

In this example, 1 ton NRG (48.7 MSCF QRG) is equivalent to 4.44 MWh electric power.

Comment

This case study emphasizes the need to standardize energy generation and consumption.

	Company A: imports electrical power			Company B: generates electrical power		
	MWh	QRG		MWh	QRG	
		t	MMscf		t	MMscf
Electrical power consumption	1200			1200		
Import electrical power	1200	270	13.2	0	0	0
Fuel Gas for generation of power	0	0	0	3871	270	13.2
Total	1200	270	13.2	3871	270	13.2

17.8 GREENHOUSE GASES[3]

THE CONCEPT OF CO_2 EQUIVALENT

Greenhouse gas (GHG) emissions are required to be reported in CO_2 equivalent (CO_2e). This is measured in terms of the Global Warming Potential (GWP), which describes the radiative forcing impact of one mass-based unit of a given GHG relative to an equivalent unit of carbon dioxide over a given time horizon. The GWPs of the various GHG are presented in Table 17.2.

Table 17.2 Global Warming Potentials for Greenhouse Gases

S	Greenhouse Gas (GHG)	Chemical Formula	Global Warming Potential (GWP) for 100 Year Time Horizon
1	Carbon dioxide	CO_2	1
2	Methane	CH_4	21
3	Nitrous oxide	N_2O	310

Courtesy IPCC Working Group 1 (WG1), 4th Assessment Report; 2007

Sources of CO_2 emissions from combustion installations and processes include the following:

- boilers
- burners
- turbines
- heaters
- furnaces
- incinerators
- engines
- flares
- scrubbers (process emissions)
- any other equipment or machinery that uses fuel, excluding equipment or machinery with combustion engines that is used for transportation purposes

Trading of GHG emissions is discussed in Section 21.8, Carbon Credits.
Trading is in Certified Emission Reductions (CERs): 1 CER = 1 metric ton of CO_2e.

17.9 COMPETITIVENESS, EFFICIENCY, AND EFFECTIVENESS

Competitiveness, Efficiency, and Effectiveness is discussed in Section 28.6, Performance Focus.

17.10 SOFT RELATIONSHIPS

The effect of soft relationships has often been grossly underestimated. Both positive and negative interaction can occur, while relationships reinforce performance improvement or could be counterproductive.

POSITIVE INTERACTION OVER A LONG PERIOD: PACESETTER STATUS

The long-term retention of highly competent staff has a major effect on all top Key Performance Indicators (KPIs) of a Business Unit.

Example:

Box 20.6, Competencies: Motivation, describes a pacesetter gas plant with very low staff turnover with an indirect positive reflection on other performance indices.

NEGATIVE INTERACTION OVER A LONG PERIOD: THE "TIPPING POINT"

Example:

An arbitrary reduction in maintenance expenditure (good performance in the short term) reduces reliability of the plant (bad performance in the long term), possibly resulting in reliability, reaching an eventual "tipping point."

Examples of Companies reaching a "tipping point" are given in Box 9.5, Accountability, Responsibility, and Discipline: BP and Box 9.6, The Jewel in the Crown: The Rise and Fall of South Africa's Electricity Supply Commission.

CROSS-LINKAGE AND CASCADING

The System Sponsor and Owner have the most influence over KPIs in a System, although others can also influence these KPIs. This cross-linkage and cascading of indicators needs to be recognized and the positive influence promoted.

An example of the cross-linkage is shown in Box 17.7.

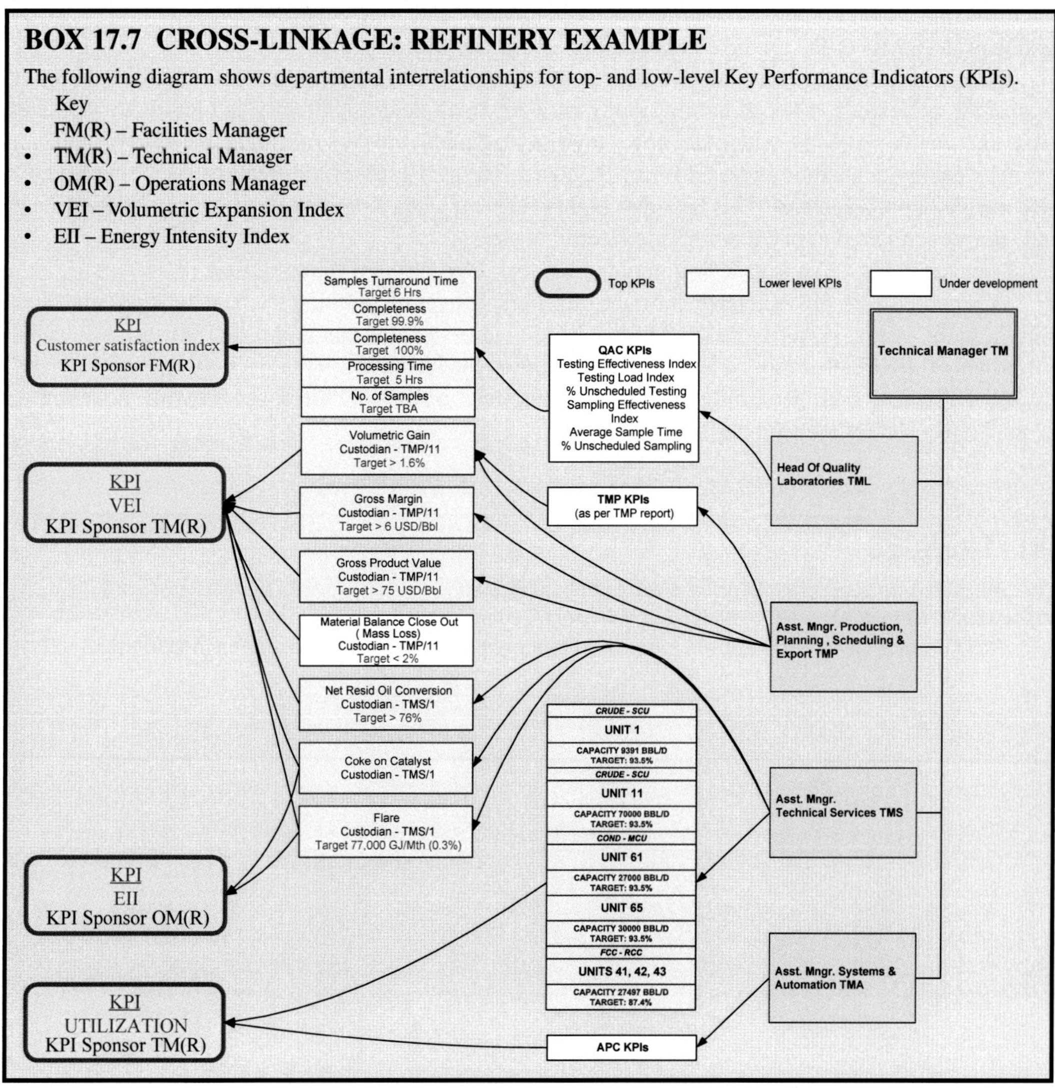

17.11 SUMMARY

Relationships between KPIs could reinforce performance improvement or be counterproductive.

In Benchmarking, size and complexity are the most important to compare "apples with apples."

Mass, volume, and heating value are based on hydrocarbons and are expected to balance. Refinery hydrocarbon balance always requires an increase in volume and decrease in mass (within accepted limits). In gas plants, all losses (mainly flaring) have to be carefully accounted for since the heating value determines the cost of the salable product.

High Utilization cannot take place without high Availability, which is dependent on planned and unplanned down days. Maximum sustainable capacity needs to be based on the real capability of the plant over an extended period without damage to the assets.

For comparison purposes, the same basis of monetary valuation is needed, whether it is the current value of the assets or the predicted expenditure over a number of years.

Personnel indices are also required for comparison. However, there are limitations to these with respect to exclusions. A benchmarking study may only include those directly and indirectly associated with the salable products, with other company staff excluded, thus distorting the personnel index.

The energy evaluation is based on a "Standard Energy Cost" with different bases established by Benchmarking Consultants. This becomes highly relevant when a process plant generates its own electric power compared to one that imports electric power.

The concept of CO_2e is used to determine the total GHG from different gases.

Competitiveness and Efficiency have the dual perspectives of "looking from the outside" and "looking from the inside." Effectiveness is more related to those actions that will have the most influence on improving the KPIs that most affect the bottom line of the business.

As KPIs have both hard and soft links, the interrelationship is best observed with modeling and comparative studies (benchmarking).

REFERENCES

1. Solomon Associates (SA). https://www.solomononline.com/.
2. Philip Townsend Associates (PTAI). http://www.ptai.com/.
3. *IPCC Working Group 1 (WG1), 4th assessment report*. 2007. https://www.ipcc.ch/report/ar4/wg1/.

PART

ASSET PERFORMANCE MANAGEMENT

5

OVERVIEW

WHY IS IT IMPORTANT TO ENSURE ASSETS PERFORM AT OPTIMAL LEVELS OVER AN EXTENDED PERIOD FOR MAXIMUM OVERALL BUSINESS PERFORMANCE?

In the process industry, the life of the physical asset is normally in excess of 30 years. An optimal level of physical asset maintenance and reinvestment is required to ensure attainment of maximum production levels over its lifetime and thus maximize profits. Human and financial assets, with the support of information, ensure this happens. However, if one reduces maintenance activity and costs before obtaining a high level of reliability, maximal long-term production levels will not be achieved.

THE OBJECTIVE

The objectives of Part 5 are the following:

1. Explain the role of different assets and their relationships with each other and the business
2. List the primary constraints within which assets need to operate

3. Embed the need for long-term integrity of the physical assets
4. Provide an understanding of tools and processes to minimize unplanned business outages through effective turnaround management

PICTORIAL VIEW

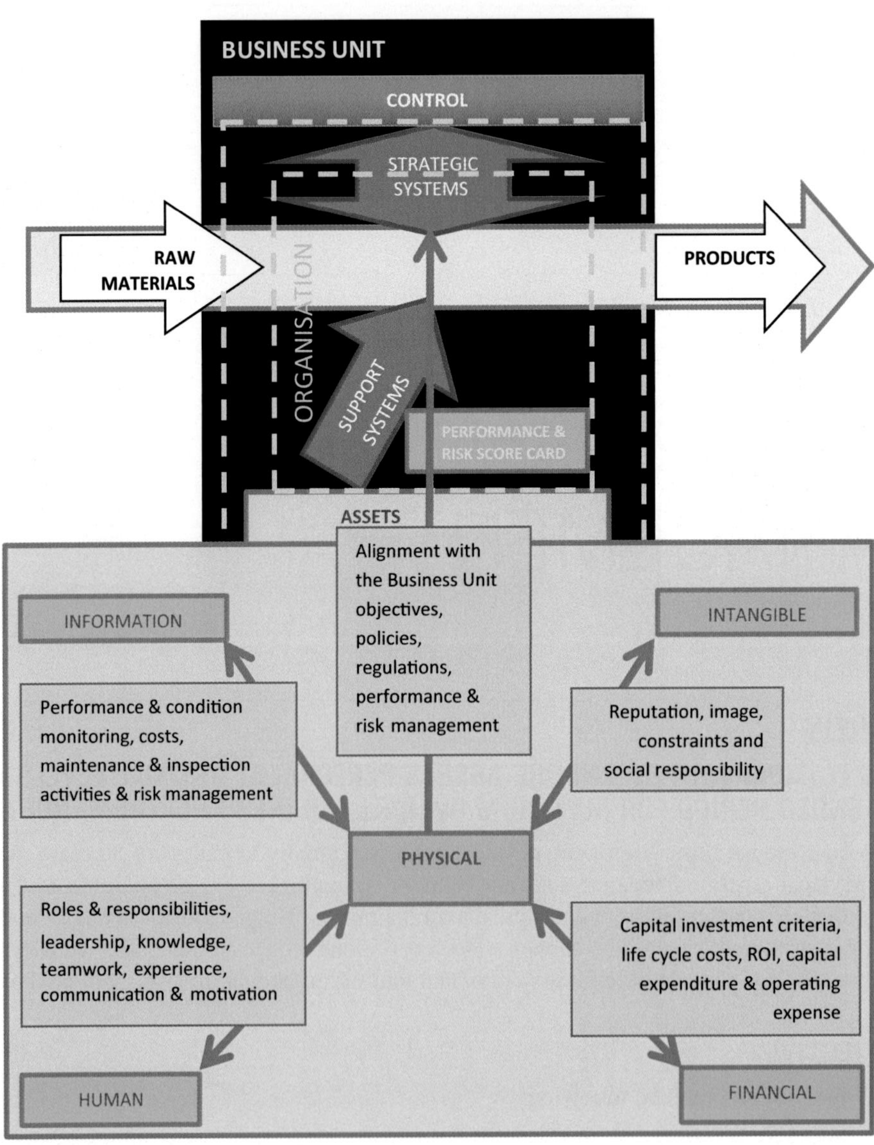

OUTLINE

- Business assets are listed with their interrelationships.
- Required regulations, standards, and codes are outlined.
- The use of a "third party" inspection authority to ensure compliance with design and construction standards and codes is summarized.
- Tools and processes, needed for the sustainability of physical assets, are detailed.
- System maturity issues, with respect to physical asset integrity, are outlined.
- The importance of maintaining and enhancing human asset competence levels, from generation to generation, over the life of the physical assets, is discussed.
- Processes and tools that ensure the required levels of competencies are outlined.
- Empowerment through behavior-based safety (BBS) is identified as the primary tool for long-term asset integrity.
- The need for secure and reliable information to ensure effective decision-making is explained.
- The requirement of strict financial management to ensure the smooth operation of the process plant, as well as optimal capital investment, is outlined.
- Tools, techniques, and processes critical to a successful turnaround, as well as the agreed run length (with minimal downtime), are explained.
- Suitable key performance indicators for physical, people, information, and financial assets are listed, as well as those required for optimal turnaround management.

CHAPTER

TYPES OF ASSETS AND APPLICABLE STANDARDS

18

18.1 INTRODUCTION

This chapter discusses the different types of assets found in the process industry, how they interact, and what restraints or standards within which they have to perform.

18.2 SUMMARY OF TYPES OF ASSETS

The assets required to add value encompass the following:

1. Information
   Definition: where value is in the content rather than the medium in which it is held
2. Human
   Definition: competencies of staff: experience, skills, knowledge, and attitudes
3. Financial
   Definition: monetary items such as cash, credit, deposits, share capital, and loan capital
4. Intangible
   Definition: value, although real, is not easily measurable, e.g., goodwill, **reputation**, morale, intellectual property, knowledge, contracts, and agreements.
5. Physical
   Definition: "hardware": mineral beneficiation, hydrocarbon/petrochemical production, or power-producing facilities as well as offices, furniture, computing, and communications equipment, etc.
6. Stock
   Definition: consumed or converted as resources during the execution of activities
7. Natural
   Definition: occurring in the natural environment, e.g., mineral resources or hydrocarbons
8. External
   Definition: used but not owned or under direct control, e.g., concession areas

The assets of an oil and gas company are generally depicted as shown in Fig. 18.1.

Performance Management for the Oil, Gas, and Process Industries. http://dx.doi.org/10.1016/B978-0-12-810446-0.00018-9

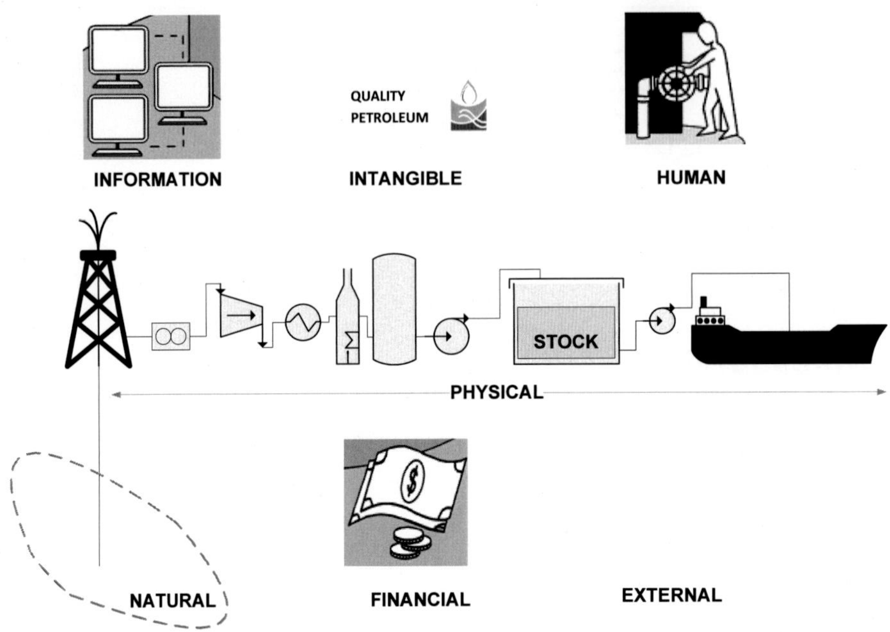

FIGURE 18.1

Oil and gas company assets.

18.3 PRIMARY RELATIONSHIPS

The **physical assets** are the center of the business, adding value to the raw material to produce the finished product. Its primary relationships include the following:

1. BUSINESS UNIT

- Alignment with the Business Unit objectives, policies, regulations, and performance and risk management requirements

2. INFORMATION ASSET

- Performance and condition monitoring, costs, maintenance and inspection activities, and risk management

3. HUMAN CAPITAL ASSET

- Roles and responsibilities [with emphasis on health, safety, and environment (HSE) and physical asset integrity], leadership, knowledge, teamwork, experience, communication, and motivation
- Physical Asset Unit operators:
    - operate and maintain the assets
    - maintain asset integrity
    - exercise cost management
    - monitor and report asset performance

4. FINANCIAL ASSET

- Capital investment criteria, life cycle costs, Return on Investment, capital expenditure, and operating expense

5. INTANGIBLE ASSET

- Reputation, image, constraints, and social responsibility

These primary relationships are depicted as shown in Fig. 18.2.

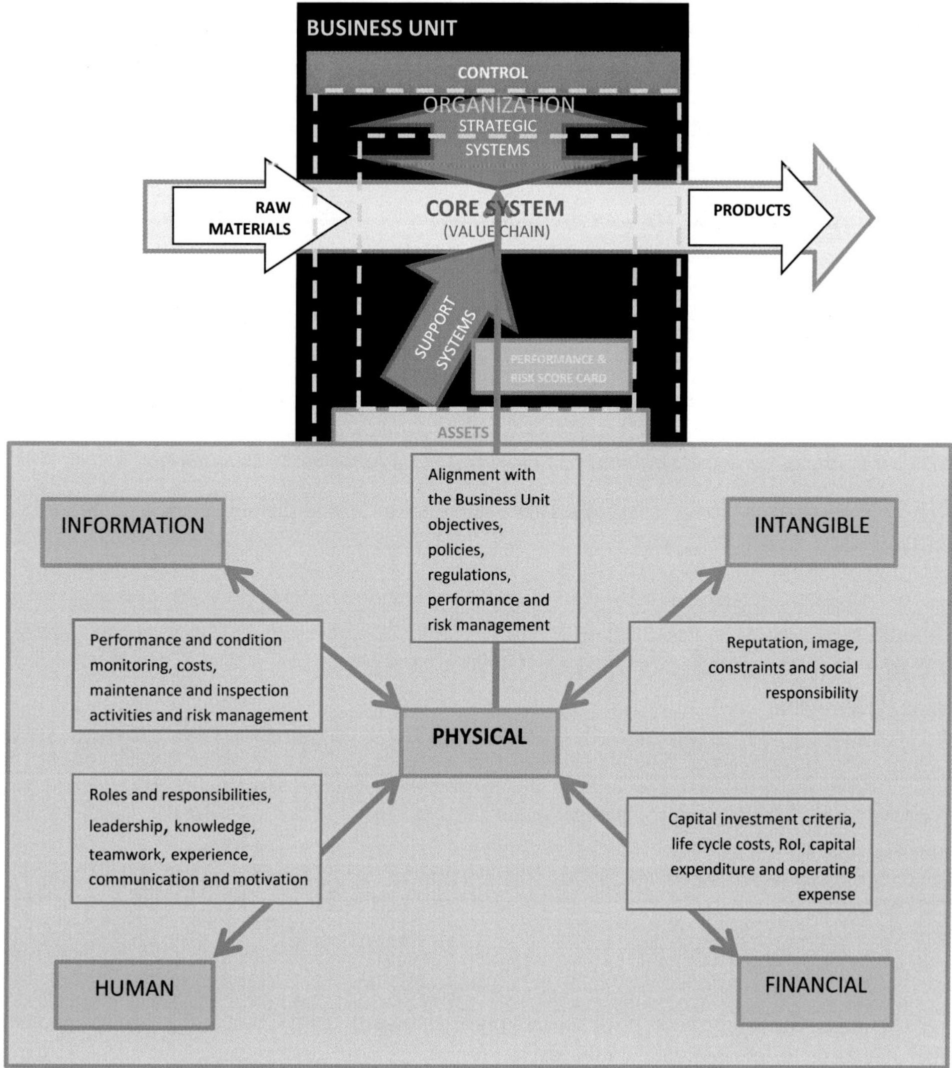

FIGURE 18.2

Asset relationships.

Box 20.6, Competencies: Motivation, discusses identified soft links between Asset Key Performance Indicators (KPIs).

18.4 STANDARDS

All the primary assets have to operate within the external restraints of standards and regulations.

A number of standards follow the Plan Do Check Act (PDCA), which is a continual improvement process based on ISO 9001.

HOW STANDARDS ARE ESTABLISHED

Industry members often collaborate to create a standard for manufacturing, maintenance, operations, etc. A government might then establish the standard nationally and then, should the issues be of global significance, the international community could form an international standard.

Example 1:

ISO 9001[1] *evolved from the UK military requirements for the manufacture of armaments. It then evolved into a British Standard, issued by the British Standards Institute, and then later was amended and adopted by the International Standards Organisation (ISO).*

Example 2:

The Project Management Body Of Knowledge[2] *was developed by the Project Management Institute, which then became an American National Standards Institute Standard (ANSI 99-001-2008). It is now used worldwide and has been translated into many languages.*

Box 18.1 describes how a compressor manufacturer established a new standard for its compressors.

BOX 18.1 REQUIREMENT FOR A NEW STANDARD

Potential Seal Failure

During a risk study carried out on a gas plant, one of the risks identified was the use of single shaft seals on gas compressors. Single shaft seals give only one layer of protection to prevent seal failure and a subsequent gas leak. Although the manufacturer stated that these seals had never failed, the operator still considered it a high risk, and requested the manufacturer to retrofit all their compressors with double seals.

Comment

The outcome was that double seals are now standard on gas compressors supplied by this manufacturer.

Fig. 18.3 gives the progression of development of standards from Companies through Industry and National standards to International standards.

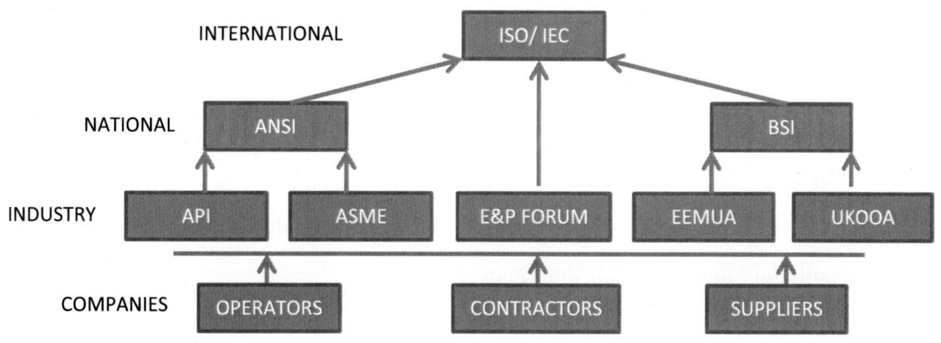

FIGURE 18.3

Standardization bodies – relationships. *ANSI*, American National Standards Institute; *BSI*, British Standards Institute.

APPLICATIONS

The following standards are grouped by application. Those in **bold** show that they follow the ISO 9001 **PDCA process** format.

GENERAL STANDARDS FOR ALL ASSETS IN THE BUSINESS UNIT

The following standards are applicable to the whole **Business Unit:**

- **ISO 9001, "Quality Management"**

  ISO 9001 is introduced in Section 2.6, ISO certification.

- ISO 31000, "Risk Management"

  ISO 31000 is introduced in Section 11.3, ISO 31000, Risk Management.

- ISO/PAS 22399, "Guideline For Incident Preparedness and Operational Continuity Management"

  ISO/PAS 22399 is introduced in Section 14.3, Applicability of Standards.

INFORMATION ASSET

The following standard is applicable to the **information asset,** no matter where it resides in the business.

- **ISO/IEC 27001, "Information Security Management"**

  The application of this standard is discussed in Chapter 19, Information.

HUMAN ASSET AND ENVIRONMENT

The following standards are applicable to both the **human capital asset** and the environment.

- **ISO 45001 (formerly OHSAS 18001), "Occupational Health and Safety Management"**
- **ISO 14001, "Environmental Management"**

The application of these standards is discussed in Chapter 20, Human Capital.

FINANCE ASSET

The following is the globally accepted standard for management of the **financial asset**.

- IFRS, International Financial Reporting Standards

The application of this standard is discussed in Chapter 21, Finance.

PHYSICAL ASSETS

Constraints on Physical assets are numerous. The application of the following standards, grouped for easy reading, is discussed in Chapter 22, Physical Asset Performance Management.

ISO PLAN, DO, CHECK, ACT (PDCA) STANDARDS

The following **PDCA standards** are applicable to the **physical assets**.

- ISO 50001, "Energy Management"
- ISO 55001, "Asset Management"

Box 18.2 promotes the use of ISO 50001, "Energy Management."

BOX 18.2 ISO 50001, "ENERGY MANAGEMENT SYSTEM PROMOTION"[3]

British Standards Institute (BSI) quote:

"Effective energy management isn't just good for business, it's also becoming a requirement. And the best way to achieve it is with ISO 50001. The international standard outlines energy management practices that are considered to be the best, globally. Energy management experts from more than 60 countries developed the standard and now we can help you tap into that expertise, every day, to help you save energy, cut costs and meet environmental requirements.

What are the benefits of ISO 50001 Energy Management?

- Identify and Manage the risks surrounding your future energy supply
- Measure and monitor energy use to identify where to improve efficiency
- Improve overall performance to cut energy consumption and bills
- Reduce carbon emissions and meet government reduction targets
- Demonstrate environmental credentials to increase tender opportunities."

Box 18.3 promotes the use of ISO 55001, "Asset Management."

BOX 18.3 ISO 55001, "ASSET MANAGEMENT PROMOTION"[4]

British Standards Institute (BSI) quote:

"ISO 55001 is a framework for an asset management system that will help your business to pro-actively manage the lifecycle of your assets, from acquisition to decommission. This system helps you to manage the risks and costs associated with owning assets, in a structured, efficient manner that supports continual improvement and on-going value creation.

What are the benefits of ISO 55001?

An asset management system provides a structured, best practice approach to managing the lifecycle of assets.

- Reduced risks associated with ownership of assets – anything from unnecessary maintenance costs and inefficiency to accident prevention (explosions at gas plants for example)
- Improved quality assurance for customers/regulators – where assets play a key role in the provision and quality of products and services
- New business acquisition - stakeholders gain confidence from the knowledge that a strategy is in place to ensure assets meet the necessary safety and performance requirements
- Supports international business growth – demonstrating that the requirements of an internationally recognised asset management system are being met."

OTHER PHYSICAL ASSET STANDARDS

Integrity

The following are guidelines for the integrity of the complete physical asset:

- Oil and Gas Producers (OGP) Report No 4, "Asset Integrity"
- Center For Chemical Process Safety (CCPS), "Guidelines For Risk Based Process Safety"

Inspection

The following is the basis for Risk Based Inspection:

- API RP 580, "Risk Based Inspection"

Maintenance

The following are the key requirements for Reliability Centered Maintenance:

- IEC 61511 and IEC 61508, "Safety Instrument Systems"
- SAE STD JA1011, "Evaluation Criteria For Reliability Centered Maintenance (RCM) Processes"
- ISO 14224, "Petroleum, Petro- Chemical and Natural Gas Industries – Collection and Exchange Of Reliability and Maintenance Data For Equipment"
- UK HSE RR509, "Plant Ageing: Management Of Equipment Containing Hazardous Fluids Or Pressure"

Assessment Methods

Chapter 12, Risk Control Mechanisms, introduces Root Cause Analysis. The following describe assessment methods for the process industry:

- IEC 61822, "Hazards and Operability Application Guide"

- ISO 17776, "Petroleum and Natural Gas Industries – Offshore Production Installations – Guidelines On Tools and Techniques For Hazard Identification and Risk Assessment"

Key Performance Indicators

The following give guidance for development of appropriate KPIs for reporting the condition of the **physical asset**:

- CCPS, "Leading and Lagging Metrics"
- UK HSE Guide 254, "Developing Process Safety Indicators"
- ANSI/API 754 RP, "For Process Safety Indicators For The Refining and Petrochemical Industries"
- EI Research Report, "Human Factors Performance Indicators For The Energy and Related Process Industries"

DESIGN ENGINEERING PRACTICES AND STANDARDS

International Oil Companies (IOCs) have internal Design Engineering Practices (DEPs) linked to national or international standards. These have associated Standard Drawings and Materials and Equipment Standards and Codes (MESC). DEPs have been developed by the companies over many years.

Example: Piping classes

The DEP contains predesigned piping systems for a variety of services. These have a direct link to the MESC materials catalog for purchasing.
This results in a large reduction in engineering and procurement effort, improves integrity control, promotes company-wide standardization, and reduces a variety of spares holding.
A collection of standardized piping components (caps, elbows, tees, reducers, branch fittings, check valves, gate valves, globe valves, ball valves, and butterfly valves) that are compatible and suitable for a defined service at stated pressure and temperature limits are specified for each class of pipe.

As the cost of maintaining these standards are high, the trend is to convert IOC standards to Industry, National, and International standards[5] (the progression is depicted in Fig. 18.3).

18.5 REPUTATION: AN INTANGIBLE ASSET

The benefits of a good reputation for organizations include the following:

1. retaining good staff and attracting the best employees;
2. attracting new business partners, customers, and suppliers;
3. allowing easier entry to new markets and brand extensions;
4. enabling successful mergers, acquisitions, and strategic alliances;
5. enhancing relationships with NGOs, or **mitigating against corporate activities that can potentially harm the business;**
6. **helping to protect the organization and brands during a crisis; and**
7. **securing future cash flows and investments.**

An example of a detrimental effect on **5 to 7 listed previously** is given in Box 18.4.

BOX 18.4 THE BATTERING OF BRITISH PETROLEUM'S REPUTATION[6]

After a series of incidents around the world, British Petroleum's reputation has taken a hard knock. The biggest of these being:

- *Grangemouth Refinery: explosion in 2000*
- *Texas City Refinery: explosion in 2005*
- *Alaska Pipeline: leaks in 2010 and 2011*
- *Exploration in the Gulf of Mexico (Deepwater Horizon): explosion and leak in 2010*

Box 34.4, Inadequate Shareholder Oversight Resulting in a Major Reputational Risk, gives an example of reputation risk related to a Joint Venture.

18.6 THIRD-PARTY COMPLIANCE CHECKS

To ensure compliance with a required standard, assurance measures have to be implemented. Section 11.9, Assurance Services, discusses these measures. However, distinct services relating to the quality of materials for construction of the physical asset requires specific attention.

Third-Party Authorities are used as independent assessors of compliance with required standards. This is a prerequisite for investment loans and also for insurance purposes.

Lloyd's of London[7] started as a marine classification society in 1760 and then went on to establish standards and employ staff to ensure compliance with their standards. Today, Lloyd's has a worldwide network of certified insurance assessors who ensure compliance with all relevant standards for a particular industry.

The following are other well-known authorities:

- Det Norske Veritas Germanischer Lloyd (DNV GL), DNV founded 1864 and GL founded 1867[8]
- Bureau Veritas, founded 1828[9]
- American Bureau of Shipping, founded 1862[10]
- Hartford Steam Boiler Inspection and Insurance Company, founded 1866[11]

It is vital that the Third-Party Authority needs to be appointed and their services paid for by the client or insurer, and **NOT** the contractor carrying out the work. Box 18.5 discusses the disastrous outcome when Standards are compromised.

BOX 18.5 THIRD-PARTY COMPLIANCE CHECKS: PETROBRAS RIG COLLAPSE[12]

A Case of Shortcutting Standards

In March 2001, a series of explosions sank Petrobras Platform 36.

> *"the project successfully rejected ... prescriptive engineering, onerous quality requirements, and outdated concepts of inspection ..." A Petrobras executive, prior to the accident, on delivering superior financials.*

Proximate Cause:

- leakage of volatile fluids burst a shutdown emergency drain tank and set off a violent chain of events

Underlying Issues:

- **a corporate focus on cost-cutting over safety;**
- poor design of individual parts (with regards to a system safety context);
- component failure without sufficient backups; and
- lack of training and communication.

18.7 SUMMARY

Physical assets are not the only assets. Human, Information, Financial, and Intangible are all assets that are relevant to all process industries. Their optimum interface with the Physical assets is critical for maximum value addition.

Standards and regulations are the external restraints within which the process industry is required to work. As there are so many, it is best to place them in asset or general business categories.

A number of standards follow the PDCA for continual improvement process, which is based on ISO 9001. These include ISO 27000, "Information Security Management"; ISO 45001, "Occupational Health and Safety Management"; ISO 14001, "Environmental Management"; ISO 50001, "Energy Management"; and ISO 55001, "Asset Management."

Business risk and continuity management are so important that the standards for these (ISO 31000 and ISO 22399) are also becoming a requirement.

A number of the Physical Asset standards should ideally be strictly adhered to by the process industry so as to ensure integrity of the assets in order to guarantee that production can continue at design capacity for the life of the investment.

Guidance in establishing leading (predictive) indicators to mitigate integrity risk is provided by US, UK, and other national standards.

It is difficult to apply KPIs directly to the Reputation Asset. However, loss of reputation can be highly detrimental to the success of a company.

Third-Party Authorities are used as independent assessors of compliance with required standards. This is mandatory for investment loans and also for insurance purposes.

REFERENCES

1. ISO 9001:2008, *Quality management.*
2. ANSI/PMI 99-001-2013. *A guide to the Project Management Body Of Knowledge (PMBOK).* 5th ed.
3. Energy management system promotion: BSI quote from www.bsigroup.com/en-GB/iso-50001-energy-management/.
4. Asset management promotion: BSI quote from http://www.bsigroup.com/en-GB/Asset-Management/Getting-started-with-ISO-55001/.
5. ISO standards for use in oil & gas industry poster. IOGP. http://www.iogp.org/portals/0/standards/standards-poster.pdf.
6. Steffy LC. *Drowning in oil: BP and the reckless pursuit of profit.* McGraw-Hill; 2011.
7. Lloyds of London. www.lr.org.
8. Det Norske Veritas Germanischer Lloyd (DNV GL). www.dnvgl.com .
9. Bureau Veritas (BV). www.bureauveritas.com .
10. American Bureau of Shipping (ABS). http://ww2.eagle.org/content/eagle/en.html .
11. Hartford Steam Boiler Inspection and Insurance Company. http://www.munichre.com/HSBBII/en/home/index.html.
12. Petrobras P36 rig collapse. https://sma.nasa.gov/docs/default-source/safety-messages/safetymessage-2008-10-01-lossofpetrobrasp36.pdf?sfvrsn=4 .

FURTHER READING

1. ISO 31000:2009. *Risk management principles & guidelines.*
2. ISO 31010:2009. *Risk management – risk assessment techniques.*
3. ISO 22399:2007. *Societal security – guideline for incident preparedness and operational continuity management.*
4. ISO 45001:2016. *Occupational health & safety management systems.*
5. ISO 14001:2004. *Environmental management systems.*
6. ISO 50001:2011. *Energy management systems.*
7. ISO 55000:2014. *Asset management – overview principles & terminology.*
8. ISO 55001:2014. *Asset management – requirements.*
9. ISO 55002:2014. *Asset management – guidelines for application of ISO 55001.*
10. BS OHSAS 1802:2008. *Annex a correspondence between 18001, 14001, 9001.*

CHAPTER

INFORMATION

19

19.1 INTRODUCTION

> Knowledge and information are recognized as two of the most important strategic resources that any organization manages.
>
> **Anonymous**

Good, quality information is critical for the optimum operation of the physical, personnel, and financial assets. Information becomes knowledge when it is applied in decision-making and further action.

DEFINITION OF "KNOWLEDGE"

> Organizational knowledge is the collective experience accumulated through systems, routines, and activities of sharing information across the organization.

The retention of organizational knowledge, from generation to generation of employees, is critical for the success of the business. Information management is thus essential for transfer of knowledge. Maintaining a balance between protecting information related to the company's intellectual property and displaying information for effective decision-making is ongoing. This chapter discusses all aspects of information management that affect value addition in line with the company direction and customer satisfaction.

Box 19.1 discusses information technology (IT) in a historical perspective.

BOX 19.1 THE EVOLUTION OF IT IN THE OIL AND GAS INDUSTRY

In the 1970s, we had pneumatic process control and mechanical metering for critical flows. With pneumatic control, the control room had to be in the center of the process plant—potentially, a very dangerous situation in a refinery. Mainframe computers such as IBM 360s supplied IT support, and refinery product mix was calculated using linear programming on the mainframe. We would provide a weekly update of turnaround progress by putting punch cards into the mainframe and getting a long printout. Accounting and other services were also carried out on the mainframe.

In the 1980s, mainframe applications evolved into enterprise resource planning such as SAP and electronic imaging for storage of drawings and other technical documents with the introduction of the likes of CIMAGE. PCs started to be used for drawing with the introduction of Autocad. Process control went electronic with data being transmitted to a control room outside the center of the process area. Languages, such as RTL2, evolved to control the process running on Digital Equipment Corporation PDP11 computers. This was the introduction of distributed control systems (DCSs).

Continued

Performance Management for the Oil, Gas, and Process Industries. http://dx.doi.org/10.1016/B978-0-12-810446-0.00019-0

BOX 19.1 THE EVOLUTION OF IT IN THE OIL AND GAS INDUSTRY—cont'd

In the 1990s, DCSs were linked to advanced process control to optimize the process. Process control started to be dominated by the likes of Honeywell and Yokogawa. PCs were linked on local area networks (LANs) where the market was dominated by companies like Novell. Some mainframe applications were replaced. For turnaround critical path method planning, Artemis on the mainframe was replaced by Open-Plan and Prima-Vera on a Local Area Network (LAN).

In the 2000s, the Internet flourished with its sister, the Intranet. Process plants are now able to be controlled from the other side of the world. Computer systems "talk" to each other. The downside is the phenomenon of cyberattacks on process plants.

We now need to think very carefully about applying IT to get the best information possible for strategic advantage. With the diminishing number of employees with long-term experience, we need to focus on storing the past experiences for future generations. An excellent example is process simulation, where all the possible scenarios for process operations can be stored so that operators can practice without blowing up the plant. Also, for turnarounds with intervals of 4 or more years, we can keep the models of past turnarounds and copy and "tweek" the best for the next turnaround, improving each time.

IT is an evergreen asset that the company needs to nurture and treasure, especially when another key asset, the human asset, has a limited lifespan.

19.2 ORGANIZATIONAL CAPABILITY: THE INFORMATION ASSET ASPECT

The information aspect of organizational capability is discussed here. Organizational capability is introduced in Section 5.2, Strategic Planning (Strategy formulation). The progression of conversion of data into knowledge is also discussed in Section 15.3, Discussion of Approach.

Knowledge is the principal productive resource of an organization. Knowledge exists in both **tacit** and **explicit** forms. Tacit knowledge is especially important as it is not easily transferred because it is acquired and stored within individuals. Specifically, production process plant operators require a wide array of knowledge, both tacit and explicit.

The essence of organizational capability is the **integration of the individual's specialized knowledge to create value through transforming input into output**.

Knowledge is created by transforming tacit into explicit knowledge, and vice versa, as well as transferring knowledge between individuals and groups, within and across organizations (e.g., benchmarking).

Organizations have to be concerned not only with knowledge creation but also with knowledge application. Organizational capability affects the following:

1. Alignment–connecting with the business unit (vertical and horizontal)
2. Innovation–bringing ideas into the business unit
3. Relationships–partnering with affiliates and
4. **Performance**–achieving high-quality business results

19.3 CONTROL OF INFORMATION

In all businesses, information is collected, stored in records (electronic or manual) in an information system, and then used (exploited). The level of security of the process is dictated by the sensitivity of the information for competitive advantage.

The control of information consists of the following facets:

- Collection
- Security

- Information system
- Information
- Utilization or exploitation

Fig. 19.1 depicts the control of information.

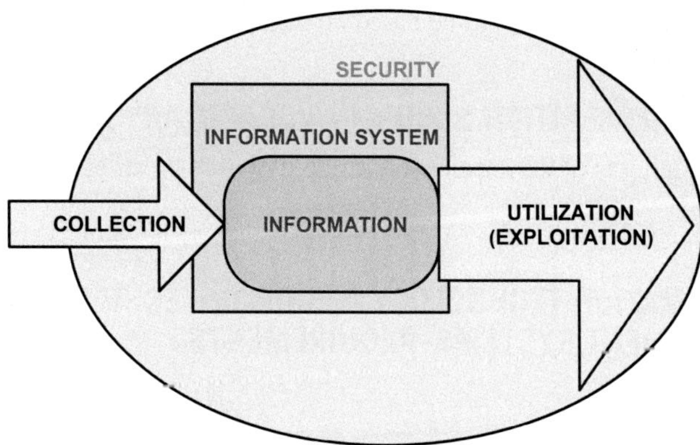

FIGURE 19.1

The control of information.

Information asset holders are responsible for the following:

1. Defining the security classification of their information
2. Maintaining the confidentiality, accessibility, and integrity
3. Promoting information sharing and rationalization

19.4 INFORMATION GOVERNANCE POLICY AND SUGGESTED FORMAT

Best practice is the enactment of an information governance policy that oversees all information technology–related activities. The suggested content of such a policy is as follows:

1. Internal information is owned by the company.
2. All information will have a defined custodian who, as the authorized agent of the company, will be responsible for its management and for making it available to those who need it.
3. Information will be managed to **support business systems and processes rather than organizational hierarchies**.
4. Information will be managed, kept accurate and up-to-date, and readily accessible to those who need it.
5. Information will not be retained or distributed unnecessarily.
6. A consistent approach to managing information will be adopted across the whole company and will cover the lifecycle of information (creation, indexing, storage, retrieval, revision, archiving/disposal).

7. Methods of information management will give due attention to security, protection, legal, environmental, and cost issues.

19.5 INFORMATION SECURITY

Information security management should be based on the ISO 27000 suite of standards.

ISO 27000-2005: INFORMATION SECURITY VOCABULARY AND DEFINITIONS

- Vocabulary and definitions of the specialist terminology used by all of the ISO standards in the 27000 framework.

ISO 27001: INFORMATION TECHNOLOGY–SECURITY TECHNIQUES–INFORMATION SECURITY MANAGEMENT SYSTEMS–REQUIREMENTS

- Scope of the standard:

> This international standard specifies the requirements for establishing, implementing, operating, monitoring, reviewing, maintaining, and improving a documented Information Security Management System (ISMS) within the context of the organization's overall business risks. It specifies requirements for the implementation of security controls customized to the needs of individual organizations or parts thereof.
>
> The ISMS is designed to ensure the selection of adequate and proportionate security controls that protect information assets and give confidence to interested parties.

- The standard is formatted in the same way as ISO 9001 Quality Management and Other Related Standards (see Chapter 18: Types of Assets and Applicable Standards). Thus, companies are formally certified compliant against this standard (see http://www.iso27001security.com/html/27001.html).

ISO 27002: CODE OF PRACTICE FOR INFORMATION SECURITY MANAGEMENT

- Previously known as BS 7799-1 and then ISO/IEC 17799, ISO 27002 describes a set of information security control objectives and a menu of generally accepted good-practice controls. The rationale here is that not all controls apply to all organizations. The menu, therefore, allows organizations to select and adopt the controls applicable to them.

ISO 27003: IMPLEMENTATION GUIDE FOR THE ISO 27000 STANDARDS FRAMEWORK

- This is the guide for applying the whole suite of standards.

ISO 27004: INFORMATION SECURITY MANAGEMENT MEASUREMENT STANDARD

- This is aimed at providing metrics for improving the effectiveness an organization's ISMS.

ISO 27005: INFORMATION SECURITY RISK MANAGEMENT (THIS REPLACES BS 7799 PART 3)

1. This includes the following process steps:
   a. Context establishment
   b. Risk assessment
      i. Risk analysis
         - Risk identification
         - Risk estimation
      ii. Risk evaluation
   c. Risk treatment
   d. Risk acceptance
   e. Risk communication
   f. Risk monitoring and review

ISO 27006: GUIDE TO ISMS CERTIFICATION

- This provides guidance on the certification process for accredited certification or registration bodies.

ISO 27007: GUIDELINES FOR INFORMATION SECURITY MANAGEMENT SYSTEMS AUDIT

- This is the guide for auditing ISMS.

Box 19.2 gives an outline of ISO 27002.

BOX 19.2 ISO 27002 CODE OF PRACTICE FOR INFORMATION SECURITY MANAGEMENT OUTLINE

4. Risk management
   - Risk assessment
   - Risk analysis
   - Risk mitigation
5. Policy
   - Principles and axioms
   - Policies
   - Standards
   - Guidelines & procedures
6. Organization
   - Structure
   - Reporting
   - Liaison
7. Asset management
   - Inventory
   - Classification
   - Ownership
8. HR security
   - Joiners
   - Movers
   - Leavers
   - Awareness, training & education
9. Physical & environmental
   - Physical access
   - Air conditioning
   - Fire & water
   - Power
10. Comms & ops management
    - Archives
    - Backups
    - Logging & alerting
    - Patching
    - Monitoring
    - Configs
11. Access control
    - Physical
    - Network
    - Systems
    - Applications
    - Functions
    - Data
12. Software development
    - Requirements
    - Design
    - Develop / acquire
    - Test
    - Implement
    - Maintain & support
13. Incident management
    - Prepare
    - Identify
    - React
    - Manage & contain
    - Resolve
    - Learn
14. Business continuity
    - Resilience
    - Disaster recovery
15. Compliance
    - Audit, SOX & Basel II
    - Policies
    - Laws & regulations
    - 3rd parties

19.6 INFORMATION SYSTEMS

Information systems exist to facilitate the decision-making process. Typical information systems in an asset business unit are as follows:

1. Executive decision support
2. Raw material supply planning
3. Product demand planning
4. Capacity planning
5. Operating plan integration/optimization
6. Operations scheduling
7. Process unit management
8. Oil movement and inventory control (refineries)
9. Custody receipts and shipments
10. Laboratory management
11. Quality management
12. Process shift worker scheduling
13. Energy/utility management
14. Operations performance analysis
15. Maintenance management overview
16. Work order planning and scheduling
17. Labor availability
18. Turnaround scheduling
19. Equipment records
20. Reliability management
21. Materials management
22. Contracted services management
23. Financial planning, analysis, and reporting
24. Accounting systems
25. Payroll
26. Personnel and training
27. Health and safety
28. Environmental management
29. Hazardous materials tracking
30. Engineering document management

In any asset business unit, these listed systems are present at different levels of computerization and integration.

Computerized information systems underpin management systems. Section 4.8, Underlying Computer Systems gives examples.

Computerized systems are either process plant IT systems or business IT systems. Generally, the information flow should be from the process plant IT systems to the business IT systems. Flow in the opposite direction requires extensive firewalling to ensure there is no disruption to production. Box 14.4, Resilience Example: The 2012 Rasgas Cyber-Attack gives an example.

BUSINESS IT SYSTEMS

Business IT systems tend to revolve around Enterprise Resource Management (ERM) systems. Examples include Oracle, SAP, and J.D. Edwards.

Box 19.3 gives an example of the application of SAP in a national oil company.

BOX 19.3 BUSINESS IT SYSTEMS EXAMPLE—NOC IMPLEMENTATION

Focus Area	SAP Category	Module
FA1 strategy	BW analytics	SEM business consolidation
FA6 asset management	Plant maintenance	Technical systems
		Maintenance order management
		Preventive maintenance planning
	Plant maintenance services	Service request processing
		Service execution and monitoring
	Mobile asset management	Fire truck inspection
		Fire and safety equipment inspection
	Materials management	Purchasing
		Inventory management
		Warehouse management
		Quality management
		Materials related master data
	External services	Contract planning, tendering and award
		Contract execution
		Contract variation
		Contract close out
		Service related master data
	Projects	Project initiation and approval
		Studies (feasibility and optimization)
		Engineering (FEED)
		Execution (EPIC/EPC)
		Planning and costing
		Close out
		E-service integration
FA2 HSSE	Environment, health and safety	Incident/accident management
		Waste management
		Hazardous substance management
		Risk assessment
		E-service integration
FA5 finance	BW analytics	Financial accounting
		Treasury management
		Funds management
FA7 personnel	Human resources	Organisation management
		Payroll
		Time management
		Recruitment
		Training and event management
		Personnel administration

BOX 19.3 BUSINESS IT SYSTEMS EXAMPLE—NOC IMPLEMENTATION—cont'd

Focus Area	SAP Category	Module
		Travel management
		E-service integration
FA9 shipping and commercial	Sales and distribution	Port services
		Land lease/rental services
		Utility services
		Other services
	IS—oil	Production and reconciliation
		Purchasing
		Tank and inventory management
		Sales and transfer
		Shipment planning and loading
		Billing

BW, Business Warehouse; *EPC*, engineer and procure contract; *EPIC*, engineer procure install and construct (contract); *FA*, focus area; *FEED*, front end engineering and design; *HSSE*, health, safety, security, and environment; *IS*, information system; *NOC*, National Oil Company; *SEM*, strategy enterprise management.

PROCESS PLANT IT SYSTEMS

The following are examples (mainly applied in refineries).

Administration

- Electronic document management system (EDMS): web-based

Strategic Planning

- Refinery-wide simulation
  - Petro-Sim
- Process evaluator
  - Icarus
- Refinery performance monitoring
  - SA Profile II

Technical Information

- Supply and demand planning
  - Orion PIMS
- Unit performance monitoring
  - HYSYS/Profimatics
- Technical Document Management System (TDMS)—drawing records system
  - CIMAGE

Plant Operation Information System (POIS)

- Blending
  - Oil Movement and Supply (OM&S)
- Product balancing and accounting
  - Product Balancing Management System (PBMS)
- Multivariable control
  - Advanced Process Control (APC)
- Refinery information
  - Real Time Information System (RTIS)
  - Laboratory Information Management System (LIMS)

Process Database

- Process Information (PI)
  - Real Time Information System (RTIS)/Process Historian Database (PHD)

Process Operation

- Distributed Control
  - Distributed Control System (DCS)
- Fail-safe Control
  - Emergency Shutdown System (ESD)
  - Fire and Gas (F&G)

Box 19.4 gives a graphic example of the integration of process plant IT systems and business IT systems.

BOX 19.4 EXAMPLE OF REFINERY IT INTEGRATION

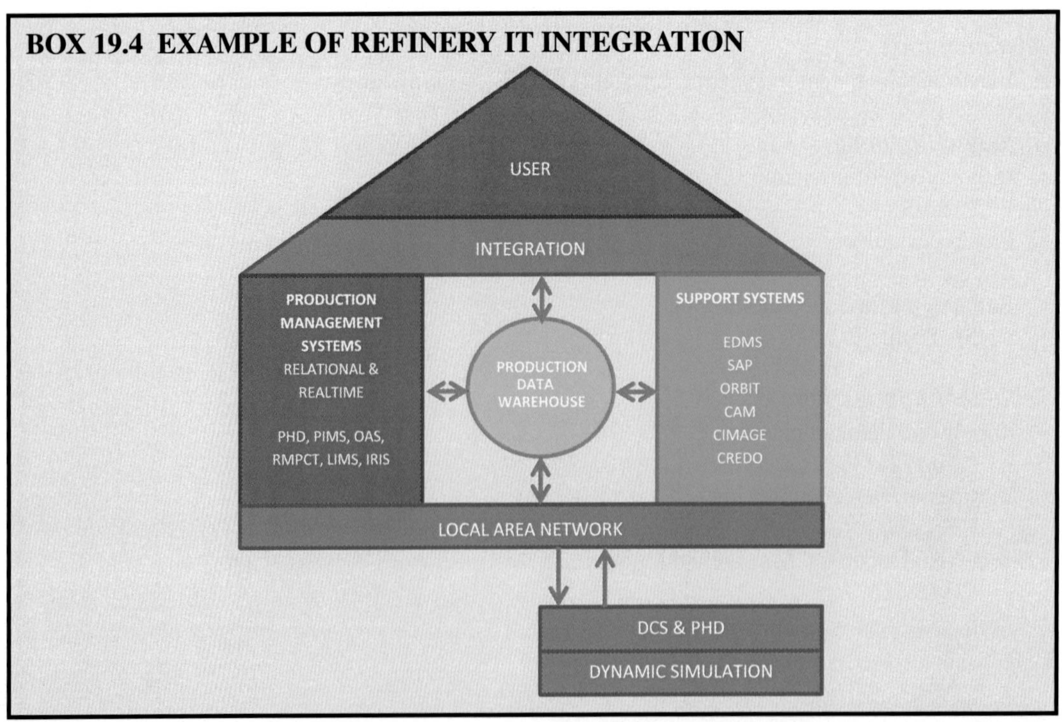

19.7 COLLECTION AND UTILIZATION OF INFORMATION

In the oil and gas industry, the flow of hydrocarbon accounting information from the process to the business is the most important process. This is described in detail in Chapter 36, Hydrocarbon Accounting.

The petrochemical industry relies heavily on international market statistics to determine production levels for specific product lines. Ongoing demand and pricing studies for petrochemical products are undertaken by organizations such as follows:

- Chemical Manufacturers Industry Association (CMIA). which is part of IHS[1]
- Strategic Business Analysis Ltd[2]

Other process industries, such as electricity and cement, use long-term national growth statistics as well as short-term (monthly) demand statistics. Short-term (monthly) demand statistics could be weather and season related. Electricity generation also has to look at daily demand fluctuations. Box 19.5 gives an example of electric power prediction.

BOX 19.5 POWER DEMAND PREDICTION MODELING

In the early 1970s, the University of Cape Town Mechanical Engineering Department developed a "power demand" computer model for the City of Cape Town Electricity Generation Department.

Generation was as follows:

1. Nuclear: steady baseload
2. Coal: variable baseload with ramp up time delay—*cheapest*
3. Hydro pump-storage: consumption at night to pump water to the upper dam and generation at peak load times from flow to the lower dam
4. Gas turbine (diesel fired)—instant load at peak times—*most expensive*

Power consumption would peak twice a day: early in the morning and in the evening at dinner time. Winter demand was also vastly different to summer demand. The model for each month of the year was built from historical records of demand for that particular month and the expected increase for the same month in the next year.

The model was used to anticipate when to use the different power sources, based on availability and cost of generation.

Comment

In many countries, electricity generation has been privatized. Environmentally friendly generation options, such as wind, solar, and wave energy, are now available, and households can even pump power back into the grid and get energy credits. (Many British homes now have solar photo-voltaic panels on the roof.) In some countries, customers can even select their generating company.

Performance data also need to flow from the process data records to strategic planning systems for analysis and reporting. A knowledge management example is shown in Box 19.6.

BOX 19.6 SHELL KNOWLEDGE MANAGEMENT SYSTEM

After an investment of $1.5 million in a knowledge management system using off-the-shelf collaborative software, Shell began developing what has evolved into 13 Web-based "communities of practice." This is now regularly tapped into by more than 10,000 Shell employees worldwide to share technical data and pose queries. Giving workers an easy way to share their wealth of knowledge has paid off. In its first full year of operating, Shell's system rang up business benefits totaling $200 million, in both decreased costs and new revenue.

19.8 THE MATURITY OF INFORMATION SYSTEMS

The maturity progression for knowledge management is described as follows:

CATEGORY 1: LAISSEZ-FAIRE

The importance and management of knowledge assets are recognized but action is spontaneous. There is no explicit strategy or resource allocation for information management.

CATEGORY 2: CONTENT BASED

Organized efforts to leverage knowledge are designed and formal systems for managing knowledge assets are built. Policies for entry, retrieval, storage, sharing, and utilization of information are designed. Focus is on internal transfer of information (within a system).

CATEGORY 3: NETWORK BASED

A move is made toward a global (internal within the business unit) knowledge management system. There is also a shift toward knowledge creation and innovation. Not only internal knowledge is leveraged but external knowledge is also leveraged (e.g., affiliate and benchmarking knowledge). Centers of excellence are recognized between affiliates.

CATEGORY 4: PROCESS BASED

There is true integration of the entire value chain (Core Systems). Continuous innovation is a key asset. There is a systematic design of intellectual capital rights (part of the reputation asset) to facilitate knowledge transfer between affiliates so as to allow affiliates to gain power by exchanging knowledge.

Box 19.7 gives an example of an assessment of the level of maturity of the knowledge management in a National Oil Company. The example shows that this company has some way to go to achieve integration of their entire value system.

BOX 19.7 NATIONAL OIL COMPANY KNOWLEDGE MANAGEMENT MATURITY STUDY

Green, implementation; **amber**, implementation in progress; **red**, no implementation.

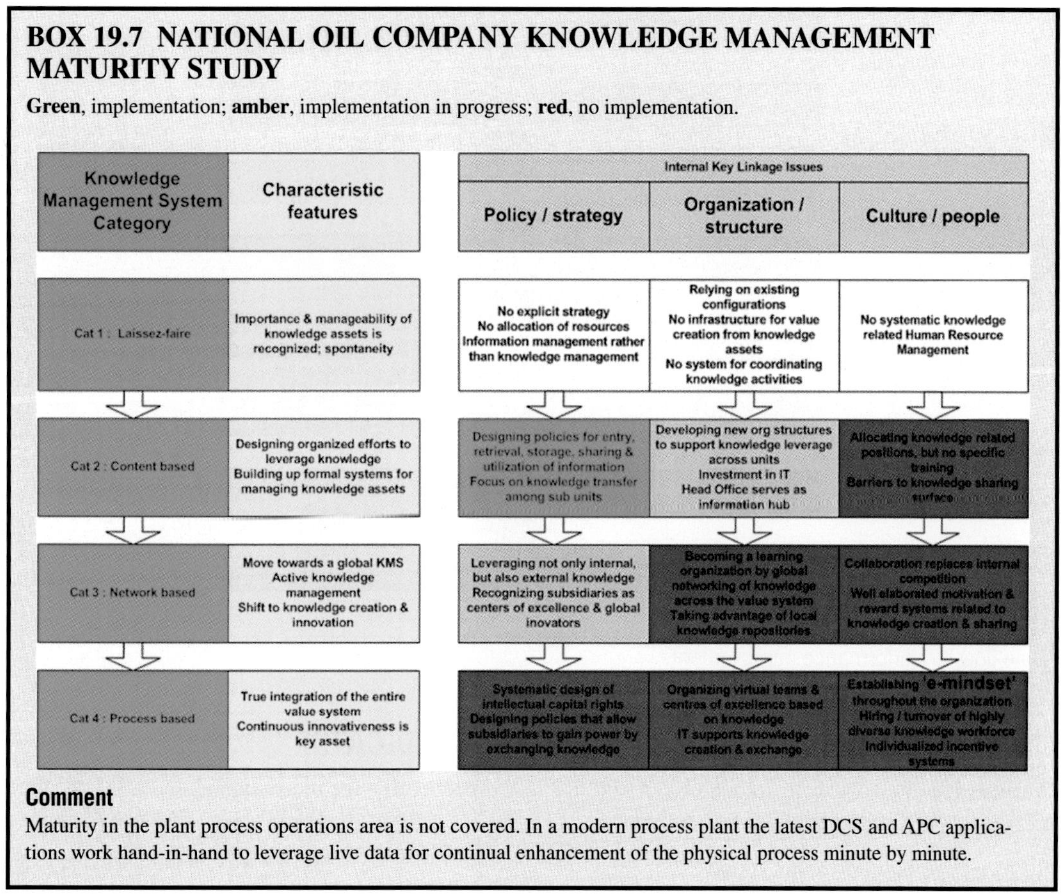

Knowledge Management System Category	Characteristic features	Internal Key Linkage Issues: Policy / strategy	Internal Key Linkage Issues: Organization / structure	Internal Key Linkage Issues: Culture / people
Cat 1 : Laissez-faire	Importance & manageability of knowledge assets is recognized; spontaneity	No explicit strategy No allocation of resources Information management rather than knowledge management	Relying on existing configurations No infrastructure for value creation from knowledge assets No system for coordinating knowledge activities	No systematic knowledge related Human Resource Management
Cat 2 : Content based	Designing organized efforts to leverage knowledge Building up formal systems for managing knowledge assets	Designing policies for entry, retrieval, storage, sharing & utilization of information Focus on knowledge transfer among sub units	Developing new org structures to support knowledge leverage across units Investment in IT Head Office serves as information hub	Allocating knowledge related positions, but no specific training Barriers to knowledge sharing surface
Cat 3 : Network based	Move towards a global KMS Active knowledge management Shift to knowledge creation & innovation	Leveraging not only internal, but also external knowledge Recognizing subsidiaries as centers of excellence & global inovators	Becoming a learning organization by global networking of knowledge across the value system Taking advantage of local knowledge repositories	Collaboration replaces internal competition Well elaborated motivation & reward systems related to knowledge creation & sharing
Cat 4 : Process based	True integration of the entire value system Continuous innovativeness is key asset	Systematic design of intellectual capital rights Designing policies that allow subsidiaries to gain power by exchanging knowledge	Organizing virtual teams & centres of excellence based on knowledge IT supports knowledge creation & exchange	Establishing 'e-mindset' throughout the organization Hiring / turnover of highly diverse knowledge workforce Individualized incentive systems

Comment

Maturity in the plant process operations area is not covered. In a modern process plant the latest DCS and APC applications work hand-in-hand to leverage live data for continual enhancement of the physical process minute by minute.

Chevron has progressed to Category 4, where web-based Centers of Excellence have been created for affiliates to access. Box 39.3, Chevron "Best Practice" Resource Map – Baldrige Award Criteria shows the outline.

19.9 DATA FLOW FOR PERFORMANCE IMPROVEMENT

Computerization of data ensures timely, accurate, and user-friendly access. The conversion of data into information that can be used for decision-making is critical for performance improvement. Information flow is summarized as shown in Fig. 19.2.

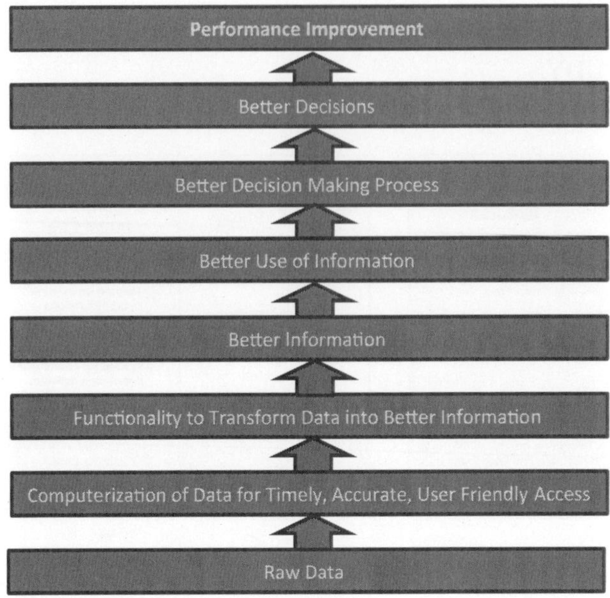

FIGURE 19.2

Data flow for performance improvement.

19.10 INFORMATION KEY PERFORMANCE INDICATORS

Various Key Performance Indicators (KPIs) are used specifically for the maintenance of IT systems. However, the fundamentals are as follows:

1. Backup planned versus actual
2. High-risk IT issues outstanding (number)
3. Audit findings outstanding beyond agreed implementation dates

The consolidation of data to obtain meaningful information for the production of various KPIs is the challenge.

19.11 SUMMARY

Good, quality information is critical for the optimum operation of the physical, personnel, and financial assets. Information becomes knowledge when it is applied in decision-making and further action. Knowledge is the principal productive resource of an organization. The essence of organizational capability is the integration of the individual's specialized knowledge to create value through transforming input into output. Best practice is the enactment of an information governance policy that oversees all IT related activities.

The control and security of information are critical. Compliance with ISO 27001 should ensure the selection of adequate and proportionate security controls that protect information assets and give confidence to interested parties. Computerized information systems underpin management systems.

We are now in the "knowledge age," where we expect to have all relevant knowledge available for effective decision-making. The move toward common platforms and simple interchange protocols is necessary, but with secure firewalls to ensure business continuity. KPIs are necessary to measure the health of the information systems. The maturity of the information systems is reflected in how well information is used for effectively managing management systems.

REFERENCES

1. IHS Chemical Market Advisory Service: Global Plastics & Polymers. https://www.ihs.com/products/chemical-market-plastics-polymers-global.html.
2. Strategic Business Analysis Ltd. www.sba-cci.com.

FURTHER READING

1. ISO/IEC 27001:2008. *Information technology – Security techniques – Information security management systems – Requirements.*
2. ISO/IEC 27005:2008. *IT – Security techniques – Information security risk management.*
3. *The risk IT practitioner guide*. ISACA; 2009.
4. Integrated Refinery Information System. *ChevronTexaco – Pembroke Refinery ERTC Computing Conference: June 10th–12th, 2002, Barcelona, Spain.*
5. PAS 77:2006. *IT service continuity management code of practice.*
6. Association for Information and Image Management (AIIM). http://www.aiim.org/.

CHAPTER

HUMAN CAPITAL

20

20.1 INTRODUCTION

Staff are regarded both as assets and resources. As their experience builds on the particular process units managed, they become more of an asset to the company. It becomes vital to document "required competencies" and establish formal training in these competencies should experience levels dwindle.

20.2 ORGANIZATIONAL CAPABILITY: THE HUMAN CAPITAL ASPECT

Organizational Capability is defined as:

Ability and capacity of an organization expressed in terms of the following:

1. **Human assets/**resources: their number, quality, skills, and experience
2. **Information assets/**resources: pool of knowledge, databases
3. **Financial assets/**resources: money and credit
4. **Physical assets**: plant and equipment, land, buildings
5. **Intangible/intellectual** resources: copyrights, designs, patents, etc.
6. **Stock**: consumed or converted as resources
7. **Natural** resources: oil, gas, minerals, water, wind, solar, etc.

Human Organizational Capability is the company's ability to manage people to gain competitive advantage. It is having the **right people** with the **right skills** available at the **right time** to meet current and future requirements to achieve the company objectives.

Retention and development of core competencies for running and growing the company is identified as a **Business Improvement Priority**: People and Capability Development (see Section 16.3: Approach to Selection).

Human Organizational Capability includes the following:

1. **Recruitment and retention** of the required skills
2. **Training** and **competence** maintenance
3. **Succession** planning
4. **Employee performance management across all levels** [including top-down cascaded alignment of personal SMART (Specific, Measurable, Achievable, Repeatable, Timely) objectives with higher-level objectives and the company Direction Statement]

Performance Management for the Oil, Gas, and Process Industries. http://dx.doi.org/10.1016/B978-0-12-810446-0.00020-7

Human Organizational Capability:

- **Focuses** on internal Systems and Processes for meeting customer needs
- **Ensures** that employee skills and efforts are directed toward achieving organizational goals and strategies
- **Creates** organization-specific competencies that provide **competitive advantage** since they are unique.

Traditional Sources of **Competitive Advantage** are as follows:

- **Economic/financial capability**: able to produce goods or services at lower cost than competitors
- **Technological capability**: products or services that customers receive are customarily innovative, high-quality, state-of-the art, in how they are built or delivered
- **Strategic/marketing capability**: products or goods that differentiate a company from its competitors, typically by "adding value" or "product–portfolio mix"

Two criteria for **Competitive Advantage**:

1. Adding perceived value to the customer
2. Offering uniqueness that cannot be easily imitated by a competitor

Strategic Planning (see Chapter 5: Business Cycle) generally focuses on **Resource Allocation** to leverage **Competitive Advantage** based on the previously mentioned **Capabilities**.

Human Organizational Capability focuses on achieving goals through employee commitment and competence.

Human Organizational Capability enhances perceived customer value in three ways:

1. **Responsiveness**: the ability of the business to understand and meet customer needs more quickly than competitors
2. **Relationships**: the ability of a business to develop an enduring relationship between customer and employee
3. **Service quality**: the ability of business to design, develop, and deliver service that meets or exceeds customer expectations

Human Organizational Capability **enhances uniqueness** because it is difficult to imitate. Imitation requires changing the way people think, act, and interact.

Social engineering of complex social processes such as culture, teamwork, and leadership are neither well understood nor easily replicated.

Four critical elements of capable organizations are as follows:

1. Shared mindset/values
2. Effective management practices
3. Capacity for change through understanding and managing organizational Systems
4. Leadership at all levels in the organization

These are expanded as follows:

1. Shared mindset/values
   - common understanding of objectives or goals (strategies) and means (Systems, Processes, activities)

- congruence between customer and employee expectations
- **embedded in corporate direction statement** (see Chapter 10: Alignment)

2. Management practices
   - policies, programs, operating procedures, and traditions that guide work
   - transform individual behavior to create customer satisfaction and consistency in how customer is treated
   - **complement and integrate with one another** to create common expectations, behaviors, and goals
3. Capacity for change
   - ability to reduce cycle time of all activities
   - Four Principles:
     - symbioses: ability to cope with external change; a bridge between internal action and external conditions
     - reflexiveness: ability to learn from past experiences; self-assessment and continuous learning
     - **alignment**: ability to integrate tasks, structures, Systems, and Processes, with political, technical, and cultural aspects of the company
     - self-renewal: ability to change over time successfully when needed
4. Leadership
   - passionately owns a Vision, which is promoted both within and outside the organization
   - translates external conditions into **Vision for organization** and how employees need to act to attain Vision
   - **empowers individuals** at all levels within the organization to act within his or her domain

Human Organizational Capability is at the heart of Richard Branson's book, "The Virgin Way."[1]

20.3 ORGANIZATIONAL EFFECTIVENESS AND LEARNING

Organizational Effectiveness and Learning measures human organizational capability and embeds continuous improvement into the mindset of employees. An appropriately worded company Value Statement would support this mindset. Section 10.5, The Development of a Policy or Direction Statement discusses Core Values.

ORGANIZATIONAL EFFECTIVENESS

Organizational Effectiveness is the measurement of Human Organizational Capability in aspects of the business.

- Most of the factors that individuals can influence so as to ensure Continuous Improvement are listed as follows:
  - leading versus managing
    - teamwork and cooperation
    - leadership role modeling
    - expanded leadership team/mobilization of all staff

- clear vision
- open and honest communication
- behavior-based performance management with focus on safety
- fact-based decision-making
- results versus activity focus
- effective meetings
- proactive versus reactive
- bottom-up–top-down dialogue
- willingness to take risks within the company risk appetite; no blame culture

- change: a sense of urgency; "a burning platform"
  - ownership of change initiatives
  - acceptance of change
  - short-term wins
  - short interval controls
  - sustainability of the change
- skills to do the work
  - performance reviews
  - adherence to job descriptions
  - competence training
- new Systems and Processes
  - sharing of Best Practices
  - eliminate silo mentality
  - Root Cause Analysis (RCA)
  - required data support

LINE OF SIGHT

Effectiveness is directly related to line of sight. Alignment is required for all performance indicators from the individual employee to shareholder aspirations. This can be cascaded as follows:

- Strategic Performance
- Operational Performance
- Team Performance
- Individual Performance

"SMART" is a useful acronym to focus on, which is as follows:

1. **S**imple
2. **M**easurable
3. **A**ligned
4. **R**epeatable
5. **T**imeliness

This is depicted in Fig. 20.1.

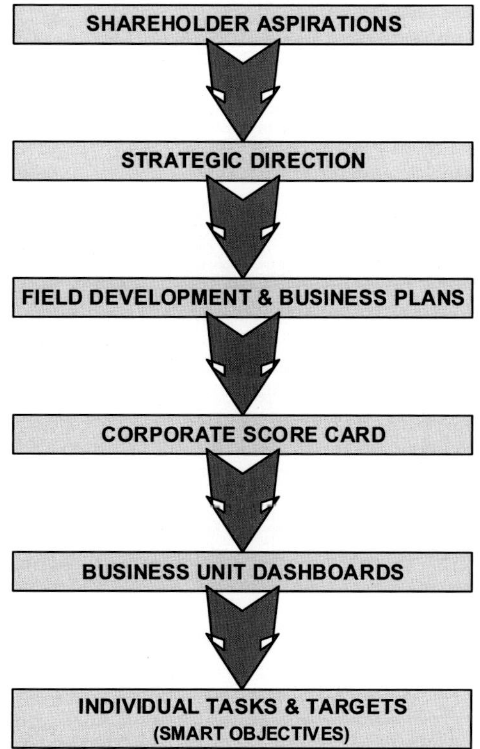

FIGURE 20.1

Vertical alignment example for an oil and gas company.

An example of the loss of effectiveness, where staff growth is unrelated to declining production, is given in Box 20.1.

ORGANIZATIONAL LEARNING

The central tenets of organizational learning are as follows:

- Managers **facilitate** rather than direct.
- Information flows and relationships between people are lateral as well as vertical (up and down).
- Organizations are pluralistic, where conflicting ideas and views are welcomed, surfaced, and become the basis of debate.
- Experimentation is the norm, so ideas are tried out in action and in turn become part of the learning process.

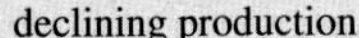

BOX 20.1 LOSS OF ALIGNMENT: NATIONAL OIL COMPANY DECLINING PRODUCTION AND SKYROCKETING STAFF

National Oil Companies (NOCs) are normally under pressure from the government to create employment for nationals and are thus generally overstaffed with nationals. The following graph is an example of the growth of staff in an NOC against declining production.

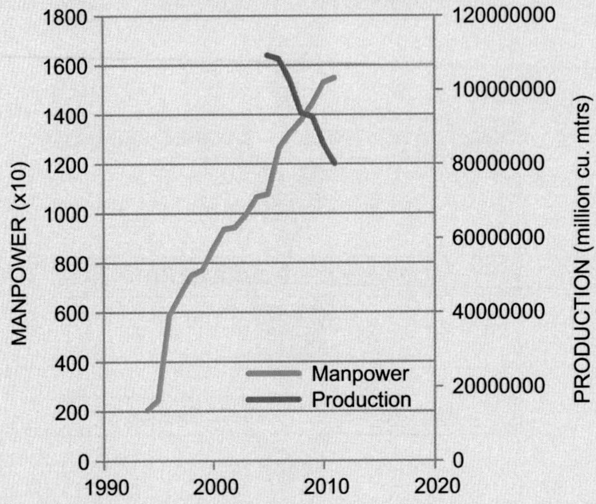

Comment

The question that should be asked of each staff member is:

"How does your job align with the strategic direction of the company, and what personal performance indicators align with the company Key Performance Indicators (KPIs)?"

(Soon after this graph was published, this company drastically reduced staff numbers by 4500, all at once.)

Box 6.5, High-Risk Processes, provides an example where challenging is not encouraged.

One also has to beware of the dangers of "group think." Box 20.2 gives an example of "group think" that led to a disaster.

BOX 20.2 GROUP THINK: CHALLENGER SPACE SHUTTLE O-RING FAILURE

On January 22, 1986, the Challenger space shuttle exploded shortly after liftoff.

The investigation into the disaster revealed that NASA scientists and engineers had been overly keen to start the mission because there had been a number of delays due to predicted bad weather (which had not actually materialized).

Prior to the launch, a component in a booster rocket called an "O-ring" had been identified by one engineer as a potential problem, but after a number of conference calls to discuss the problem, the decision was made to launch anyway.

Comment

This group exhibited a number of the conditions of groupthink. They were under pressure to launch the shuttle as soon as possible, and so they ignored any warnings that would compromise their goal.

There was also an apparent feeling of invulnerability because, up to that point, NASA had an excellent safety record and so they failed to fully examine the risks associated with their decision. This led to the tragic loss of the astronauts, a very expensive space shuttle, and the resulting loss of reputation of a near-perfect safety record.

20.4 COMPETENCIES

DEFINITION

Competence is a person's ability to accurately and reliably meet the performance requirements for a defined role.

Competence includes the skills and knowledge necessary to perform the required tasks successfully, the ability to recognize personal limits and so seek physical help or input from others when appropriate, and the conscientious application of skills and knowledge every time they are used.

Competence thus includes a behavioral element, i.e., ability to apply personal skills and knowledge in typical workplace situations.

Competence for each role should be managed as follows:

1. Identifying the required competencies
2. Providing relevant training (knowledge and skills)
3. Assuring and verifying these competencies (ability to apply knowledge and skills)
4. Refreshing competencies as appropriate

REQUIREMENTS OF STANDARDS

ISO 45001 and ISO 14001 are introduced as relevant to Human Capital in Section 18.4, Standards. Each of these standards has a section on "competence, training, and awareness."

Specific competencies are required in line with other standards, such as American Society of Mechanical Engineers (ASME) and American Petroleum Institute (API) standards for welders and nondestructive testing (NDT) inspectors.

The UK nuclear industry is required to have registers of Suitably Qualified and Experienced Persons (SQEP registers), which identify key safety-related roles and their training and experience requirements.

SPECIFIC TECHNICAL COMPETENCIES

Specific required technical competencies include:

Plant Process

The analogy of an aircraft shown in Section 22.4, Loss Of Primary Containment: Analogy Between an Aircraft and a Process Plant, is very relevant to process panel operators who are required to do simulator training, just as pilots are required to do flight simulator training. Box 20.3 discusses **inadequate** operator competency training. Competencies required for abnormal operation must be practiced on a simulator on a regular basis.

BOX 20.3 PLANT PROCESS COMPETENCIES: THE SIMULATOR

Modern refineries are built with a simulator room where operator training commences even before construction is complete.

Standard practice is to have all operators undertake simulator training for their respective process units before allowing them to work on the live units. After working competently on the panel of the live unit under supervision, they qualify as panel operators for their specialist process unit. Periodic retesting is undertaken on the simulator to expose operators to experiences that, hopefully, may never happen in reality. Cross-training also takes place on the simulator before operators qualify for operating other process units. Regular refresher training on the simulator is recommended.

Continued

BOX 20.3 PLANT PROCESS COMPETENCIES: THE SIMULATOR—cont'd

A technical audit in a particular refinery revealed that operators were initially trained on the simulator and only returned to the simulator should they need to be trained on another process unit.

The audit recommended that operator simulator training be repeated at regular intervals for the same process unit.

Analogy: Annual pilot simulator training as discussed in Section 22.4, Loss of Primary Containment: Analogy Between an Aircraft and a Process Plant

Comment

The installation of a complete control room simulator is also common practice when building a nuclear power station.

Maintenance

Maintenance staff, who are required to calibrate and test safety and critical process measuring instruments and equipment, need to be qualified by an external authority.

Welders, specifically, need to be qualified to a standard such as ASME IX. Tests are required and documented, and qualifications are valid for a limited period before retesting and recertification.

Inspection

Inspectors have to be certified to certain industry standards. X-ray certification is an example.

Health, Safety, and Environment

All staff require basic Health, Safety, and Environment (HSE) training. Offshore operation requires enhanced safety and escape training.

Maintenance and Inspection Staff for Turnarounds

Additional staff, with high levels of competence, are required for turnarounds. These are sourced from competent turnaround contractors, experienced in the type of turnaround to be undertaken. International Oil Companies tend to draw secondees from other affiliates for particular turnarounds since they know the required company standards and procedures.

The impact of inadequate competencies on a turnaround could be devastating to the company. An example is discussed in Box 9.6, The Jewel in the Crown: The Rise and Fall of South Africa's Electricity Supply Commission: Koeberg Nuclear Power Station.

THE VALUE OF SECONDEES

Secondees boost the technical "know-how" of the company. Industry good practice is that, besides receiving secondees, an affiliate should also send employees to affiliated companies in other countries as secondees.

Typically, in Joint Ventures (JVs), the shareholders make complementary contributions to the JV. Generally, one partner to the venture has a higher level of technical expertise who supplies a number of secondees at the beginning of the project to ensure a flawless startup and initial operation.

When cutting costs, shareholders tend to look at "expensive" secondees. An example of using higher numbers of secondees for a "new technology" startup is shown in Box 20.4.

> **BOX 20.4 SECONDEES: GAS TO LIQUID PLANT**
>
> A recently commissioned Gas to Liquid (GTL) plant had a higher number of secondees than associated LNG and sales gas plants in which a particular shareholder had interests.
>
> This shareholder wanted to pressure the company to reduce secondees from the other Joint Venture partner. The counter argument was that the GTL plant involved new technology and required more specialist GTL operating experience.
>
> At the time of reporting, secondees in the GTL joint venture have remained constant.

Secondee roles and responsibilities are discussed in Section 34.7, The Value of Secondees.

STRUCTURED COMPETENCY TRAINING

Apprenticeships, where trainees work under experienced staff over an extended period, have become more difficult as older and more experienced staff go into retirement. To address this problem, the oil and gas industry established a competency-based system for training in specialist oil and gas skills in 2001. This is outlined in Box 20.5.

> **BOX 20.5 COMPETENCIES: PETROSKILLS[2]**
>
> BP, Shell, and other oil and gas companies created the PetroSkills Alliance in 2001 to accomplish a mission of vital importance for the future of the industry.
>
> The Alliance provides high-quality, competency-based training that is critically important for all oil and gas companies, but not unique to any particular company, allowing Alliance members to more strategically leverage their investments in training, competency development, and competency assurance.

COMPETENCY AND PERFORMANCE MANAGEMENT SYSTEMS

As discussed in Section 20.3, a structured approach is required to ensure alignment between the individual and the shareholders' aspirations. A Personnel Performance Management System, designed to motivate all staff, is thus required. Box 20.6 gives an example of the potential for high performance of a company.

> **BOX 20.6 COMPETENCIES: MOTIVATION**
>
> A certain "pacesetter" LNG producer has performance and competency systems that transcribe into highly motivated staff who have been employed by the company for a long time.
>
> In recent benchmarking exercises, they have consistently shown a very low annual staff turnover of 2.5%, which is the best in the industry.
>
> This reflects indirectly in all other performance indicators. Well-structured competency and personnel performance systems are well-established, and safety and maintenance Key Performance Indicators (KPIs) are also among the best in the industry.
>
> See also Box 22.7, System Maturity: Gas Plants.

COMPETENCIES AND ORGANIZATIONAL KNOWLEDGE MATURITY

Further to the discussion on the subject of maturity of knowledge management/information systems in Chapter 19, Information, the level of maturity of staff competencies to manage this knowledge is important. The categories could be depicted as follows:

Category 1: Laissez-Faire

- No systematic knowledge-related Human Resource Management

Category 2: Content-based

- Allocating knowledge-related positions but not providing specific training
- Barriers to knowledge-sharing surface

Category 3: Network-based

- **Collaboration** replaces internal competition
- Well-elaborated motivation and reward systems related to knowledge creation and sharing

Category 4: Process-based

- Individualized incentive schemes

Box 20.7 gives an example of a system that does **not** encourage **collaboration** as discussed previously.

BOX 20.7 PERSONNEL MANAGEMENT SYSTEMS: FORCED RANKING

A National Oil Company (NOC) developed a Personnel Performance System based on Managerial/Behavioral and Technical Skills. Individual targets for the year were agreed and set, based on cascaded targets linked to the business objectives of the company. Midyear self-assessments were carried out, and annual assessments resulted in varying annual bonuses for staff.

However, the system started to unravel when Technical competencies were dropped from the assessments and forced ranking of individuals was undertaken. The spread was as follows:

Rating 1	Rating 2	Rating 3	Rating 4	Rating 5
10%	20%	62%	6%	2%

The forced ranking was calibrated to ensure adherence to the distribution curve guidelines. In other words, the numbers of staff in a department are distributed by percentage between each rating.

As staff were not rated on their Technical competencies, they just focused on pleasing their manager and did not offer critical or objective advice. Teamwork was thus affected as individuals were after good assessments for themselves.

Comment

Oil and Gas companies must have the highest levels of Technical competencies and have to encourage teamwork. The alternative is an elevated risk to the company operations.

20.5 PROCESS SAFETY MANAGEMENT AND SAFETY CULTURE

To quote from the Baker Report:

> "The passing of time without a process safety accident is not necessarily an indication that all is well."

Box 13.4, BP and the Oil Industry After the Texas City Refinery Disaster, discusses this issue.

In the past, Process Safety Management (PSM) focused on the management of major risks, while Occupational Health and Safety focused on protecting the Human Assets of the company and other Stakeholders from the potential negative impact that the physical assets could have on people and the environment.

As a result of the Texas City Refinery Disaster and other major incidents, the focus has now changed. Primary focus areas are as follows:

1. Corporate **Safety Culture**
2. Process Safety/Integrity Management **Systems**
3. **Performance** Evaluation, Corrective Action, and Corporate Oversight

Two key questions:

1. How well do we ensure the integrity of the operation?
2. How will we **know** we are doing it?

The relationship between Health and Safety and the Integrity of the physical asset is a complex one. The common denominator is the Human, as indicated in the following guidelines.

PROCESS SAFETY MANAGEMENT FRAMEWORKS

The Energy Institute (EI)[3] has produced a high-level framework for PSM, "**PSM Framework**," and the EI Process Safety Survey, "**EIPSS**."

The PSM framework is a comprehensive PSM framework that captures industry good practice in PSM and provides the energy industry with a consistent and effective approach to answer the two key questions with absolute confidence.

The EIPSS is an in-depth survey based on the PSM framework, allowing companies to assess their process safety arrangements and benchmark performance against industry good practice.

Box 20.8 gives the focus areas and elements of the PSM framework.

BOX 20.8 ENERGY INSTITUTE PROCESS SAFETY MANAGEMENT FRAMEWORK

EI PSM consists of 4 focus areas and 20 elements as follows:

Process Safety Leadership

1. Leadership commitment and responsibility
2. Identification and compliance with legislation and industry standards
3. Employee selection, placement, competency, and health assurance
4. Workforce involvement
5. Communications with stakeholders

Risk Identification and Assessment

6. Hazard identification and risk assessment
7. Documentation, records, and knowledge management

Risk Management

8. Operating manuals and procedures
9. Process and operational status monitoring and handover
10. Management of operational interfaces

Continued

BOX 20.8 ENERGY INSTITUTE PROCESS SAFETY MANAGEMENT FRAMEWORK—con'd

11. Standards and practices
12. Management of change and project management
13. Operational readiness and process startup
14. Emergency preparedness
15. Inspection and maintenance
16. Management of safety critical devices
17. Work control, permit to work, and test risk management
18. Contractor and supplier, selection, and management

Review and Improvement

19. Incident reporting and investigation
20. Audit, assurance, management review, and intervention

The US equivalent is the US Center for Chemical Process Safety guidelines,[4] which is summarized in Box 20.9.

BOX 20.9 *CENTER FOR CHEMICAL PROCESS SAFETY GUIDELINES FOR RISK BASED PROCESS SAFETY, 2007*

The guidelines are based on four pillars:

1. Commit to process safety
2. Understand hazards and risk
3. Manage risk
4. Learn from experience.

Subjects discussed include:

- Process Safety Culture
- Compliance with Standards
- Process Safety Competency
- Workforce Involvement
- Stakeholder Outreach
- Process Knowledge Management
- Hazard Identification and Risk Analysis
- Operating Procedures
- Safe Work Practices
- Asset Integrity and Reliability
- Contractor Management
- Training and Performance Assurance
- Management of Change
- Operational Readiness
- Conduct of Operations
- Emergency Management
- Incident Investigation
- Measurement and Metrics
- Auditing
- Management Review and Continuous Improvement
- Implementation and the Future

CCPS.

PSM is linked to Asset Integrity. This relationship is discussed in detail in Section 22.5, Asset Integrity and Process Safety Management.

CORPORATE SAFETY CULTURE

A primary element of PSM is Corporate Culture.

Corporate Culture can be defined as:

> shared values (what is important) and beliefs (how things work), which interact with a company's structure and control systems (our focus) to produce behavioral norms (the way we do things around here).[5]

Corporate Safety Culture is an integral part of Corporate Culture, which, in turn, is determined by the Company's Core Values. Section 10.5, The Development of a Policy or Direction Statement, discusses Core Values.

The "human factor" in HSE is introduced in Section 12.2, Accident/Incident Causation Theory.

Falling incident rates are discussed in Section 15.10, Health, Safety, Environment, and Integrity: Change of Focus.

The focus now is on the promotion of a Safety Culture.

HOW DO WE CHANGE A CULTURE?

The steps are no different to discussions in Section 30.6, Change Management, but require special emphasis, as a safety culture change affects all staff **and** contractors.

The same questions are asked as in Chapter 1, What Is Performance Management?:

1. Where are we?
   - Audits, site visits and inspections, hazard reporting, incident analysis, and RCA
2. Where do we want to be?
   - Benchmark HSE performance against others, external and internal surveys, statistical and trend analysis
3. How do we get there?
   - Gap analysis, establish resources to bridge the gaps, understand the culture of the company, in particular its readiness to change, develop an action plan, and implement and track changes

For successful change management, two issues are particularly important: leadership and readiness to change. A literal "burning platform" is often the trigger for change to a safer operation. Change of attitude to safety is described in Box 20.10.

BOX 20.10 BEHAVIOR-BASED SAFETY: STOP CARDS

Single Point Mooring Maintenance Incident[7]

A routine maintenance on a Single Point Mooring (SPM) was planned by a contractor. It would appear that the workboat staff uncoupled the line without testing for gas. Gas subsequently escaped and was ignited, probably by the workboat engine. The ensuing explosion **killed 7 people** on the workboat, including the captain.

Lesson Learned

Deaths apparently occurred as a result of unsafe practices.

Empowerment Initiative

Subsequent to this incident, the Operating Company initiated a Behavior-Based Safety (BBS) program where all employees and contractors are empowered to **stop** and record unsafe acts using a **"STOP"** card. The "number of STOP cards issued" is now a Key Performance Indicator (KPI) and required to be reported to the company shareholders every quarter.

Section 30.6, Change Management, discusses the change management process.

ASSESSING A SAFETY CULTURE[6]

A measure of safety culture has been developed in five levels:

1. Pathological: no one knows or cares about safety.
2. Reactive: improvements are only made after a serious incident.
3. Calculative: complex management systems are used to encourage and monitor safe working practices.
4. Proactive: people try to avoid problems occurring and exist in a constant state of awareness.
5. Generative: safety is integral to everything we do.

These five levels are depicted in Fig. 20.2.

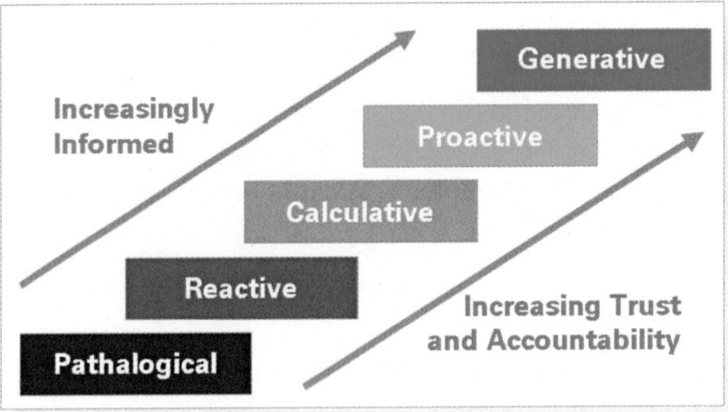

FIGURE 20.2

Five levels of safety culture.

In a survey of Oil and Gas Producer member companies, these levels were used to assess the levels of safety culture using the 20 elements shown in Box 20.11.

BOX 20.11 OIL AND GAS PRODUCER 20 ELEMENTS[6]

1. Benchmarking, Trends, and Statistics
2. Audits and Reviews
3. Incident/accident reporting, investigation, and analysis
4. Safety reports
5. Who causes accidents in the eyes of management?
6. What happens after an accident? Is the feedback loop being closed?
7. How do safety meetings feel?
8. Work planning, including Permit To Work (PTW)
9. Contractor management
10. Standards setting and by whom
11. Competency/training: Are workers interested?

BOX 20.11 OIL AND GAS PRODUCER 20 ELEMENTS[6]—cont'd

12. Worksite hazard management techniques
13. Who checks safety on a day-to-day basis?
14. Balance between safety and profitability
15. Is management interested in informing workforce about safety issues?
16. Commitment level of workforce and level of care for colleagues
17. How do you get new/improved procedures?
18. What is the purpose of procedures?
19. What is the size/status of the Health, Safety, and Environment (HSE) department?
20. What are the rewards of good safety performance?

A top-class safety culture would include the following:

1. Benchmarking outside the industry.
2. Searching for nonobvious problems with self and cross-audits.
3. Aggregating data across business functions to look for trends and issues that need to bc addressed.
4. Involvement of Senior management on a routine basis, which includes setting reporting goals.
5. Management looking to themselves to assess what could be improved, and taking a broad view looking at the interaction of systems and people.
6. Visible involvement of top management in public activity after an accident, and employees taking accidents involving others personally.
7. Informal safety meetings, which feel like an opportunity for communication.
8. Polished planning processes with anticipation and review of work progress with employees trusted to do most planning: less paper, more thinking, and processes are well-known and disseminated.
9. Integrated workforce of contractor and company staff including: shared information leading to integration of policies, procedures, and practices; postponement of work if no contractor meets the HSE requirements; joint training and competency programs.
10. The standards regulator being influenced by the company in the setting of higher standards with standards defined by the workforce.
11. Attitudes becoming as important as knowledge and skills. Training seen as a process rather than an event. Needs identified and methods of training suggested by the workforce, who are seen as an integral part of the process rather than just passive receivers.
12. Job safety analysis revised regularly in a process. People (workers and supervisors) not afraid to tell each other about hazards
13. Everybody checking safety, looking out for themselves and their workmates. Supervisor inspections largely unnecessary. There is no problem with insisting on shutdowns (of operations) for safety reasons.
14. The balance between safety and profit becoming a nonissue. The company accepting delays to get contractors up to standard in terms of safety. Management believing that safety makes money.
15. A transparent two-way communication process in which managers get more information back than they provide.
16. Contractors included in commitment to safety and care for their fellow workers from day one. Levels of commitment and care very high and driven by employees.
17. Procedures developed by the workforce and reviewed constantly.

18. Employees are trusted. Noncompliance goes through recognized channels. Procedures refined for efficiency.
19. HSE staff minimal.
20. Recognition for good safety performance is itself seen as high value. Tokens are not given, as the workforce knows they perform well. Evaluation is process-based.

IMPLEMENTING BEHAVIOR-BASED SAFETY PROGRAMS

An important component of PSM is the establishment of Behavior-Based Safety (BBS) Programs, where all staff are empowered to intervene when unsafe acts are committed.

After the Texas City Disaster, BP introduced a BBS program, which is outlined in Box 20.12 (the progression is depicted as shown in Fig. 20.2).

BOX 20.12 BP'S ORGANIZATIONAL CULTURE FOCUS

Assessment and improvement tools developed and applied globally:

1. **People Assurance Survey (PAS)**
   a. "Culture pulse check"
2. **Safety culture survey**: broadly applicable web-based tool and coaching support
3. **Specific assessment and improvement tools**, e.g.:
   a. Refinery HRO assessment tool for in-depth operations review
   b. Incident and Injury Free (IIF) programs
4. **Leadership**

20.6 HUMAN ASSET KEY PERFORMANCE INDICATORS

Those indicators most affecting relationships with other assets are as follows:

- Staff turnover
- Critical positions open for more than 3 months
- Assessment of technical competencies
- Technical competencies development plans
- Implementation and maintenance of a Safety Culture
- Fatalities
- Lost Time Incidents
- Major environmental spills
  - air
  - water
- Total Greenhouse Gas emissions (direct and indirect)

It is important to note that Key Performance Indicators (KPIs) must not be looked at in isolation. Leading indicators that are produced elsewhere **may preempt an incident**. KPIs and the safety mindset of staff go hand in hand.

Examples are:

- *Increased emergency work orders*
- *Increased panel alarms* per *hour*
- *Increased number of STOP cards issued*

Appendix B, Key Performance Indicators, lists possible KPIs for the Human Capital Asset and Health and Safety KPIs that people are able to influence and possible Environmental KPIs that people can influence.

20.7 SUMMARY

Human Organizational Capability is the company's ability to manage people to gain competitive advantage. It is having the **right people** with the **right skills** available at the **right time** to meet the current and future requirements to achieve the company objectives. Human Organizational Capability includes recruitment and retention of the required skills, training and competence maintenance, succession planning, and employee performance management across all levels.

Organizational effectiveness and learning measures the human organizational capability and embeds continuous improvement into the mindset of employees.

Competence is a person's ability to accurately and reliably meet the performance requirements for a defined role. Motivation and competency are essential, and so systems need to be designed to maximize the potential of every individual. Each staff member needs objectives and targets that align with the company Vision and Mission (SMARTs). Specific competencies are required in line with other standards, such as ASME and API standards for welders and NDT inspectors.

Secondees boost the technical "know-how" of the company.

Should experience levels drop, it is important to have structured competency-based training to operate and maintain the process within accepted risk levels. Petroskills is an industry "cooperative" that ensures competencies (knowledge and skills) in the process industry are retained for future generations.

The relationship between Health and Safety and the Integrity of the physical asset is a complex one. The common denominator is the Human. BBS needs to be embedded into each individual's responsibilities and accountabilities, so they are encouraged to be open to report and discuss safety issues with anyone, even up to the CEO.

Top Personnel KPIs are considered as important as the top Physical Asset KPIs.

REFERENCES

1. Branson R. *The virgin way – how to listen, learn and lead.* Penguin Random House; 2014.
2. PetroSkills Competency System (PCS). www.petroskills.com.
3. Energy Institute (EI) Process Safety Management (PSM) framework. https://www.energyinst.org/technical/PSM/PSM-framework.
4. *CCPS Guidelines for Risk Based Process Safety*. 2007.
5. International Association of Oil & Gas Producers (OGP). *A guide to selecting appropriate tools to improve HSE culture. Report No: 435*. March 2010.

6. International Association of Oil & Gas Producers (OGP). 20 elements of safety culture. http://www.iogp.org/human-factors#2643395-20-elements.
7. Qatar incident. http://www.hazardexonthenet.net/article/50279/Tugboat-explosion-kills-seven-off-Qatar.aspx.

FURTHER READING

1. Johnson, Scholes, Whittington. *Exploring corporate strategy*. Pearson; 2008.
2. Burke R, Barron S. *Project management leadership*. Burke Publishing; 2007.
3. *The report of the BP US refineries independent safety review panel*. January 2007. [Baker Report].
4. Gladwell M. *What the dog saw. Chapter: 'The talent myth'*. Penguin; 2009.
5. *Human factors performance indicators for the energy & related process industries*. Dec 2010. [EI Research Report].
6. ISO 45001:2016. *Occupational health & safety management systems – requirements*.
7. *Guide to petroleum professional certification for petroleum engineers*. SPE; 2005.
8. Hudson P. Centre for Safety Research, Leiden University, The Netherlands. *Safety management and safety culture the long, hard and winding road*. http://www.caa.lv/upload/userfiles/files/SMS/Read%20first%20quick%20overview/Hudson%20Long%20Hard%20Winding%20Road.pdf.
9. Senge P, et al. *The fifth discipline field book*. London: Nicholas Brealey; 1994.

CHAPTER

FINANCE

21

21.1 INTRODUCTION

This chapter discusses finance as an asset. The annual operating budget and the capital budget are the two key elements.

21.2 STANDARDS

The primary standards for managing financial assets are the International Financial Reporting Standards (IFRS).[1]

WHAT ARE INTERNATIONAL FINANCIAL REPORTING STANDARDS?

Accounting provides companies, investors, regulators, and others with a standardized way to describe the financial performance of an entity. Accounting standards present preparers of financial statements with a set of rules to ensure standardization across the market. Companies listed on public stock exchanges are legally required to publish financial statements in accordance with the relevant accounting standards.

IFRS is a single set of accounting standards, developed and maintained by the **International Accounting Standards Board (IASB)** with the intention of those standards being applied on a globally consistent basis—whether by developed, emerging, or developing economies—thus enabling comparison of the financial performance of publicly listed companies on a like-for-like basis with international peers.

IFRS are now mandated for use by more than 100 countries, including the European Union and by more than two-thirds of the G20.

IFRS are developed by the **International Accounting Standards Board (IASB)**, the standard-setting body of the IFRS Foundation—a public-interest organization with award-winning levels of transparency and stakeholder participation. The IASB's 14-member board is appointed and overseen by 22 trustees from around the world, who are in turn accountable to a monitoring board of public authorities.

21.3 FINANCIAL RISKS AND UNDERLYING COMPUTER SYSTEMS

Section 11.4, Categories of Risk, discusses four categories of risk, with finance/reporting being one of the categories.

Project financial risks are managed within the project risk management system during design and construction and up to handover of the completed plant to the operator. During the project phases, there are two primary risks: project cost and startup date [when the plant will begin to produce product for sale and thus start to give a return on investment (ROI)]. Increased costs or late startup will increase the risk of not getting the required ROI. During the project phase, there are also generic financial risks such as fluctuating exchange and interest rates.

Performance Management for the Oil, Gas, and Process Industries. http://dx.doi.org/10.1016/B978-0-12-810446-0.00021-9

Section 33.4, Investment Risk Assessment and Section 33.5, Project Risk Mitigation, discusses project risk in detail.

Once the project has been completed, the asset is registered in the company asset register and handed over to the operator. The ROI is then tracked for the operating life of the investment. Various computer systems are used to assess project risk and manage project expenditure.

Generic financial risk categories could include the following:

- Currency exchange
- Credit
- Interest rates

Revenue and expenditure management is managed within an Enterprise Resource Management (ERM) system such as SAP[2] or Oracle. See Fig. 21.1, showing the rolling up of major financial risks.

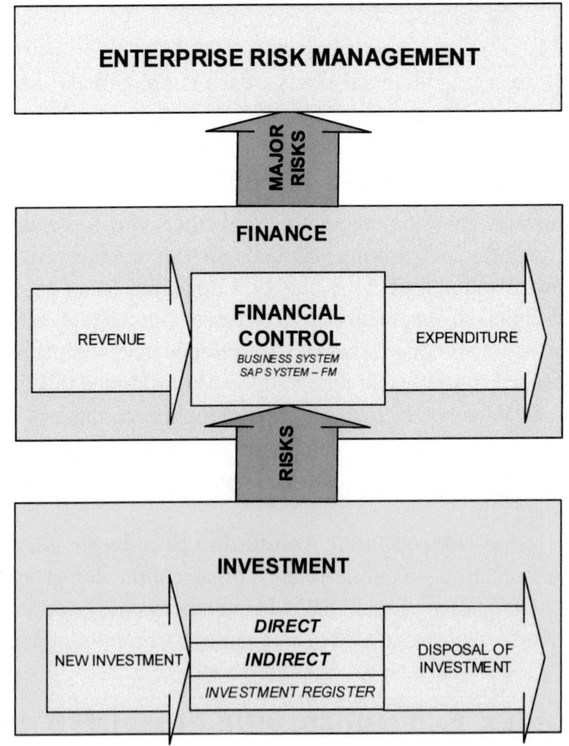

FIGURE 21.1

Finance model with risk rollup.

21.4 OPERATING EXPENSE

Activity-based costing (ABC) identifies the least profitable systems. Costs are directly or indirectly allocated to particular Core Systems. Core Systems be a gas processing train, a refining unit, a petrochemical product train, or a power plant.

The cost categories need to be grouped for comparative studies (benchmarking). Table 21.1 shows the required contents that are typical in the oil and gas and petrochemical industries. This is normally completed in the form of a spreadsheet.

Table 21.1 Operating Expense Categories

		Variable Costs
A.		Purchased Energy costs
B.		Purchased Non-Energy Utilities costs
C.		Process Materials costs
D.		Other Variable Costs

		Fixed Costs
E.		Own Personnel salary related costs
	1	Operators
	2	Maintenance and Engineering staff
	3	Support staff
	4	Trainees
F.		Non-Maintenance Contract Services costs
G.		Maintenance and Engineering costs for Surface facilities
	1	Maintenance Labour – Contractors
	2	Maintenance Materials - supplied by contractors
	3	Maintenance Materials - own supply
	4	Capitalised maintenance costs
H.		Property charges
I.		Insurance costs
J.		Environmental costs
K.		Other Fixed costs

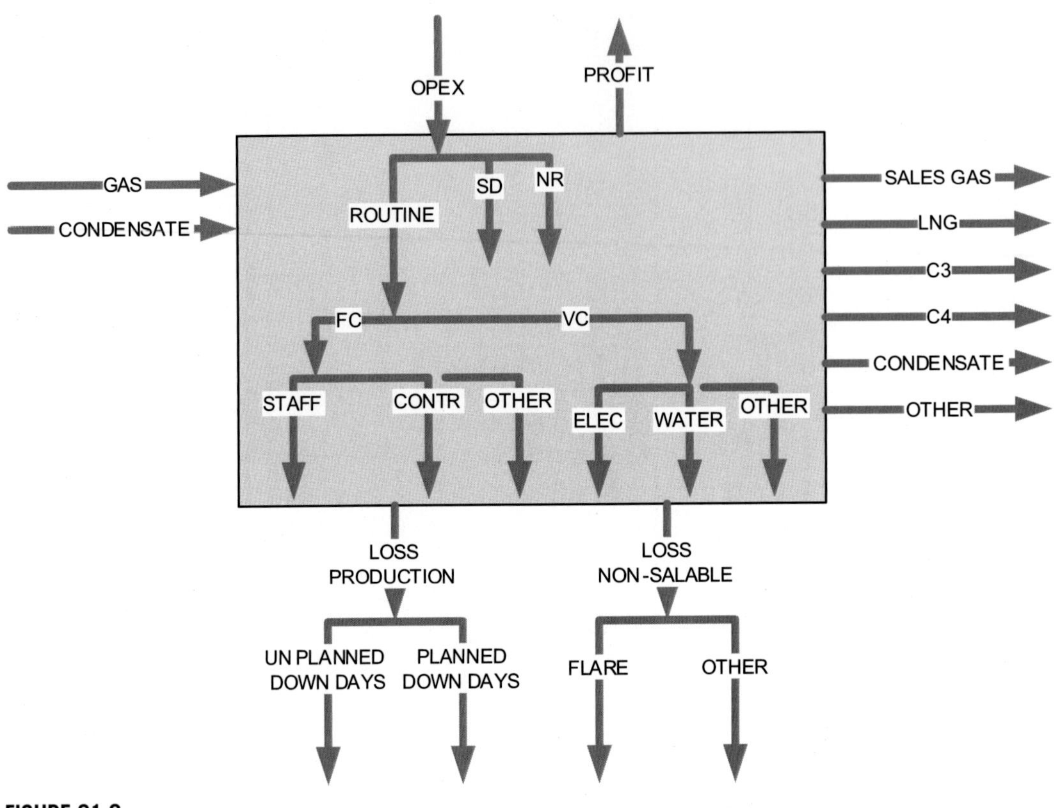

FIGURE 21.2

Gas plant Opex model.

Fig. 21.2 is an example of an operating expense (Opex) model for a gas plant. A refinery or petrochemical plant would be similar but with different raw materials and products.

Planned shutdowns (SDs) or turnarounds and other nonrecurring (NR) costs are normally separated from the routine operating expense.

Box 25.4, Example of Cost Comparison Using Replacement Value (RV) as Denominator gives an example of cost comparison using the model in Fig. 21.2.

21.5 CAPITAL EXPENSE/INVESTMENT

Capital is a measure of all the cash that has been deposited in a company over its lifetime, irrespective of its financing sources. Funding of minor capital projects is generally from the company revenue account, on condition that there is a strong cash flow or regulatory/integrity requirement and the project value is within certain limits. Major capital expense (capex) could be funded from loans and/or joint venture partners.

Motivation for capital expense could be one or more of the following:

1. ROI
2. Regulatory requirement (health, safety, environment, or other)
3. Asset integrity

An example of ROI and environmental justification for investment is discussed in Box 21.1.

BOX 21.1 CAPEX JUSTIFICATION—ENVIRONMENTAL COMPLIANCE AND RETURN ON INVESTMENT

During the past 15 years, the LNG industry has grown exponentially with significant advances in technology (see also Box 33.16: Qatar's Partnering Strategy—Exxon-Mobil). The flaring of boil-off gas (BOG) when loading LNG tankers has become a major environmental issue due to the massive increase in export of LNG. Rasgas and Qatargas, two large integrated gas plants in Qatar, joined forces to invest in a BOG recovery system. The scale of the project made it partially economically viable, with the recycling of the BOG to the process units. It is anticipated that carbon credits for the project may also be traded in the future, thus making it even more financially feasible (see Section 21.8).

Capital expense is discussed further in Chapter 33, The Opportunity Life Cycle.

21.6 FINANCIAL STATEMENTS—THE BASICS

QUALITATIVE CHARACTERISTICS OF FINANCIAL STATEMENTS

Qualitative characteristics of financial statements include the following:

- Understandability
- Reliability
- Comparability
- Relevance

ELEMENTS OF FINANCIAL STATEMENTS

The financial position of a company is primarily provided in the statement of financial position. The elements include the following:

1. **Assets**: an asset is a resource controlled by the enterprise as a result of past events from which future economic benefits are expected to flow to the enterprise.
2. **Liabilities**: a liability is a present obligation of the enterprise arising from past events, the settlement of which is expected to result in an outflow from the company's assets.
3. **Equity**: equity is the residual interest in the assets of the company after deducting all the liabilities under the historical cost accounting model. Equity is also known as owner's equity. Under the units of **constant purchasing power model** equity is the **constant real value** of shareholders' equity.

Fig. 21.3 shows the value of the physical asset with equity and liabilities superimposed on it.

FIGURE 21.3

Assets and liabilities—example.

The financial performance of an enterprise is primarily provided in an income statement or profit and loss account. The elements of an income statement or the elements that measure the financial performance are as follows:

1. **Revenues**: increases in economic benefit during an accounting period in the form of inflows or enhancements of assets, or decrease of liabilities that result in increases in equity. However, it does not include any contributions made by the equity participants, i.e., proprietor, partners and shareholders.
2. **Expenses**: decreases in economic benefits during an accounting period in the form of outflows, or depletions of assets or in currencies of liabilities that result in decreases in equity.

Revenues and expenses are measured in nominal monetary units under the historical cost accounting model and in units of constant purchasing power (inflation-adjusted) under the units of constant purchasing power model.

21.7 COST ESCALATION

Comparison of costs becomes difficult when different approaches to cost escalation are used, in particular for comparing budgets for similar joint ventures and the determination of replacement value for benchmarking purposes.

MONEY OF THE DAY AND REAL TERMS

Annual budgets require cost projections for the next 5–7 years. Some companies' reports are based on money of the day (MOD), while others are in real terms (RT). The basis for cost projections needs to be clearly stated for complete clarity.

Section 17.5, Money of the Day (MOD) and Real Terms (RT) discusses the relationship.

REPLACEMENT VALUE

Replacement value is normally estimated based on the escalation from the original invested value, using baseline costs, or based on the cost of recent similar projects. Replacement value is usually required for insurance purposes. However, benchmarking consultants derive this calculation in different ways for their particular purposes.

COST ESCALATION INDICES

A reliable source for cost escalation in the refining industry is the *Oil and Gas Journal*, which publishes Nelson-Farrar Cost Indices for Refinery Construction and Refinery Operations.[3]

Box 21.2 discusses examples of cost escalation for benchmarking purposes.

BOX 21.2 COST ESCALATION FOR BENCHMARKING

The two most popular benchmarking consultants for oil and gas and petrochemicals are Solomon Associates and Philip Townsend Associates International (PTAI) (see Chapter 24: The Benchmarking Process and Consultant Selection). Both consultants use asset replacement value (RV) as a denominator for their benchmarking performance indicators, although they use different methods to arrive at this value.

Solomon Associates

The Solomon RV is based on the US Gulf Coast figures for typical unit size and then adjusted for the following:

a. Reported size
b. Location
c. Inflation (to date of benchmarking period)
d. Sulfur

Statistics for this calculation are based on those reported in the *Oil and Gas Journal*.

Philip Townsend Associates International

PTAI uses the current insured value based on an insurance assessor's report. However, this could vary enormously from assessor to assessor.

21.8 CARBON CREDITS[4]

INTRODUCTION

To address the issues of global warming, the United Nations Framework Convention on Climate Change (UNFCCC) was adopted in 1992, with the objective of limiting the concentration of greenhouse gases (GHGs) in the atmosphere. Subsequently, to supplement the convention, the Kyoto Protocol came into force in February 2005. This sets limits on the maximum amount of emissions allowed by countries. At the time of writing, the Kyoto Protocol committed 41 countries to reduce their GHG emissions by at least 5% below their 1990 baseline emission within the commitment period of 2008–12. As per the Kyoto Protocol, developing and least developed countries are not bound to reduce their GHG emissions.

MECHANISMS TO MEET EMISSION REDUCTION TARGETS

The Kyoto Protocol provides three market-based mechanisms to facilitate the countries and, in turn, the local businesses and entities, to meet the emission reduction targets as follows:

Joint Implementation

Under Joint Implementation (JI), a developed country with a relatively high cost of domestic GHG reduction can set up a project in another developed country that has relatively low cost and earn carbon credits that may be applied to their emission targets.

Clean Development Mechanism

Under Clean Development Mechanism (CDM), a developed country can take up a GHG reduction project activity in a developing country where the cost of GHG reduction is usually much lower and the developed country would then be given carbon credit for meeting its emission reduction targets.

Entities in a developing or least developed country can also set up a GHG reduction project, get it approved by UNFCCC and earn carbon credits. Such carbon credits generated can be bought by entities of developed countries with emission reduction targets.

International Emission Trading

Under International Emission Trading (IET), developed countries with emission reduction targets can simply trade on the international carbon credit market. This implies that entities of developed countries exceeding their emission limits can buy carbon credits from those whose actual emissions are below their set limits. Carbon credits can be exchanged between businesses or bought and sold on the international market at the prevailing market prices. IET serves the objectives of both developed countries with emission reduction targets who are buyers of carbon credits, as well as developing and least developed countries with no emission targets who are the sellers of carbon credits.

POSITION OF INTERNATIONAL ACCOUNTING STANDARDS

At the time of writing, there is no accounting standard or interpretation within the International Financial Reporting Standards (IFRS) that deals with the accounting of Certified Emission Reductions (CERs): 1 CER = 1 metric ton of CO_2 equivalent.

STAGES IN GENERATION AND SALE OF CERTIFIED EMISSION REDUCTIONS

To be eligible for CDM benefits, the proposed project must have the feature of "additionality." In other words, the CDM project must provide reductions in emissions that are additional to that which would occur in the absence of the project.

Example:

A company can generate CERs under CDM if it installs a waste heat boiler (WHB) that saves energy.

Registration/Accreditation of a Clean Development Mechanism Project

A company desirous of setting up a CDM project needs to register the project with the CDM Executive Board of the UNFCCC.

Monitoring, Verification, and Issuance of Certified Emission Reductions

Once the project is registered and becomes operational, the performance is monitored and verified periodically to determine whether emissions reductions have taken place before the CDM Executive Board can issue CERs.

Sale/Trade of Certified Emission Reductions

The CERs obtained by the company can also be traded.

Box 21.3 discusses a CDM project that did not receive CERs.

> **BOX 21.3 CARBON EMISSION REDUCTION—FLARE REDUCTION PROJECT**
>
> A Production Sharing Agreement (PSA) contractor implemented a flare reduction project on its offshore oil platform. Regrettably, the planned CERs were not obtained as the proper process was not followed for registration/accreditation of a clean development mechanism project with the CDM Executive Board of the UNFCCC.
>
> **Comment**
>
> It is advisable to stick strictly to the requirements of UNFCCC when implementing a CDM project. As this story illustrates, a lot of time and money can be needlessly wasted by not following the registration process.

21.9 PROFIT AND INVESTMENT

In oil refining terms, gross margin is the difference between the raw material price and the selling price per barrel of oil. Net margin, or profit per barrel of oil, is the gross margin minus the operating expense. Box 21.4 discusses the effect of the type of crude on profit, based on the products a refinery can extract using its existing assets.

> **BOX 21.4 REFINING MARGINS EXAMPLE—CRACK SPREAD[5]**
>
> Crack spread is a term used in the oil industry and futures trading for the differential between the price of crude oil and petroleum products extracted from it—that is, the profit margin that an oil refinery can expect to make by "cracking" crude oil (breaking its long-chain hydrocarbons into useful shorter-chain petroleum products).
>
> For simplicity, most refiners wishing to hedge their price exposures have used a crack ratio usually expressed as X:Y:Z where X represents a number of barrels of crude oil, Y represents a number of barrels of gasoline, and Z represents a number of barrels of distillate fuel oil, subject to the constraint that $X=Y+Z$. This crack ratio is used for hedging purposes by buying X barrels of crude oil and selling Y barrels of gasoline and Z barrels of distillate in the futures market.
>
> Widely used crack spreads have included 3:2:1, 5:3:2, and 2:1:1. As the 3:2:1 crack spread is the most popular of these, widely quoted crack spread benchmarks are the "Gulf Coast 3:2:1" and the "Chicago 3:2:1."

Further to discussion in Section 2.7, Integration, large integrated complexes that extend the supply chain tend to be more profitable. As the profits pour in, the tendency is to reinvest further down the supply chain.

Some integrated complexes can convert low-grade crude oil into highly profitable products such as gasoline and diesel, as well as various petrochemicals.

Box 21.5 gives an example of extraordinary gross margins and continuing investment.

BOX 21.5 RELIANCE REFINERY—PROFITS AND INVESTMENT[6]

Reliance Refinery, in India, is the largest integrated refinery in the world (as at the time of writing) and is owned by Reliance Industries Limited (RIL). Profits are high, relative to industry norms. Reliance Refinery is among a handful of global refiners with the ability to process low-grade crude into high-value products and switch between fuels depending on market prices.

The following is an extract of the gross refinery margins over nine quarters:

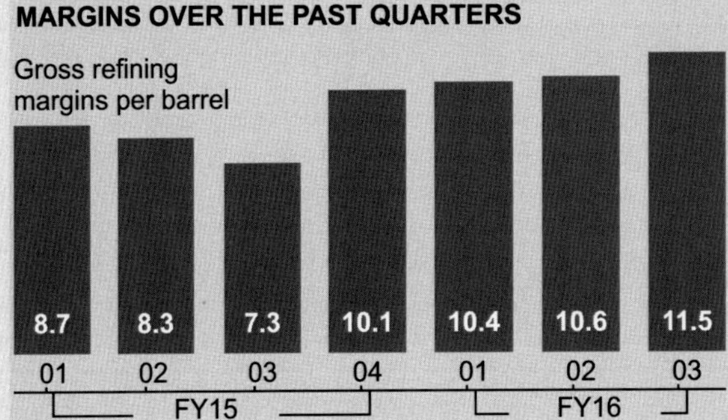

However, the company continues to expand its petrochemical production. As of early 2016, RIL is investing $16 billion in expanding petrochemical production. Of this, $4.6 billion is for a project to convert captive petcoke to synthetic gas; $4.5 billion for a refinery off-gas cracker to extract ethane, ethylene, propylene, butanes, and propanes; $5 billion to expand polyester production; and $1.5 billion to import ethane from the United States to replace costly propane imports and naphtha. See also Box 27.7, Reliance Refinery Benchmarking Performance.

21.10 FINANCE KEY PERFORMANCE INDICATORS

Finance Key Performance Indicators (KPIs) can be divided as follows:

A. Business
B. Benchmarking
C. Operations (Opex)
D. Projects (Capex)
E. CDM credits

A. BUSINESS

There are a number of standard business KPIs, but we will concentrate on just two:

1. Profit
2. ROI

Profit

Profit is reported in many forms. The more common forms are as follows:

- Net profit ($)
- Net profit margin (%)
- Gross profit margin (%)
- Operating profit margin (%)
- Earnings before interest and tax (EBIT; $)
- Earnings before interest, tax, depreciation and amortization (EBITDA; $)

Investment

Investment is reported in many forms. The more common forms are as follows:

- ROI: used to assess potential investments

$$\text{ROI} = (\text{Gain from investment} - \text{Cost of investment}) / \text{Cost of investment}.$$

- Return on capital employed (ROCE)

$$\text{ROCE} = \text{EBIT/Total Capital employed}.$$

ROCE is regarded by some as the best measure of historical capital productivity. This can be related to the **efficiency** of the investment as discussed in Section 15.8, Performance Gap. An example of the use of ROCE is discussed in Box 21.6.

> **BOX 21.6 ROCE APPLICATION—EXXONMOBIL[7]**
>
> ExxonMobil has used a modified version of the ROCE for many years. It regards this as the best measure for historical capital productivity in such a long-term capital-intensive industry, both to evaluate management's performance and demonstrate to shareholders that capital has been used wisely over a long term.

B. BENCHMARKING

KPIs are normalized and are reported in

1. $ per Mechanical Unit (MUc)
   Or
2. $ per Normal Shift Position (NSP)
   Or
3. $ per Equivalent Distillation Capacity (EDC)
   Or
4. $ per Equivalent Gas Processing Capacity (EGPC)

C. OPERATING EXPENSE

The common measurement is cost per amount of product. Examples are as follows:

- Power plants generally report in $ per KWH.
- Gas plants normally report in $ per 1 million BTUs for flowing gas and $/ton for salable products.
- Refineries report in cost per barrel of crude oil into the refinery.

As indicated, reporting for refineries is normally based on the operating expense per barrel of crude oil. However, when the market price of the product decreases yet the crude price remains the same, profits are squeezed. Fig. 21.4 shows the relationships.

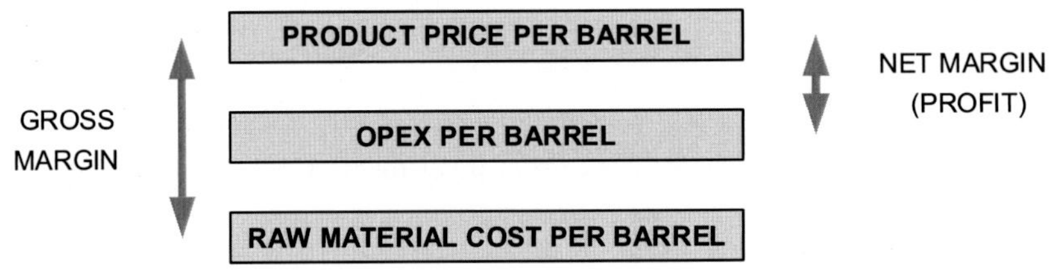

FIGURE 21.4

Refining gross and net margins.

D. CAPITAL EXPENSE[8]

The following are used for specific projects.

- ROI
- Actual versus budget
- Cost variance (CV): the difference between the value of work completed to a point in time, usually the data date, and the actual costs to the same point in time
  - CV = Earned Value (EV) − Actual Cost (AC) to date
- Estimate at completion (EAC): the expected total cost of completing all work expressed as the sum of all actual cost to date and the estimate to complete. If future work will be accomplished at the planned rate the following equation is applicable.
  - $EAC = AC + [(BAC - EV)/(CPI \times SPI)]$

where BAC is the budget at completion, CPI is the cost performance index and SPI is the schedule performance index.

E. CLEAN DEVELOPMENT MECHANISM CREDITS

Actual CERs versus total CERs, where total CERs = actual CERs + potential CERs.

21.11 SUMMARY

International Financial Reporting Standards are the standards for financial reporting throughout most of the world.

The financial risks throughout the lifecycle of the investment, in both the project and operating phases, are rolled up into the enterprise risk management system.

Standard operating expense (Opex) budget templates are useful for comparison to previous budgets and the budgets of similar operations (benchmarking).

Capital expense (Capex) must always have a justification and, where possible, a return on investment (ROI). Other than economic benefit (ROI), justification includes categories such as regulatory requirements, health and safety, and environment.

The financial position of a company is primarily provided in the statement of financial position, which includes assets, liabilities, and equity.

A standard approach to cost escalation calculations is required for benchmarking (for replacement value at time of benchmarking), budgets (for predictions of 5–7 years), as well as major investment projects taking a number of years (for cost escalation throughout the project period).

Carbon credits can be revenue generators when the process for applying for certified emission reductions is strictly adhered to.

Primary business finance KPIs are net profit after tax and return on capital used. However, the underlying Opex and Capex KPIs are required to understand the basis of the performance of the business. Benchmarking KPIs are required for comparison to the competition.

In summary, control and monitoring of the performance of the operating and capital budgets are critical for optimum value addition.

REFERENCES

1. International Financial Reporting Standards (IFRS). http://www.ifrs.org/pages/default.aspx.
2. Enterprise-wide financial control SAP Business Planning and Consolidation application. www.sap.com/solutions/performancemanagement.
3. Oil and Gas Journal Nelson-Farrar Cost Indices. http://www.ogj.com/articles/print/volume-114/issue-3/processing/nelson-farrar-cost-indexes.html.
4. United Nations Framework Convention on Climate Change (UNFCCC). http://unfccc.int/essential_background/convention/items/6036.php.
5. Crack spread. Wikipedia: https://en.wikipedia.org/wiki/Crack_spread.
6. Pathak K. *Reliance refinery margins to remain strong*. Mumbai: Business Standard; January 20, 2016. http://www.business-standard.com/article/companies/why-does-ril-enjoy-unmatched-refinery-margins-116012000645_1.html.
7. Marr B. *Key performance indicators*. Pearson; 2012.
8. ANSI/PMI 99-001-2013. *A Guide to the Project Management Body of Knowledge (PMBOK)*. 5th ed. [Chapter 7: Project Cost Management].

CHAPTER

PHYSICAL ASSETS

22

22.1 INTRODUCTION

This chapter discusses those factors that ensure that the physical asset performs at design capacity for the life of the investment, without major loss of primary containment (LOPC).

WHAT IS ASSET MANAGEMENT?

BS PAS 55: 2008 definition[1]

> "systematic and coordinated activities and practices through which an organization optimally and sustainably manages its assets and asset systems, their associated performance, risks and expenditures over their life cycles for the purpose of achieving its organizational strategic plan."

22.2 ASSET MANAGEMENT SYSTEM STANDARDS

The following are the three standards in the ISO 55000 suite:

1. ISO 55000:2014
   - provides an overview of asset management, its principles and terminology, and the expected benefits of adopting asset management. It can be applied to all types of assets and by all types and sizes of organizations. It also provides the context for ISO 55001 and ISO 55002.
2. ISO 55001
   - specifies requirements for an asset management system.
3. ISO 55002
   - gives guidelines for the application of ISO 55001.

These standards supersede BS PAS 55: 2008.

BENEFITS OF ADOPTING THE ISO 55000 SUITE OF STANDARDS

The adoption of this suite of standards enables an organization to achieve its objectives through the effective and efficient management of its assets. The application of an asset management system provides assurance that those objectives can be achieved consistently and sustainably over time.

Performance Management for the Oil, Gas, and Process Industries. http://dx.doi.org/10.1016/B978-0-12-810446-0.00022-0

22.3 ASSET PERFORMANCE MANAGEMENT AND INTEGRITY

There is a direct link between an optimized operating cycle and the profitability of the business, and this link is critical for the survival of the business. The integrity of the plant is tested as the boundaries of performance are pushed. The key is to MAXIMIZE **long-term availability** to operate at design capacity.

The focus must be **on (hydrocarbon) containment** for the duration of the investment while **optimizing throughput**. Personnel safety and prevention of damage to the environment are integral parts of this.

Minimizing downtime (planned and unplanned) and Loss Of Primary Containment (LOPC) over the long term leads to maximum availability, plant integrity, and sustainability. Fig. 22.1 depicts the relationships.

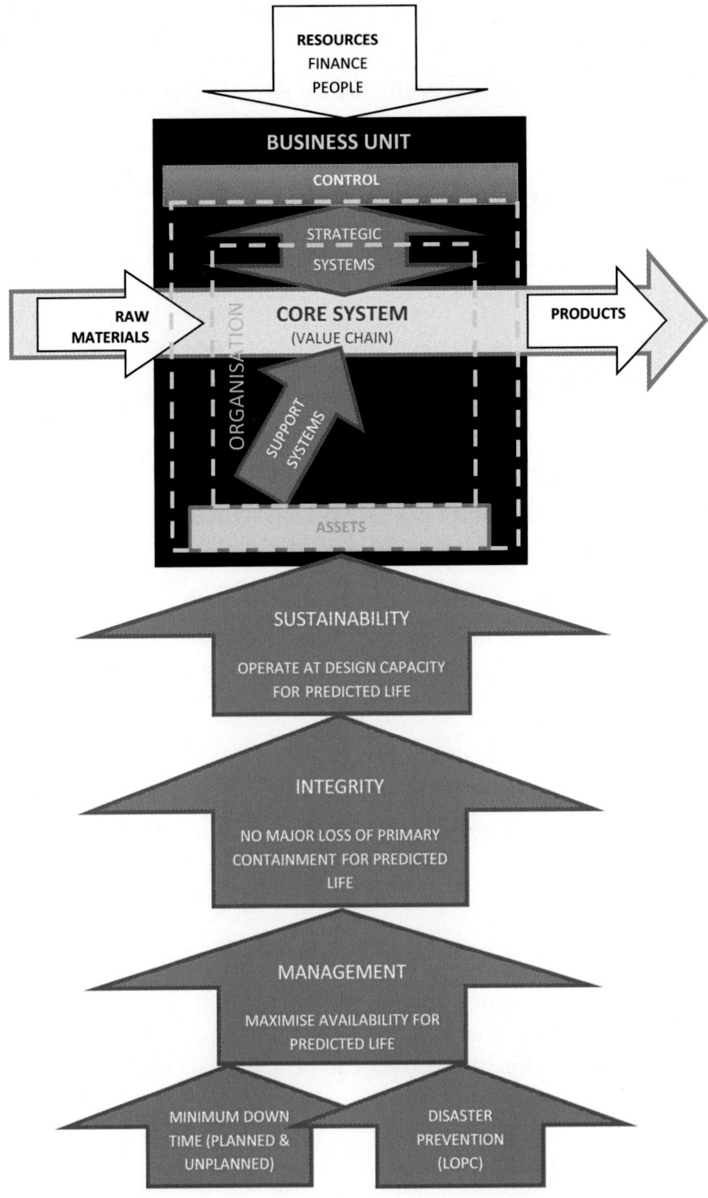

FIGURE 22.1

Performance management/integrity relationship.

22.4 LOPC: ANALOGY BETWEEN AN AIRCRAFT AND A PROCESS PLANT

Lessons can be learned from other industries. The similarities between the aircraft industry and the process industry are discussed here. A passenger aircraft is simply a pressure vessel. This is depicted as shown in Fig. 22.2.

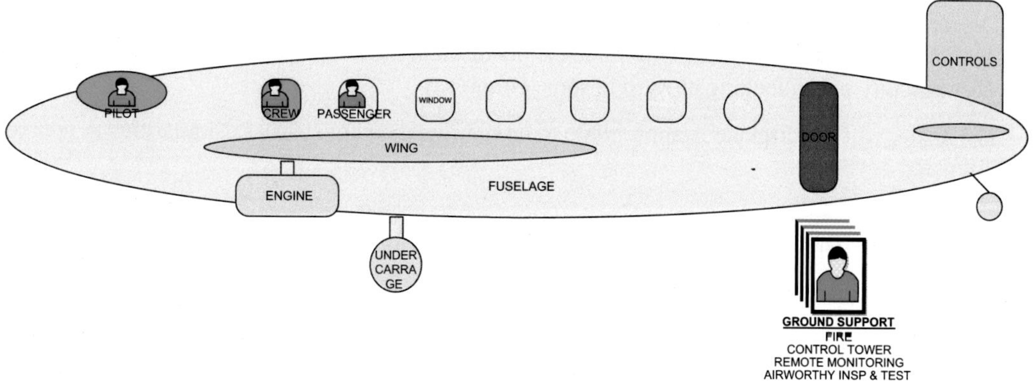

FIGURE 22.2

Passenger aircraft analogy.

The primary containment of a process plant is no different. This can be depicted as shown in Fig. 22.3. Both figures are color coded for identification of **pressure containment**, **operation**, and **maintenance**.

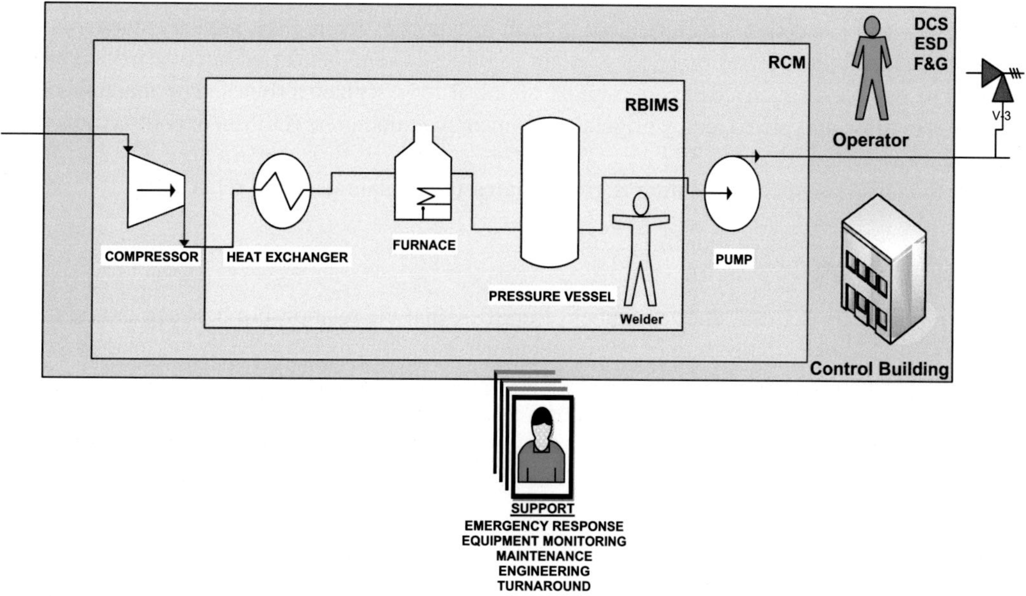

FIGURE 22.3

Process plant primary containment.

In each case, should there be a major LOPC, lives could be lost and the asset destroyed.

Section 14.12, Accidents and Knowledge Transfer: Do We Learn From Mistakes? identifies the following for discussion:

- **Learning from other industries**
- Refresher and simulator training
- Documenting systems and processes to counter loss of corporate memory
- Application of the highest standards and best practices in the industry
- The difficulty in challenging upward in some cultures

The analogy of the airplane is appropriate to emphasize the above issues. Similarities in approach are discussed as follows:

FAIL-SAFE DESIGN

Aircraft doors and steam boiler oval manways both swing inward, preventing any chance of bursting outward.

OPERATION

Simulator training is undertaken on a regular basis for both pilots of aircraft and console operators of process plants so as to practice eventualities that, hopefully, may never happen. Before each flight the pilot performs a physical inspection of the airplane. Similarly, the operator tours his or her area of responsibility at the start of the shift.

MAINTENANCE AND INSPECTION

Maintenance and inspection are carried out in both instances to the highest industry standards in line with documented industry best practices. Certain activities have to be carried out by certified persons.

Aircraft engines in many cases are almost identical to gas plant turbines. The maintenance and inspection of these are carried out by original equipment manufacturers (OEMs) or contractors certified by them.

In both industries, special attention is given to pressure containment and relief systems.

HUMAN FACTORS

The aircraft industry learned, after a number of crashes, that the relationship between the captain and his first officer is crucial. It needs to be an **equal** partnership. The process industry has learned from this with their "behaviour-based safety programs" where anyone can challenge a safety issue and even take the issue to the chief executive of the company.

The psychological aspect of working under extreme stress in an emergency situation is also an area for study in both industries.

22.5 ASSET INTEGRITY AND PROCESS SAFETY MANAGEMENT

An integrated approach to safety management has evolved because it has become apparent that good Occupational Health and Safety (OHS) performance does not guarantee major incident prevention. In contrast to occupational injuries, large losses are typically the result of the failure of multiple safety barriers, often with complex scenarios. These are difficult to identify using just a simple experience-based hazard identification and risk assessment process.

Process safety management (PSM) is introduced in Secion 20.5, Process Safety Management (PSM) and Safety Culture. The relationship between PSM and Asset Integrity Management (AIM) overlaps and is a critical one.

Example: An asset integrity Key Performance Indicator (KPI) can foresee a process safety incident.

The Shell categorization of safety and integrity, showing overlaps, is given in Box 22.1.

BOX 22.1 SHELL SAFETY AND INTEGRITY CATEGORIZATION

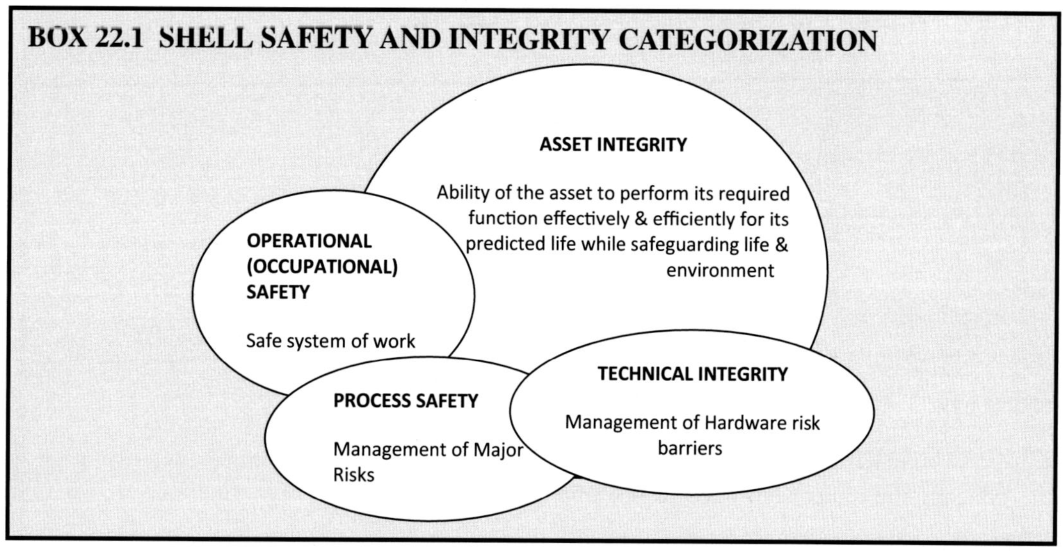

The "Swiss cheese" model in Fig. 22.4 shows the total integration of process safety with asset integrity embedded in each slice of cheese with **varying levels of contribution**. The "slices" are as follows:

1. Prevention: containment
2. Detection: gas/flame
3. Control: emergency shutdown (ESD)
4. Mitigate: fire protection, deluge

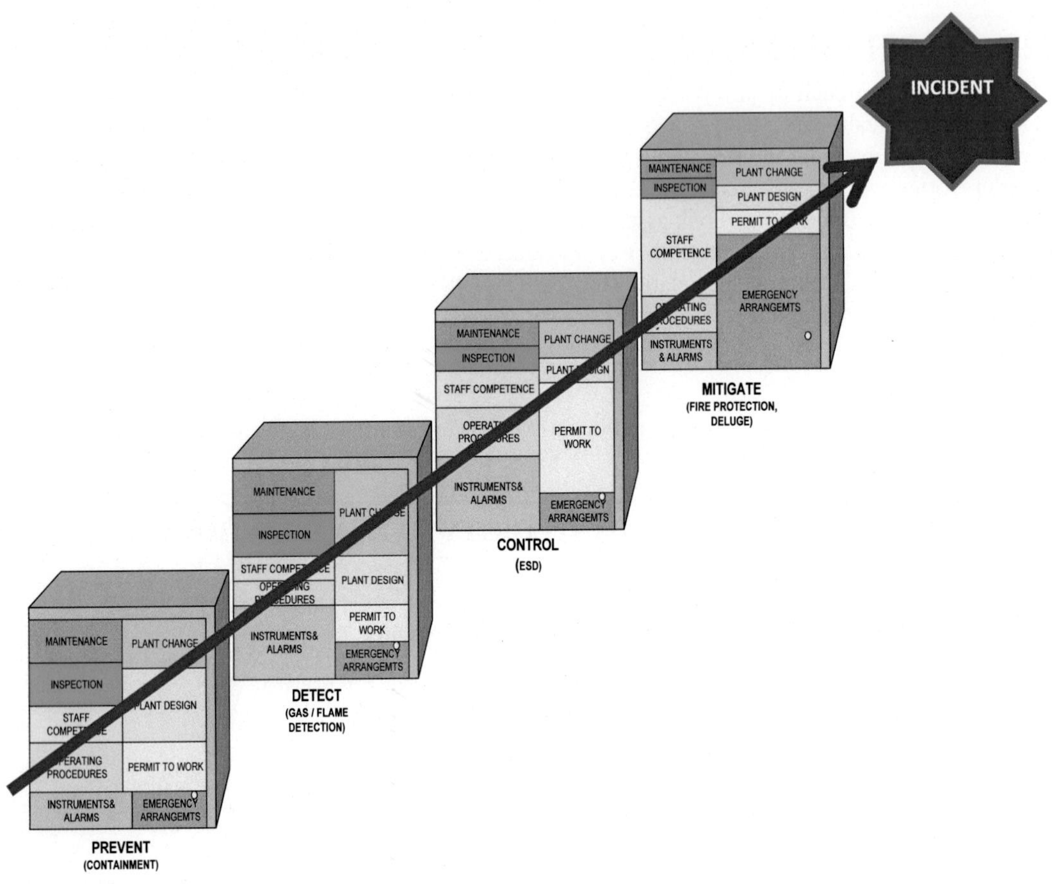

FIGURE 22.4

Swiss cheese with process safety/asset integrity integration.

Typically, the components of each slice are as follows:

- Maintenance
- Inspection
- Staff competence
- Operating procedures
- Instruments and alarms
- Plant change
- Plant design
- Permit to work
- Emergency arrangements

Asset integrity is the outcome of good design, construction, and operating practices. It is achieved when facilities are structurally and mechanically sound, perform the required processes and produce the products for which they were designed. Emphasis is on **preventing unplanned hydrocarbon releases that may**, either directly or via escalation, **result in a major incident**.

Key elements of **asset integrity** are as follows:

1. **Assurance** that there will be **no "loss of containment"** (by maintaining the pressure envelope)
2. Ensuring that all **safeguarding systems** act in the desired manner at the proper time
3. Ensuring that the equipment is being **operated within predefined limits** that are necessary to ensure continued operation from one major turnaround to the next
4. **Preventing unexpected equipment failures**
5. Ensuring that all **staff are trained, motivated, and fully competent** for their jobs

A "cradle to grave" ("asset lifecycle") risk management approach should be adopted for all of the company's ***asset business units***.

Sustainable integrity management consists of the following:

1. Procedures
2. Management Of Change (MOC)
3. Operating conditions
4. Training

The integrity and reliability management objective or goal should include "sustainable top performance in reliability and maintenance at optimum cost." An asset integrity guideline could consist of the following headings:

1. Introduction
2. Asset integrity risk management process (ISO 31001)
3. Barriers
4. Integrity throughout the lifecycle of the asset
5. Human factors
6. Competencies
7. Monitoring and review

The EI PSM Framework (see Box 20.8: EI PSM Framework) and CCPS Guidelines for Risk-Based Process Safety (see Box 20.9: CCPS Guidelines for Risk Based Process Safety, 2007) give guidelines specifically for the process industry, and the Oil and Gas Producers (OGP) Report no 415 "Asset Integrity"[2] gives guidelines specifically for the oil and gas industry.

Note that there are overlaps between these guidelines and ISO 55000 "Asset Management." A cross-reference matrix may help address the specific requirements of each standard/guideline.

Tools for risk assessment are discussed in Section 12.8, Tools for Operational Risk and Performance Analysis and Evaluation and Chapter 29, Identification, Analysis, and Evaluation of Gaps.

Further to discussion in Section 9.4, The Board: The Fulcrum of Business Performance, the board of directors responsibility must include the long-term sustainability of the assets.

BP's focus on cost reduction in past years resulting in long-term damage to its assets is discussed in Box 9.5, Accountability, Responsibility and Discipline: BP and Box 24.2, Integration Into the Business Example 1: Cost Reduction.

22.6 APPLICATION OF ASSET PERFORMANCE MANAGEMENT

PRIMARY FOCUS

Primary focus areas are as follows:

1. Corporate safety **culture**
2. Process safety/integrity management **systems**
3. **Performance** evaluation, corrective action, and corporate oversight

KEY ELEMENTS

The following are key elements of physical asset performance management that require addressing:

1. RCM (Reliability-Centered Maintenance)
2. SIF (Safety Instrument Function)
3. RBI (Risk-Based Inspection)
4. Turnaround optimization
5. Initiative activity ranking
6. Priority setting of corrective maintenance
7. Defect elimination

RCM, SIF, and RBI are discussed in the following paragraphs. Turnaround optimization is discussed in Chapter 23, Turnarounds.

Initiative activity ranking, priority setting of corrective maintenance, and defect elimination are embedded in each of these elements.

22.7 MAINTENANCE

PRIMARY FOCUS

The primary focus must be on safety-critical plants and equipment. These are plants and equipment that are relied on to ensure safe containment of hazardous chemicals and stored energy and for continued safe operation. This will typically include the following items in a plant's preventative maintenance program:

- Pressure vessels
- Piping systems
- Relief and vent devices
- Instruments
- Control systems
- Interlocks and emergency shutdown systems
- Mitigation systems
- Emergency response equipment

STANDARDS

Some useful standards are as follows:

- **SAE STD JA1011**: Evaluation Criteria for Reliability Centered Maintenance (RCM) processes
- **UK HSE RR 509**: Plant Aging—Management of Equipment Containing Hazardous Fluids or Pressure
- **ISO 14224**: Petroleum, Petro Chemical and Natural Gas Industries—Collection and Exchange of Reliability and Maintenance Data for Equipment

MAINTENANCE PRINCIPLES

The primary maintenance principles are as follows:

1. **Defect elimination**
2. Optimal reliability and integrity
3. Optimal work volume
4. Maximal efficiency of execution

APPLICATION OF THE PRIMARY MAINTENANCE PRINCIPLES

Defect elimination reduces:

1. **number of breakdowns** (emergency maintenance) increasing reliability and integrity,
2. **reactive work** (corrective maintenance) optimizing work volume, and
3. **high priority work** increasing efficiency or work.

Reliability and integrity improvement also improves optimization of work volume, which in turn has a knock-on effect on improved efficiency of execution.

Active **defect elimination** is, thus, a key element of RCM and RBI.

AREAS OF EFFECTIVE MAINTENANCE

Areas of effective maintenance can be grouped as follows:

- Maintenance management
- Turnaround management
- Projects
- Other
  - Risk-based decision-making
  - Initiative management
  - Performance management

EQUIPMENT CRITICALITY ANALYSIS

Criticality of equipment is directly related to the effect on production throughput.

Criticality is generally classified as follows:

A: Primary process stream equipment shutdown causes production loss.

Example: If there is only one pump in the primary process stream and it fails, the process stops and production is lost. This pump would be regarded as a "criticality A" item of equipment.

B: Primary process stream equipment with parallel full capacity equipment.

Example: Two crude feed pumps are installed in parallel, each capable of handling 100% of the refinery design capacity. One is electrically powered while the other is steam turbine driven. When power is lost to the plant, the steam-driven pump automatically cuts in. These pumps are regarded as "criticality B" items of equipment.

C: Equipment that could cause production loss.

Example: Three air compressors supply the plant air system. 100% of the air requirements are covered by two compressors. Failure of more than one air compressor would shut part of the plant down. These compressors are regarded as "criticality C" items of equipment.

Spares for criticality A equipment are always[a] retained in stores. These are referred to as insurance spares and would include items such as spare compressor and turbine rotors for criticality A equipment.

Examples of equipment criticality are discussed in Box 22.2.

BOX 22.2 EQUIPMENT CRITICALITY

Compressors

Compressed air is critical for the smooth running of instruments that keep a refinery operating.

Two air compressors, each with 100% capacity, supplied instrument air for a particular refinery. One of these broke down and could not be repaired immediately since there were no spares in stock. As luck would have it, the other compressor also failed, causing the entire refinery to shutdown. Production stopped completely for a number of hours, before a portable compressor was hired to start up and keep the refinery on line until the broken compressors were fixed.

Comment

Critical spares were not stocked since the compressors were rated as criticality B. Nonetheless, the lost production was significant. A service agreement with the compressor supplier, where spares are required to be supplied within hours, would have avoided the problem.

Relief Valves

A gas plant experienced an increase in the backlog of repairs to relief valves. (Relief valves and their components are criticality A items.) An investigation was initiated as to the reasons for the backlog, and it was found that there was an unacceptably long delivery time for these spares.

Outcome

Service agreements were established with the agents of all relief valve suppliers, whereby they guaranteed supply of spares, on demand, within a number of hours.

As a result the backlog was cleared.

---

[a]Box 9.6, The Jewel in the Crown: The Rise and Fall of South Africa's Electricity Supply Commission—Koeberg Nuclear Power Station discusses the replacement of a spare 900-MW steam turbine rotor that is regarded as a "criticality A" insurance spare. However, the probability of failure of this was extremely low and so the company took the decision not to stock a spare, with resulting disastrous consequences.

PLANNED MAINTENANCE INTERVALS

Planned maintenance (PM) intervals are determined in the following order:

1. OEMs' recommendations
2. Owner experience of this equipment based on RCM and RBI
3. Staff experience of this type of equipment

Standards such as described in Box 22.4 also assist in determining PM intervals.

RELIABILITY CENTERED MAINTENANCE PROCESSES

Failure analysis needs to be carried out on equipment that has failed to prevent repetitive failure and thus move toward **eliminating defects**.

Mechanical

Vibration analysis and monitoring of rotating equipment are key activities. Major equipment on primary systems requires online vibration measurement. Lubrication oil analysis should be undertaken on a regular basis for key machines.

Critical success factors (CSFs) for determining run time of critical rotating equipment are as follows:

- Type categories
- Vibration analysis
- Lubrication oil analysis
- Failure analysis
- Categorization of mean time between failures (MTBF)

Electrical

Compliance with various International Electrotechnical Commission (IEC) Standards is required.

CSFs for determining electrical inspection intervals for equipment requiring process shutdown are as follows:

- List of equipment requiring a process shutdown
- Analysis of required equipment shutdown intervals

Key elements of good practice are as follows:

- An extensive and effective lock-out, tag-out system controlled by the permit system
- Routine thermography of all transformers and switchgear
- Electrical earth bonding circuits tested annually and post maintenance break-in
- Maintenance of steam and gas turbine generator units done in accordance with the OEMs' recommendations

Instrument

IEC 61511 covers the design and management requirements for Safety Instrument Systems (SISs) from cradle to grave. Its scope includes initial concept, design, implementation, operation, and maintenance through to decommissioning.

IEC 61508 is a generic functional safety standard, providing the framework and core requirements for sector-specific standards. Three sector -specific standards have been released using the IEC 61508 framework: IEC 61511 (process), IEC 61513 (nuclear), and IEC 62061 (manufacturing/machineries). IEC 61511 provides good engineering practices for the application of safety instrumented systems in the process sector.

In line with **IEC 61511** and **IEC 61508** standards on safety instrument systems, the primary focuses are as follows:

- ESD
- Fire and Gas (F&G)

Safety instrument function (SIF) is required to be carried out at regular intervals as determined by these standards.

CSFs for determining periods between inspection of instrument systems are as follows:

- List of equipment requiring process shutdown
- Analysis of required equipment shutdown intervals

CSFs for determining periods between inspection for F&G and Emergency Shutdown (ESD) systems are as follows:

- F&G and ESD systems requiring process shutdown
- Analysis of required ESD intervals

COMPUTERIZED MAINTENANCE MANAGEMENT SYSTEMS

These are now well established in the process industry, either as part of the enterprise resource planning (ERP) system, such as SAP or Oracle, or have strong links to the company ERP system. Maintenance modules are linked to stores/purchasing and accounting modules so that the planning and scheduling of maintenance activities are streamlined.

The following are some advantages achieved with computerized maintenance management systems (CMMS):

- Maintenance human resource restraints identified and leveled
- Material and tools availability identified from the materials module
- Financial (budget) restraints identified
- Categorization of work orders: emergency, corrective, and planned
- Prioritization based on equipment criticality
- Standard planned maintenance (PM) activities and intervals embedded
- Multidiscipline job preparations done using detailed task breakdowns
- Efficiency improvement by monitoring "hands-on tool" time and waiting time for permit issue
- Easy monitoring of maintenance KPIs

BASIC EQUIPMENT CARE

Analogy: If you own a car and you personally drive it and check the oil, water, and tire pressure on a regular basis, it is very likely that you will spot any maintenance issues that arise even before your vehicle mechanic is aware of them.

Similarly, operators are more familiar with the performance of the specific equipment that they operate than are maintenance staff, as they are there listening, seeing, and feeling on an hourly basis.

They also understand the criticality of each item of equipment. They could thus carry out minor work on the equipment such as attending to small leaks, general cleaning, clearing blockages, and other duties so as to release maintenance staff to do more skilled work. However, it is important to note that the tasks allocated to operators must be clearly defined, interruptible, and of short duration, and operators must be properly trained to perform these tasks. Ideally, a feeling of ownership needs to be inculcated into the way operators manage and care for their equipment.

The trend in some critical plant processes is to have maintenance staff on shift as a front line to give a faster response when maintenance expertise is needed.

The outcome is that plant reliability improves as problems are identified and attended to at an early stage before they grow to be much bigger complications.

22.8 INSPECTION

API RP 580 is the basis for establishing RBI in the hydrocarbon industry.

SCOPE OF RBI

RBI relates to the whole pressure envelope and is grouped as follows:

1. Pressure vessels—all pressure-containing components
2. Boilers and heaters—pressurized components
3. Heat exchangers (shells, floating heads, channels, and bundles)
4. Rotating equipment—pressure-containing components
5. Process piping—pipe and piping components
6. Storage tanks—atmospheric and pressurized
7. Pressure-relief devices

ADVANTAGE OF RBI

The primary advantage of RBI is that uninspectable risks are reduced (see Fig. 22.5).

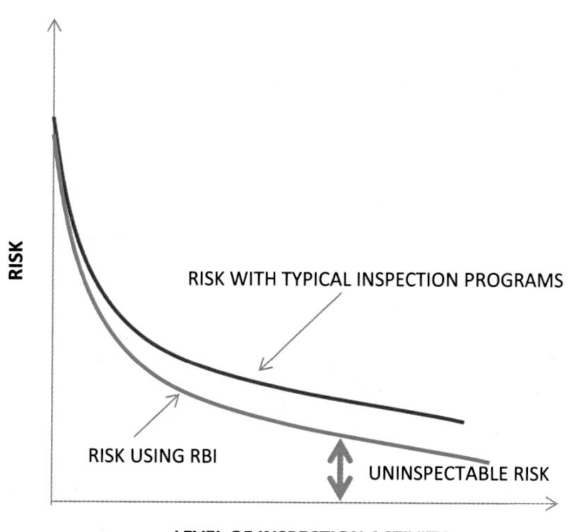

FIGURE 22.5

Advantage of risk-based inspection.

CODES

RBI relies on the application of a number of codes. The U.S. API and ASME codes for pressure containment of vessels, pipelines, and tanks are depicted in Fig. 22.6.

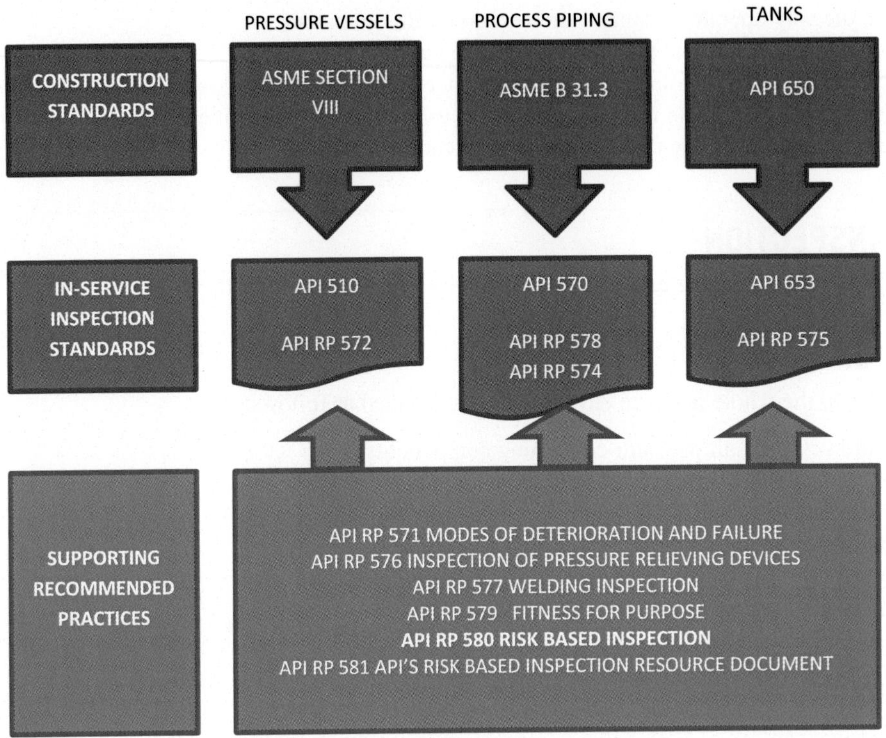

FIGURE 22.6

API and ASME codes involved with risk-based inspection.

PROCESSES

Remaining Life Calculations

Regular thickness measurement of all primary pressure systems and the use of risk-based prediction software tools can predict the extent of the remaining life of that system or item of equipment.

Nondestructive Testing

Nondestructive testing covers all tools and methods used to assess the condition of the pressure systems. These include X-ray, ultrasonic thickness (UT) testing, thermography, etc.

Destructive Testing

Destructive testing includes the testing of corroded/eroded items removed from the pressure systems when replacement takes place. An analysis may initiate a change of material for a particular service.

Corrosion Management

Corrosion management includes the injection of corrosion inhibitors, impressed current, and use of corrosion coupons to monitor and minimize corrosion of pressure systems and structures.

OUTPUT OF RBI PROCESSES

The output of the inspection planning process, conducted according to API 580 RBI guidelines, should be an inspection plan for each equipment item analyzed that includes the following:

1. Type of inspection methods that should be used
2. Extent of inspection (% of total area to be examined or specific locations)
3. Inspection interval or next inspection date (timing)
4. Other risk mitigation activities
5. Residual level of risk after inspection and other mitigation actions have been implemented

An RBI application is discussed in Box 22.3.

BOX 22.3 INSPECTION—RBI APPLICATION

An asset technical integrity audit revealed an excellent risk-based inspection management system (RBIMS) for all pressure containment within the gas plant process areas. However, none of the pipelines leading across the country to the terminal and downstream processing units had a system for assessing risk and inspection frequency.

Spreadsheet listings of cross-country pipelines and respective inspection intervals were improved with the intent to adopt a suitable risk modeling tool for each pipeline.

Comment

It stands to reason that risk assessment of all pressure containment within the black box must be addressed. However, RBI, as defined by API 580, does not cover cross-country pipelines since these are covered by various modeling tools and techniques.[6]

Box 22.4 outlines **BP RP 32-3,** which became the de facto unofficial standard for many oil and gas companies and was one of the foundations of later RBI standards.

BOX 22.4 BP RP 32-3 INSPECTION AND TESTING OF IN-SERVICE CIVIL AND MECHANICAL PLANT—MANAGEMENT PRINCIPLES OCTOBER 1998

Background

Industry is continually grappling with the pros and cons of extending run length between turnarounds using RBI. This standard sets intervals for plant based on experience.

Scope

This is applicable to oil refineries, petrochemical and chemical plants, onshore and offshore production facilities, transmission pipelines, and storage and distribution facilities.

Value

Ensuring a high standard of plant integrity based on economical safe practice.

CSFs for determining periods between inspections of static equipment:

- List of equipment requiring process shutdown
- Remaining life calculation
- Recommended inspection intervals

CSFs for determining periods between inspections of mechanical safety devices:

- List of devices requiring process shutdown
- Recommended inspection intervals

22.9 ASSET PERFORMANCE MANAGEMENT SOFTWARE

Asset performance management software is becoming a necessity in the process industry.

Section 22.12 discusses the requirement for this, and Box 22.7 clearly shows the impact on other KPIs.

Asset performance management software should cover all assets in alignment with recognized and generally accepted good engineering practices as per the following standards or similar:

- RCM: SAE standard JA1011, "Evaluation Criteria for Reliability Centered Maintenance (RCM) Processes"
- SISs: ISA 84/IEC 61511 and IEC 61508
- RBI: API 580
- Hazard Analysis and Operability Studies (HAZOP/HAZAN): IEC 61882 "Hazards and Operability Studies Application Guide"

FOCUS AREAS

Strategy Management

- Asset strategy manager
- Asset strategy optimization

Strategy Execution

- Calibration management in line with ISO 9000 and API Manual of Petroleum Measurement Standards
- Inspection management
- Operator rounds
- Thickness Measurement Location (TML) management including monitoring and thickness calculation according to ASME and API industry standards for piping, tanks and pressure vessels

Strategy Evaluation

- Generation management
- Metrics and scorecards
- Production loss accounting
- Reliability analytics
- Root cause analysis
- Vibration analysis
- Asset criticality analysis

Meridium is the most popular software covering these focus areas. Meridium installations include Saudi Chemical, Bapco, ChevronTexaco, Qatargas, Rasgas, SABIC, Qatar Chemical, Saudi Aramco, and ExxonMobil. This is outlined in Box 22.5.

BOX 22.5 MERIDIUM ASSET INTEGRITY SOFTWARE

The Meridium software model consists of the following five APM work processes:

1. APM foundation
   a. APM framework
   b. Asset health indicators
   c. EAM integration (SAP, Maximo, Oracle)
   d. On-line analytical interfaces (AMS, OSI-PI, OPC)
2. Failure elimination
   a. Production loss accounting
   b. Root cause analysis
   c. Metrics and scorecards
   d. Reliability analytics
   e. Generation management
3. Asset strategy
   a. Asset strategy management
   b. Asset strategy optimization
   c. RCM/FMEA
   d. Operator rounds
   e. Lubrication management
   f. Calibration management
4. Mechanical integrity
   a. RBI
   b. Inspection Management
   c. Thickness monitoring
5. Asset safety
   a. Hazards analysis
   b. SIS management

22.10 ENERGY MANAGEMENT

Box 18.2, ISO 50001 Energy Management System promotion outlines the **ISO 50001 Energy Management** standard that is applicable.

Energy efficiency is established at the design stage, whereas energy management is totally dependent on good operating and maintenance practices and directly affects environmental emissions.

Energy review is the heart of the energy planning process as follows:

1. Analyze energy use and consumption
2. Identify areas of significant energy use and consumption
3. Identify opportunities for improving energy performance

An energy model of the process plant is thus required. This was introduced in Section 3.3, Mass and Energy Balance Models.

Energy review is depicted in Fig. 22.7.

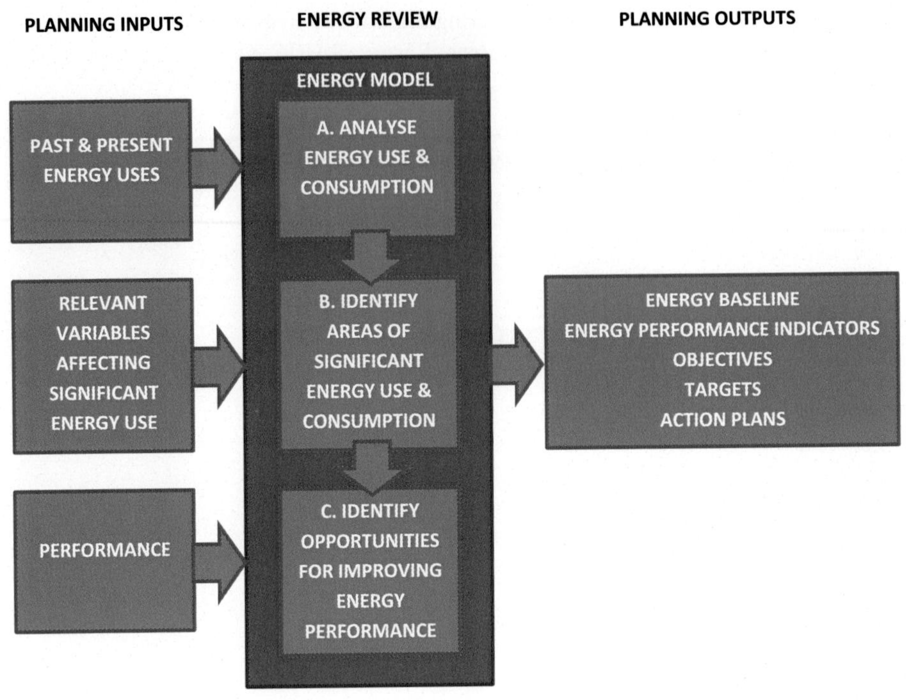

FIGURE 22.7

Energy review.

Examples of energy performance are discussed in Chapter 36, Hydrocarbon Accounting.

22.11 REVIEWS AND BENCHMARKING

Regular reviews of the asset management system are required. ISO 55000 requires internal (peer) reviews and certification audits (if certified by an external authority). However, performance relative to the competition also needs to be assessed. Reliability and Maintenance (RAM) benchmarking is carried out by various benchmarking consultants as discussed in Chapter 24, The Benchmarking Process and Consultant Selection.

Various internal tools to ensure integrity over the life cycle of the asset have been developed. An example is the **Asset Integrity Toolkit** described in Box 22.6.

BOX 22.6 ASSET INTEGRITY TOOLKIT

In support of asset integrity inspections undertaken by the Health and Safety Executive, UKQQA established the Installation Integrity Working Group (IIWG) in 2004. Its aim was to focus industry efforts not just on improving asset integrity over the prescribed period, but also to promote a sustainable set of KPIs for the long-term future of the industry.

The work group helped develop and collate a collection of good practice techniques and guidelines aimed to assist operators in their efforts to maintain and enhance asset integrity.

The toolkit is a collection of these good practices and provides a central reference for managers, supervisors and the workforce.

BOX 22.6 ASSET INTEGRITY TOOLKIT—cont'd

The scope of the Asset Integrity Toolkit is to provide a practical framework of "observed good practice," checklist and tools to facilitate and enable review of Asset Integrity Management throughout the "Six Stages of The Lifecycle" of any offshore installation.

The tools are as follows:

- Tool No.1 Assurance and Verification
- Tool No. 2 Assessment/Control and Monitoring
- Tool No. 3 Competence
- Tool No. 4 Planning
- Tool No. 5 Maintenance
- Tool No. 6 Quality and Audit

22.12 SYSTEM MATURITY

As maintenance and inspection systems evolved, from the implementation of CMMS with associated materials modules, through the development of RBI and RCM tools, to having an integrated asset management system, maintenance costs have come down and reliability of the equipment has gone up. However, with each step of improvement there is a learning curve where costs may go up due to extra staffing and training before settling at a lower cost level. Reliability may also decrease as the reliability boundaries are pushed before settling at a higher reliability level. Box 22.7 gives an example of three similar gas plants at different stages of maturity.

BOX 22.7 SYSTEM MATURITY—GAS PLANTS

A 2013 analysis of three similar gas plants determined how system maturity affects KPIs.

Two of the gas plants had implemented Meridium asset integrity software. Gas plant 1 had had the software for a while and was the pacesetter in the industry with, among other KPIs, the lowest maintenance and inspection costs, and the lowest staff turnover. Gas plant 2 had recently purchased Meridium and was still in the learning stages with higher staff and other costs. Gas plant 3 did not have any integrated asset management but was believed to be pursuing certification to ISO 55000 "Asset Management." It was already certified to ISO 9001, ISO 14001, and ISO 45000 within its integrated management system. The other two gas plants did not intend, at that stage, to certify to ISO 55000.

This is graphically shown here:

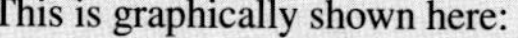

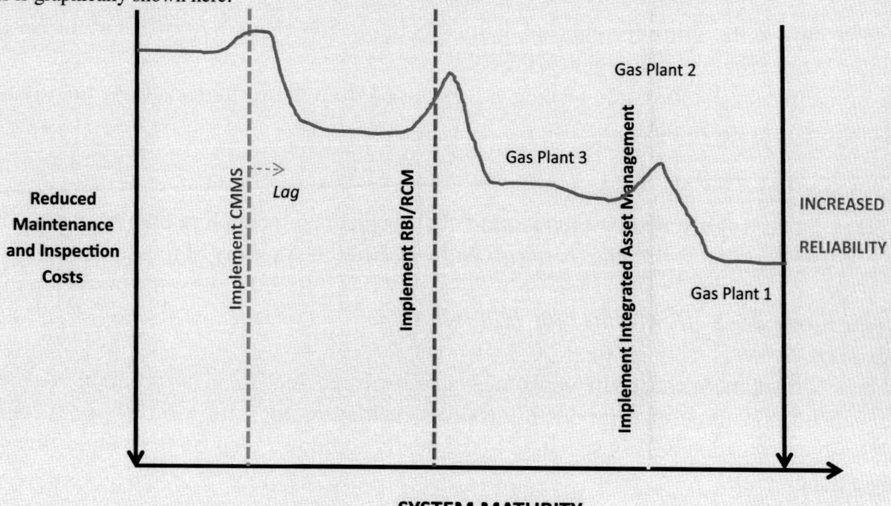

Also see Box 20.6, Competencies Motivation.

22.13 SUGGESTED KPIs

There should be a focus on leading KPIs so as to prevent a major LOPC.

The following give guidance for the development of appropriate KPIs for the reporting of the condition of the **physical asset** from the aspects of process safety and integrity:

1. Center for Chemical Process Safety: Leading and Lagging Metrics[3]
2. UK HSE Guide 254 Developing Process Safety Indicators[4]
3. ANSI/API 754: RP for Process Safety Indicators for the Refining and Petro Chemical Industries[5]
4. EI Research Report: Human Factors Performance Indicators for the Energy and Related Process Industries[7]

The following are samples of those KPIs that have the most influence on the performance of the physical asset.

AVAILABILITY/RELIABILITY KPIS

Mechanical Availability

Availability is primarily dictated by maintenance downtime in days. This is split into **unplanned** and **planned** and is normally referred to as mechanical availability (see Section 17.4: Utilization and Availability). The hierarchy is as follows:

- Availability
- Reliability
- Planned down days
- Unplanned down days

Reliability

$$\text{Reliability} = (\text{Total Days} - \text{Shutdown Days} - \text{Unplanned Down Time}) \times 100/(\text{Total Days} - \text{Shutdown Days})$$

Shutdown days in the above formula are defined as those days that are regarded as turnaround days in the annual Work Program & Budget (WP&B).

Note: Sometimes planned shutdown days are excluded from the formula as these are often annualised and could distort the indicator.

Planned Down Days for Maintenance

This is referred to as a **shutdown or turnaround**. This is planned from more than a year in advance to ensure a run-length in line with industry norms and to restrict the number of unplanned down days.

Unplanned Down Days (Downtime) for Maintenance

These are *unpredicted.*

Unplanned downtime is caused by unplanned shutdowns that lead to an interruption of a production train. Unplanned downtime of a production train is defined as the time (in days or parts of days) required for shutdowns that are not specified in the yearly shutdown plan (in the annual Business Plan and Budget document).

Unplanned downtime related to unreliability can be caused by maintenance, process upsets (e.g. process plant trips, heat exchanger fouling), consequential downtime caused by unplanned outages of upstream equipment (e.g. wells, slug-catcher) and equipment failure or breakdown (such as leakages and fire) and by external events such as power failure. Shutdowns that may result from exceedances of flaring or emission levels as defined in the "permit to operate" agreed with the state environmental authority, also fall under this category.

In reporting unplanned downtime, the following guidelines are recommended:

1. Downtime caused by an unforeseen event that can be deferred and scheduled in the next monthly or quarterly shutdown schedule, should still be reported as unplanned downtime. Only when it can be deferred and included into the next year annual plan, can it be classified as planned downtime.
2. Any time that a shutdown takes longer than defined in the annual plan for causes mentioned above shall be classified as unplanned downtime.

Depending on the industry, downtime is usually measured in fractions of days.

INTEGRITY KPIS

Development of integrity KPIs is introduced in Section 16.7, Physical Asset Integrity KPIs – An Introduction.

Process safety events (loss of primary containment or potential for LOPC) generally in line with **API RP 754 Guide to reporting process safety events**, are categorized as follows:

Tier 1.

- **LOPC with severe consequences**:
  - A fatality and/or LTI or
  - A fire or explosion that causes $50,000 or more of direct cost or
  - Officially declared community evacuation and/or on site shelter-in-place

OR

- **Release in kg in 1 h** of:
  - 5 kg toxic material or
  - 500 kg flammable gas or flammable vapor arising from flammable liquid or
  - 1000 kg flammable liquids or
  - 2000 kg combustible liquids

Tier 2.

- **LOPC with limited consequences**:
  - A recordable injury or
  - A fire or explosion that results in damage that must be repaired as a result of the fire or explosion

OR

- **Release in kg in 1 h** of:
  - 0.5 kg toxic material or
  - 50 kg flammable gas or flammable vapor arising from flammable liquid or
  - 100 kg flammable liquids or
  - 200 kg combustible liquids

Tier 3.

- **Challenges to safety systems**
  - Any Loss of Primary Containment (LOPC) other than Tier 1 and 2
  - Other fires not related to process units
  - Central operator panel alarm priority one exceedences
  - Demand on safeguarding system (process shutdown: FandG, ESD etc.)
  - **Overdue inspection recommendations**

Tier 4.

- **Operating discipline and management system performance indicators**
  - Bypass override (minimum number)
  - Alarm management (number per hour per panel - target 5)
  - **Emergency exercises (including a minimum of 1 full exercise per annum)**
  - **Planned maintenance (PM) compliance (target 90%)**
  - **Safety critical elements PM compliance (target 100%)**
  - **Safety critical elements test exceedances (min number)**
  - **Inspection schedule compliance (equipmt/piping/corrosion inhibitor program etc.)** (%)
  - Compliance to process safety audit schedule (%)
  - Overdue audit items (process safety field observations - number).

Leading and lagging indicators related to LOPC tiers are indicated in Fig. 22.8. It could be said that the more there are of the tier three and four indicators the better the prediction.

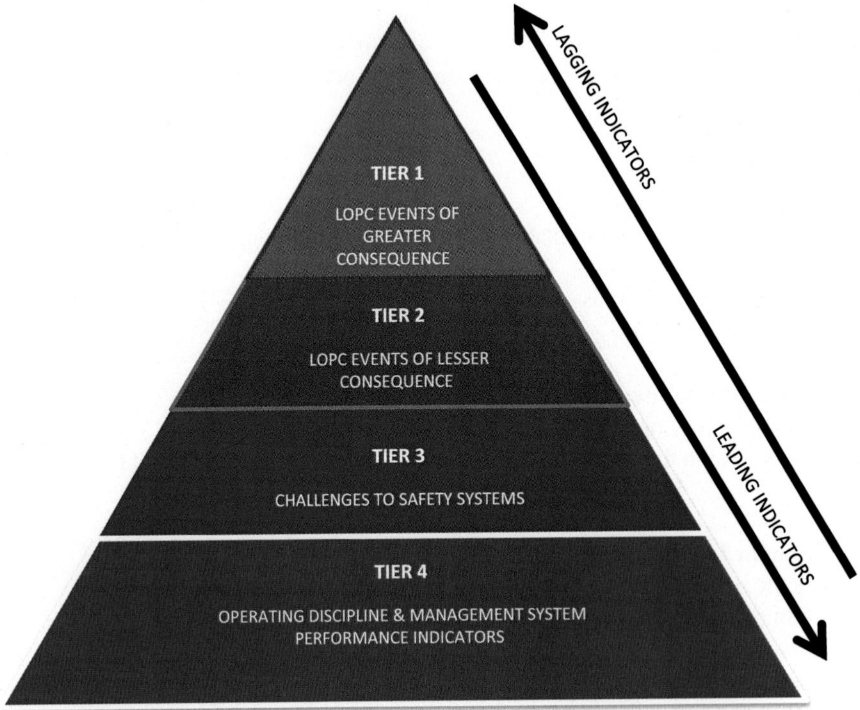

FIGURE 22.8

Leading and lagging indicators related to loss of primary containment tiers.

MAINTENANCE KPIS

1. **Items listed in LOCP Tier 4 above**
2. Majority of work is planned (proactive >80%)
3. Ratio of proactive to reactive work is 80:20
4. < 5% of work orders classified as urgent
5. Compliance to schedule is >85%
6. Work order completions are fully documented
7. Complete formal training for planners and schedulers

INSPECTION KPIS

1. **Item listed in LOCP Tier 3 above**
2. **Items listed in LOCP Tier 4 above**
3. Compliance to schedule

ENERGY KPIS

- Energy losses: weight% of intake
- GHG equivalent: % of intake

JOINING THE DOTS

The interrelationship between the above indicators and others such as cost and competency need to be appreciated (see Box 20.6: Competencies – Motivation).

22.14 SUMMARY

Physical asset performance management is dependent on the integrity of the asset being sustained for the life of the investment, in order to perform at design capacity without a major LOPC.

The adoption of the ISO 55000 suite of standards enables an organization to achieve its objectives through the effective and efficient management of its assets.

Integrity/process safety has five key elements: no loss of containment assurance, ensuring safeguarding systems operate properly, operating equipment within predefined limits, ensuring no unexpected equipment failures and ensuring that staff is trained, motivated and fully competent.

Asset performance management has focus elements for success. These are: implementation of risk-based inspection and reliability centered maintenance, application of safety instrument system principles for instruments, turnaround optimization, corrective maintenance prioritization, initiative activity ranking and defect elimination.

Performance indicators in the areas of availability, integrity, maintenance, inspection, energy, competencies and cost interact both negatively and positively to result in overall optimization of asset performance. The focus should be on leading KPIs so as to prevent a major loss of primary containment.

Minimizing holes in each slice of Swiss cheese (mitigate, control, detect, and prevent) is essential for the overall integrity for the life of the investment.

Asset performance management software should cover all assets in alignment with recognized and generally accepted good engineering practices. Maturity of the asset integrity systems shows in the KPIs, where more mature systems are reflected by better performance indicators.

In summary, physical asset performance is dictated by the health of the equipment. If appropriately inspected and maintained, by competent staff, using the latest tools available, a high level of performance is likely to be achieved for the life of the asset.

REFERENCES

1. BSI PAS 55–1. *Asset management (superseded by ISO 55000:2014)*. 2008.
2. International Association of Oil and Gas Producers (OGP). *Asset integrity: the key to managing major incident risks*. Report No 415. December 2008.
3. *Process safety leading and lagging metrics*. CCPS; 2008.
4. UK HSE guide 254. *Developing process safety indicators*. HSE Books; 2006.
5. API RP 754. *RP for process safety indicators for the refining and petro chemical industries*. 2010.
6. Dey PK, Gupta SS. Risk-based model aids selection of pipeline inspection, maintenance strategies, *Oil Gas J* July 09, 2001;**99**(28).
7. *Human factors performance indicators for the energy & related process industries*. EI Research Report; December 2010.

FURTHER READING

1. SAE STD JA 1011. *Evaluation criteria for reliability centered maintenance (RCM) processes*. 2009.
2. HSE Research Report RR509. *Plant ageing: management of equipment containing hazardous fluids or pressure*. HSE Books; 2006.
3. ISO 14224:2006. *Petroleum, petrochemical and natural gas industries - collection and exchange of reliability and maintenance data for equipment*.
4. BP RP 32–3. *Inspection and testing of in-service civil and mechanical plant – management principles*. October 1998.
5. International Electrotechnical Commission (IEC). http://www.iec.ch/.
6. IEC 61508:1998. *Functional safety of electrical/electronic/programmable electronic safety-related systems*.
7. IEC 61511:1998. *Functional safety - safety instrumented systems for the process industry sector*.
8. IEC 61882:1998. *Hazards and operability studies application guide*.
9. ISA 84. Safety instrumented systems certificate programs.
10. API 580. *Risk based inspection*. November 2009.
11. API 581. *Risk based inspection technology*. September 2008.
12. RR 363. *Best practice for risk based inspection as a part of plant integrity management*. UK: HSE; 2001.
13. *Managing ageing plant: a summary guide*. UK: HSE. http://www.hse.gov.uk/offshore/ageing/ageing-plant-summary-guide.pdf.
14. Meridium website. www.meridium.com.
15. *Asset integrity toolkit*. UK: Offshore Operators Association/HSE. https://www.stepchangeinsafety.net/safety-resources/publications/asset-integrity-toolkit.
16. *Guidance on safety performance indicators*. OCED; 2003.
17. Venn C. Chevron step change – integrity management, 2005.
18. D'Aquino R, Berger S. Baker report: analyzing BP's process safety program. *Chem Eng Prog* Feb 2007.
19. ISO 17776:2000. *Petroleum and natural gas industries - offshore production installations - guidelines on tools and techniques for hazard identification and risk assessment*.
20. ISO 13702:2015. *Petroleum and natural gas industries - control and mitigation of fires and explosions on offshore production installations - requirements and guidelines*.

CHAPTER

TURNAROUNDS

23

23.1 INTRODUCTION

This chapter discusses the elements that are key to the success of turnarounds (TAs). TAs apply **people** and **financial assets** to the inspection and repairing of the **physical assets** so as to optimize the next process run length. They are critical to maximizing production over the lifetime of the asset.

23.2 DEFINITION

A turnaround is a shutdown that has been planned from more than a year in advance to ensure a run length in line with industry norms with minimal unplanned down days.

The TA skeleton plan and budget is normally included in the company business plan and budget (BP&B).

TAs are complex projects. Principles of project management should be applied as per the Project Management Body Of Knowledge (PMBOK),[1] an American National Standard that is applied worldwide.

23.3 WHY EFFECTIVE TURNAROUND MANAGEMENT IS CRITICAL

TAs directly affect the **availability** of the plant to achieve maximum production. **Days lost in a TA are lost forever**.

If a TA is not done effectively, unplanned down days increase and the interval to the next TA may be below industry norms and so result in losing competitive edge. Section 28.6, Performance Focus details how TAs relate to the full cycle of operation of the process plant and a key effectiveness Key Performance Indicator (KPI)—availability. Also, Section 30.5, Mechanical Availability Drill-Down, discusses the important contribution of a TA in optimizing the full operating cycle.

Performance Management for the Oil, Gas, and Process Industries. http://dx.doi.org/10.1016/B978-0-12-810446-0.00023-2

23.4 BASIC REQUIREMENTS FOR OPTIMIZATION OF THE TURNAROUND

A project management approach is essential for effective management of a TA. The following elements are especially critical:

1. Ensure a clear alignment of TA statement of commitment with the corporate vision and mission
2. Carry out structured phased planning and execution
3. Undertake structured scope screening
4. Apply front end loading (FEL)
5. Use Critical Path Modeling (CPM) tools
6. Choose the right experienced contractors and incentivize them appropriately
7. Ensure staff are competent in both technical and management fields
8. Use lessons learned for the next TA

All of the these actions should be well documented in a TA Manual (or TARMAP: Turn ARound MAnagement Process) or TA Framework. This is summarized in Fig. 23.1.

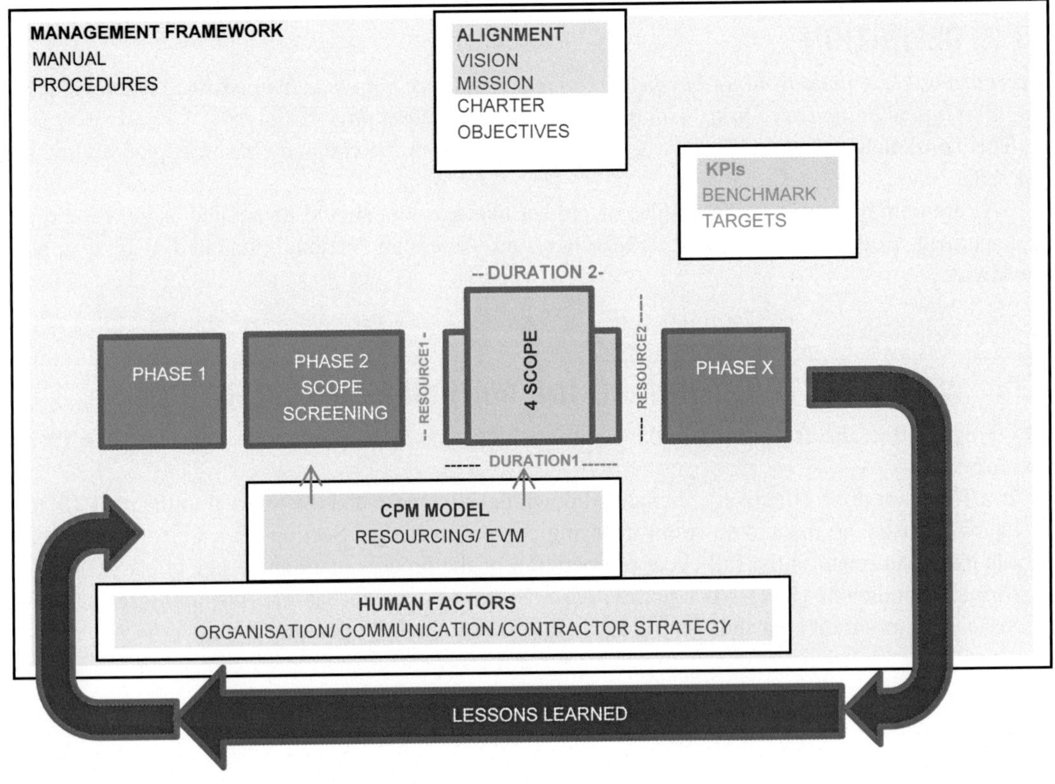

FIGURE 23.1

Turnaround management framework.

These elements are discussed in detail as follows:

1. ALIGNMENT HIERARCHY

The following hierarchy is imperative. The statement of commitment serves to keep the TA team focused throughout the TA.

1. Corporate vision and mission
2. TA vision and mission (all TAs)
   a. In introduction to TA Manual
3. Project charter (upcoming TA)
4. Statement of commitment (upcoming TA)
   a. Includes measurable project objectives and related success criteria
      i. Commitment to achieve the scope, duration, and cost, with zero incidents, etc.
   b. Signed by all team leaders (staff and contractors)

2. OPERATING CYCLES AND SHUTDOWN/TURNAROUND PHASES

Definitions

Operating Cycle[a]

This is the **run length** or time to the next TA and is determined from reviews of benchmarking, reliability-centered maintenance, and risk-based inspection data (see Section 23.7).

Full Cycle[b]

This is the time from startup (online producing on-spec products) to startup after the next TA. This is used to annualize the TA planned down days.

Shutdown Cycle[c]

This is the time from initiation of the next TA project to completion of the project (project duration). The shutdown cycle is divided into clear phases with each phase consisting of activities and targets for completion.

As in most projects, a project review or audit needs to take place at the end of each phase in the project, the most important being the readiness review/audit at the end of the detailed planning phase.

Front End Loading

See Element 4, Apply Front End Loading section.

Freeze Date

This is the date when the scope is frozen and the model **baseline** is set. This is done at least 6 weeks before the TA. Actual progress is then measured against this **baseline** plan.

Fig. 23.2 depicts the previously mentioned cycles, Front End Loading and Freeze Date, as well as the typical phases of a TA project.

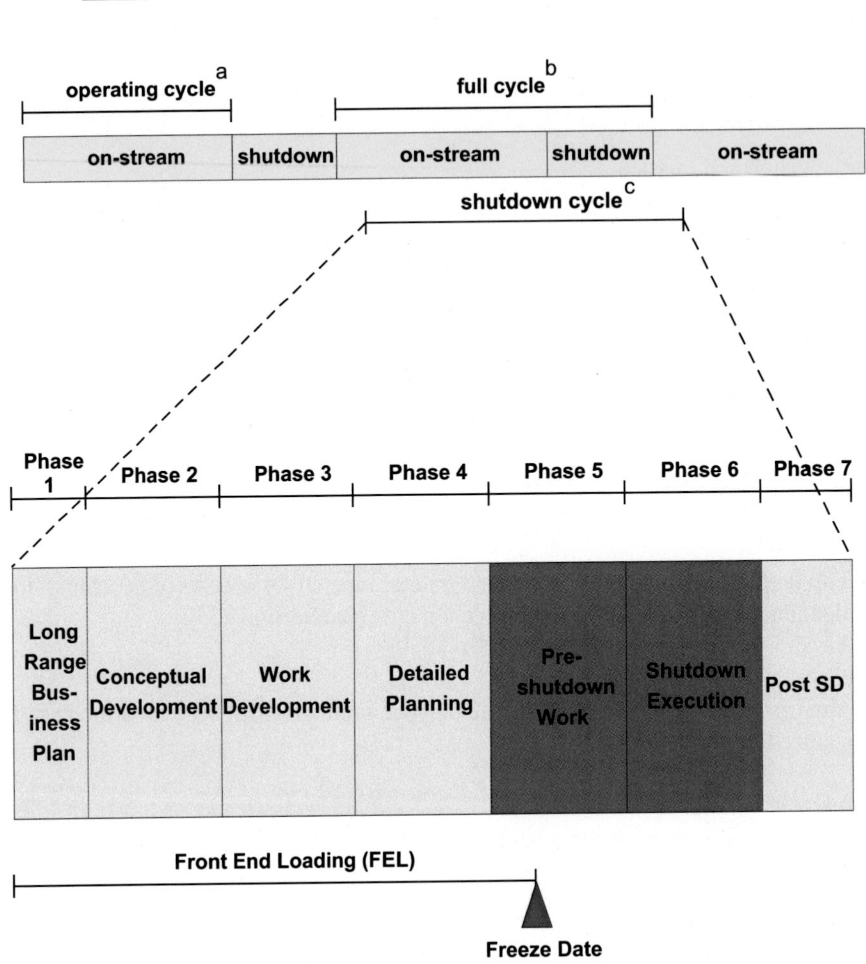

FIGURE 23.2

Cycles and phases.

3. SCOPE SCREENING

Scope screening, or challenge, is **critical** to ensure that **only those items that cannot be done online are carried out in the shutdown.**

This is done through a series of scope challenge workshops with cross-functional participation.

Any work **scope additions** or **significant scope growth after the freeze date** should be approved by the appropriate authorities before field execution. This additional work must have the following before approval:

1. Detailed work sequence and plans
2. Estimated work hours and resource requirements

3. Material requirements and availability
4. Estimated cost impact
5. Work category priority
6. Schedule analysis and impact

An example of scope screening, resulting in a significant reduction in TA time, is described in Box 23.1.

BOX 23.1 SCOPE CHALLENGE—GAS PLANT TURNAROUND

A gas plant typically had TAs in excess of 5 weeks, which was way above the industry norm. This was unacceptable to management.

Shell Global Solutions (SGS) was called in to facilitate a series of "scope challenge" workshops, and as a result, the TA plan was reduced to 3 weeks. In fact, the actual duration turned out to be even less than the agreed plan.

Consequently, TA staff who were trained in these workshops went on to advise other gas plants within the company on the optimization of TAs resulting in improvements in TA durations elsewhere in the company.

4. FRONT END LOADING (FEL)

An FEL index is a useful tool to assess performance in preparation for the TA. This index is defined as follows:

A quantitative measure of the level of definition for a TA.

TA definition includes defining the following:

- **What** will be done
- **How** it will be done
- **Who** will do it
- **When** it will be done

Ratings are suggested as follows:

1. No evidence—zero
2. Poor—25%
3. Fair—50%
4. Good—75%
5. Best—100%

Each area of assessment shown below is given a rating:

Area A: Scope Definition

1. TA goals determined
2. Stakeholder buy-in obtained
3. Work lists for maintenance, inspection, and capital projects completed
4. Scope control measures established
5. Work orders/engineering packages completed

Area B: Execution Strategy

1. Roles and responsibilities defined
2. Contracting strategy agreed
3. Schedule development completed
4. Risk mitigation and contingency planning actioned
5. Lessons learned incorporated

Area C: Planning Status

1. TA start date set
2. Availability of resources established
3. Materials deliveries on schedule
4. Shutdown, start-up, and operations procedures completed
5. Detailed plans, job plans, rigging plans, scaffolding plans completed

Alternatively, these subjects could be included in a more detailed readiness audit.
FEL is graphically described in Fig. 23.3.

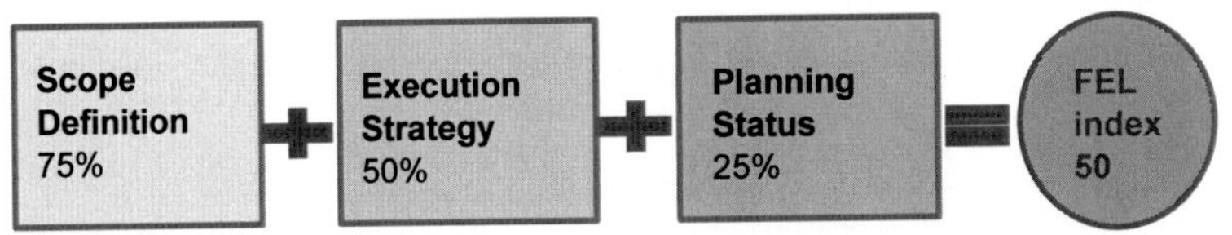

FIGURE 23.3

Front End Loading example.

5. CRITICAL PATH MODELING

Optimum downtime is determined by modeling all activities and determining and minimizing the **longest path of activities with optimum use of resources** required for the shutdown. A software Critical Path Model (CPM) such as Oracle Prima-Vera or Deltek Open-Plan is essential.

Analysis of the model must include thorough evaluation of the critical path and near critical path activities. The number of near critical path activities can affect the probability of achieving the planned completion date. If the model is well structured, probability analysis of the completion date could be undertaken. Prima-Vera Risk Analysis (formerly Pertmaster) is a useful tool for this purpose.

OBS, WBS, CBS, and resourcing must be integrated and include the following:

- Organizational Breakdown Structure (OBS)
  - Includes all TA project supervision and management (staff and contractors)
  - OBS chart to be distributed at start of TA for posting on bulletin boards and walls so that information about who is responsible for what is visible and easy to access
  - Also, 48-h "look ahead" printouts are easily available for those responsible for actions
- Work breakdown structure (WBS)
  - All work must be under this hierarchy by area, discipline, and system

- Cost breakdown structure (CBS)
  - Every activity must have a cost attached
- Contractor breakdown structure (could be part of OBS or separate)
- Resourcing
  - Detailed man-hours and equipment hours for all resources to achieve each activity

Advantages of Critical Path Modeling

1. Models can be archived and copied for next TA.
2. Work scope and resourcing can be produced quickly for emergency shutdowns from archived model(s), and opportunity work can be easily added to the emergency work.
3. Negates the need for building each TA model from scratch.
4. Progressive improvement of the model for each successive TA.

It is important to have the final CPM model agreed by all interested parties involved with a TA.

Box 23.2 gives an example of an instance where buy-in was not obtained from the top management and so corrective action was not initiated, resulting in a major loss to the company.

BOX 23.2 CRITICAL PATH MODELING

A major refinery TA was planned using an Open-Plan Professional (OPP) CPM model. All units were integrated into the master model including significant engineering work. Fluor, Foster–Wheeler, and CBI were the major engineering contractors, and they prepared their plans in Prima-Vera or Open-Plan for consolidation into the master plan, which was managed by the operating company.

After a full review, analysis, and amendment of the model, a 6-week TA was predicted and subsequently agreed on by management. On approval, scope was frozen and the TA commenced. The model was updated daily with actual progress and after 2 weeks was predicting a startup of 2 weeks later than planned. Management was in denial and refused to believe the results of the model. No corrective or contingency action was taken and the model continued to predict an overrun of about 2 weeks.

As predicted, actual start-up was just over 2 weeks late. Being a regional refinery with no other refineries in the area meant that this refinery had to stockpile product for the duration of the TA. As a result of the late startup, the stockpile ran dry and product had to be railed in from other refineries at great cost. Unsurprisingly the company made a major loss that year. Not only the company was affected, though, the fuel shortage also had a detrimental impact on the local economy.

Lessons Learned

This was a complicated shutdown with extensive engineering work, raising the probability of not completing the work on time. Probability tools were not used at that stage.

Communication: It was clear that management did not trust the model. The situation might have been somewhat allayed had the planners explained "the build" of the model in more detail to management before the shutdown in order to build trust.

Finally, it should be noted that a situation such as this should be declared an incident, and entered into an incident register, as it resulted in a major loss to the company.

6. CONTRACTOR STRATEGY

The business operator must ensure that the contractors chosen are very experienced in the type of shutdown to be undertaken.

The most effective approach is to incentivize the contractors. The incentivization must be focused on carrying out the required work in the specified time at the required quality standard. Further details are discussed in Section 13.2, Procurement Risk and Performance.

It is important to remember that contract values are a fraction of the cost of lost production, and **lost downtime can never be recovered**.

7. STAFF COMPETENCE AND MOTIVATION

Five different types of knowledge are required for TA management. These are as follows:

1. Local—knowledge of the history, layout, and peculiarities of the plant
2. Work management—knowledge of tasks to be undertaken
3. Craft—knowledge of details of work methods, tools and equipment required, and time for completing the tasks
4. Specialist—knowledge of specialist activities such as those relating to rotating machinery, fiscal metering, and licensed equipment
5. Management of discrete projects—project management knowledge and skills

Key staff must know the history of the asset and have been trained on technical aspects of maintenance/inspection and the project management of shutdowns.

Box 23.3 gives an example of a change in **management competence** and **contracting strategy** as a result of change in ownership.

BOX 23.3 COMPETENCE

A refinery changed ownership in 2000. Shortly thereafter, a crude unit TA was being undertaken with a duration planned at 40 days. At the time Solomon Associates benchmarking estimate for the duration of this type and size of crude unit was 30 days.

The question was raised as to whether the duration could not be reduced to 30 days, and the answer was: "We have always done it in this duration and cannot do better." [The TA contract was for a single TA contractor based on a bill of quantities, which does not motivate a contractor to reduce duration. On the contrary, a contractor is motivated to maximize the amount of work].

However, a search of old records found that under the previous ownership this particular crude unit TA had, in fact, been completed in 30 days. Even when confronted with this information, the refinery leadership was still not willing to change.

Lesson Learned

The new management should have been led through a paradigm shift using a facilitated change management process and an independent facilitator.

It was clear that the following was not evident, and was required:

1. Adequate full-time TA planning staff with CPM modeling experience
2. A structured scope challenge process
3. An integrated CPM model that is archived after each TA and copied and improved on for the next TA
4. A contracting strategy to employ very experienced TA contractors and motivate them to complete the TA in the required duration

Shell Global Solutions could have possibly offered the service as an independent facilitator as was done in the case described in Box 23.1.

Comment

Resistance to change is greatest when people think they are performing well but, in fact, are not. The leadership of this refinery clearly thought they were doing fine and were not willing to change. Only when a leader understands that the competition is doing better by taking a different approach, can things start to change.

8. APPLICATION OF LESSONS LEARNED

Lessons learned need to be documented and applied as follows:

a. The TA report must be compiled in such a way that makes it easy to extract lessons learned for application to the next TA.
b. The 80:20 rule should be applied to focus on primary issues for improvement in carrying out the next TA.
c. The planned and actual activities from the completed TA must be saved for future reference and improvement.

23.5 TURNAROUND RISK PROFILE

This profile is applicable to process plants containing hydrocarbons operating under pressure.

During normal operation, without plant upsets, the risk profile remains steady. However, risk increases with the opening of the pressurized hydrocarbon-containing vessels, equipment, and pipelines. Once all of the hydrocarbon-containing vessels, equipment, and pipelines are gas free, the risk profile reduces. The highest risk profile is at startup when hydrocarbons are reintroduced into the pressure system, which is when the potential for leaks is highest. Flawless startup (no incidents, **no leaks**) is an important KPI.

A TA risk profile is identified in Fig. 23.4.

		SHUTDOWN PERIOD			
	NORMAL OPERATIONS	**PREPARATION**	**MAINTENANCE WINDOW**	**START-UP**	**NORMAL OPERATIONS**
Risk Drivers	Pressurized hydrocarbons and hazardous chemicals Risk of hydrocarbon release.	Isolating/ depressurizing equipment Continued risk from hydrocarbons and work Risk slightly elevated than normal operations.	Running multiple simultaneous jobs for maintenance, inspection and plant upgrades Minimized risk from hydrocarbons.	Recommissioning Increased risk profile from hydrocarbons and work **Highest risk profile.**	Pressurized hydrocarbons and hazardous chemicals Risk of hydrocarbon release.

FIGURE 23.4

Turnaround risk profile.

23.6 INNOVATIVE IDEAS FOR REDUCTION IN TURNAROUND DURATION

Once the scope of the TA is agreed, the TA duration has to be reduced as much as can be safely achieved. Innovative solutions are sometimes needed when the critical path turns out to be excessively long. Some ideas are as follows:

FCCU RISER INTERNAL INSPECTION

In a particular case, scaffolding was initially planned but finally ab-sailing was used, after inspectors were trained in rope techniques. Inspection duration was drastically reduced.

SPADING AND DESPADING OF FLANGES

A flange spreading tool such as that produced by Equalizer[2] is used for the spreading of flanges for spading and aligning flanges for bolt up. Spading and despading time is considerably reduced using this tool.

Prior to a TA, operations and maintenance staff identify spading points for the isolation of process pipe and other areas in preparation for opening. A tag is placed at each location, identifying what spade, and its pressure rating, is required. The process of swopping a tag for a spade, and displaying all tags/spades on a tag/spade board for each work package, reduces the risk of leaving a spade in the plant during startup. When the spaded line is safe to work on, the board displays only tags. At startup, the board only display spades.

Example of Unaccounted Spade

On one occasion in the Bahrain Refinery, a spade was accidently left in a vapor line on a column at startup after a crude unit TA, thus causing a startup delay of hours. This was a very large spade!

HEAT EXCHANGER BUNDLE EXTRACTION, CLEANING, AND HYDRO-TESTING

The use of a heat exchanger bundle puller cradle, for extracting large tube bundles, speeds up the heat exchanger cleaning and hydro-testing cycle. However, the use of a single cradle could cause a bottleneck when all the bundles are installed at the same time toward the end of the TA.

Very high pressure (as opposed to high pressure) water cleaning speeds up the bundle cleaning process. Adequate numbers of these very high pressure machines are required to suit the number of exchangers which need to be cleaned.

Example of Removing the Complete Exchanger From the Process Area

Bahrain Refinery has the practice of removing the complete exchangers to the workshop for cleaning and hydro-testing, thus decongesting the area of work in the process plant, and increasing productivity by creating a mini production line in the workshop. The refinery has, of course, been designed for this sort of practice.

CHEMICAL CLEANING

Chemical cleaning could be applied to “hard to remove” deposits. The cleaning cycle of spading, pump-around, passivating, and despading needs to be built into the preparation phase of the TA. Only those professional cleaning companies that have successfully completed similar work before should be considered. Box 23.4 gives an example of chemical cleaning gone wrong.

BOX 23.4 CHEMICAL CLEANING GONE WRONG—POLYETHYLENE TERAPHTHALATE PLANT

During a TA, a company employed a contractor to chemically clean the nitrogen plant air compressor heat exchangers. (Nitrogen is required to flush the process piping at start-up and blanket the process vessels during operation.)

The cleaning contractor performed the task but, on startup, it was found that the compressors could not produce air for the nitrogen plant and so the startup was delayed.

Pinhole leaks in the air compressor heat exchangers, caused by incorrect cleaning methods, were discovered. This resulted in all air compressor heat exchangers having to be replaced. Luckily, the compressor manufacturer (Atlas Copco) had spares, but more than a week's production was lost due to this incident.

Lessons Learned

1. Ensure the contractor is qualified to perform this specialized sort of chemical cleaning.
2. Visit sites where the contractor has done similar work to be assured of their capabilities, before awarding the contract.
3. Request and verify their process for cleaning and monitor their progress to ensure that they work to the process.
4. Ensure that the contractor's appointed supervisor has done many similar cleaning operations before.

MODULAR REPLACEMENT

On a small scale, mechanical seals can be removed and replaced with reconditioned seals, on condition that the removed seals are sent to an original equipment manufacturer qualified seal repair contractor for reconditioning between TAs.

On a much larger scale, a gas turbine could be swopped out for an identical rebuilt turbine. The turbine that has been removed is then sent to the vendor for rebuilding for the next TA.

For example, Qatargas has a maintenance agreement with Al Shaheen GE Services Company for the exchange of gas turbines.

23.7 STRATEGIC PLANNING AND RUN LENGTH DETERMINATION

The following questions need to be raised before each TA:

1. What are the legal/insurance/industry standards requirements?
2. When are the required upstream shutdowns to take place?
3. What are the predicted market conditions? (to time the shutdown for minimum lost profit)
4. How are the units performing—mechanically and from a process point of view?
5. What other major shutdowns are planned in the country? (so as not to overload available contractor resources)
6. Should the whole plant be shut down, or just complexes or units?
7. What plant improvements/modifications are required during the shutdown?
8. What run length is being envisaged?

Factors that affect extension of TA periods are as follows:

- Process severity issues
  - Catalyst life
  - Fouling

  - Corrosion
  - Coking
- Resistance to change by the leadership team and specialists.

Different approaches for determining the run length (operating cycle) are as follows:

1. Benchmarking
2. Reliability-centered maintenance
3. Risk-based inspection

These are shown in Table 23.1.

Table 23.1 Operation Cycle Determination

	Benchmarking	Reliability-Centered Maintenance	Risk-Based Inspection
Approach	Top-down generic	Bottom-up specific equipment	Bottom-up specific pressurized systems
Focus	Unit—hierarchical	Equipment—process flow	Pressurized systems
Comparison	Industry norms for type of unit	Statistical run time for equipment	Industry norms for pressure containing equipment
Application	Immediate	Long term	Long term
Data source	Last benchmarking study	Equipment vendor, industry and Reliability Department historical data	Plant historical data
Methodology	Solomon model (profile II), EDC weighted	FMECA, etc.	Risk assessment software (e.g., Lloyds-Capstone[3])
Targets	Easily and quickly determined from industry norms	Determined after extensive data analysis	Set from output of risk assessment software
Objective	**Achieve industry norms for mechanical availability**	**Achieve optimal level of reliability for required process unit run length**	**Optimize period between internal inspections**

23.8 TURNAROUNDS AND PROFIT RELATIONSHIP

Maintenance and inspection competence within a mature maintenance and inspection system are the foundations for determining good KPIs and strategies. This and a well-structured TA framework and competent TA management staff ensure an optimal TA duration for an optimum operating cycle or run length. This in turn determines optimum production over the long term, thus maximizing long-term revenue generation.

As reliability improves with the optimization of TA duration and operating cycle, the maintenance and inspection costs come down, thus reducing operating expense and increasing profit.

Fig. 23.5 depicts these relationships.

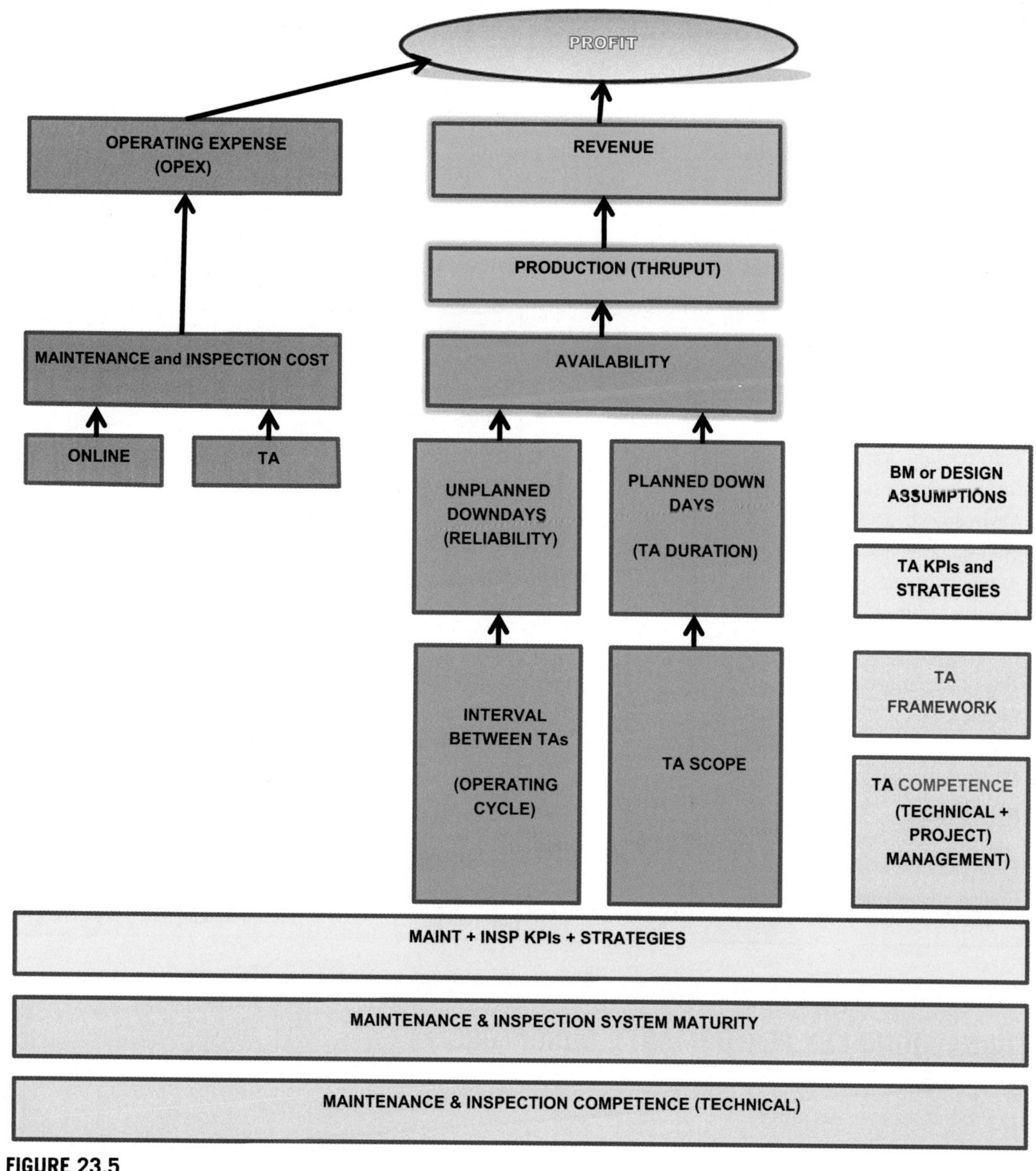

FIGURE 23.5

Turnarounds and profit relationships.

Box 22.7, System Maturity—Gas Plants discusses the effect of system maturity on the increase in reliability and reduction of maintenance costs for three gas plants.

23.9 TURNAROUND CRITICAL SUCCESS FACTORS AND KEY PERFORMANCE INDICATORS

Critical success factors (CSFs) determine measurement areas on which to focus. KPIs can then be determined more easily. Table 23.2 identifies the primary CSFs and related measurement areas.

Table 23.2 Critical Success Factors and Measurement Areas

Critical Success Factor	Measurement Area
Specific Turnaround	
Duration	Oil out to on-spec product: days and annualized days
Turnaround cost	Actual and annualized
Predictability (Variance)	Actual versus planned: Man-hours, duration and cost
Safety incidents	From oil out to on-spec: number and rate
Environmental incidents	From oil out to on-spec: number—total and major
Start-up incidents	Number and days lost
Additional work	Actual versus contingency
Leak free start-up	Number of leaks causing delay in start-up
Other	
Total maintenance cost	Turnaround and routine maintenance annualized
Frequency	Run length in months
Unscheduled shutdowns	Days lost during the run between turnarounds
Mechanical availability	Time available %—annualized
Initiatives	Time and cost savings—decontamination, modularization etc.
Front end loading	Preparedness for the next turnaround
Comparative statistics	Mechanical availability, turnaround duration (days), turnaround cost, run length (months), unplanned down days

TURNAROUND KEY PERFORMANCE INDICATORS

Categories

1. Predictability (**variance**)
   a. Measures a company's performance versus the **company's estimates and targets** set by the TA team
   b. Applies only to a specific TA
   c. Short term: target achievable with current tools/strategies
2. Absolute (**efficiency**)
   a. Measures a company's effectiveness versus that of other companies (**benchmarked**)
   b. Applies to all TAs
   c. Long term: normally involves a paradigm shift by those involved for **step improvement**

Next Turnaround

1. Schedule
   a. Behind/ahead of critical path: hours
   b. **Schedule variance** (SV): actual/planned duration
   c. Schedule performance index (SPI)—alternative to SV
      i. Deviation from planned to date: earned value/planned value
2. Cost
   a. **Cost variance** (CV): actual/planned costs
   b. Cost performance index (CPI)—alternative to CV
      i. Deviation from budget to date: earned value/actual cost
   c. Deviation from budget: estimate at completion/budget at completion
   d. Additional work: actual versus contingency %
3. Scope
   a. Emergent work
      i. Man-hours as % of total TA man-hours: % *(minimize)*
4. HSE
   a. Lost time incidents (LTIs): Number *(target: zero)*
      i. Monitor lower-level HSE indicators to prevent even one incident
   b. Flawless start-up (no incidents and **no leaks**): Number *(target: zero)*
      i. Measure from "hand back" to "on-stream at required quality"

Future Turnarounds

1. Long-term TA duration target: **schedule efficiency**
   a. Shutdown duration relative to others with similar scope and complexity
2. Plant availability **(benchmarked)**
   a. Maximal interval between TAs with
   b. Minimal unplanned down days and
   c. Minimal planned down days
      i. Annualized
3. Future runtime target
4. FEL
   a. Preparedness indicator

23.10 SUMMARY

A TA is a shutdown that has been planned from more than a year in advance to ensure a run length in line with industry norms with minimal unplanned down days. It is a complex project to be managed in line with the PMBOK.

The objectives of TA management are: to have an optimal TA cycle and minimal unplanned down days, in order to maximize long-term availability of the plant.

Pacesetter TA management requires the following:

1. Clear alignment of TA statement of commitment with the corporate vision and mission
2. Structured, phased planning and execution

3. Structured scope screening
4. Front End Loading
5. Critical Path Modelling tools
6. Appropriate, experienced, and incentivized contractors
7. Competent staff (both technical and management)
8. Use of "lessons learned" for the next TA

The risk level for a TA is highest in the startup phase.

Innovative ideas for TA duration reduction are required.

Run length is determined from three sources:

1. Benchmarking
2. Reliability-centered maintenance
3. Risk-based inspection

Optimum TA duration for an optimum operating cycle is critical for the long-term profitability of the plant.

Performance indicators have two focuses:

1. **Variance** from the company's targets for a specific shutdown
2. **Efficiency**, which entails comparing performance of all TAs to other companies

REFERENCES

1. Singh R. *World class turnaround management*. Everest Press; 2000.
2. Equalizer International Group. www.equalizerinternational.com.
3. Lloyds Capstone RBI software. http://rbi.lrenergy.org/capstone-rbmi-software/.

FURTHER READING

1. Lenahan T. *Turnaround management*. Butterworth-Heinemann; 1999.
2. *Shutdown survey. IIR 5th Southern African Annual Symposium*. July 2002.
3. Hey RB. *The best turnaround ever*. Hydrocarbon Processing; April 2001.
4. *Practical management for turnarounds* Chapter 1: plant turnaround. www.reliabilityweb.com.
5. Oliver R. Complete planning for maintenance turnarounds will ensure success. *Oil Gas J* April 29, 2002.
6. Oliver R. *Turnarounds, an integral component of asset performance management*. World Refining; March 2003.
7. TAR fighter shutdown project simulation. www.tacook.com/tar.
8. Open Plan website. www.deltek.com/openplan.
9. Primavera website Oracle. https://www.oracle.com/applications/primavera/index.html.
10. Singh R. *Executive leadership essential to ensure world class turnarounds*. Global Turnaround and Maintenance, Hydrocarbon Processing; 2012.
11. Singh R. *Achieve excellence and sustainability in your next turnaround*. Hydrocarbon Processing; January 04, 2014.
12. Operations integrity management system (OIMS). ExxonMobil. https://www.exxonmobil.com/UK-English/about_integrity_oims.aspx.
13. Levitt J. *Managing maintenance shutdowns and outages*. Industrial Press; 2004. https://www.amazon.com/Joel-Levitt/e/B001IU2Q9Y/ref=dp_byline_cont_book_1.
14. Sahoo T. *Process plants: shutdown and turnaround management*. CRC Press; 2013.

PART

BENCHMARKING

6

OVERVIEW

WHY IS IT IMPORTANT TO BENCHMARK?

It is vital for a business to know where it stands and whether it is losing or gaining ground relative to the competition, so as to ensure it remains competitive. Details of shortcomings need to be identified before improvement initiatives can be undertaken.

THE OBJECTIVE

The objective of benchmarking is to find examples of superior performance and understand the processes and practices driving that performance. Companies then improve their performance by tailoring and incorporating these best practices into their own operations.

The objectives of Part 6 are the following:

1. Outline the need for using established benchmarking consultants
2. Explain the value and limitations of benchmarking
3. Show the benchmarking data verification process to ensure confidence in the data
4. Establish how to use benchmarking outputs effectively

PICTORIAL VIEW

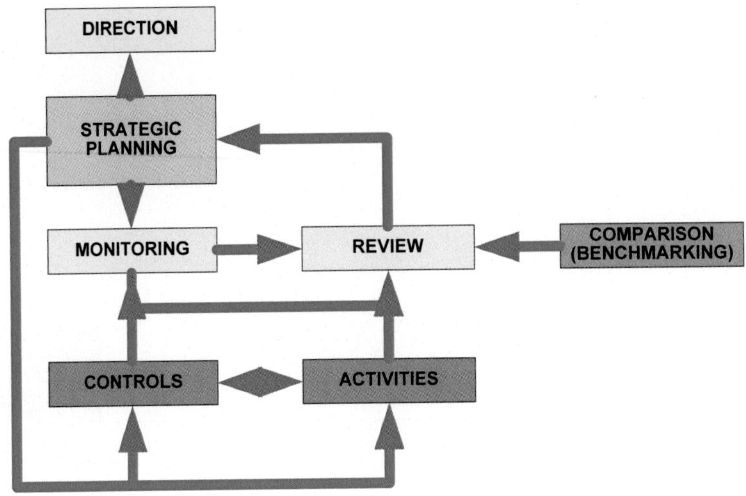

OUTLINE OF PART 6

- The necessity of benchmarking is explained.
- The generic benchmarking process is outlined.
- The improvement process, resulting from benchmarking, is illustrated.
- The defining features of benchmarking are listed.
- The benchmarking code of conduct is laid out.
- The advantages of benchmarking are listed.
- The importance of selecting established benchmarking consultants is explained.
- Definitions of industry leaders or "pacesetters" are defined.
- Process plant categories for specific comparative studies are listed.
- Peer grouping for improved comparison is explained.
- Project and turnaround benchmarking concepts are discussed.
- Common denominators (proprietary and generic) and their relationships are illustrated.
- Proprietary indexes are explained.
- Data verification methods and tools, from input to assessment, are discussed.
- The need for evaluating the implications of benchmarking envelopes is discussed.
- Leads for using the results are outlined, including business integration issues.
- Types of output formats are demonstrated.

CHAPTER

THE BENCHMARKING PROCESS AND CONSULTANT SELECTION 24

24.1 INTRODUCTION

DEFINITION

Benchmarking is the process of comparing one's business processes and performance metrics to industry bests and/or best practices from other industries. Benchmarking is a powerful tool to assess where the company is relative to the competition and is a key element in the continuous improvement process. Benchmarking supplies information for management to ascertain the following:

1. Where the business **was**
2. Where the business **is now**
3. Where the **competition is**

Management then knows **confidently** where the business **is going**.

Focus should be on those elements that most affect the commercial success of the business.

24.2 INTEGRATION INTO THE BUSINESS

In the past, a benchmarking consultant would be employed to complete the benchmarking study and, at the end of the process, the beautifully bound benchmarking report would be put on the manager's bookshelf, never to be referred to again.

Today, this report is in many forms and enables detailed analysis by the company's performance experts, before appropriate plans and actions are undertaken as part of the business strategy.

Fig. 24.1 shows the integration of benchmarking into the business.

Benchmarking is a process that measures the business performance and compares it with its peers. In other words, benchmarking identifies where the company stands or rates on a particular issue relative to the competition. Action for step improvement of particular issues can then be undertaken.

Performance Management for the Oil, Gas, and Process Industries. http://dx.doi.org/10.1016/B978-0-12-810446-0.00024-4

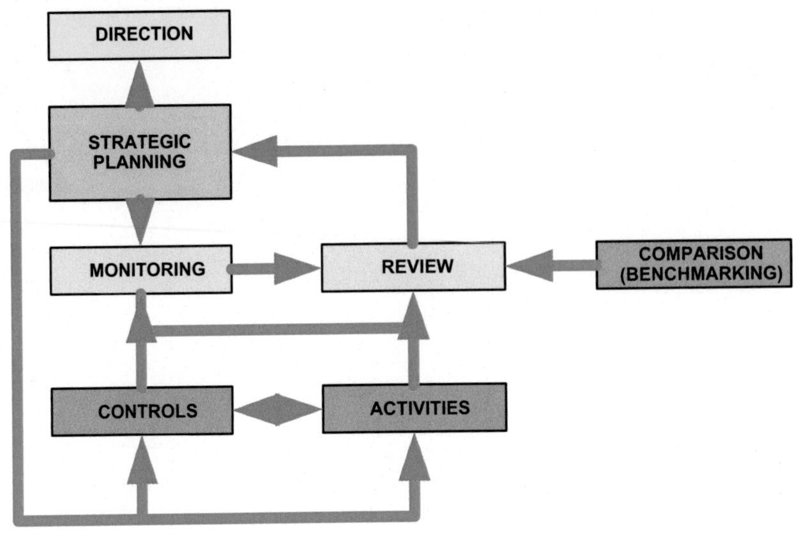

FIGURE 24.1

Integration of benchmarking into the business.

The overriding objective of benchmarking is to identify best practices. The **purpose** of benchmarking is to generate improvement action to enhance the value of the business unit to its stakeholders. Benchmarking helps to:

1. gain a detailed understanding of current and potential performance;
2. quantify the gaps;
3. prioritize initiatives; and
4. close the performance gaps.

The defining features of benchmarking are as follows:

1. Purpose of benchmarking—specific to the business unit
2. Externally focused—comparison with competitors
3. Measurements based—using industry standard measurements for proper comparison
4. Information intensive—extensive and appropriate data
5. Objective—facts based
6. Action generating—clear actions, responsibilities, and deadlines

Benchmarking **should** show:

1. time-related "before and after" or "trend over time" situations; **and**
2. frozen-in-time comparison of "apples" versus "apples" at a particular moment in time.

Time-related benchmarking identifies convergent or divergent performance over time relative to the competition. Section 31.4, The Use of Graphical Displays, gives an example of time-related benchmarking.

Frozen-in-time means that all data are collected at the same time and, consequently, no time-related corrections are required. However, some functional benchmarking has to be time adjusted.

Example: For project benchmarking, comparable projects are projects that have been completed.

Benchmarking **should** facilitate the establishment of high-level performance targets for the company. Top **Key Performance Indicators (KPIs)** thus need to be common in the industry for comparison.

Benchmarking is not:

- *only* competitive analysis;
- "number crunching";
- "industrial tourism";
- just "copying" or "catch up";
- spying or espionage;
- quick and easy.

Benchmarking is an ongoing process:

- To understand your own Controls, Assets, Organization, and Systems and Processes and
- To compare your own Controls, Assets, Organization, and Systems and Processes
  - Within your organization
  - To world class organizations
    - In your industry
    - Outside your industry
    - In your region or worldwide

Benchmarking ensures a single improvement system that encompasses the following:

1. Continuous improvement ***and***
2. Breakthrough or step improvement

Examples:

Continuous improvement: Energy Index gap—

- *Create a burner management program to ensure clean combustion.*

Breakthrough or step improvement: turnaround duration gap—

- *Change turnaround contracting strategy.*

24.3 WHY BENCHMARK?

The following are the primary reasons for benchmarking:

1. Improve:
   a. Profits
   b. Effectiveness
   c. Efficiency
2. Accelerate and manage change
3. Set stretch goals/targets

4. Achieve breakthroughs/innovations
5. Create a sense of urgency
6. Overcome not invented here (NIH) complacency or arrogance
7. See "outside the box"
8. Understand world class (pacesetter) performance

24.4 GENERIC PROCESS

Benchmarking is undertaken between industry peers, that is, similar business units or functions.

The basic process is as follows:

1. Plan and establish the particular benchmarking exercise as a project
2. Collect and normalize data
3. Analyze results
4. Identify strategies and actions
5. Report to decision-makers
6. Implement improvements

In the case of having an experienced benchmarking consultant on board, the following occurs:

1. Financial and operational data are submitted to the benchmarking consultant in a standard format where the consultant can enter the data into the benchmarking model.
2. Data are validated, checked for consistency, and normalized in the model.
3. Quantitative results are produced for each client showing their data against the competitions' grouped data.

The benchmarking process is shown in Fig. 24.2.

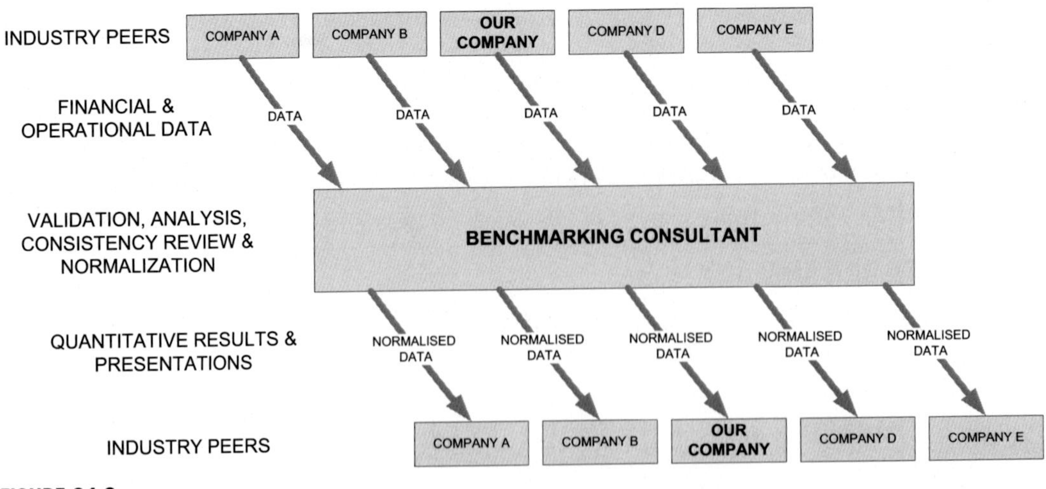

FIGURE 24.2

The benchmarking process.

Once the benchmarking report has been completed, the real work starts. Industry-specific reports are very comprehensive and one needs to look for the factors that have the most influence on overall improvement. This is covered in Chapter 29, Identification, Analysis and Evaluation of Gaps.

24.5 BENCHMARKING CATEGORIES

Benchmarking is categorized as follows:

1. Company
2. Functional
3. Operational/competitive
   a. Internal
   b. External

1. COMPANY BENCHMARKING

Scope

Company benchmarking includes the following:

A. Corporate governance
B. Organizational structures

Frequency

This is required only when a major restructuring or change in business system is anticipated and would therefore be carried out on an ad-hoc basis.

Methodology

Selection, data gathering, and reporting processes need to be standardized and documented.

2. FUNCTIONAL BENCHMARKING

Scope

Functional benchmarking includes the following:

A. Procurement
B. Finance
C. Human resources
D. Information technology
E. Turnarounds (planned shutdowns)
F. Capital projects

Frequency

Functional benchmarking should be carried out periodically at intervals of up to 5 years or when a major change or disruption in operations occurs.

Methodology
Functional benchmarking methodology is determined by the selected benchmarking consultant.

3. OPERATIONAL/COMPETITIVE BENCHMARKING

Scope
This type of benchmarking is always performed in the same industry within which the company operates. Operational benchmarking includes the following:

1. Refining
2. Gas operations
3. Oil/gas supply chain
4. Petrochemicals
5. Power
6. Aluminum
7. Cement
8. Paper
9. Steel

Frequency
Operational benchmarking should be done at a maximum of 3-year intervals.

Methodology
Internal

Internal benchmarking is carried out between similar business units of the same company. Confidentiality is not required so there should be an open flow of information, making the benchmarking easy to implement.

> *Example:*
>
> *Benchmarking between two similar coal-fired power stations (business units) of an electricity-generating company.*

External

The consultant determines the methodology. For refining, gas operations and petrochemicals, the methodology is based on the consultant's benchmarking model.

24.6 BENCHMARKING CONSULTANT SELECTION

Consultants are selected based on their knowledge of a particular industry and their extensive benchmarking database of this industry. The partnership with a benchmarking consultant **should** be a long-term relationship to ensure time-related trending against industry peers.

1. COMPANY BENCHMARKING

Benchmarking should be carried out against similar industries (industry peers). Section 7.7, Governance Surveys, discusses company governance benchmarking.

2. FUNCTIONAL BENCHMARKING

These should be carried out with consultants with extensive databases related to the particular function. The following consultants are specifically suggested:

- Independent Project Analysis (IPA)[1] or Solomon Associates[5] for:
  - all planned major (i.e., process unit or complex) process plant turnarounds
- IPA for:
  - major oil and gas capital investments
- Towers Perrin,[2] etc., for:
  - salary and benefits surveys
- Gartner[3] for:
  - major IT implementation projects
- International Oil and Gas Producers Association[8]
  - oil and gas HSE

3. OPERATIONAL BENCHMARKING

These should be carried out with consultants with extensive databases related to the particular type of business unit. The following consultants are specifically suggested:

- Refining: Solomon Associates
- Gas operations: Phillip Townsend Associates Inc. (PTAI),[6] Solomon Associates,[5] or Juran Institute[4]
- Oil and gas supply chain (transmission): Phillip Townsend Associates, Solomon Associates, or Juran Institute
- Offshore oil and gas production: Juran Institute, McKinsey[7]

Their respective primary expertise (with some overlap) is as follows:

SOLOMON ASSOCIATES

- Fuels refineries (>90% of fuels refineries worldwide)
- Specialist processes:
  - Fluidic catalytic cracking units (FCCUs)
- Lube refineries
- Gas plants
- Gas transmission
- Petrochemicals
- Liquid pipelines
- Power generation
- Terminals
- Functional benchmarking:
  - Reliability and maintenance
  - Refinery turnarounds
  - Process control (automation)
  - Laboratories/quality control

PTAI

- LNG plants (>70% of LNG plants worldwide)
- LNG terminals and shipping
- Basic chemicals
- Intermediate chemicals
- Polymers
- Functional benchmarking:
  - Reliability and maintenance
  - Energy
  - Site personnel index
  - Process safety

JURAN

- Production and processing facilities
- Marine terminals
- LNG receiving terminals
- Gas transmission systems
- Liquid and gas pipelines
- Underground Gas storage facilities

McKINSEY

- Oil and gas supply (value) chain
- Offshore

IPA

- Major projects
- Turnarounds

INTERNATIONAL OIL AND GAS PRODUCERS (IOGP)

- Health and safety (for members)

24.7 PEER GROUPS

As benchmarking does not permit identification of primary competitors, the concept of peer groups helps identify performance relative to these primary competitors. Thus, a critical factor in benchmarking is the selection of the peer group against which the participating plant compares its performance.

Why use more than one peer group?

- No single peer group will enable a reasonable and relevant comparison of the full range of comparison metrics. For example, a single peer group for both high-profitability and low-cost analyses usually results in missed opportunities.
- In performing comparative analysis, it is common to make the same analysis across a number of peer groups to confirm messages.
- Users of the analysis must feel confident that they are comparing apples with apples. The usual defense against accepting comparison conclusions is that "we are different."

The following are sample peer groups for each type of process plant.

REFINING

1. Equivalent Distillation Capacity (EDC) (see Section 25.3: Consultants Proprietary Common Denominators for definition)
2. Market supply—supply corridors
3. Geography—responsive to the same regional regulatory and taxation bodies, face the same demands for product quality, and have common pricing structures and market forces
4. Ownership—International oil company, National Oil Company, independents
5. Three sequential fuels studies
6. Pacesetters

LNG

1. Equivalent gas processing capacity (EGPC) (see Section 25.3: Consultants Proprietary Common Denominators)
2. Corrected mechanical units (MUc) (see Section 25.3: Consultants Proprietary Common Denominators)
3. Normalized Shift Positions (NSP)/Site Personnel Index (SPI) (see Section 25.3: Consultants Proprietary Common Denominators)
4. Market supply—supply corridors

PETROCHEM

1. Source of feedstock
2. Market supply—worldwide

POWER

1. Energy source—nuclear, coal, hydro, gas (turbine), wind, solar, etc.
2. Market supply—regional, limited by the power grid

ALUMINUM

1. Primary raw material source—bauxite
2. Source of cheap power—hydro, gas, or other
3. Market supply—worldwide

CEMENT

1. Source of raw materials—limestone, shale, iron ore, gypsum
2. Source of cheap fuel—coal, gas, or other
3. Market supply—regional

PAPER MILL

1. Source of timber
2. Market supply—worldwide

STEEL

1. Source of iron ore
2. Source of cheap fuel—coal, gas, or other
3. Market supply—supply corridors

RELIABILITY AND MAINTENANCE

For reliability and maintenance benchmarking, the two dominant peer groups are as follows:

1. Geographical region—for reasons of prevailing hourly labor costs, employment and contracting practices, safety and environment protection norms, climate, management culture, and the industrial development level of the region
2. Process unit family—for reasons of equipment types, process severity (corrosivity, toxicity, etc.), and maintenance complexity

   Examples of process unit families are as follows:

   - *Fuels refineries*
     - *CDU: atmospheric Crude Distillation Unit*
     - *FCCU: Fluidic Catalytic Cracking Unit*
   - *Chemicals*
     - *Primary olefin plant petrochemical products: ethylene, propylene, butadiene, mixed aromatics*
     - *Polyolefin thermoplastics: low-density polyethylene, linear low-density polyethylene, high-density polyethylene, polypropylene*
   - *Pulp and paper*
     - *Pulp continuous digester*
     - *Paper machine: cellulose*

24.8 PROCESS CATEGORIES FOR BENCHMARKING

Benchmarking studies are undertaking in the following categories.

UPSTREAM OIL AND GAS PRODUCTION

- Onshore
- Offshore

REFINERIES

1. Topping
2. Catalytic cracking
3. Hydro cracking
4. Coking
5. Lube oil

GAS PLANTS

1. LNG
2. Natural gas liquefaction: pipeline/flowing/sales gas
3. Gas-to-liquid

Example:

An LNG complex includes trains dedicated to pipeline/flowing/sales gas. For benchmarking purposes these are given equivalent LNG capacity.

PETROCHEMICAL PLANTS

- Olefins
- Low-density polyethylene
- High-density polyethylene
- Vinyl chloride monomer
- Ammonia
- Urea
- MTBE
- Methanol
- Benzene
- Polyethylene terephthalate
- Polypropylene
- And so on

POWER PLANTS

- Nuclear steam turbines
- Coal-fired steam boilers/turbines
- Oil-fired steam boilers/turbines
- Hydroelectric
- Gas turbine: open cycle
- Gas turbine: combined cycle
- Wind
- Solar
- And so on

ALUMINUM

- Alumina refining
- Smelting

CEMENT PLANT

- Wet process
- Dry process

PULP AND PAPER MILLS

- Pulp—batch digester
- Pulp—continuous digester
- Paper—cellulose
- Paper—fine paper
- Paper—Kraft

STEEL MILLS

- Blast furnace/basic oxygen furnace
- Smelt reduction/basic oxygen furnace
- Direct reduced iron/electric arc furnace
- Scrap/electric arc furnace

24.9 DEFINITION OF *PACESETTER*

SOLOMON ASSOCIATES

Achieve first or second quartile accomplishments in all the major performance areas for the entire trend period (three consecutive studies—6 years)

The major performance areas are as follows:

1. Energy Intensity Index
2. Maintenance Index
3. Operating expense (opex)
4. Return on Investment
5. Work hours per EDC
6. Utilization
7. Mechanical Availability

The pacesetter comparison could be done on regional, size, or other basis within a process category.

Example:

A fuels refinery chose to target being a "pacesetter in the Middle East"— compared against all sizes and refinery process categories.

PTAI

The average of the two sites nearest to the top of the comparison group.

PTAI's definition only applies to comparison groups.

Example:

Pacesetter for LNG

Some benchmarking exercises use the more subjective "world class" definition for the top performers. Others use the term "best in class" for peer comparisons.

24.10 PROJECT BENCHMARKING

As part of the **business risk mitigation** process, projects greater than a certain value should be benchmarked by parties outside the project team. Typically, this is over \$500 million. Benchmarking is primarily a check on the cost and schedule.

The time to benchmark is generally in the definition phase of a project. This is depicted in Fig. 24.3.

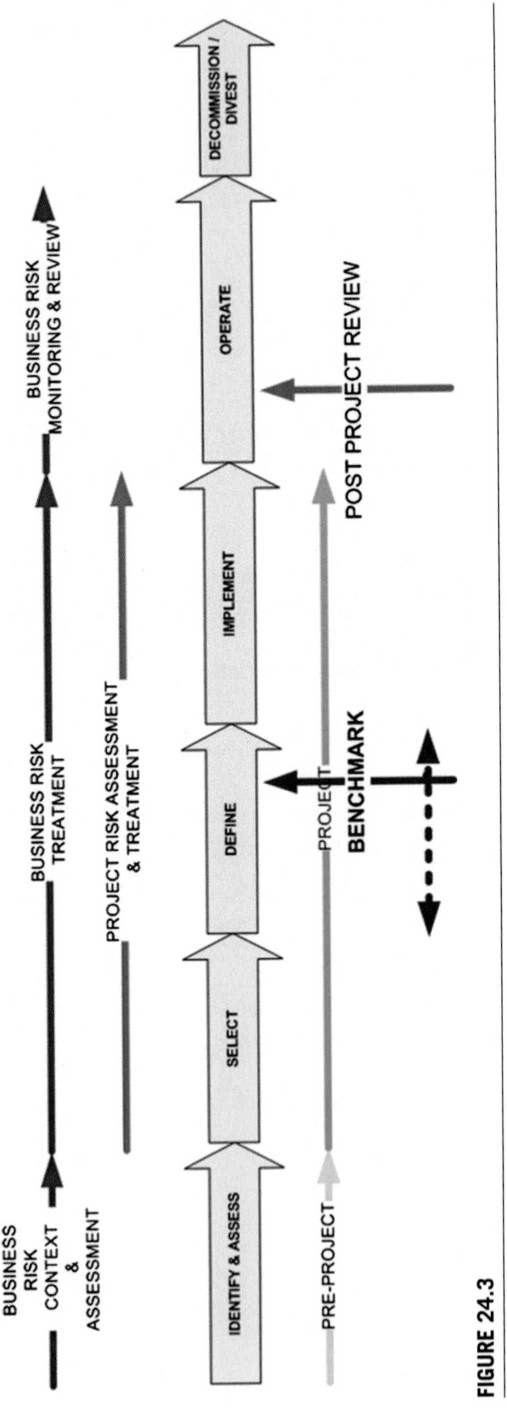

FIGURE 24.3

Lifecycle risks.

IPA performs project benchmarking using their database of specific process units recently constructed. They have more than 200 projects in their $1 billion-plus category and more than 8000 projects in their $5 million-plus category.

Solomon Associates performs a refinery project review based on their operating refinery database. Section 29.10, Assessment by Consultants, outlines this.

24.11 TURNAROUND (TA) BENCHMARKING

The objective of turnaround benchmarking is to understanding the effectiveness in the company's practice of planning, defining, and executing turnarounds. The purpose of turnaround analysis is to:

- measure turnaround target setting and performance of a company's turnaround management system relative to industry average;
- measure turnaround target setting and performance of an individual turnaround relative to the best turnarounds;
- quantify the level of definition or readiness to achieve "best in class" performance;
- identify gaps in targeted performance and gaps that may prevent "best in class" performance; and
- develop a set of actionable recommendations for the teams to implement.

Solomon Associates undertakes turnaround benchmarking specific to refining while IPA undertakes benchmarking on all refining and petrochemical industry turnarounds.

IPA has a database of more than 250 turnarounds.

The metrics provided to the client by IPA are shown in Box 24.1.

BOX 24.1 IPA BENCHMARKING REPORTS

Prospective Analysis: 4 to 6 months Out	Readiness Review: 6 weeks Out	Retrospective: Closeout	System Benchmarking
Current front end loading (FEL) Estimated FEL Schedule performance risk Cost performance risk Estimated team integration Estimated use of value improvement processes (VIPs) Recommendations	Progress since earlier evaluation Actual FEL and rating Updated schedule performance risk Updated cost performance risk Updated team integration Updated use of VIPs Updated recommendations	Actual FEL Schedule performance Cost performance Total schedule predictability Startup schedule predictability Cost predictability Team integration assessment Safety performance Lost time incidents Recordable incidents Environmental releases Calculated based on turnaround size Work order growth Lessons learned	Metrics similar to retrospective metrics except that the system is evaluated as well as consistency (or lack thereof) within the system

24.12 BENCHMARKING CODE OF CONDUCT

The following is accepted practice for participants in benchmarking:

1. Keep it legal.
2. Be willing to give what you get.
3. Respect confidentiality.
4. Keep information internal.
5. Do not refer without permission.
6. Be prepared from the start.
7. Understand expectations.
8. Act in accordance with expectations.
9. Follow through with commitments.

24.13 BUSINESS IMPROVEMENT PROGRAM (BIP) CATEGORIES FOR OPERATIONAL BENCHMARKING

Typical categories for a **balanced** Business Improvement Program (BIP) resulting from an operational benchmarking exercise are as follows:

- Mass balance
- Energy balance
- Utilization
- Mechanical availability
- Maintenance costs
- Opex
- Profit
- Return on investment
- Human resources
- Health and safety
- Environment

Gaps in these categories are identified and targets and actions are allocated to the relevant business systems, with clear ownership for implementation. Improvement is monitored before the next round of benchmarking. This is depicted as shown in Fig. 24.4.

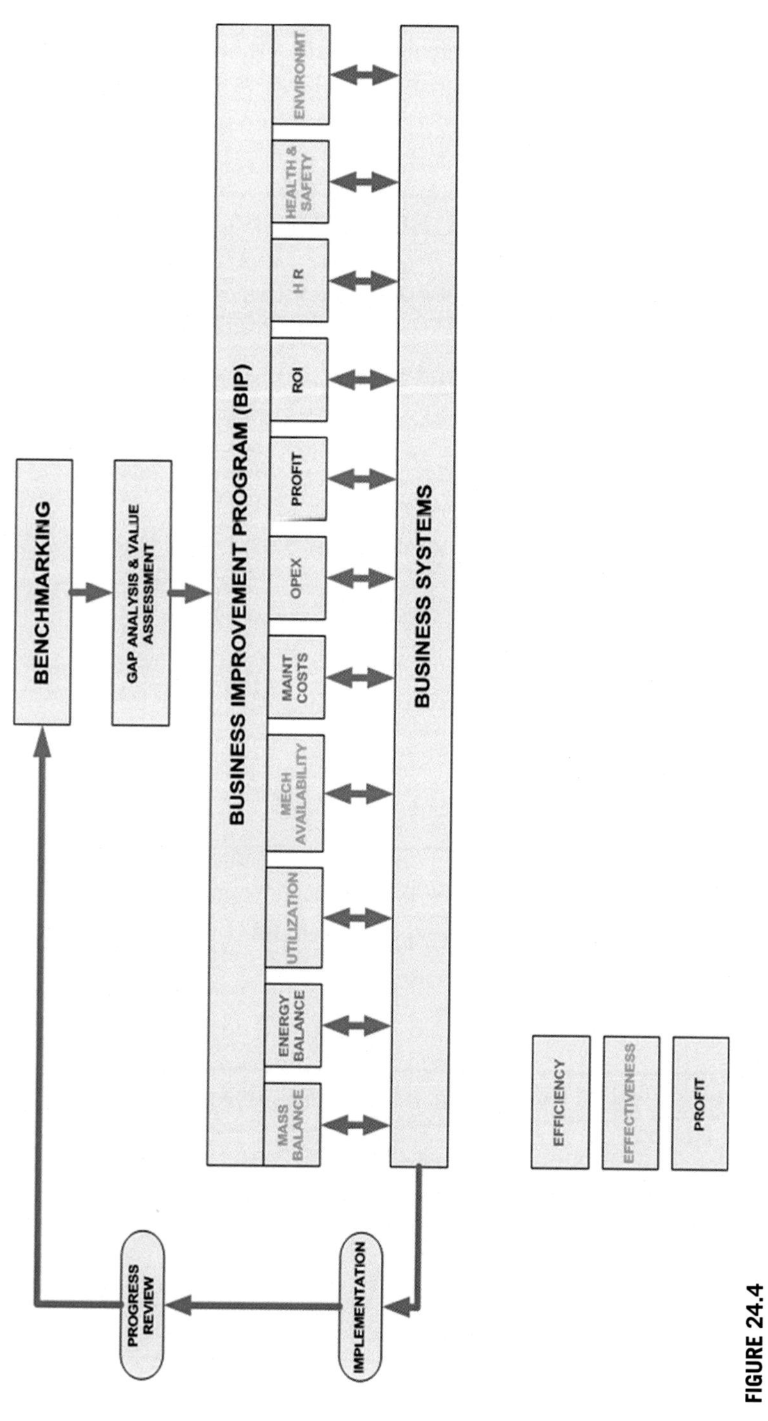

FIGURE 24.4
The business improvement program.

Chapter 29, Identification, Analysis and Evaluation of Gaps, details the assessment process.

It must be kept in mind that business KPIs are interrelated. A short-term improvement in one could have a detrimental effect on others in the long term. Box 24.2 gives an example of excessive focus on short-term indicators.

BOX 24.2 INTEGRATION INTO THE BUSINESS—EXAMPLE 1: COST REDUCTION

Focusing on short-term profit, at the expense of long-term reliability and integrity, **will** be detrimental to the company's health. BP's obsession with focusing on driving costs down resulted in major incidents.[11] The BP Texas City Refinery disaster in March 2005 killed 15 and injured 180 people.

Quote From the Baker Report[10] Published in January 2007

The Panel does not believe that the absolute level of funding is a deciding factor with respect to the state of process safety performance in a refinery. Many of the Solomon Associates Pacesetters with respect to reliability-centered maintenance—the industry leaders—frequently are not the biggest spenders on maintenance. Spending more money does not necessarily improve process safety ***if the money is not spent effectively.***

Quote From US Chemical Safety and Hazard Investigation Board (CSB) Report[9] Published in March 2007

...The presentation indicated that a new 1000-day goal had been added to reduce maintenance expenditures to 'close the 25 percent gap in maintenance spending' identified from Solomon benchmarking...

According to Solomon benchmarking (prior to 2005), the BP Texas City refinery was spending more on maintenance than other refineries of similar size and complexity. The 1000 Day Goal was established to close that gap.

Comment

This example clearly shows focus on a short-term target for maintenance costs without assessing the underlying issues of reliability and integrity. **Interpretation** of benchmarking results **is very important**. It can be perilous to focus on each KPI separately— they are interactive and should be viewed together.

Box 24.3 gives an example of a **balanced long-term reliability focus that automatically results in reduced costs.**

BOX 24.3 INTEGRATION INTO THE BUSINESS—EXAMPLE 2: RELIABILITY FIRST

A **long-term reliability focus** accordingly reduces costs. A particular gas operating company has the ideal combination of low costs, high utilization and availability with low staff turnover. By ensuring that their targets are complementary, they have struck a happy medium that has earned them pacesetter status.

Comment

Maintenance costs cannot be reduced before an excellent reliability and asset integrity record has been established. Costs first tend to increase in order to establish robust reliability and asset integrity systems, and only then do they come down (see also Section 27.5: Competitiveness, Efficiency and Effectiveness).

24.14 BENEFITS OF BENCHMARKING

PRIMARY

- Improved reliability
  - Optimized reliability and maintenance
  - Decreased risk of incidents
- Increased profits
  - Assessment of performance gaps in financial terms
  - Substantial portion of gap translated into increased profit

OTHER

- Understanding and documenting what "good" performance is
- Identifying performance gaps and lost revenue opportunities
- Gaining a valuable planning tool to assist in:
  - Setting realistic, measurable goals and performance targets
  - Allocating resources and CAPEX efficiently
- Accelerating chosen performance improvement actions
- Pinpointing problem areas and determining frequency of recurring events
- Increasing net cash margin
  - Reducing operating expenses
  - Lowering maintenance costs
- Measuring industry trends

Box 23.3 gives an example of the benefits of benchmarking a turnaround.

24.15 SUMMARY

Benchmarking tells you where you were, where you are, and where the competition is. It tells you how you have improved since the last benchmarking exercise and where you currently stand relative to the competition. It does **not** tell you how to get there.

Benchmarking is specific to a business operation (e.g., comparing various fuels refineries) or function (e.g., comparing specified turnarounds) and uses industry-standard performance indicators. It is information intensive and fact based.

Benchmarking takes a slice in time to compare all similar businesses.

The concept of peer groups helps identify performance relative to primary competitors.

Benefits include clear identification of those gaps that can significantly improve performance and supports the sense of urgency to improve specific deficiencies. Pacesetters (the best of the best) give an indication of what can be achieved and thus set stretch targets for the company.

Effective benchmarking consultants have large historical databases of many similar businesses. The leading benchmarking consultant for refineries is Solomon Associates, while for gas and process plants, it is Philip Townsend.

The definition for pacesetter varies. Philip Townsend defines it as the average of the top two performers in a performance indicator, whereas Solomon Associates has a more complex definition of all top indicators being in the top two quartiles for six consecutive years.

Project benchmarking has both generic components and components unique to a particular process plant. Independent Project Analysis undertakes benchmarking of most major oil and gas projects as well as turnarounds.

The benchmarking code of conduct ensures confidentiality of data by not permitting disclosure of information of specific competitors.

The primary benefits of operational benchmarking are improved reliability and increased profits.

REFERENCES

1. Independent Project Analysis (IPA). http://www.ipaglobal.com/.
2. Audit Analytics. *Compensation Consultant Analysis Briefing. August 2008.* http://www.auditanalytics.com/doc/Compensation_Consultant_Analysis_9_08.pdf.
3. Gartner. http://www.gartner.com/technology/home.jsp.
4. Juran's Benchmarking Approach. https://juranbenchmarking.com/our-approach/.
5. Solomon Associates. https://www.solomononline.com/.
6. Philip Townsend Associates Inc. http://www.ptai.com/.
7. McKinsey. http://www.mckinsey.com/.
8. International Oil and Gas Producers Association. http://www.iogp.org/.
9. US Chemical Safety and Hazard Investigation Board (CSB). *Investigation report: refinery explosion and fire. BP Texas City March 23, 2005. Report no. 2005-04-I-TX*. March 23, 2005. https://www.propublica.org/documents/item/csb-final-investigation-report-on-the-bp-texas-city-refinery-explosion.
10. *The report of the BP US refineries independent safety review panel (Baker report)*. January 2007. http://www.documentcloud.org/documents/70604-baker-panel-report.
11. Steffy LC. *Drowning in oil: BP and the reckless pursuit of profit*. McGraw-Hill; 2011.

FURTHER READING

1. *Benchmarking: pure and simple: a quick reference guide to benchmarking*. American Productivity and Quality Center. https://www.apqc.org/knowledge-base/documents/benchmarking-pure-and-simple.
2. McNair CJ, Liebfried KHJ. *Benchmarking—a tool for continuous improvement*. John Wiley; 1992.
3. *Manderstam's experience in benchmarking in hydrocarbon processing*. http://www.manderstam.com/benchmarking/Experience_Benchmarking.pdf.
4. Kelessidis V. *Report produced for the EC funded project INNOREGIO: dissemination of innovation management and knowledge techniques, Benchmarking;* January 2000.
5. RISI – the leading information provider for the global forest products industry. http://www.risiinfo.com/.
6. Worrell E. *Benchmarking energy efficiency in the iron and steel industry*. Paris: IIP Advisory Panel; June 2011.

CHAPTER

COMMON DENOMINATORS AND INDICES USED IN BENCHMARKING

25

25.1 INTRODUCTION

Benchmarking entails comparing different sizes, complexities, designs, and operating modes within the same operations or functions, thus requiring mechanisms to "level the playing field" for purposes of comparison.

The value of common denominators and indices is that they compare "apples with apples."

Common denominators normalize the numerator so that the numerator can be used for comparison exercises in spite of differences in size and complexity of the competition.

Example:

- *The cost of maintenance in a small, simple refinery can be compared to that of a larger and more complex refinery.*

Indices compare actual against a standard where the standard is determined by the best in efficient design and effective operation.

Examples:

- *The design of one heater can be more efficient than another.*
- *The use of one catalyst can be more effective than another.*

25.2 PRIMARY COMMON DENOMINATORS

The following are the primary common denominators used:

GENERIC

These are as follows:

1. Replacement Value (RV)
2. Shift Positions
3. Topside weight (for offshore)

These are normally clear, easy to determine, and indisputable. However, RV is defined more specifically by Solomon Associates (SA) and Philip Townsend Associates International (PTAI).

Performance Management for the Oil, Gas, and Process Industries. http://dx.doi.org/10.1016/B978-0-12-810446-0.00025-6

COMPLEX

These are used for the benchmarking consultant's proprietary indicators and have usually been developed and refined by the consultant over many years. They often have an empirical basis.

These are as follows:

Solomon Associates

- Equivalent Distillation Capacity (EDC)
- Equivalent Gas Processing Capacity (EGPC)

Philip Townsend Associates International

- Normalized Shift Positions (NSP)
- Corrected Mechanical Units (MUc)

Juran

- Juran Complexity Factor (JCF)

These proprietary common denominators are discussed in detail in the following section.

25.3 CONSULTANTS' PROPRIETARY COMMON DENOMINATORS

The effect of size and complexity is explained as follows, based on a particular consultant's approach.

SOLOMON ASSOCIATES

Refineries: Equivalent Distillation Capacity

EDC evolved from Nelson complexity factors initially developed by Dr. Nelson over 60 years ago (see Box 25.1).

> **BOX 25.1 EQUIVALENT DISTILLATION CAPACITY**
>
> **An approach to Size and Complexity Comparisons**
>
> In articles published in The Oil and Gas Journal more than 60 years ago, Dr. W.L. Nelson presented papers on correlation techniques that could be used to compare staffing levels and operating costs between refineries of diverse processing techniques and capacities. In a 1973 paper, he presented a process complexity factor related exclusively to unit investment levels, concluding that coking plants cost 5.5 times more per barrel than crude units.
>
> In 1980, Solomon Associates (SA) conducted its first Comparative Performance Analysis (CPA) of fuels refineries based primarily on Dr. Nelson's complexity factors, and, after many years of refinement, SA, who have more than 500 fuels and lube refineries on their database, still use Equivalent Distillation Capacity (EDC) as the basis of their studies.

EDC is based on process unit capacity and a configuration factor: the ratio of the particular process unit investment costs to the standard crude unit investment costs per bbl.

EDC is built up from a Configuration factor (CF). This starts with *a standard crude unit with a CF of 1.*

Examples:

- *Residual Catalytic Cracker: CF 10*
- *Sulfur Recovery Unit (SRU): CF 300*

The CF is multiplied by the stream day capacity of the process unit, modified using a multiplicity factor as required, and added together with all units in the refinery to get the EDC.

Example:

A 130,000 barrel a day Refinery with a Catalytic Cracker could have an EDC of 1 million barrels a day.

Box 25.2 gives a graphical representation of a comparison over time using EDC.

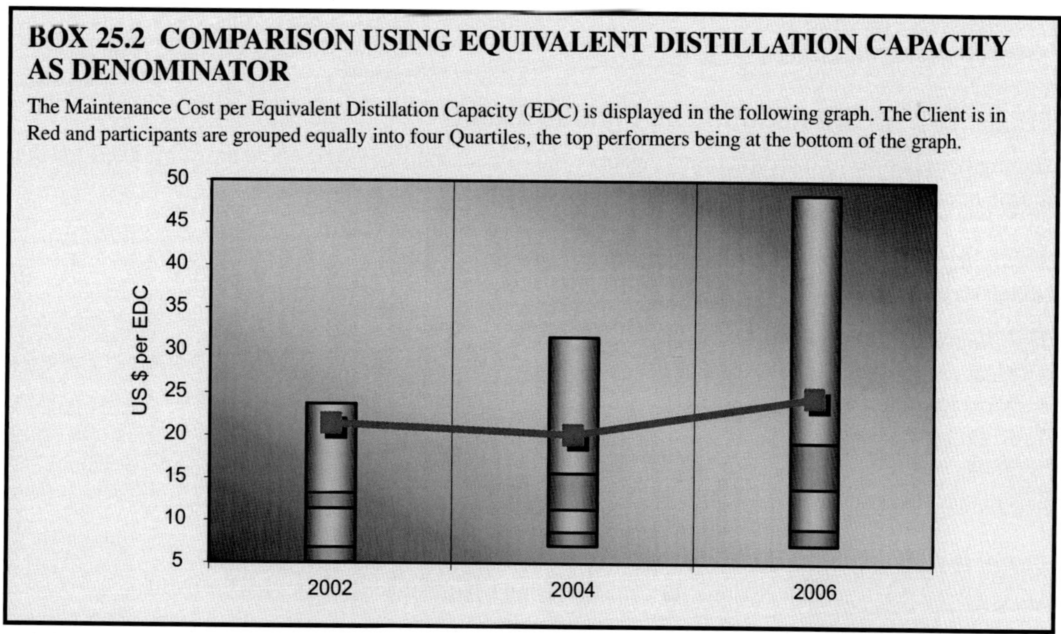

BOX 25.2 COMPARISON USING EQUIVALENT DISTILLATION CAPACITY AS DENOMINATOR

The Maintenance Cost per Equivalent Distillation Capacity (EDC) is displayed in the following graph. The Client is in Red and participants are grouped equally into four Quartiles, the top performers being at the bottom of the graph.

Gas Plants: Equivalent Gas Processing Capacity

In a similar manner to Refinery EDCs, EGPCs are built for Gas Plants using the CFs for standard units common to refineries and gas plants (such as SRUs), and those for specialized gas processing units.

Box 25.3 gives a graphical representation of a comparison using EGPC.

BOX 25.3 EQUIVALENT GAS PROCESSING CAPACITY (EGPC) APPLICATION

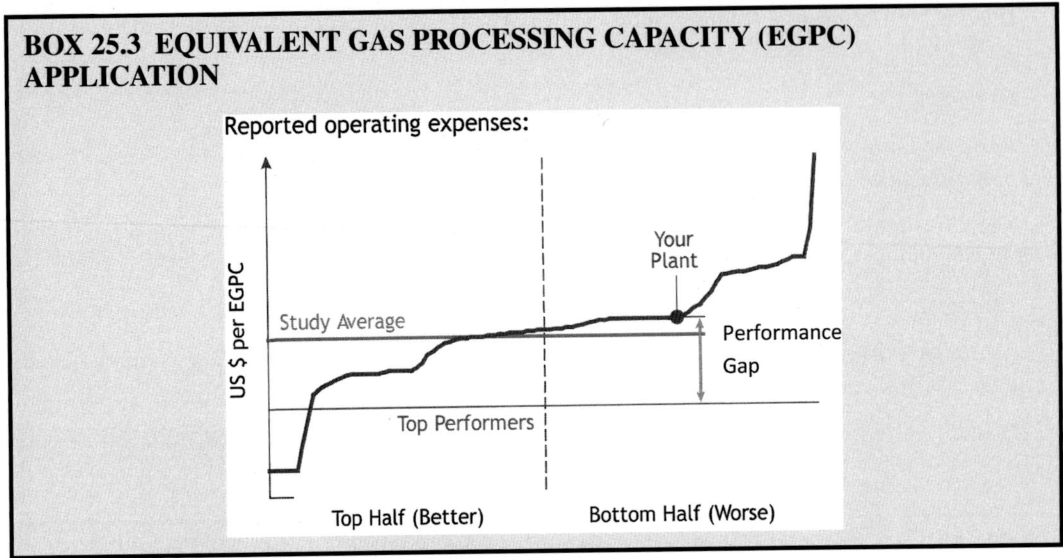

PHILIP TOWNSEND ASSOCIATES INTERNATIONAL

Having bought Shell's benchmarking database, PTAI uses methodologies developed by Shell for their internal comparative studies.

Mechanical Units

MU: Mechanical Units are categorized as follows:

1. Furnaces/boilers
2. Heat exchangers
3. Vessels/columns
4. Rotating equipment
5. Tanks
6. Others

Corrected Mechanical Units

MUc: Corrected Mechanical Units for capacity using Maintenance Scaling Factor (MSF)

MUc = MU × MSF

A simple example demonstrating correction for size and complexity is shown in Fig. 25.1.

Normalized Shift Positions/Site Personnel Index

These are discussed in detail in Section 17.6, Human Capital.

All maintenance hours are adjusted using the MSF.

Examples:

- *A crude distillation unit has NSP of 1.15.*
- *A large natural gas liquefaction/sales/flowing gas plant could have an NSP of 25 with an MSF of 1.5.*

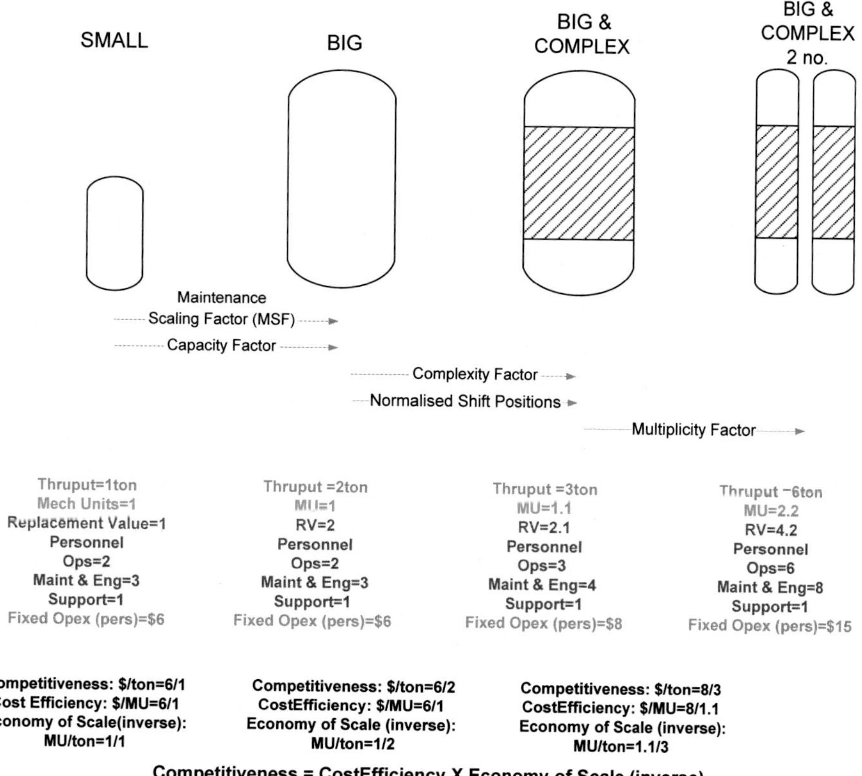

FIGURE 25.1

Correction for Size and Complexity.

- *A mega-train LNG plant or Gas to Liquid plant could have an NSP of about 70 with an MSF between 1.5 and 1.7.*

JURAN

Juran Complexity Factor

Juran has developed the JCF to be able to apply a uniform approach to all oil and gas exploration, transportation, processing, and storage assets. It is derived from a library of Weight Factors (WF) for standard modules of equipment. Each module is given a specific value of WF based on the aggregate of the worldwide operation and maintenance hours. Each **WF** value equates to **1 man-year operational and maintenance effort**. For each facility, each module is identified, and the arithmetic total of the WFs gives the final Complexity Factor (CF).

The JCF is of particular use for value-chain benchmarking that may highlight inefficiencies in coordination between Business Units along the value chain. Section 34.2, Value Chain Oversight and Changing Risk, discusses the value chain risk profile.

REPLACEMENT VALUE

RV is usually the current insured value. However, this method is not always very reliable since insured values could differ markedly.

Example:

The shareholder of three similar gas plants queried the insured value of one plant, which was markedly different to the other two. The following year, there was better alignment.

Nonetheless, PTAI uses the RV approach based on the insured value, with some reservations.

To standardize Refinery RVs, SA uses published escalated US Gulf Coast values[1] for typical refining unit size and adjusted for the following:

1. Reported size
2. Location
3. Inflation (to date of benchmarking period)
4. Sulfur

To standardize RVs for Reliability and Maintenance benchmarking, SA uses the Chemical Engineering "Plant Cost Index" to adjust the construction value to the value in the benchmarked year. This is published by the "Chemical Engineering"[2] periodical on an annual basis.

Box 25.4 gives an example of a cost comparison using RV as defined as the current insured value. The cost data input sheets used for the example were as defined in Section 21.4, Operating Expense.

BOX 25.4 EXAMPLE OF COST COMPARISON USING REPLACEMENT VALUE (RV) AS DENOMINATOR

Graphical output of operating costs for five Operating Companies, managing 22 gas trains.

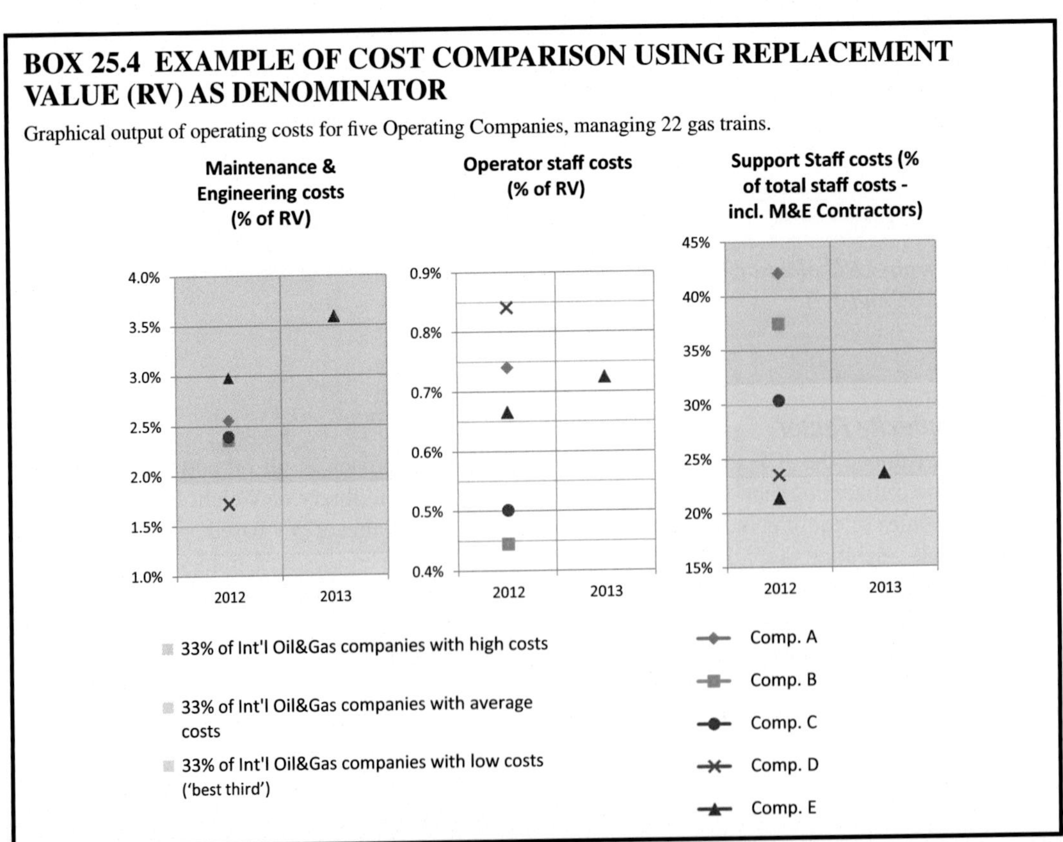

25.4 COMMON DENOMINATOR RELATIONSHIPS

Mechanical Units, NSP, and RV have a strong correlation and linear relationship. For any particular plant, cross-checking the values of MUc, NSP, and RV could determine a spurious value for one of the three and thus initiate correction.

Example:

A gas plant operator queried its assigned MUc value with the benchmarking consultant, compared to a similar sized sister plant. The NSP and RV values were similar for the two plants, but the MUc value was different. The consultant corrected the client's MUc value.

Relationships are shown in Fig. 25.2.

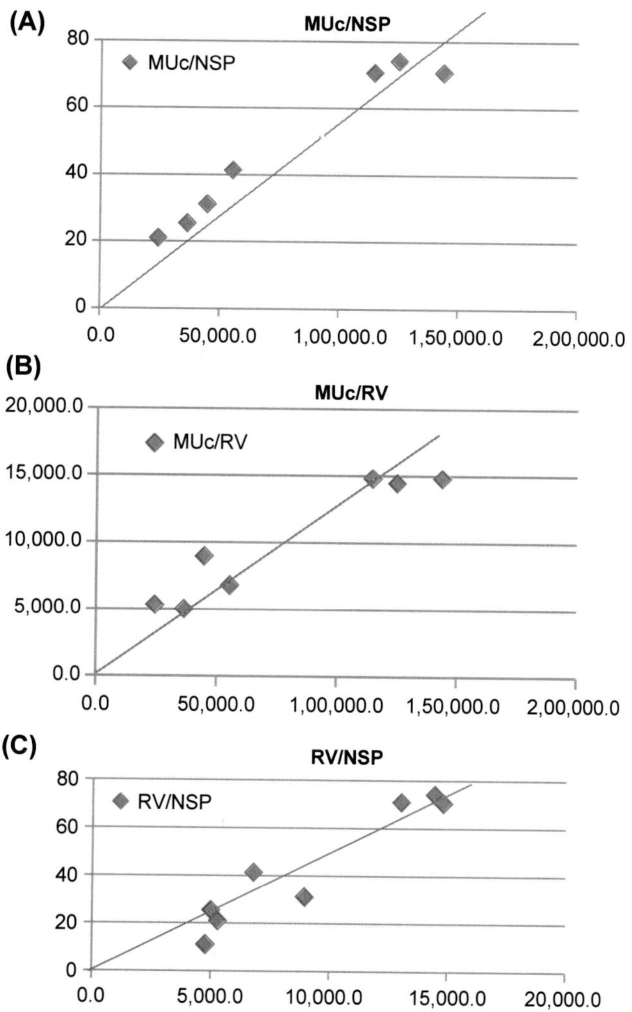

FIGURE 25.2

Mechanical Units, Normalized Shift Positions (NSP), and Replacement Value (RV) relationships: (A) Equipment Count (corrected Mechanical Units; MUc) versus Complexity (NSP); (B) Equipment Count (MUc) versus Investment Replacement Value (RV); (C) Investment Replacement Value (RV) versus Complexity (NSP).

25.5 PROPRIETARY INDICES

INTRODUCTION

Each consultant has a set of indices that have been developed based mainly on empirical information. Specific consultants' indices are outlined.

Personnel indices are discussed in Section 17.6, Human Capital. PTAI's Site Personnel Index (SPI) is discussed further in Section 25.3.

SOLOMON

SA has a number of indices, the most important being:

Volumetric Expansion Index

SA developed this index in 1984 to measure the ability of a refinery to produce and recover salable products. A high Volumetric Expansion Index (VEI) is the result of obtaining good yields and good recovery of process output as salable liquid products and fuel gas.

Energy Intensity Index

SA developed the Energy Intensity Index (EII) in 1981 to compare energy consumption among fuels refineries, since complex refineries require more energy than simple ones. The approach is similar to that described for Corrected Energy and Loss Index (CEL) as follows.

Maintenance Index

SA's Maintenance Index (MI) represents the overall cost of maintenance and is the sum of the non-Turnaround (Routine) Maintenance Index and the Turnaround Maintenance Index. The common denominator used is EDC, and costs are annualized.

PHILIP TOWNSEND ASSOCIATES INTERNATIONAL

The following is PTAI's primary index:

Corrected Energy and Loss Index

This is a measure of processing efficiency with regards to energy usage and hydrocarbon losses. It compares actual site energy to a standard. The actual energy used is converted to tons of "Standard Reference Fuel," which is defined as a fuel with a lower heating value of 40,500 kJ/kg. The calculated theoretical energy allowances correspond to a level of energy consumption commensurate with an energy efficient design and steady operation, the lower the index the better.

Section 17.7, Energy, discusses SA and PTAI approaches, and Box 17.6, Gas Plant Energy Balance, emphasizes the need for a standard energy unit for comparison.

JURAN

Juran Hydrocarbon Index[3]

This index is a cumulative indicator of the overall competitiveness of an asset, based on efficiency and effectiveness of its performance. It is a numerical value representing the percentage deviation of performance, relative to the corresponding industry or peer group average. The components are shown in Box 25.5.

BOX 25.5 JURAN HYDROCARBON INDEX (JHI) COMPONENTS

Efficiency Components	Effectiveness Components
1. Business Overhead	1. Utilization
2. Operation	2. Emission
3. Maintenance	3. Waste
4. Annualized Maintenance	4. Incidents
5. Health, Safety, Security, and Environment	5. Absenteeism
6. Technical Support	6. Availability
7. Energy	7. Reliability
8. Facility Specific	8. Service Level
SUM (8)	SUM (8)

25.6 SUMMARY

Common denominators are essential for benchmarking, where the value of a common denominator is such that it enables comparison of "apples with apples."

Primary generic common denominators are RV, Shift Positions, and Topside Mass.

Proprietary common denominators, based on the consultants' extensive experience which has been gained over many years, as well as vast databases, are EDC and EGPC for SA, NSP and MUc for PTAI, and JCF for Juran.

There is a strong correlation between RV, NSP, and MUc.

Indices compare process unit performance to a standard performance. Proprietary indices include VEI, EII, and MI for SA, CEL for PTAI, and Juran Hydrocarbon Index (JHI) for Juran.

REFERENCES

1. Nelson-Farrar cost indices. *Oil and Gas Journal*. http://www.ogj.com/articles/print/volume-114/issue-3/processing/nelson-farrar-cost-indexes.html.
2. Chemical Engineering plant cost index. http://www.chemengonline.com/pci-home.
3. Single index measures operational performance of hydrocarbon facilities. *Oil and Gas Journal* October 10, 2005. http://www.ogj.com/articles/print/volume-103/issue-38/processing/single-index-measures-operational-performance-of-hydrocarbon-facilities.html.

FURTHER READING

1. *Global industrial energy efficiency benchmarking: an energy policy tool working paper*. United Nations Industrial Development Organization; November 2010.

CHAPTER

BENCHMARKING DATA VERIFICATION

26

26.1 INTRODUCTION

All too often, when poor performance results are submitted by the consultant to the client, the client's "knee-jerk" reaction is: "Ah, but we are different." It is therefore very important that all data have a strong factual basis that cannot be challenged. Data verification is thus vital to build confidence, especially for buy-in when major improvement is required.

Each benchmarking consultant has a standard frame of reference to ensure each participant supplies data based on the same definitions. The focus of the study is generally limited to those elements that most affect the commercial success of the business. These elements use only generic terms that are clearly defined, in detail, to ensure consistent submission by each client. In addition, all performance data is usually based on the same time frame.

26.2 MONTHLY PERFORMANCE MODELING

Tools for performance modeling were introduced in Chapter 3, Models.

In some cases, data input from different sources can be used to cross-check authenticity.

Example:

Solomon's Profile II is used to crosscheck production figures by generating a Volumetric Expansion Index (VEI) and Energy Intensity Index (EII). The VEI has to be positive and within range, and the EII has to be negative within range. Incorrect input data is flagged by "out-of-range" *VEI and EII.*

Box 17.1, Profile II check on Mass Balance, gives a case where incorrect production figures were identified.

In the case of using Solomon's Profile II, the data accumulated in the model over 2 years can be uploaded into the benchmarking model, thus saving some time in completing data input sheets for the benchmarking exercise. Errors due to manual transposition of data are also reduced.

Performance Management for the Oil, Gas, and Process Industries. http://dx.doi.org/10.1016/B978-0-12-810446-0.00026-8

26.3 BENCHMARKING DATA INPUT TABLES

Data input tables are usually in a spreadsheet format.

CELL DATA RANGE

Data entry is barred for out-of-range data. Alternatively, a conditional entry is accepted with an explanation in a separate field.

Example:

LNG plant Greenhouse Gas emissions range: 18%–30% woi (weight on intake).

CELL ENTRY VALUE

Cell format is fixed by the consultant to only accept text, numeric, date, or a prescribed (pull-down) value.

CROSS-VALIDATION BETWEEN SPREADSHEETS

An entry in one spreadsheet is cross-validated against an entry in another spreadsheet. This normally entails some calculation.

Example 1:

Data entry could be "number of operators" *in one spreadsheet and* "shift positions" *in another spreadsheet. From the* "shift positions" *entered, a macro calculates the total number of operators based on the number of shifts and gives an error message if this is significantly different to the* "number of operators."

Example 2:

The operating cost input sheet is cross-checked against the personnel input sheet. The hourly rates are calculated for various disciplines and compared against an expected range.

VARIATION FROM THE PREVIOUS YEAR'S BENCHMARKING DATA SUBMISSION

If the physical assets have not changed, then variations are not expected to exceed a certain percentage.

Out-of-range error reports are generated as the tables are populated (Solomon). There is usually a maximum acceptable list of errors in a residual error report before submission of the Data Input Tables. All errors require explanation.

EXCLUSIONS

The scope of the benchmarking exercise needs to be strictly adhered to. Section 27.2, Benchmarking Envelopes, gives examples of different scopes.

Specific areas of attention are as follows:

Personnel

"Standard Activities" only include normal operations, maintenance, administration, and management of the facility. Box 26.1 gives an example where the personnel to be included in the study are those undertaking "standard activities" only. The study input data sheet does not force the client to account for other employees. Thus the client could "play down" their staff numbers to produce a more rosy report.

> **BOX 26.1 EXCLUSIONS: BODY COUNT**
>
> A company producing only LNG and associated products has about 2700 company personnel. However, for benchmarking purposes, only 2000 persons are taken into account, as these undertake "Standard Activities" associated with the production of LNG and associated products. The balance, being 25% of the total number of employees, undertake "Nonstandard Activities."
>
> "Other" activities of the company include the following:
>
> - Exploration
> - Major projects
> - Medical center
> - School
> - Staff village
>
> **Comment**
>
> The benchmarking guidelines need to be changed so as to challenge the company about how the salaries of the remaining 700 staff are generated.

Maintenance Operating Expense

The line drawn between maintenance and capital expense is defined differently by consultants.

Philip Townsend Associates International

For maintenance costs, Philip Townsend Associates International (PTAI) includes plant changes involving moving, rearranging, or otherwise changing existing property, plant, and equipment where the cost involved is less than $400,000 per plant change.

Solomon

Solomon employs a decision tree to determine the split between maintenance operating costs and capital investment costs. This is shown in Fig. 26.1.

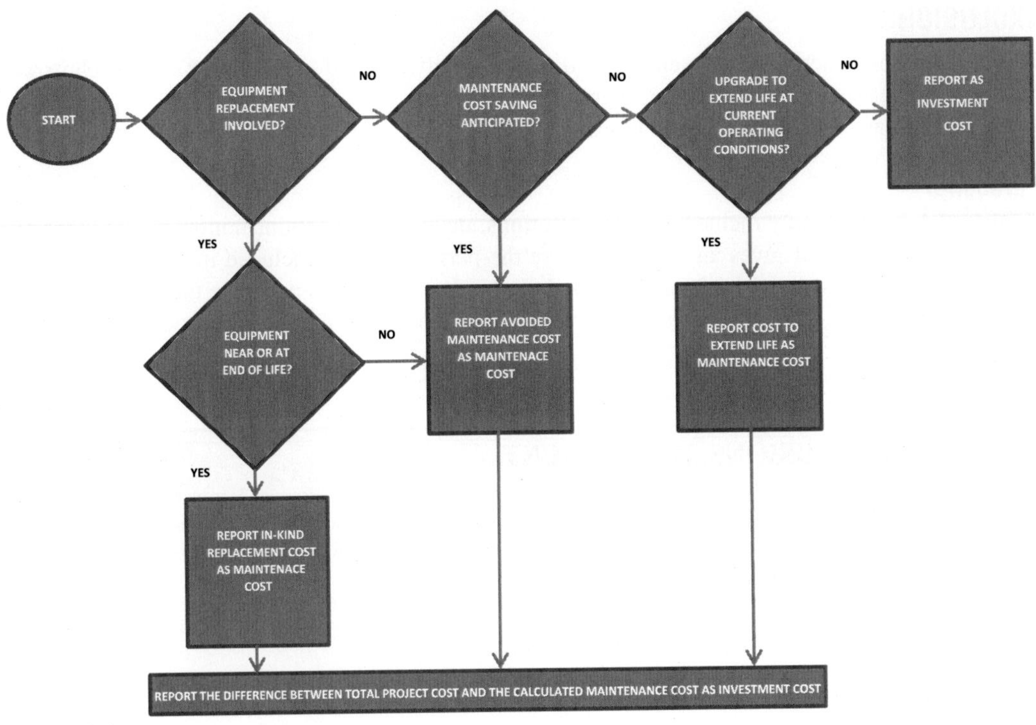

FIGURE 26.1

Solomon Associates maintenance operating expense decision tree.

INTERPRETATION OF DEFINITIONS

Maximum Sustainable Capacity

Maximum Sustainable Capacity is clearly defined by the consultants. This is discussed in detail in Section 17.4, Utilization and Availability. However, if the client has chosen not to change the "nameplate" maximum design capacity, he may continue to report 100% utilization and therefore not be able to analyze any potential improvements.

Box 26.2 gives an example of differing views.

BOX 26.2 INTERPRETATION OF DEFINITIONS: MAXIMUM SUSTAINABLE CAPACITY

A shareholder has interests in two operating companies, each of which manage a number of identical LNG mega-trains.

The maximum Design Capacity for each LNG mega-train was calculated to be 7.7 Million Tons Per Annum (MTPA), and this was fortunately achieved after startup.

Trials were undertaken by one of the two operating companies a year or so after commissioning. This time, Maximum Sustainable Capacity of 8.5 MTPA was achieved for each train, and the one operator changed to this new capacity for calculation of **Utilization.**

BOX 26.2 INTERPRETATION OF DEFINITIONS: MAXIMUM SUSTAINABLE CAPACITY—con'd

However, the second operator continued to use 7.7 MTPA as its Maximum Sustainable Capacity for each train. This had always been achieved, as a dip in production was made up by exceeding the Maximum Sustainable Capacity over the short term.

The shareholder thus has difficulty in gauging the performance of the second operator's mega-trains relative to that of the first operator.

At the time of writing, the second operator had not undertaken trials to determine if the Maximum Sustainable Capacity should be changed.

Comment

The second operator is masking potential improvements in performance by retaining the lower Maximum Sustainable Capacity and thus not registering all Lost Capacity. This practice potentially does not align with benchmarking requirements. For benchmarking with PTAI, the following categories are registered:

- Planned
- Unplanned
- Consequential
- Opportunity maintenance
- Project installation/revamp
- Economics and scheduling
- Other/unidentified

Annualization of Turnaround Costs

Accepted practice is to take the last turnaround cost and divide it by the interval between the last two turnarounds. This is graphically shown in Fig. 26.2.

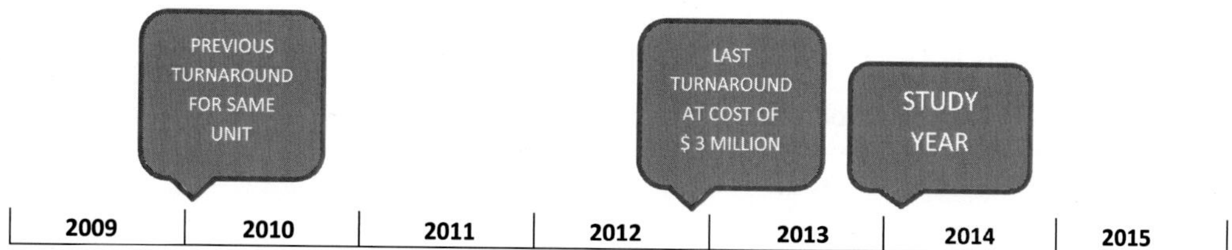

FIGURE 26.2

Annualized turnaround cost calculation.

If an insufficient number of turnarounds have occurred, the predicted turnaround interval in the design specification should be used. It is not considered good practice to use an estimated interval.

26.4 BENCHMARKING MODEL

The benchmarking model is initially populated with the client's unique configuration.

Example: Refinery: Solomon Associates

Each process unit's number, type, and capacity is entered into the consultant's model.

Example: Gas Plant: PTAI

Each item of equipment and its capacity are entered into the consultant's model.

The consultant employs an experienced industry analyst who reviews each submission for reasonableness and consistency by physically checking data input and residual error reports. He raises queries for clarification from the client before collating data from all clients in the benchmarking model. Additional out-of-range error reports could still be generated after the data input sheets are entered into the benchmarking model.

26.5 BENCHMARKING CONSULTANT AUDIT

Some benchmarking consultants undertake site audits.

Box 26.3 gives an example of an audit.

> **BOX 26.3 DATA REPORTING: SOLOMON AUDIT**
>
> Once a refinery benchmarking cycle has been completed, Solomon Associates selects approximately 10% of participating sites where a team spends a number of days at each site to carry out an audit to ensure data reporting accuracy and consistency. Refineries are selected randomly, but are generally drawn from the top half of the overall performance indicators.

26.6 ASSESSMENT

At the start of the assessment, process data requires validation. The context in which the data is used also needs to be stated.

DIFFERENT SOURCES

Cross-checks are required between sources to check for different values on the same basis or different basis.

Example:

COST/TON salable product. Cost could exclude or include nonrecurring and turnaround costs, or turnaround costs could be annualized. Cost could be applicable to onshore only or both onshore and offshore.

DRILL DOWN TO IDENTIFY BASIS OF THE DATA

Drill down is required for major values (for example in mass balance) to ascertain the source of the data: distributed control systems, laboratory information management system, automatically calculated by the mass balance model, manually adjusted, etc.

DEFINITION OF THE DENOMINATOR VALUE

Check on the basis for the indicator value is required.

Example:

OPERATOR STAFF COST Percentage of Replacement Value (RV). RV is calculated differently by different benchmarking consultants: original invested value (escalated), US Gulf coast current replacement value, and local insured value.

SCOPE

The scope of different comparative studies and Business/Work Program and Budgets vary, and so the context needs to be stated.

Example:

A Key Performance Indicator (KPI) for a gas plant may be based on total operating expense (opex) for both offshore and onshore, or could be just within the gas plant fence.

Box 27.1, Benchmarking Envelopes: Gas Plant Example, gives an example of a number of benchmarking studies carried out on the same plant with different scopes.

TIME AND DAYS

The data collection period would be different for benchmarking reports and current operational performance reports. Leap years also need to be accounted for when calculating daily throughput.

DISTORTION

An indicator may be different for plants with identical outputs and size.

Example:

One plant purchases most of its power where another produces most of its own power, thereby giving the cost of power in a financial spreadsheet different KPI values.

IDENTICAL KEY PERFORMANCE INDICATOR VALUE FOR DIFFERENT COMPANY

A data input error could have occurred when data entry is for multiple Business Units/Companies. This could be attributed to a copy/paste error.

26.7 SUMMARY

All data must have a strong factual basis that cannot be challenged. Data verification is therefore vital to build confidence with decision-makers and actioners, especially for buy-in when major improvement is required.

To ensure credibility and objectivity of benchmarking data, it is essential to have a series of data validation processes and tools.

If the company has the consultant's software (e.g., Solomon's Profile II), validation can be done on a monthly basis.

Each benchmarking consultant has a standard frame of reference to ensure each participant supplies data based on the same definitions.

Input data tables are designed to:

1. limit values to acceptable ranges and notation;
2. cross-check data inputted on different sheets of the spreadsheet; and
3. check against the spreadsheets submitted in a previous benchmarking exercise.

The benchmarking consultant receives the data sheets from the client, possibly with a small number of errors and with explanations attached. A model and various other methods are used to execute a number of checks, and the client may be asked for further clarification.

After publication of the results of the benchmarking exercise, the assessment process commences. Data verification continues with cross-checks against other sources of data. Note that benchmarking is done for a particular slice of time, and other data may be from different time periods or defined differently. The benchmarking results may also have a limited scope.

FURTHER READING

1. IBNET. The International Benchmarking Network for Water and Sanitation Utilities. Data verification. http://www.ib-net.org/en/texts.php?folder_id=123&mat_id=102&L=1&S=0&ss=0.

CHAPTER

USING BENCHMARKING RESULTS

27

27.1 INTRODUCTION

The benchmark report should identify gaps that need to be addressed. Subsequent Gap Analysis/Root Cause Analysis (RCA) workshops should identify the best actions to take in bridging the gap. The actions should result in a step improvement in the specific aspect of the business that has been benchmarked. Fig. 27.1 shows the generic basic process that **should** be undertaken.

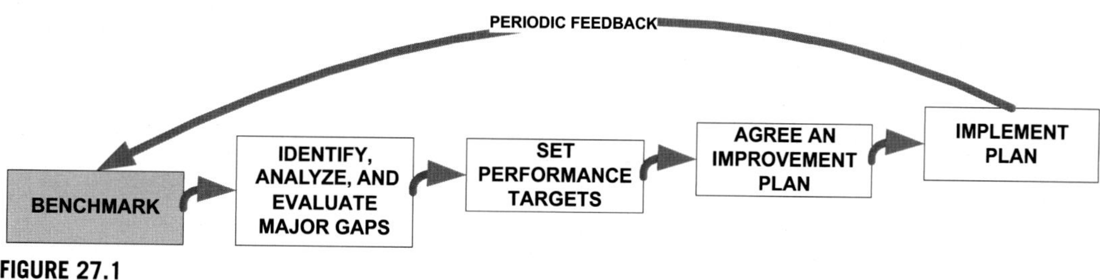

FIGURE 27.1

Basic improvement process.

27.2 BENCHMARKING ENVELOPES

In using benchmarking results, there has to be awareness of the scope (envelope) of the benchmark. Business performance Key Performance Indicators (KPIs) and targets may not necessarily line up with the benchmarking KPIs and targets. Evaluating the implications of these benchmarking envelopes is required.

Section 29.4, Assessment Methods, discusses this further.

Fig. 27.2 shows the standard envelope for a Fuels Refinery Benchmarking Study. The partial inclusions of the Power Plant and Receiving and Shipping Facilities depend on services utilized by others.

Performance Management for the Oil, Gas, and Process Industries. http://dx.doi.org/10.1016/B978-0-12-810446-0.00027-X

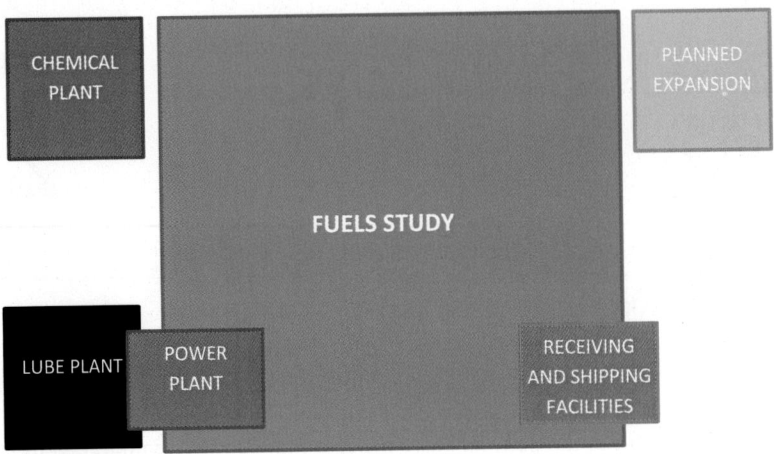

FIGURE 27.2

Fuels refinery benchmarking envelope.

Box 27.1 gives an example an integrated gas plant where a number of benchmarking studies with different envelopes have been carried out.

BOX 27.1 BENCHMARKING ENVELOPES: GAS PLANT EXAMPLE

Benchmarking (BM) and Assets envelopes:

1. Core BM: shareholder BM from Slug Catcher to Tankage
2. Philip Townsend Associates International (PTAI) BM: Gas trains from Slug Catcher to Loading
3. SH BM: Shareholder BM from Wellhead to Loading
4. Solomon BM: Refinery
5. McKinsey BM: Offshore from wellhead to slug catcher
6. Total Assets under Production Sharing Agreement (PSA) or Joint Venture (JV): from Reservoir to Destination, including Shipping

SH, shareholder; *TO*, joint terminal operations.

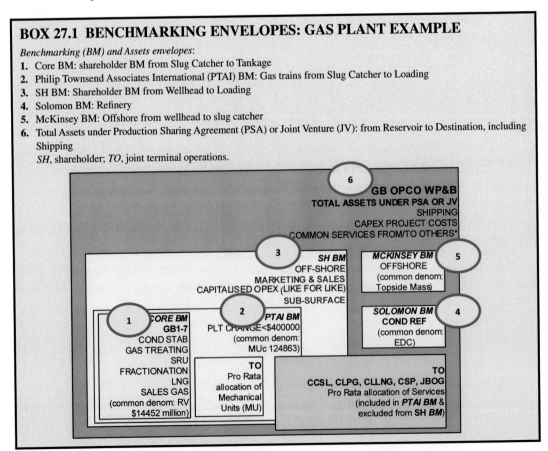

6
GB OPCO WP&B
TOTAL ASSETS UNDER PSA OR JV
SHIPPING
CAPEX PROJECT COSTS
COMMON SERVICES FROM/TO OTHERS*

3
SH BM
OFF-SHORE
MARKETING & SALES
CAPITALISED OPEX (LIKE FOR LIKE)
SUB-SURFACE

5
MCKINSEY BM
OFFSHORE
(common denom: Topside Mass)

4
SOLOMON BM
COND REF
(common denom: EDC)

1
CORE BM
GB1-7
COND STAB
GAS TREATING
SRU
FRACTIONATION
LNG
SALES GAS
(common denom: RV $14452 million)

2
PTAI BM
PLT CHANGE<$400000
(common denom: MUc 124863)

TO
Pro Rata allocation of Mechanical Units (MU)

TO
CCSL, CLPG, CLLNG, CSP, JBOG
Pro Rata allocation of Services (included in PTAI BM & excluded from SH BM)

27.3 OUTPUT FORMATS

It often takes time to understand the content of benchmarking reports. They are very complex and extensive, and one has to "mine" for information that is relevant and significant.

The following are some summary formats.

SPIDER DIAGRAM

A spider diagram can be convenient for a summary of a benchmarking study. A functional benchmarking exercise is shown in Box 27.2 using a spider diagram.

BOX 27.2 SPIDER DIAGRAM: LOGISTICS BENCHMARKING STUDY SUMMARY

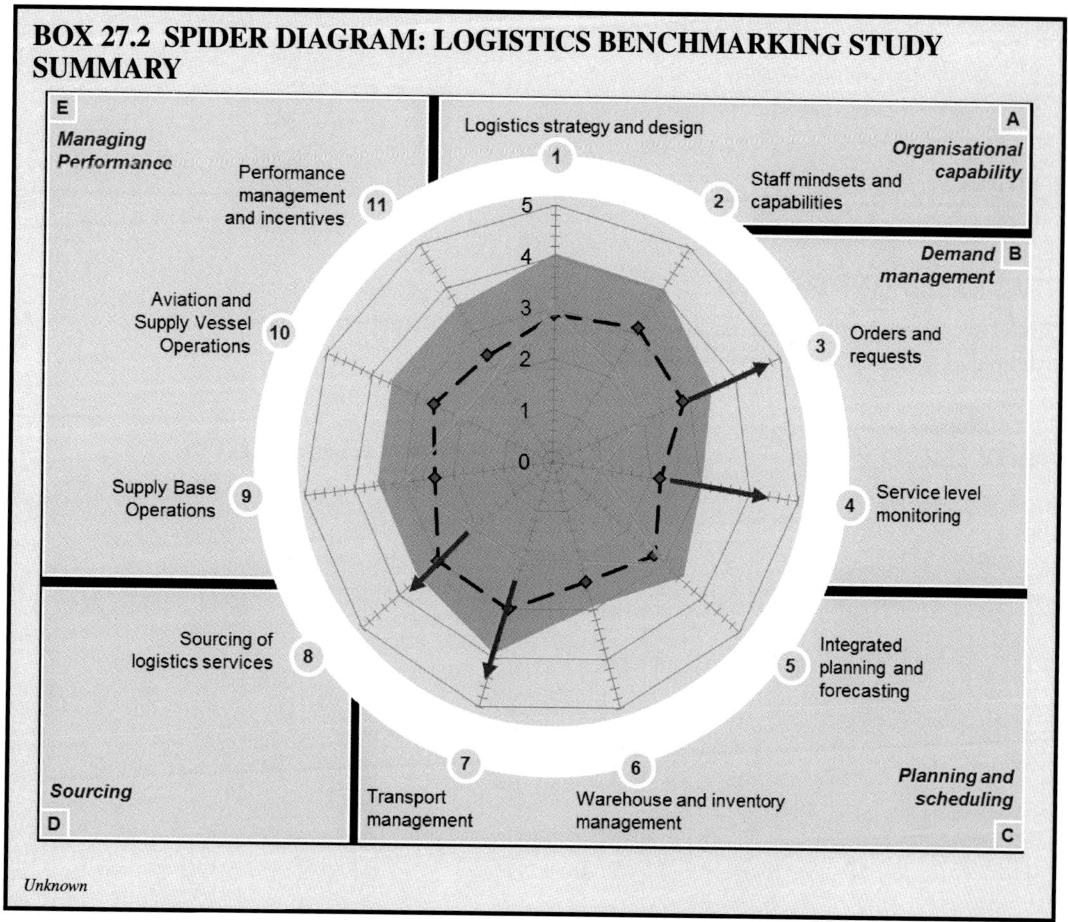

Unknown

SUMMARY BAR GRAPHS

Using a summary bar graph, a number of KPIs for a benchmarking exercise can be easily summarized. This is shown in Box 27.3.

BOX 27.3 SUMMARY BAR GRAPH: GAS PLANT BENCHMARKING STUDY

X denotes the client's position.

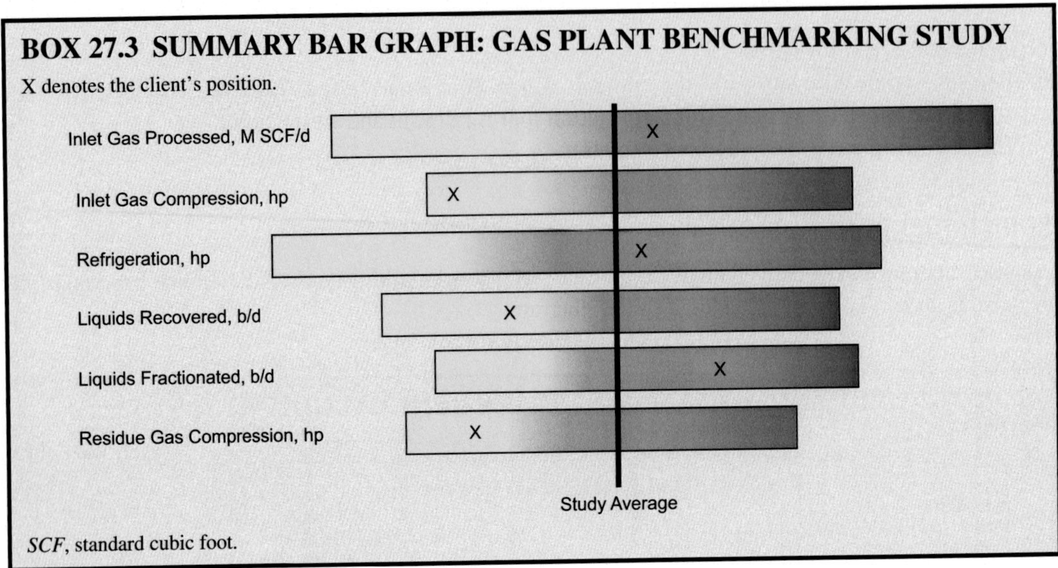

SCF, standard cubic foot.

TIME-BASED LINE GRAPHS

Time-based line graphs show relative improvement or deterioration of performance over time. An example of a time-based line graph is shown in Box 27.4.

BOX 27.4 TIME-BASED LINE GRAPH: REFINERY UTILIZATION

1Q and 4Q are the first and fourth quartiles of groupings of study participants. The red is the client's performance. The 2007 value was determined with the use of Solomon's Profile II in the year between benchmarking studies.

Utilization KPI

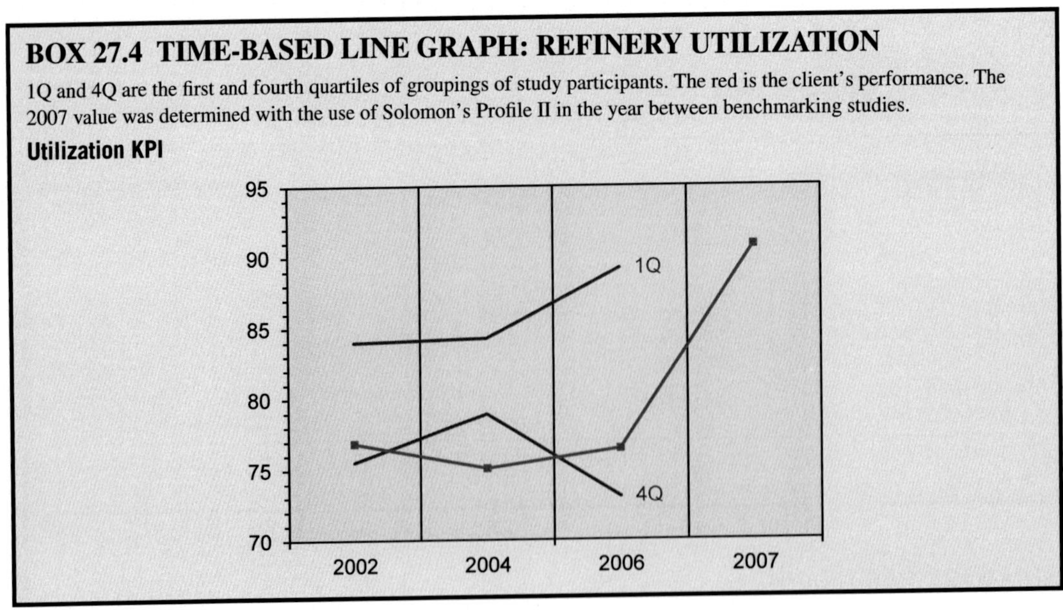

STACKED BAR GRAPHS

Stacked Bar graphs can show the breakdown of the client performance for an indicator relative to the competition. An example of a stacked bar graph is shown in Box 27.5.

BOX 27.5 STACKED BAR GRAPH: ANNUALIZED PLANT DOWNTIME

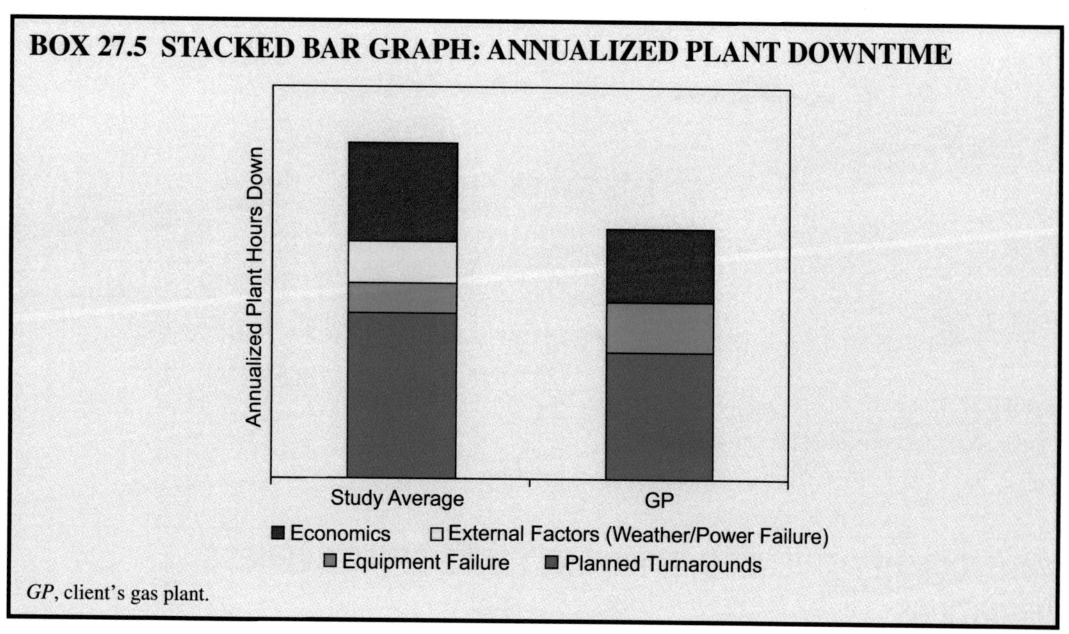

GP, client's gas plant.

COMPLEX GRAPH

Some consultants build complex graphs to visualize the benchmarking outcomes. This sometimes makes it difficult to drill down to focus on the lower level KPIs that have most influence over a top KPI.

An example of a graphical display of the Juran Hydrocarbon Index is shown in Ref. 3 of Chapter 25, Common Denominators and Indices Used in Benchmarking.

27.4 WHERE TO APPLY THE BENCHMARKING RESULTS

Ways to use the Benchmarking results include:

- determining competitive position using standard industry peer groups;
- developing gap analysis to assess specific performance shortfalls;
- formulating optimization plans to eliminate shortfalls; and
- using results as a first step to a performance improvement program.

As discussed in Chapter 5, The Business Cycle, results are typically applied at the business strategy review stage of the business cycle. This is shown in Fig. 27.3.

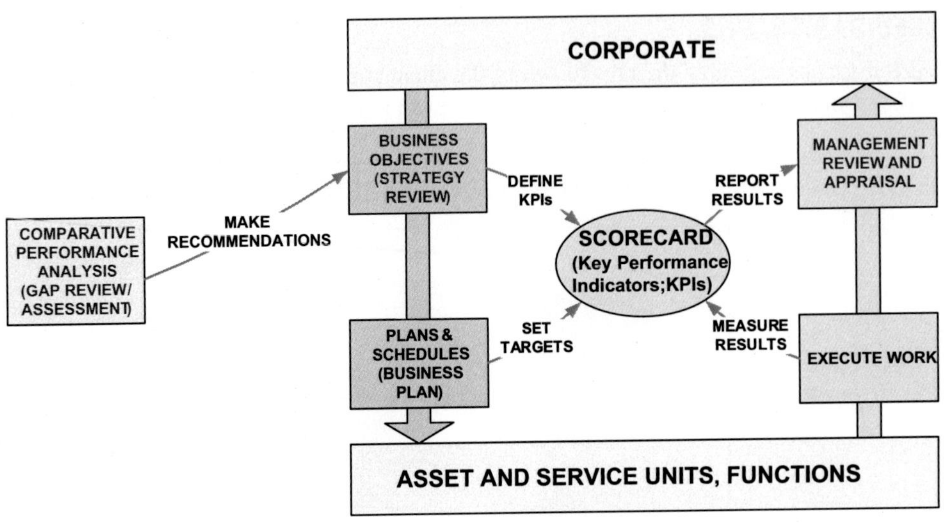

FIGURE 27.3

Application of comparative performance analysis.

27.5 COMPETITIVENESS, EFFICIENCY, AND EFFECTIVENESS

DEFINITIONS

Competitiveness

Competitiveness is looking in from the outside. "How competitive is my Asset Business Unit in the market place?"

Efficiency

Efficiency is looking from the inside. "How well is my Asset Business Unit operating with the facilities that it has today?"

Effectiveness

Effectiveness identifies focus areas for significant improvement. Pareto's 80:20 rule generally applies.

Competitiveness and Efficiency are discussed in Section 15.9, Competitiveness and Efficiency.

Section 28.6, Performance Focus, discusses Availability as the primary area of Effectiveness. This is directly linked to Utilization. Fig. 27.4 plots Utilization (Effectiveness) against Maintenance Cost Index (Efficiency), showing the correct route to Pacesetter status being the improvement of Utilization before an attempt is made at reducing costs. The danger of attempting cost reduction before ensuring improved utilization would be disastrous, as experienced by BP (see Box 24.2: Integration into the Business Example 1: Cost Reduction). Performance improvement has to be sustainable.

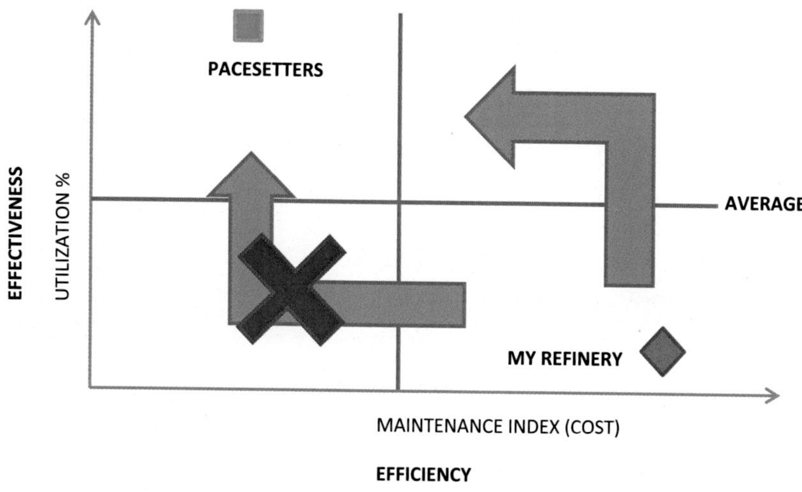

FIGURE 27.4

Effectiveness versus efficiency graph.

27.6 BENCHMARKING TOP INDICATORS

These are discussed in detail in Section 16.8, Examples of High-Level Key Performance Indicators. However, they are summarized here to show the relevance of the proprietary common denominator and indices.

SOLOMON: REFINING

Competitive

- Net Cash Margin $/bbl
- Return on Investment %
- Cash Opex $/Equivalent Distillation Capacity (EDC)
- Nonenergy Cost $/EDC
- Maintenance Index $/EDC
- Personnel Index Work hours/100 EDC
- Personnel Cost Index $/EDC
- Capital Investment Index $/EDC

Efficiency

- Gross Margin Index
- Volumetric Expansion Index
- Energy Intensity Index
- Carbon Emission Index
- Maintenance Cost Efficiency Index
- Personnel Efficiency Index
- Nonenergy Cost Efficiency Index

Box 27.6 gives an example of a benchmarking output format showing the client's position for some top KPIs.

BOX 27.6 REFINERY BENCHMARKING SUMMARY

Green (top quartile); Yellow, Amber, and Red (bottom quartile) are the Quartiles in which the participants have been placed in equal numbers. C is the position of the Client relative to all participants. Bracketed values are negative.

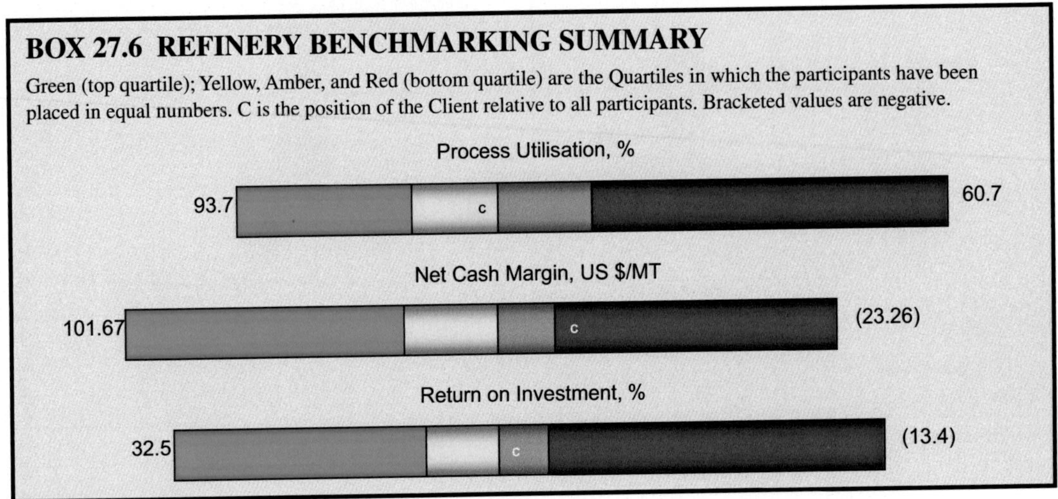

PHILIP TOWNSEND ASSOCIATES INTERNATIONAL: GAS PLANTS AND PETROCHEMICALS

Health, Safety, and Environment

- Number of Losses of Primary Containment per Normalized Shift Position
- Total Reportable Case Frequency
- Total Greenhouse Gas Emission (% weight on intake)

Margins

- Utilization (%)
- Opportunity to Increase Production (%)
- Annualized Lost Capacity for Maintenance (%)
- Annualized Unplanned Lost Capacity (%)
- Total Identified Losses (excluding Chemical losses) (% weight on intake)
- Corrected Energy and Loss Index

Costs

- Cost Index
- Fixed Costs (adjusted for Maintenance) ($/Mechanical Units)
- Annualized Maintenance Costs ($/corrected Mechanical Units; MUc)
- Annualized Maintenance Material Costs ($/MUc)
- Overall Personnel Index
- Average Personnel Costs ($/hr)

Box 27.7 gives an example of a top refinery benchmarked by Shell [prior to Philip Townsend Associates International (PTAI) taking over the benchmarking activities of Shell], and Solomon Associates.

BOX 27.7 RELIANCE REFINERY BENCHMARKING PERFORMANCE[1]

Business Standard: January 13, 2004

Reliance Jamnagar refinery ranked best

Reliance Industry Ltd's (RIL) refinery at Jamnagar has been ranked the best by Shell benchmarking magazine for the third consecutive year in 'energy and loss' performance from among 50 refineries worldwide.

The Jamnagar refinery has consistently lowered the "**energy and loss index**" **(CEL)** during the past 3 years, decreasing it from 95.6 CEL Index in 2000 to 88.7 CEL Index this year. The lowest CEL Index shows a reduction in energy loss and places the company as the world leader in energy performance.

Reliance's performance was also at number one in **operating cost, manpower cost, maintenance cost and plant utilization** in Shell Benchmarking.

Earlier this year, Reliance was ranked number one in energy performance among large complex group refineries in the Asia-Pacific region by Solomon Benchmarking of the Solomon Associates of the US. Reliance's **energy intensity index (EII)** was the best among 64 large and complex refineries in the Asia-Pacific region. Its position was at number one in **personal cost index, maintenance index, cash operating expense, net cash margin and return on investment** among 71 refineries in the region.

These energy saving efforts at the Jamnagar refinery have not only reduced energy indices, but has also resulted in Rs. 76 crore (USD 11 million) worth of fuel savings, a company statement said on Monday.

The statement said that the refinery was also awarded the **ISO 14001** certification by Lloyd's Register Quality Assurance (LRQA) for its **environment management system** and the Environment Excellence Gold Award by the Greentech Foundation for environment care.

The integrated petrochemical and refinery complex at Jamnagar was commissioned in 2000. The 27 million ton refinery accounts for 24% of India's refining capacity and is the world's largest grass-roots refinery and the fifth largest at any single location.

27.7 SUMMARY

The comparative study (benchmark) report should identify gaps that need to be addressed. Subsequent Gap Analysis/RCA workshops would then identify the best actions to take in bridging the gap.

"Benchmarking envelopes" are used to set common asset boundaries for a particular comparative study. These may be different to the **Business Unit**, but still within it.

The format of the benchmarking output summary, divided into quartiles for the top indicators in the study, clearly shows the position of the company relative to the rest of the industry. It also shows the change in performance relative to the previous two benchmarking studies (see also Chapter 31: Reporting).

Benchmarking outcomes are used for strategic planning (see Chapter 5: The Business Cycle) and input into the Assessment process (see Part 7: Assessment and Reporting) and may result in determining KPI targets for the Company.

Competitiveness, Efficiency, and Effectiveness relationships help analyze results and determine the maximum positive effect on total performance. Strategies that have the most effect on these indicators can then be applied.

The primary benefits of benchmarking are improved reliability and increased profit. These indicators are complementary. The route for improvement of these two indicators is to improve reliability first, with possible short-term increased costs, before costs reduce and profits increase.

Solomon Associates divides top indicators into "competitiveness" and "efficiency," and PTAI divide them into "HSE," "Margins," and "Costs." Top benchmarking indicators need to be integrated into the business and may need to be simplified or broken down for understanding, buy-in, and ownership.

REFERENCE

1. Reliance Refinery Benchmarking. http://www.business-standard.com/article/economy-policy/reliance-jamnagar-refinery-ranked-best-104011301078_1.html.

FURTHER READING

1. PWC. *Insights into the performance of the global shipping industry*. Global Shipping Benchmarking Survey; 2010.
2. IOGP. *Environmental performance in the E&P industry. 2007 data*. Report No. 414. December 2008.
3. IOGP. *Safety performance indicators. 2008 data*. Report No. 419. May 2009.
4. Bloch HP, Hernu M. *Maintenance/reliability. Performance benchmarking update: expectations and reality*. Hydrocarbon Processing; December 2007.

PART

ASSESSMENT, STRATEGIES, AND REPORTING

7

OVERVIEW

WHY IS IT IMPORTANT TO DO A ROUTINE FULL ASSESSMENT OF THE PERFORMANCE OF THE BUSINESS?

The strategic direction of the business needs to be reviewed prior to preparation of the annual Business Plan and Budget. Primary inputs to this review are assessment of past performance, performance against the competition and projected short & long term performance.

THE OBJECTIVE

The objectives of Part 7 are the following:

1. Give guidelines for a structured annual assessment of a business
2. Apply simple tools and methods that focus on the few strategies that have the most impact on the improvement of the overall business

3. Apply a change management process to ensure change is embedded
4. Tailor reports for understanding and effective decision-making

PICTORIAL VIEW

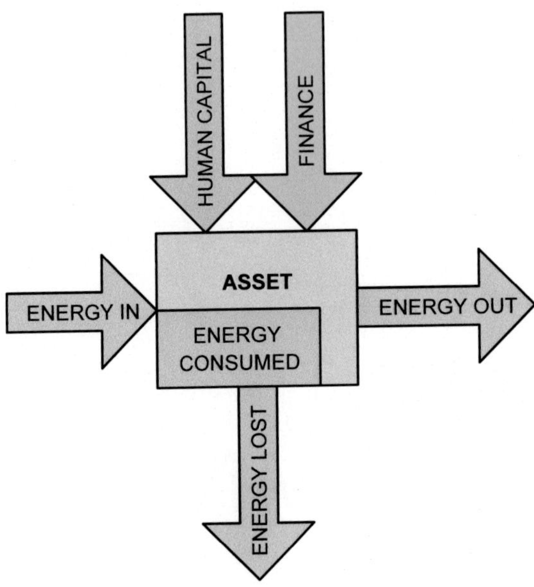

OUTLINE OF PART 7

- Assessment objectives are outlined.
- A few strategies that have the most impact on improvement are demonstrated.
- The inputs and outputs of assessment as part of strategy management are explained.
- The basic assessment cycle based on ISO 31000 is outlined.
- The primary categorization/grouping of business performance for assessment is outlined.
- Views and perspectives on performance assessment are highlighted.
- Evaluation and prioritization are outlined.
- The use of "rules of thumb" for quick assessment is mentioned.
- Assessment focus areas for performance of the whole business, from obtaining resources to delivery to the customer, are outlined.
- Approaches to assessment are outlined.
- Assessment methods are detailed.
- A process for preparation for structured assessment workshops is discussed.
- Mapping is outlined, using examples, showing how to identify the 20% high-impact actions that resolve 80% of the issues.

- Questions on the level of achievement, relative to the competition, are discussed.
- Target setting is discussed, with an emphasis on setting short-term and long-term targets for mechanical availability.
- Types of strategies and plans and their grouping are discussed, supported by examples of improvement plan reports.
- The requirement for formal change management is emphasized with detailed discussions on current practices.
- Reporting to ensure effective decision-making is discussed.
- Examples of different types of reports, relevant to the audience, are given.

CHAPTER

ASSESSMENT OVERVIEW

28

28.1 INTRODUCTION

This chapter gives an overview of the assessment process. Risk and performance assessments are to be undertaken together, where possible.

Focus categories and various assessment tools are introduced.

28.2 ASSESSMENT OBJECTIVE

The objective of undertaking assessments is to identify those few items that can be actioned and will have the biggest impact on performance improvement (Pareto's 80:20 rule). These few items are referred to as High-Impact Actions (HIAs) in this chapter. The intent is to have breakthrough or step improvement in the overall business performance.

Specific objectives are to:

1. optimize the external risks and opportunities in line with the risk appetite of the company;
2. maximize the long-term return for investors;
3. perform better than the competition;
4. maximize throughput at design capacity for the life of the investment;
5. exceed customer expectations;
6. obtain greater return on investment (ROI) than justified for investment projects; and
7. minimize the impact on the environment.

Fig. 28.1 superimposes the previously mentioned objectives on the Business Unit.

28.3 ASSESSMENT GROUPS AND GENERIC PROCESS

Assessment generally takes place in the following instances:

A. BUSINESS STRATEGY FORMULATION

1. Competitive and efficiency assessment after a benchmarking exercise
2. Business risks and opportunities, and performance assessment prior to the development of an annual business plan and budget

See Chapter 29, Identification, Analysis and Evaluation of Gaps, for details.

Performance Management for the Oil, Gas, and Process Industries. http://dx.doi.org/10.1016/B978-0-12-810446-0.00028-1

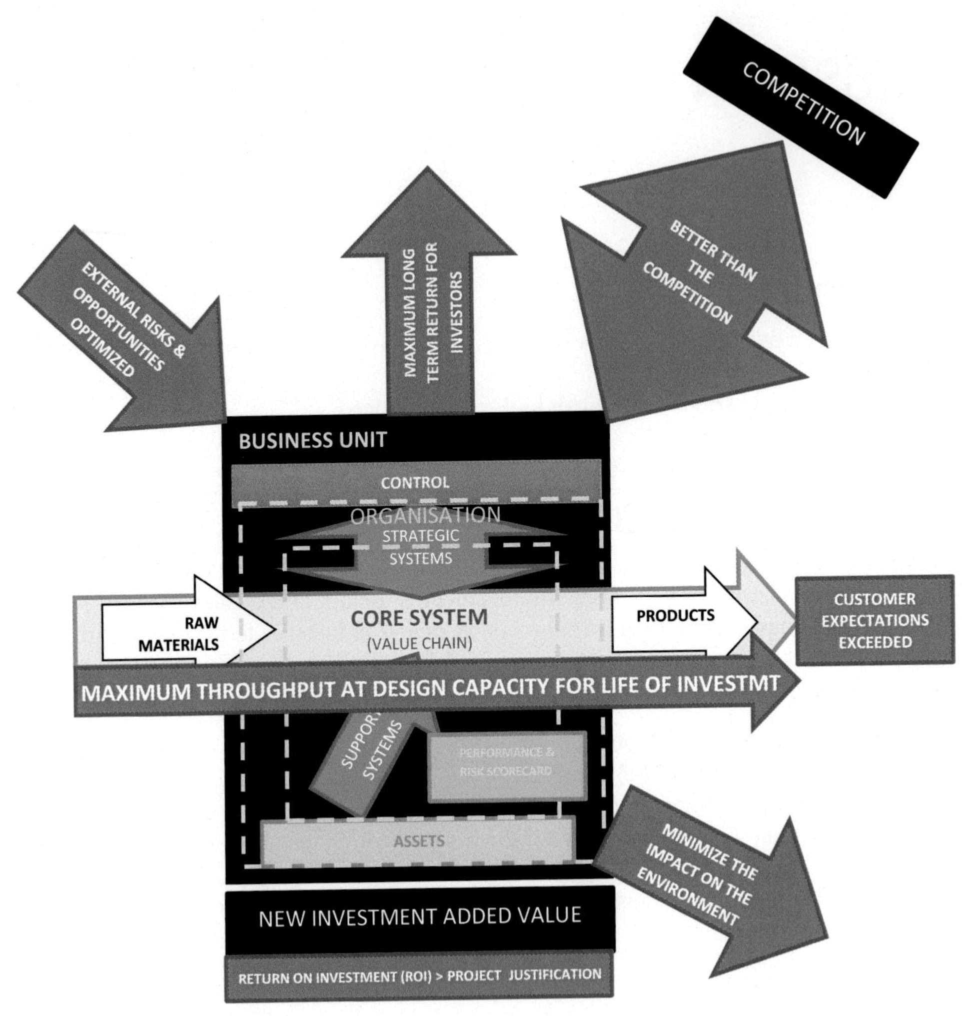

FIGURE 28.1

The Business Unit and assessment objectives.

B. BUSINESS OPERATION

1. Business risk assessment prior to establishing an Enterprise Risk Management (ERM) System
2. Incident Root Cause Analysis (RCA) after an incident

See Section 12.8, Tools for Operational Risk and Performance Analysis and Evaluation, for details.

C. INVESTMENT PROJECT

1. Investment business assessment prior to the Initial Investment Decision gate and reviewed at Preliminary Investment Decision gate and Final Investment Decision gate during the project stage of the investment cycle
2. Project risk assessment during the project phases of the investment cycle to optimize design and construction.

See Section 33.3, The Investment Cycle and Section 33.4, Investment Risk Assessment, for details.

The assessment process follows the "Assessment" component of the ISO 31000 risk management process as introduced in Chapter 11, Risk Management.

The basic process is:

1. identify the gaps (risk or opportunity);
2. quantify significant gaps;
3. prioritize as per quantified values; and
4. action the closing of the gaps.

This is depicted in Fig. 28.2.

Examples of actions that have had a major impact on the "bottom line" of the business are listed as follows.

A. STRATEGIC

- *National Oil Company (NOC): implement Enterprise Risk Management (ERM)*
- *NOC/International Oil Company refineries: obtain certification to ISO 9001, 14001, and 45001*

B. OPERATIONAL

Process

- *Gas Plant:*
  - *Change amine treatment*
  - *Install energy management system (in line with ISO 50001)*
  - *Apply advanced process control (APC) to gas trains*

Planned Shutdowns/Turnarounds

- *Gas Plant:*
  - *Apply rigorous turnaround scope challenge*
  - *Swop out gas turbine frame in a turnaround*
  - *Minimize gas train flaring/cooldown time for shutdown/startup*
- *Refinery/Gas Plant:*
  - *Implement a phased approach to turnarounds*
  - *Incentivize turnaround contractors*
  - *Assess rope access versus scaffolding for confined internal inspections during turnarounds*

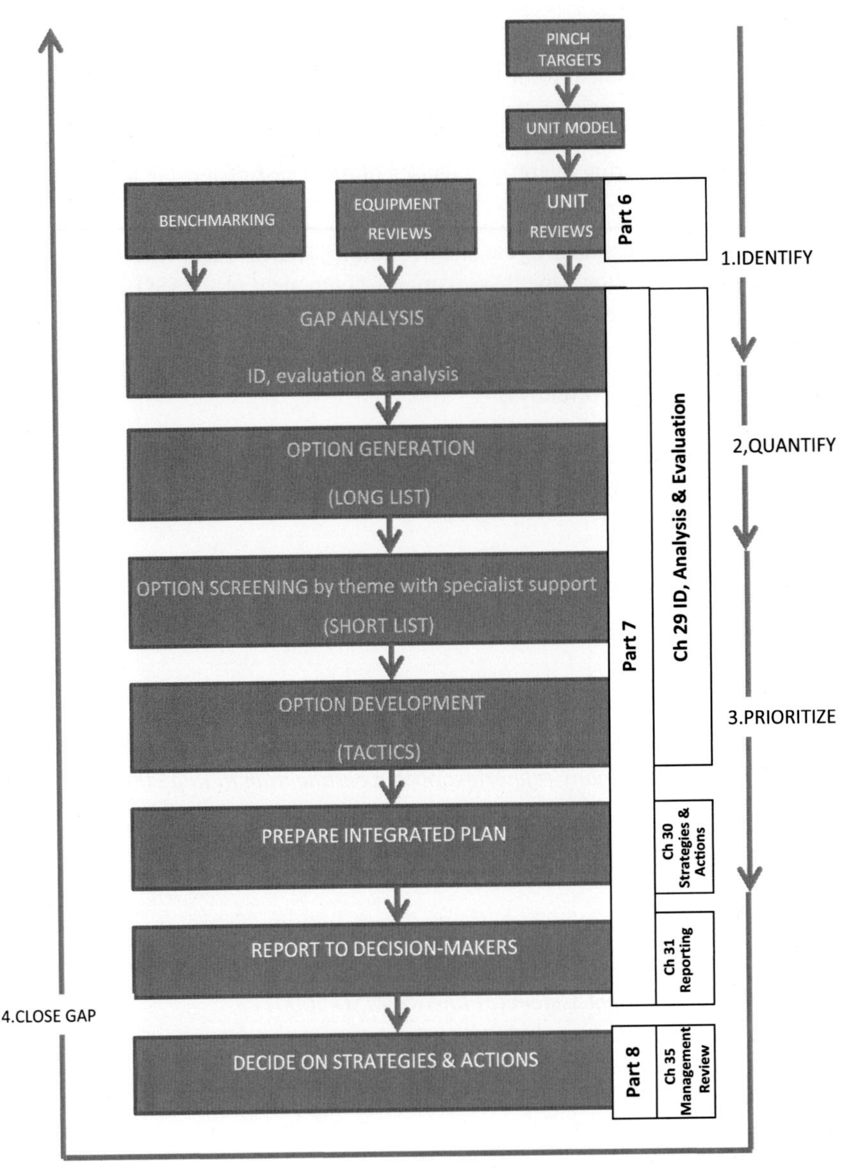

FIGURE 28.2

Basic assessment process.

- *Polyester Plant:*
  - *Assess chemical cleaning versus physical cleaning at start of turnarounds*

C. NEW INVESTMENT

- *Ensure start up on time, within budget, to design capacity and on-specification product quality*

The **full assessment process** is required at least on an annual basis.

28.4 BUSINESS RISKS AND OPPORTUNITIES AND PERFORMANCE ASSESSMENT

Chapter 5, The Business Cycle, introduces assessments as part of **Business Strategy Formulation**. These are grouped together as they formulate the inputs to the **Annual Business Planning Process**.

Fig. 28.3 depicts Assessments as part of Business Strategy Formulation.

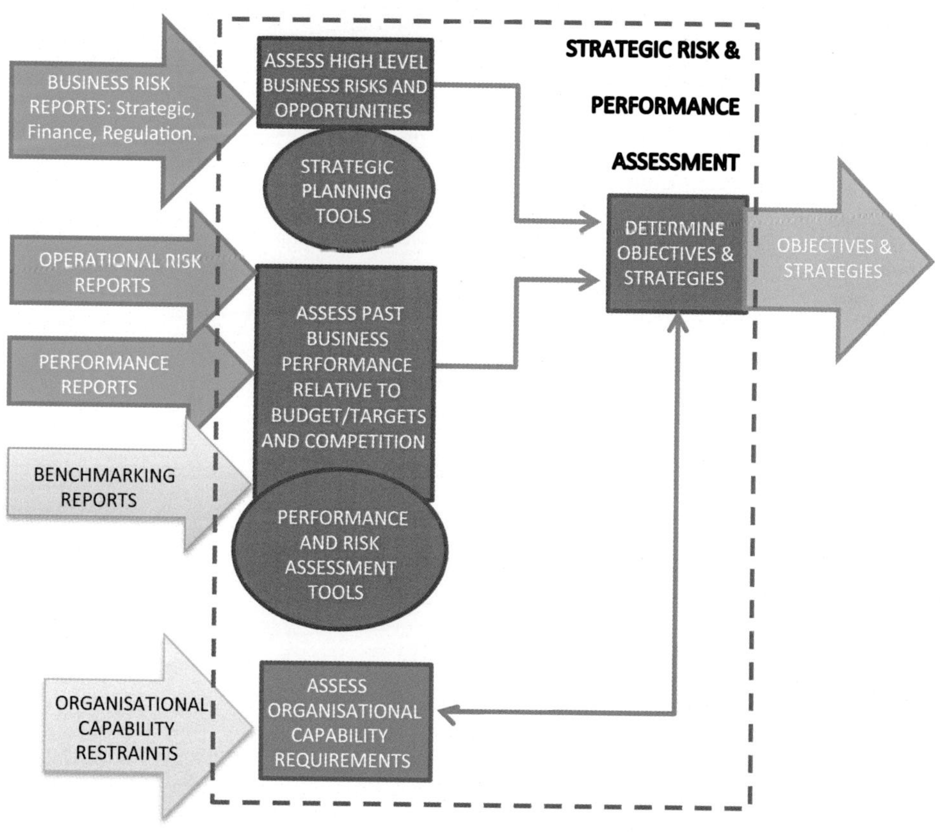

FIGURE 28.3

Assessment and Business Strategy Formulation.

Corporate strategic risk assessment generally:

- identifies and ranks the risks inherent in the company's strategy (including its overall goals and appetite for risk);
- selects the appropriate risk management approaches and transfers or avoids those risks that the business is not competent or willing to manage;

- implements controls to manage the remaining risks;
- monitors the effectiveness of risk management approaches and controls; and
- learns from experience and makes improvements.

Box 28.1 gives an example of ISO 31000 as applied to shareholder's interests in a number of integrated gas plants, totaling 22 gas trains.

BOX 28.1 GAS PLANT JOINT VENTURE (JV) ANNUAL BUSINESS PLAN AND BUDGET PREPARATION

Performance Assessment Outline

1. Gas plant shareholding interests estimated to generate one-third of the shareholding company's profit (context)
2. Quarterly performance and other reports (including benchmarking) relative to the previous approved Work Plan and Budget (WP&B) reviewed (communication and consultation)
3. Assessment:
   a. Operations performance gaps identified
   b. Root Cause Analysis (RCA) from the shareholders perspective undertaken (quantify)
   c. Evaluation undertaken and areas for significant improvement determined (prioritize)
4. Action: initiated through a shareholder "letter of wishes" to the management of the affiliate companies prior to developing affiliate WP&Bs for the subsequent year (risk treatment)
   Box 35.2, Shareholder's "Letter of Wishes" gives sample contents.

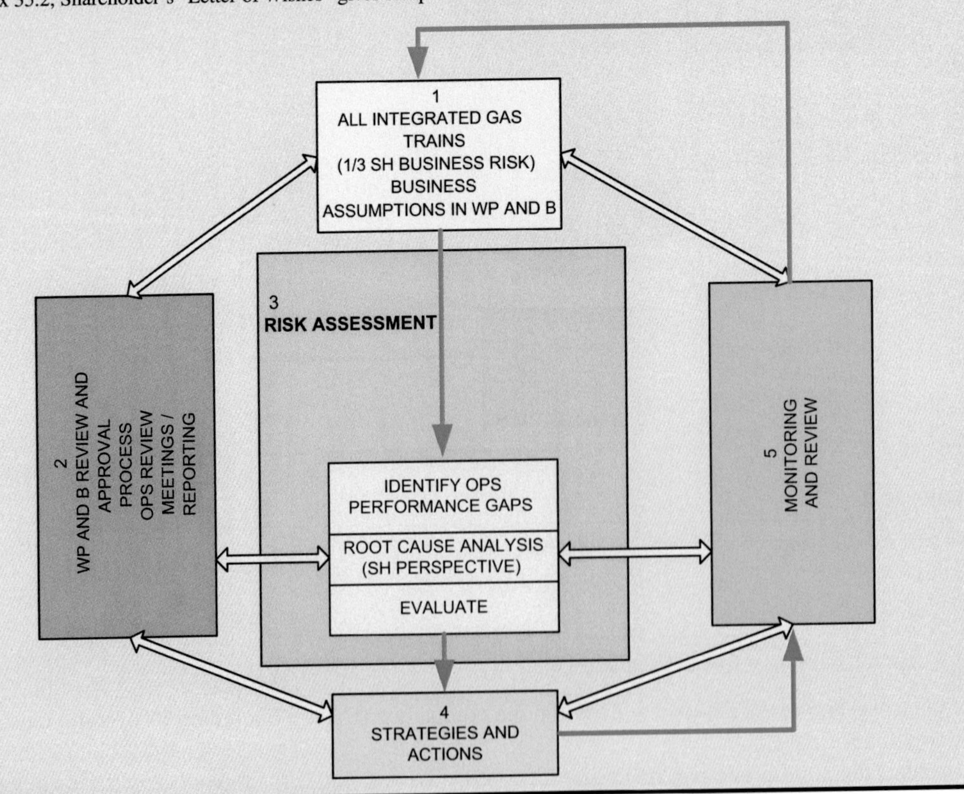

28.5 CATEGORIZATION

Assessment requires all information on related issues to be assembled for review.

Common practice is to categorize related issues as follows for assessment:

1. **E**nergy
2. **F**inancial Assets
3. **P**hysical Assets
4. **H**uman Assets

This is depicted as per Fig. 28.4.

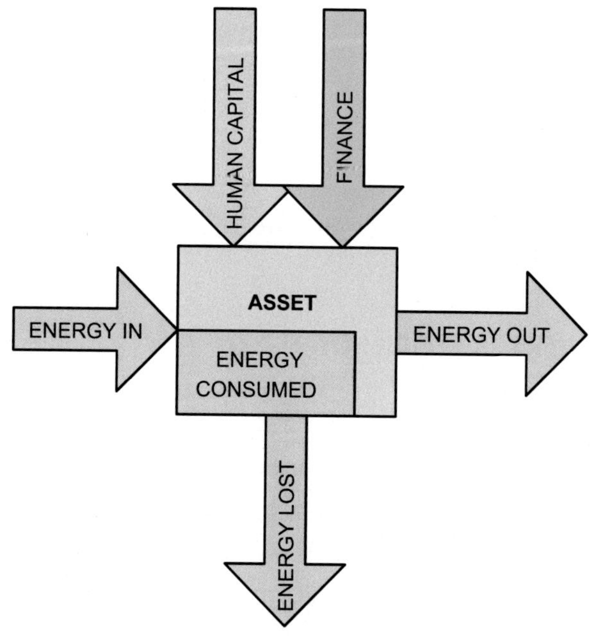

FIGURE 28.4

Energy, Finance, Physical Asset, and Human Asset (EFPH) categories of assessment.

Energy covers Production and Downtime.

Health, Safety, Security, and Environment spans both Energy and Physical Assets and is covered by them.

Categorization is covered in detail in Section 29.5, Preparation for Workshops.

28.6 PERFORMANCE FOCUS

Performance needs to focus on three fronts: competitiveness, efficiency, and effectiveness.

Competitiveness and efficiency is introduced in Section 15.9, Competitiveness and Efficiency, where competitiveness and efficiency gaps are discussed.

Section 23.3, Why Effective Turnaround Management Is Critical, introduces effectiveness related to turnarounds and how turnarounds influence improvement in two of the most effective Key Performance Indicators: availability and utilization.

Section 27.5, Competitiveness, Efficiency, and Effectiveness, discusses the relationship between competitiveness and effectiveness and the danger of reducing cost (competitiveness) before increasing reliability (effectiveness).

Assessment Focus expands on the previous discussion as follows:

- **Competitiveness**
  - Looking from the outside
  - "How competitive is my Asset Unit in the marketplace?"
  - Indicators: Profit, Utilization, Availability, etc.
- **Efficiency**
  - Looking from the inside
  - "How well is my Asset Unit operating the facilities that it has today?"
  - Indicators: Energy Efficiency, etc.
- **Effectiveness**
  - Actions that will address primary issues
  - "Which indicators have most effect on the Corporate Objectives?"
  - Application of the 80:20 rule

COMPETITIVENESS VERSUS EFFICIENCY

Shareholder focus has two components:

1. Cost per saleable product: an indication of **Competitiveness**
2. ROI: an indication of **Efficiency** (lower cost = higher return)

The four quadrants of a competitiveness/efficiency assessment are as follows:

1. **"Big and Unprofitable"**:
   - **Poor use of economy of scale and not efficient**
2. **"Niche players"**:
   - Low economy of scale and very efficient
3. **"Big and Profitable"**:
   - Good economy of scale and not efficient
4. **"Pacesetters"**:
   - **Good economy of scale and efficient**

These can be depicted as per Fig. 28.5.

Note that ExxonMobil focuses on Return On Capital Employed, and so, they would look at reducing $ per Replacement Value (RV) (see Box 21.6: Return On Capital Employed Application: ExxonMobil).

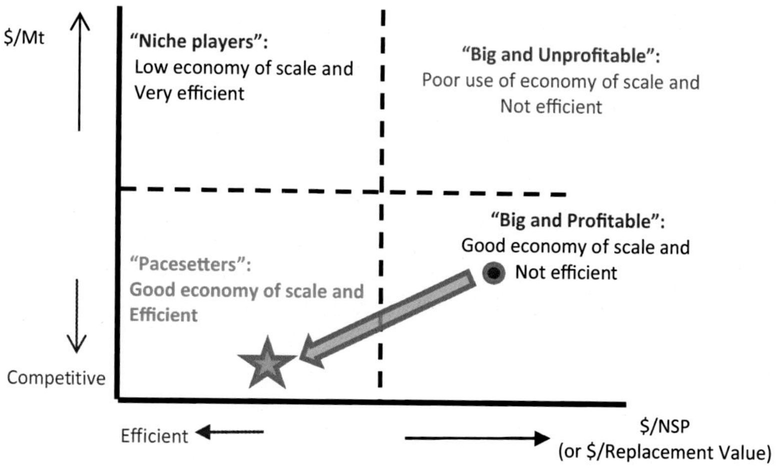

FIGURE 28.5

Competitiveness versus efficiency costs. *NSP*, normalized shift positions.

EFFECTIVENESS

Effectiveness relates primarily to Availability and Utilization. Fig. 28.6 gives a pictorial view of effectiveness. The first graph indicates what is targeted and results in a satisfied stakeholder, whereas

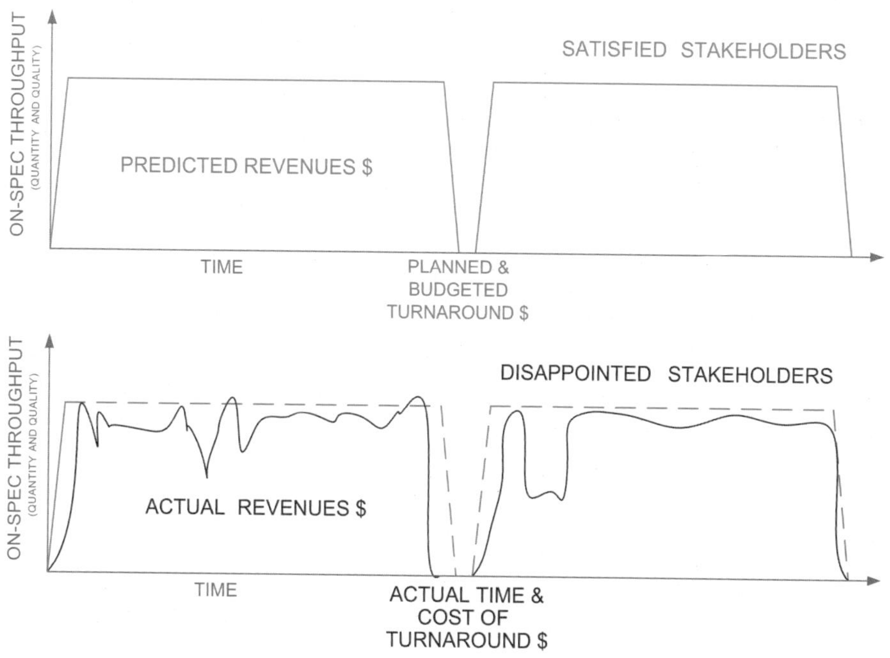

FIGURE 28.6

Effectiveness.

the second graph shows how ineffective Availability and Utilization could result in a disappointed stakeholder. Essentially, the area under the graph needs to be maximized.

Availability and Utilization definitions and relationships is discussed in Section 17.4, Utilization and Availability.

A record of past turnarounds and intervals between turnarounds is critical for the assessment of effectiveness.

For Availability to be optimized, **reliability** has to be improved, people have to be competent and motivated, and waste has to be minimized, and only then will costs come down.

Effectiveness, using the 80:20 rule and focusing on high impact actions (HIAs) in reliability, energy efficiency, and cost optimization, including the reduction of a major component of fixed cost, staff, as applied to the assessment of a gas plant, is shown in Box 28.2.

BOX 28.2 EFFECTIVENESS AS APPLIED TO A GAS PLANT ASSESSMENT

Mapping Outcome

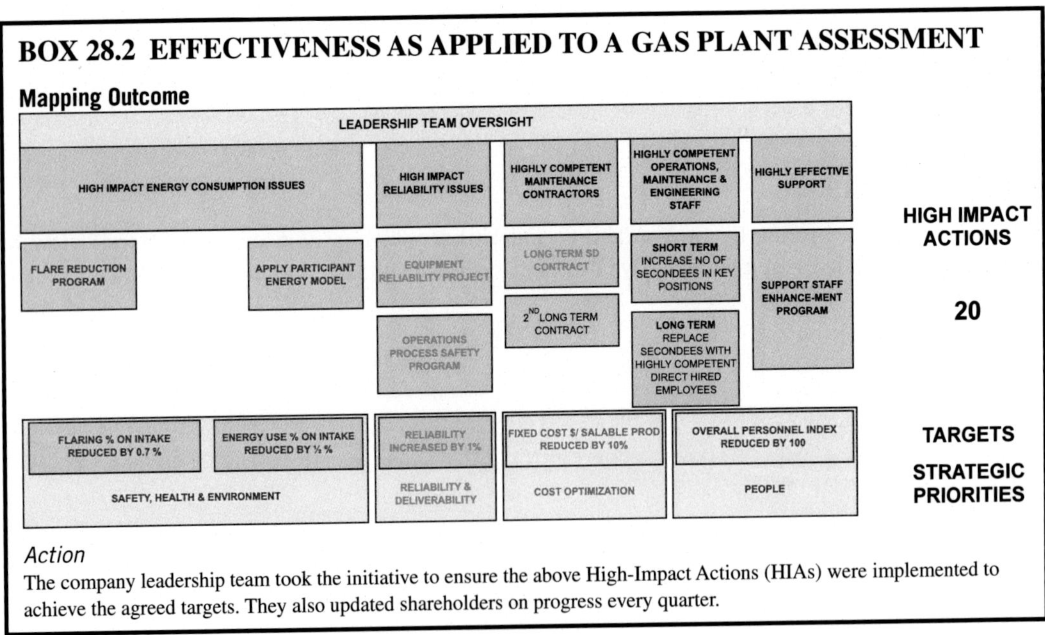

Action

The company leadership team took the initiative to ensure the above High-Impact Actions (HIAs) were implemented to achieve the agreed targets. They also updated shareholders on progress every quarter.

28.7 EVALUATION AND PRIORITIZATION

RETURN/EFFORT MATRIX

This matrix is normally used as a simple tool to prioritize plans and strategies.

Quadrants are as follows:

- Quadrant 4: High Return, Low Effort: Quick hit.
- Quadrant 3: High Return, High Effort: Assign to experts to investigate and produce a proposal.

- Quadrant 2: Assign to champions to study and action as required.
- Quadrant 1: Remove from list.

Fig. 28.7 depicts the previous.

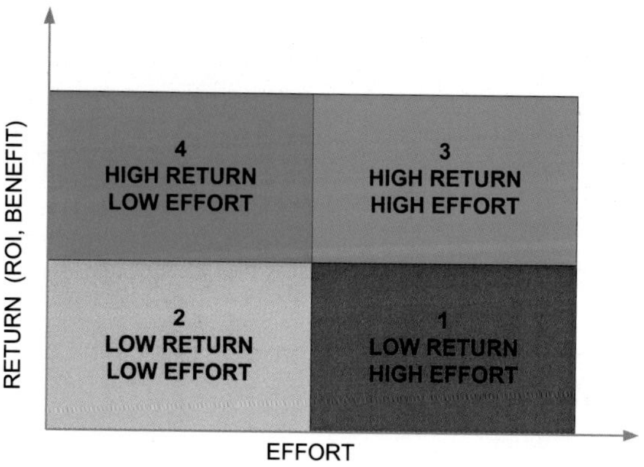

FIGURE 28.7

Return/Effort Matrix.

VOTING

Voting by participants in a brainstorming workshop could determine priorities. An example is shown in Box 28.3.

BOX 28.3 PRIORITIZATION VOTING EXAMPLE

Energy Workshop No. 1 Outcomes

Most Important + Important >3

Potential Strategies	Important	Most Important	Important + Most Important	Supporting Data	Possible Ownership	System	Possible Action
Process							
Optimize equipment capacity utilized	3	1	4		TMP	PP	
Maximize liquefied petroleum gas (LPG) Recovery	2	2	4	Excessive flaring	TMS	PE	
Improve tank insulation	4		4		MCM	MC	

Continued

BOX 28.3 PRIORITIZATION VOTING EXAMPLE—cont'd

Energy Workshop No. 1 Outcomes

Most Important + Important >3

Potential Strategies	Important	Most Important	Important + Most Important	Supporting Data	Possible Ownership	System	Possible Action
Maximum power generation package (PGP) utilization	2	2	4	Design 15 MW Operating at 8 MW	OMF	PS	
Improve power generating efficiency	4		4	Design 15 MW Operating at 8 MW	OMF	PS	

28.8 QUICK ASSESSMENTS

Quick assessments are required when:

- there is not enough time for a full assessment and where time is of the essence to resolve a problem or obtain competitive advantage; or
- the amount of data for assessment is so overwhelming that it is difficult to know where to start and focus.

RULES OF THUMB

"Rules of Thumb" give ranges or targets that should be achievable. They can be used in a quick assessment of those that are far off target, prior to a "deep dive" to find root causes of poor performance.

Some examples of "Rules of Thumb":

- Mass Balance
  - Refining: Volume gain/Mass loss +0%/–0.2%
  - Gas Plants: Mass loss/gain less than) ±0.6%
- Energy Efficiency
  - LNG mega train: 7.9% of weight on intake (woi)
- Operating expense (Opex)
  - Refinery Maintenance cost < 1.7% of RV
- Turnarounds Duration
  - Crude Unit: 30 days
- Turnarounds Interval
  - Gas trains (turbines): 5.5 years
- Reliability
  - LNG mega-train design basis: 98% reliability
    - 7 unplanned down days per year

The "rules of thumb" are also applied in Chapter 26, Benchmarking Data Verification.

Appendix C gives a full list of "rules of thumb."

CONSULTANT

Juran[1] offers quick assessments through their proprietary processes, as outlined in Box 28.4.

BOX 28.4 JURAN ASSESSMENTS

Comprehensive Organizational Review for Excellence Assessment

Juran's Comprehensive Organizational Review for Excellence (CORE) Assessment is a tool to help clients worldwide measure system-wide performance with a common methodology, similar to the National Awards for Excellence in place in many countries around the globe. It is a "**quantitative assessment**" used to measure and evaluate an organization's total business system against a set of global competencies and best practices. By conducting a CORE Assessment throughout an organization, the organization will gain a third-party review of the critical processes that impact quality, performance, and customer satisfaction, as well as identify where changes should be made to maximize performance improvement results in business operations.

Quick Strike Global Assessment of Performance

This analysis is similar to what is described in the CORE Assessment except that the scope is reduced to focus on specific elements of CORE. It is a "**qualitative assessment**." Juran gathers information through interviews and review of important company documentation. A written report will be prepared and presented to senior management. The report from a Quick Strike Global Assessment of Performance (GAP) analysis uses a narrative format rather than the scoring format utilized in the CORE to capture strengths, deficiencies, and recommendations to improve on the deficiencies and build on the strengths.

28.9 SUMMARY

The objective of the Assessment is to derive those Strategies (**HIAs**) that will have the most effect on performance improvement.

The ISO 31000 process can be used for assessment (identify, analyze, and evaluate) to determine appropriate strategies and to ensure they are carried out.

Common practice is to divide the assessment process into Energy, Finance, Physical Asset, and Human Asset categories and attempt to cover all aspects of the business within each category.

The focus needs to be on Competitiveness, Efficiency, and Effectiveness throughout the process.

The process is simply to identify, quantify, prioritize, and close the gap. The full process should be completed at least once per year.

Business Risks and Opportunities and Performance Assessment need to be carried out together.

Rules of thumb are useful to do quick assessments and cross-checks on data.

Appropriate tools include spreadsheets, matrices, and structured processes, some using software to assist (for example, Taproot RCA). Checklists ensure all items are covered. Evaluation and prioritization includes tools such as the Return/Effort Matrix.

REFERENCE

1. Juran's benchmarking approach. https://juranbenchmarking.com/our-approach/.

FURTHER READING

1. ISO 31000:2009. *Risk management – principles & guidelines.*
2. ISO 31010:2009. *Risk management – risk assessment techniques.*
3. Barr S. How to improve what matters most with the 80/20 rule. http://kpilibrary.com/topics/how-to-improve-what-matters-most-with-the-8020-rule.
4. Setting targets for measures used in performance management systems. www.2gc.co.uk.

CHAPTER

IDENTIFICATION, ANALYSIS, AND EVALUATION OF GAPS

29

29.1 INTRODUCTION

This chapter discusses the details of a performance assessment as a result of Benchmarking (BM) and for the preparation of the Business Plan and Budget (BP&B). Examples are discussed in support of the theory.

Business risks and opportunities are discussed in Section 5.2, Strategic Planning (Strategy Formulation).

29.2 FOCUS AREAS

Section 28.5, Categorization, introduces four basic categories for assessment—Energy, Finance, Physical asset, and Human asset. Focus areas cover a wider scope but are mostly grouped within these generic categories.

The following have been used for shareholder assessment:

1. STRATEGY

This relates to external business risks and opportunities and the strategic direction of the company.

2. HSSE

This relates to all health, safety, security, and environment issues. It overlaps with energy and mass balance and with physical, human, and finance asset management.

3. ENERGY AND MASS BALANCE

This relates to the overall balance between raw materials coming in to the Business Unit and products going out of the Business Unit and includes salable and nonsalable products as well as all emissions to the atmosphere, ground, and water.

Performance Management for the Oil, Gas, and Process Industries. http://dx.doi.org/10.1016/B978-0-12-810446-0.00029-3

4. PRODUCTION AND DOWNTIME

This relates to the value addition process and the maximization of salable products. It overlaps with energy and mass balance and physical asset management.

5. FINANCE ASSET MANAGEMENT

This relates to operating and capital expenditure.

6. PHYSICAL ASSET MANAGEMENT

This relates to the physical assets carrying out the value addition of the raw materials.

7. HUMAN ASSET MANAGEMENT

This relates to all aspects of human asset management, including recruitment, motivation, training, and competency assessment.

8. RAW MATERIALS

This relates to timely acquisition of all raw materials that are compatible with the design of the physical assets.

9. SHIPPING

This relates to the timely and cost-effective shipment of salable products.

10. MARKET CUSTOMER

This relates to all marketing and customer issues.

Table 29.1 gives details of assessment focus areas and the primary focus for different industries.

Colored items in Table 29.1 are the categories noted in Chapter 28, Assessment Overview, and discussed in detail later in this chapter.

Shareholder oversight focus areas, described in Section 34.8, Shareholder Oversight Focus Areas, differ as these focus on the duties of the shareholder representative. However, assessment is a primary component of shareholder oversight focus and is intertwined with other shareholder representative duties.

Table 29.1 Assessment Focus Areas

Focus Area	Industry				Assessment application	Participants
	Gas Plants	Refineries	Petro-chemical Plants	Generic		
1 Strategy	Based on E&Y Risks & Opportunities	Scale & Integration	Scale & Integration, niche markets	Scale	Business Risks & Opportunities	Leadership team
2 HSSE	LOPC, Flaring	Flaring	Toxic Emissions	Lost Time Incidents (LTI)	Benchmarking andBusiness Plan & Budget (BP&B)	Experts
3 Energy and Mass Balance	HydroCarbon Management 1. Balance Unaccounted losses 2. Royalty accounting Miss-measurement	HydroCarbon Management 1. Balance Unaccounted losses	Conversion Management	Conversion Management	Benchmarking andBusiness Plan & Budget (BP&B)	Experts
4 Prod & Downtime	Salable Products	White Oil Production	Production of Primary Product	Production of Primary Product	Benchmarking andBusiness Plan & Budget (BP&B)	Experts
5 Finance Asset Management	Opex Fixed Costs	Opex Fixed Costs	Opex Fixed Costs	Opex Fixed Costs	Benchmarking andBusiness Plan & Budget (BP&B)	Experts
6 Physical Asset Management	Reliability & Maintenance (RAM)	Reliability & Maintenance (RAM)	Reliability & Maintenance (RAM)	Reliability & Maintenance (RAM)	Benchmarking andBusiness Plan & Budget (BP&B)	Experts
7 Human Asset Management	Competencies	Competencies	Competencies	Competencies	Benchmarking andBusiness Plan & Budget (BP&B)	Experts
8 Raw Materials	Sub Surface, Reserve Replacement	Crude assay vs. design capability	Upstream/ Optimum sources	Optimum sources	Business Risks & Opportunities	Leadership team and Experts
9 Shipping	LNG tankers/ pipelines	See & land transport	See & land transport	Distribution network	Business Risks & Opportunities, Benchmarking andBusiness Plan & Budget (BP&B)	Leadership team and Experts
10 Market Customer	Supply chain management	Transport fuels marketing network	IHS Chemical Market Advisory Service	Marketing network	Business Risks & Opportunities	Leadership team and Experts

29.3 APPROACH

Approaches could be

1. Directed: Performance Analysis Workshops (PAWs) with appointed specialists/experts, or
2. General: management review by the management leadership team.

Proposed changes could either be

1. incremental or
2. step (breakthrough).

For step changes, those people required to change will, in all likelihood, need to go through a paradigm shift, with the application of a formal change process to prevent regression to the previous state. This is discussed in detail in Section 30.6, Change Management.

29.4 ASSESSMENT METHODS

Assessments can be carried out using any combination of the following:

1. Structured review techniques
2. Codes/standards
3. Checklists
4. Experience/judgment

1. STRUCTURED REVIEW TECHNIQUES: PAWS

Generic Process

This directed approach entails assembling a small group of senior staff who are totally committed to their field, have sound judgment and extensive experience, and have influence within the business.

A multidisciplinary team representing the primary areas of the business is required as follows:

a. Strategic
b. HSSE
c. Production
d. Maintenance, engineering, and inspection (asset integrity)
e. Finance
f. Human assets

The process generally is to:

1. prepare using code and standard requirements, checklists, and spreadsheets of prepared data (see data input later);
2. assess in **EFPH** categories;
3. search for data in **EFPH** categories;
4. apply risk management process 31001;
5. use brainstorming/de Bono's "Six Hats"[5] or other tools to identify gaps;
6. use risk assessment tools;
7. do effectiveness testing;
8. identify bad actors; and
9. select appropriate strategies.

Data Input

Sources of data would include the following:

- Business/Work Plan and Budgets (B/WP&Bs)
- Progress on current investment projects

- Performance reports (monthly/quarterly performance reports and other participant reports)
- Regulatory reports (production/royalty, environmental, health, and safety)
- Audits including internal, certification, HSE, etc.
- Risk management survey and asset valuation reports (as required for insurance purposes)
- Reviews including Quantitative Risk Assessments (QRAs), etc.
- Benchmarking reports
- **Mechanical Availability** history of at least the past 6 years and an assessment of the next 5 years.

A **Mechanical Availability** assessment is required to determine the long-term optimization of the full cycle of operation. The full cycle of operation is from startup, the duration of operation of the plant until the next turnaround, as well as the turnaround duration (see Fig. 23.2: Operating Cycles and Shutdown Phases). This is the basis for calculating the total annualized down days (see Section 17.4: Utilization and Availability for details). Section 30.5, Mechanical Availability drill down discusses the determination of targets for this important Key Performance Indicator (KPI).

Note that the period for which the data are extracted and the scope for which the data are valid need to be identified and assessed. Section 27.2, Benchmarking Envelopes, discusses different benchmarking scopes.

2. CODES/STANDARDS

Clear noncompliance would be identified from consent-to-operate, international standards, government regulations, industry best practices, etc.

3. CHECKLISTS

Checklists could take many forms (see Box 29.2).

Inclusions would be:

1. compliance with agreements;
2. compliance with regulations;
3. alignment with corporate vision;
4. assessment of enterprise risk management framework; and
5. identification of the parent company directors on affiliate BoD/Mancom and shareholder representatives

It is important to assess the level of maturity in applying a listed item, as this gives an indication of expected performance at that level of maturity. It also helps to set short-term targets for lower levels of maturity where it will take time to achieve higher levels, and it shows management commitment over time between assessments.

4. EXPERIENCE/JUDGMENT

It is recommended that at least one member of the performance analysis team have an understanding of the business as a whole, in addition to being a subject expert.

29.5 PREPARATION FOR WORKSHOPS

Preparation for workshops includes the creation of spreadsheets, graphs, and checklists for easy discussion and drawing of conclusions.

ASSESSMENT SUBJECT AREA SPREADSHEETS AND GRAPHS

The assessment subject areas need to cover a wider area than the assessment focus areas shown in Table 29.1, as there could be significant interaction and influence between assessment focus areas.

The categorization Energy, Finance, Physical asset, and Human asset (EFPH) mentioned in Section 28.5, Categorization, is used for the assessment subject areas, covering the primary aspects of the "black box" Business Unit.

The use of spreadsheets facilitates the production of graphs. Graphs give clear comparison of the data on the tables. Examples are given in Box 29.3.

These are discussed in more detail in the following paragraphs.

1. Energy

All production is "energy." Specifically, liquefied natural gas (LNG) is sold based on heating value. Internal consumption is significant in gas plants (see Appendix C: Process Industry Rules of Thumb, for typical values).

Any losses are of concern. All flaring, internal consumption, and "discounted product" ("give away") need to be assessed.

2. Finance

All aspects of finance in the business are to be covered. Typically, for gas plants, fixed costs are 60%–80% of the operating expense, which gives an immediate focus for assessment.

Minor capital expense (Capex) needs to be assessed with respect to items that are not improving throughput. These are "replacement in kind" and Health, Safety & Environment projects. The line to be drawn between maintenance and capital projects is discussed from the benchmarking perspective in Section 26.3, Benchmarking Data Input Tables.

Major Capex needs to be assessed as to planned versus actual plant **startup date**, when revenue will start to be generated. Changes to the original approved budget also need to be reviewed. Note that the project is approved based on return on investment, and therefore project cost and date for first revenue generation are critical.

Typically, in gas plants, the "own consumption" gas value is excluded from the finances, or is of nominal value, and is not a factor in the finance assessment. However, energy expenses for a fuel refinery are typically 45%–55% of total cash operating expense (Opex). This is thus significant for assessment of refinery Opex.

For benchmarking, the cost of internally generated power needs to be assessed against purchased power.

Example:

An LNG plant's energy cost is one-fifth the energy cost of a similar LNG plant as it generates most of its own electricity (see Box 17.6: Gas Plant Energy Balance).

An example of the finance spreadsheet for this assessment is shown in Box 29.1.

BOX 29.1 LNG PLANT ASSESSMENT SPREADSHEET—FINANCE

Focal Point	INDICATOR	MEASURE	VALUE	TARGET
A	NET PROFIT AFTER TAX (NEPAT)/ TON salable product	$/TON		
A	COST/ TON salable product	$/TON		
A	ROYALTY/ TON salable product	$/TON		
A	COST/ TON LNG loaded	$/TON		
A	FIXED COST/ TON salable product	$/TON		
A	OPS & MAINT COST/ TON salable product	$/TON		
A	OPS & MAINT COST % of TOTAL OPEX	%		
A	OPEX % OF TOTAL EXPENSE	%		
B	MAINT & ENG COST % of RV	%		
B	MAINT & ENG COST/ NSP	INDEX		
B	OPERATOR STAFF COST % of RV	%		
B	OPERATOR STAFF COST/ NSP	INDEX		
B	SUPPORT STAFF COST % of Total Staff including contractor labor	%		
B	SUPPORT STAFF COST/ NSP	INDEX		
B	TURNAROUND COST	$		
B	TURNAROUND COST (ANNUALISED)	$		
C	ROCE	%		
C	REPLACEMENT IN KIND % of CAPEX			
C	HSE % OF CAPEX			
C	AVERAGE ROI FOR BALANCE OF CAPEX (excl above)			
C	OPERATING COST INDEX	INDEX		
C	NON ENERGY COST INDEX	INDEX		
C	FIXED COSTS (adjusted for maint)	$/MU		
C	AVERAGE PERSONNEL COSTS	$/HR		
C	OVERALL PERSONNEL INDEX	INDEX		
C	PERSONNEL COST/NSP	M$/NSP		
C	UNIT OPERATING COST/ TON salable product	$/TON		
C	UNIT FIXED COST/ TON salable product	$/TON		
C	UNIT VARIABLE COST/ TON salable product	$/TON		
C	MAINT COSTS (ANNUALISED)	$/MUc		
C	MAINT MATERIAL COSTS (ANNUALISED)	$/MUc		
C	MAINT LABOUR COSTS	$/MUc		
C	WAREHOUSE OPERATING COST	$/MUc		
C	INSURANCE PREMIUM	$/MUc		
C	INSURANCE PREMIUM	$/RV		
D	LNG SHIPPING COSTS	$/ TON		
E	SIGNIFICANT AUDIT ISSUES OUTSTANDING			

Blue, mainly performance reports; Red, mainly benchmarking reports.

3. Physical Asset Integrity

The primary focus is preventing major loss of primary containment, for the life of the physical asset. Risk-based systems and processes help with predictive inspection and maintenance. As a result, intervals between planned shutdowns are being stretched, with minimal unplanned shutdowns between planned shutdowns. Leading indicators are, therefore, becoming more important. These are derived from disparate areas, ranging from "fail safe" design, health and safety, and static and rotating inspection to planned maintenance and safe operations.

4. Human Capital

Organizational capability means having the right people with the right skills available at the right time to meet current and future requirements, so as to achieve the company objectives. Assessment here tends to be more subjective than the preceding three subject areas. Nevertheless, assessment needs to cover both behavioral and technical aspects of managing the business. Recruitment and retention of highly competent personnel are essential for the success of the business.

Example:

A particular company with low staff turnover has significantly higher (better) "unrelated" performance indicators than a similar company (see Box 20.6: Competencies—Motivation and Box 24.3: Integration into the Business—example 2 Reliability first).

Technical competency is essential for the following:

- Operations
- Inspection
- Maintenance

Continual up-skilling is also required.

Example:

A panel operator is required to undertake simulator training initially and at subsequent regular intervals in the same way as an airline pilot (see Box 20.3: Plant Process Competencies—The Simulator).

COMPANY CHECKLISTS

Company checklists identify all systems and processes required for operational good practice. The level of maturity of the system or process is indicated on this checklist; this is defined in Table 29.2.

Table 29.2 System Maturity Levels

Level of Maturity	Classification
1	No formal system
2	Under development
3	Well established
4	Fully integrated

A partial checklist example is shown in Box 29.2. A full example is shown in Appendix D, Operational Performance Health Checker Example.

BOX 29.2 AFFILIATE COMPANY CHECKLIST EXAMPLE (EXTRACT)

Company:

Date:

rev 130103	Applied	LoM	Comments
FA1-STRATEGY			
ERM	?		
IMS	Y		
Top 10 risks / opportunities	?		
KPI scorecard	Y		
SH audit rights routine	Y		
ISO 9001 certification	Y		
FA2- HSSE			
Health & Safety			
OHSAS 18001 certification	Y		
RCA process using Tripod Beta, Taproot, Proact or other	Y		

Level Of Maturity	CLASSIFICATION
1	No formal System
2	Under development
3	Well established
4	Fully integrated

29.6 BENCHMARKING GAP TROUBLESHOOTING WORKSHOPS

Gap Troubleshooting Workshops (GTWs) use a variety of tools to assess the performance of the company, relative to benchmarked peer groups.

The process is similar to the generic process described earlier but is more focused. Details are completed on prepared spreadsheets derived from the benchmarking study. Diagnosis tables and graphs are presented. Some examples are shown in Box 29.3.

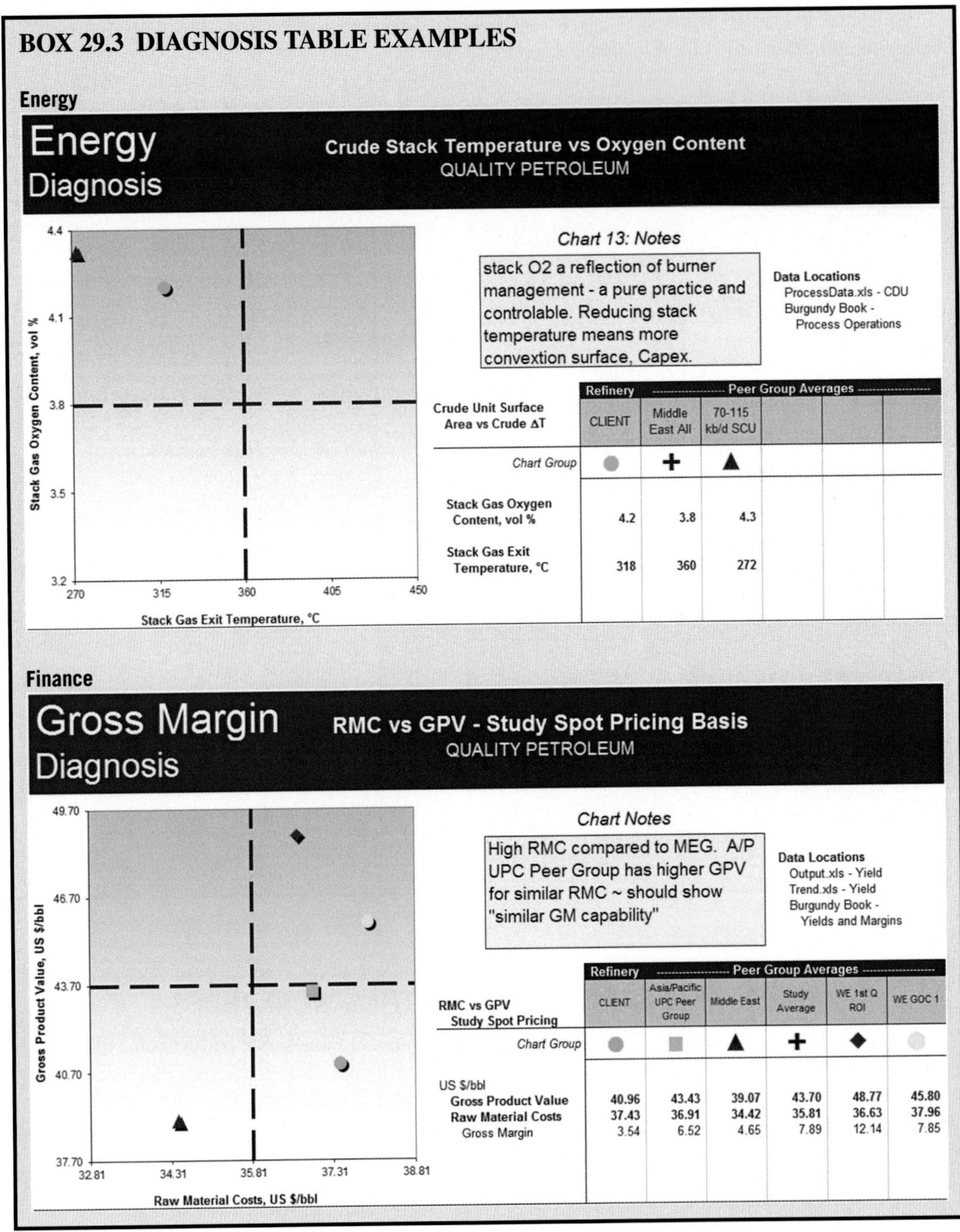

Crude Unit Surface Area vs Crude ΔT	Refinery	Peer Group Averages				
	CLIENT	Middle East All	70-115 kb/d SCU			
Chart Group	●	+	▲			
Stack Gas Oxygen Content, vol %	4.2	3.8	4.3			
Stack Gas Exit Temperature, °C	318	360	272			

RMC vs GPV Study Spot Pricing	Refinery	Peer Group Averages				
	CLIENT	Asia/Pacific UPC Peer Group	Middle East	Study Average	WE 1st Q ROI	WE GOC 1
Chart Group	●	■	▲	+	◆	●
US $/bbl						
Gross Product Value	40.96	43.43	39.07	43.70	48.77	45.80
Raw Material Costs	37.43	36.91	34.42	35.81	36.63	37.96
Gross Margin	3.54	6.52	4.65	7.89	12.14	7.85

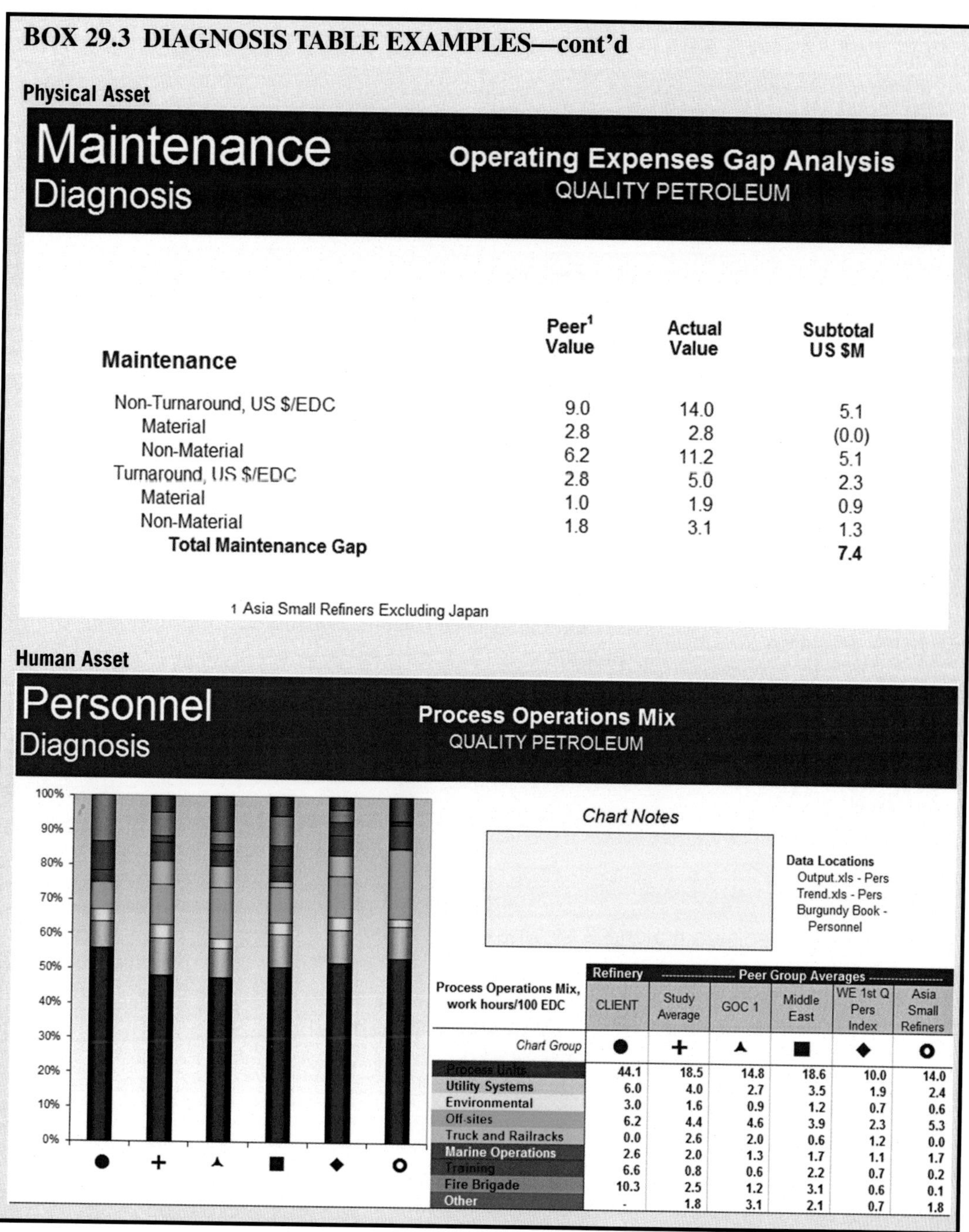

BOX 29.3 DIAGNOSIS TABLE EXAMPLES—cont'd

Physical Asset

Maintenance Diagnosis — Operating Expenses Gap Analysis — QUALITY PETROLEUM

Maintenance	Peer[1] Value	Actual Value	Subtotal US $M
Non-Turnaround, US $/EDC	9.0	14.0	5.1
Material	2.8	2.8	(0.0)
Non-Material	6.2	11.2	5.1
Turnaround, US $/EDC	2.8	5.0	2.3
Material	1.0	1.9	0.9
Non-Material	1.8	3.1	1.3
Total Maintenance Gap			**7.4**

1 Asia Small Refiners Excluding Japan

Human Asset

Personnel Diagnosis — Process Operations Mix — QUALITY PETROLEUM

Process Operations Mix, work hours/100 EDC	Refinery	Peer Group Averages				
	CLIENT	Study Average	GOC 1	Middle East	WE 1st Q Pers Index	Asia Small Refiners
Chart Group	●	+	⅄	■	◆	o
Process Units	44.1	18.5	14.8	18.6	10.0	14.0
Utility Systems	6.0	4.0	2.7	3.5	1.9	2.4
Environmental	3.0	1.6	0.9	1.2	0.7	0.6
Off-sites	6.2	4.4	4.6	3.9	2.3	5.3
Truck and Railracks	0.0	2.6	2.0	0.6	1.2	0.0
Marine Operations	2.6	2.0	1.3	1.7	1.1	1.7
Training	6.6	0.8	0.6	2.2	0.7	0.2
Fire Brigade	10.3	2.5	1.2	3.1	0.6	0.1
Other	-	1.8	3.1	2.1	0.7	1.8

After discussion, brainstorming is carried out to determine the root causes of each problem. An action plan is then produced from the brainstorming session using the return/effort matrix discussed in Section 28.7, Evaluation and Prioritization. Examples are shown in Box 29.4.

BOX 29.4 SAMPLE BRAINSTORMING AND ACTION PLAN

Maintenance
Brainstorming

Define Problem: Mechanical Availability

No.	Plant/Practice	Impact	Improvement Opportunity
1	Practice	4	Lack of decision making & slowness. Zero risk attitude (craftsmen)
2	Practice	4	24 hour TA working on critical path
3	Plant/Prac	4	Turnaround workscope control. Reduce TA jobs.
4	Plant/Prac	4	Material procurement - long lead time
5	Plant/Prac	4	Operational faults
6	Practice	4	Delay in releasing equipment
7	Practice	4	Slow work execution
8	Practice	4	KPI coordination
9	Practice	4	No radios and tools
10	Practice	4	No multi disciplinary teams to address outstanding problems
11	Practice	4	Lack of performance targets for maintenance. No cost of downtime

Maintenance
Sample Action Plan

No.	Action Item	Champion	Date
1	Lack of decision making & slowness. Zero risk attitude (craftsmen)		
2	24 hour TA working on critical path		
3	Turnaround workscope control. Reduce TA jobs.		
4	Material procurement - long lead time		
5	Operational faults		
6	Delay in releasing equipment		
7	Slow work execution		
8	KPI coordination		
9	No radios and tools		
10	No multi disciplinary teams to address outstanding problems		

29.7 BEST PRACTICE SCORECARDS

Some large companies and consultants have extensive sets of best practice scorecards that can be used to determine gaps. The methodology is shown in Fig. 29.1.

FIGURE 29.1

Best practice review methodology.

A typical scorecard format is shown in Fig. 29.2.

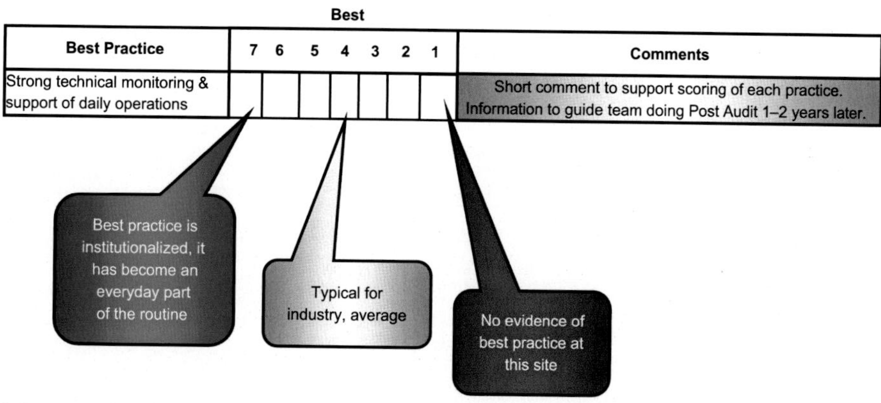

FIGURE 29.2

Scorecard format.

29.8 MAPPING USING THE 80:20 RULE

Mapping is a technique to identify all top KPIs together with actual values and targets, linked to current strategies and actions. This pictorial view is used to identify interrelationships and determine the focus on 20% of the actions to resolve 80% of the issues. The KPIs that are most likely to reflect the **High-Impact Actions (HIAs)** are to be identified with clear targets related to the **HIAs**.

Box 29.5 shows an example of an integrated gas plant performance assessment, using the 80:20 rule to focus on significant indicators and strategies.

BOX 29.5 MAPPING—INTEGRATED GAS PLANT PERFORMANCE ASSESSMENT

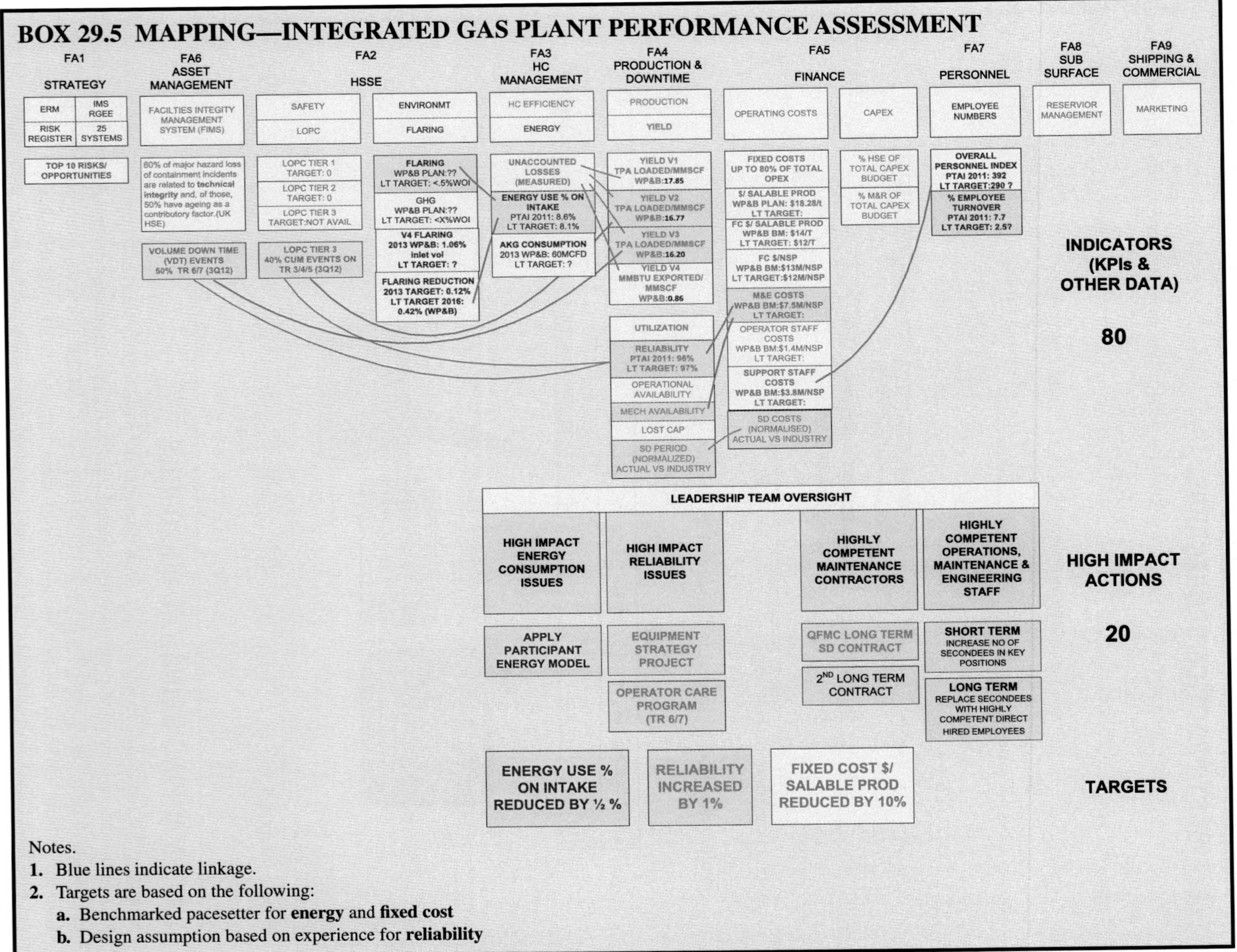

Notes.

1. Blue lines indicate linkage.
2. Targets are based on the following:
   a. Benchmarked pacesetter for **energy** and **fixed cost**
   b. Design assumption based on experience for **reliability**

Box 29.6 shows an example of a refinery performance assessment, using the 80:20 rule to focus on significant indicators and strategies. This was carried out after the refinery was benchmarked against other fuels refineries.

BOX 29.6 MAPPING—REFINERY PERFORMANCE ASSESSMENT

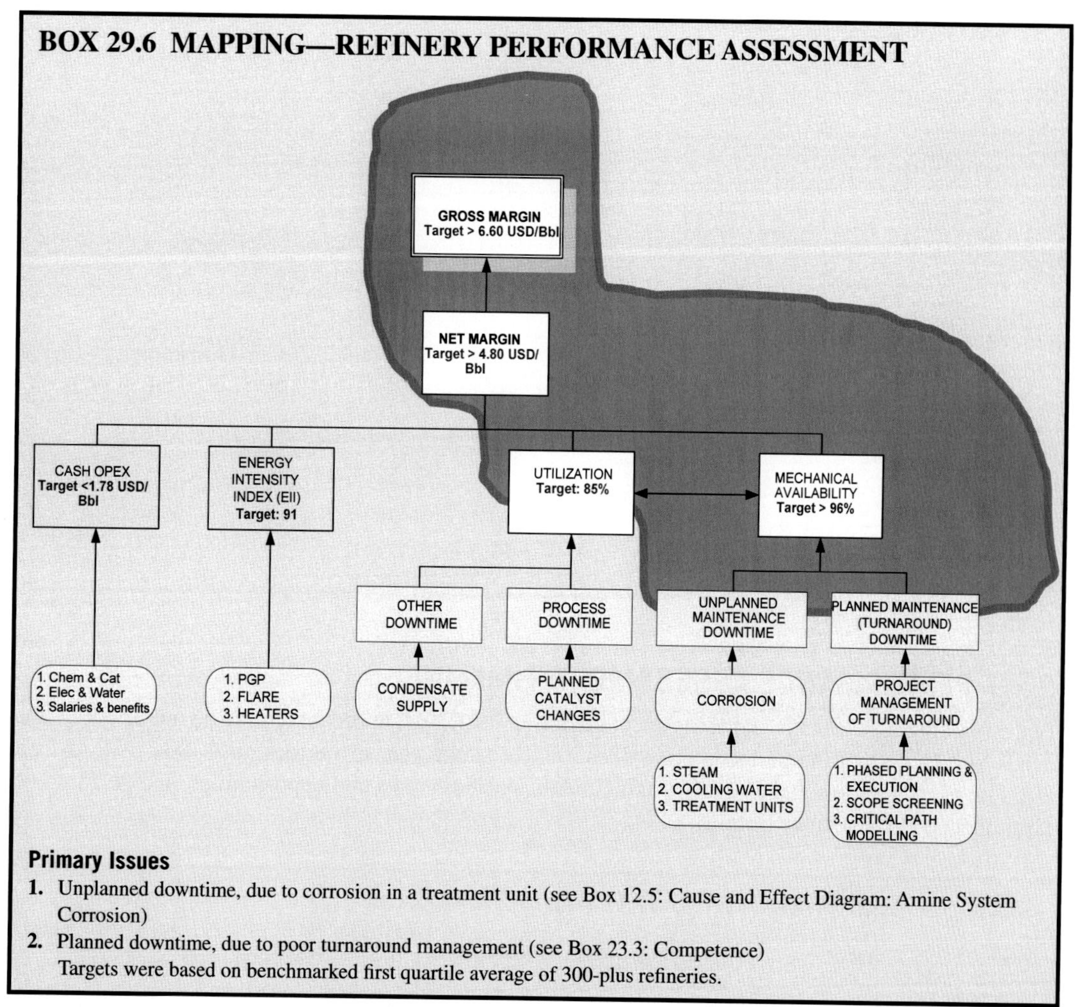

Primary Issues

1. Unplanned downtime, due to corrosion in a treatment unit (see Box 12.5: Cause and Effect Diagram: Amine System Corrosion)
2. Planned downtime, due to poor turnaround management (see Box 23.3: Competence)
   Targets were based on benchmarked first quartile average of 300-plus refineries.

Box 29.7 shows a shareholder joint assessment process for four affiliate companies with a total of 20-plus gas trains. EFPH categorization checklists and mapping, were used for this exercise. **HIAs** and targets were generated and performance health checker datasheets were completed. Box 29.5 shows the **HIAs** and targets for one of the affiliates.

BOX 29.7 INTEGRATED GAS PLANT GROUP ASSESSMENT

Assessment of four integrated gas plant (IGP) affiliates: P, Q, R, and S.

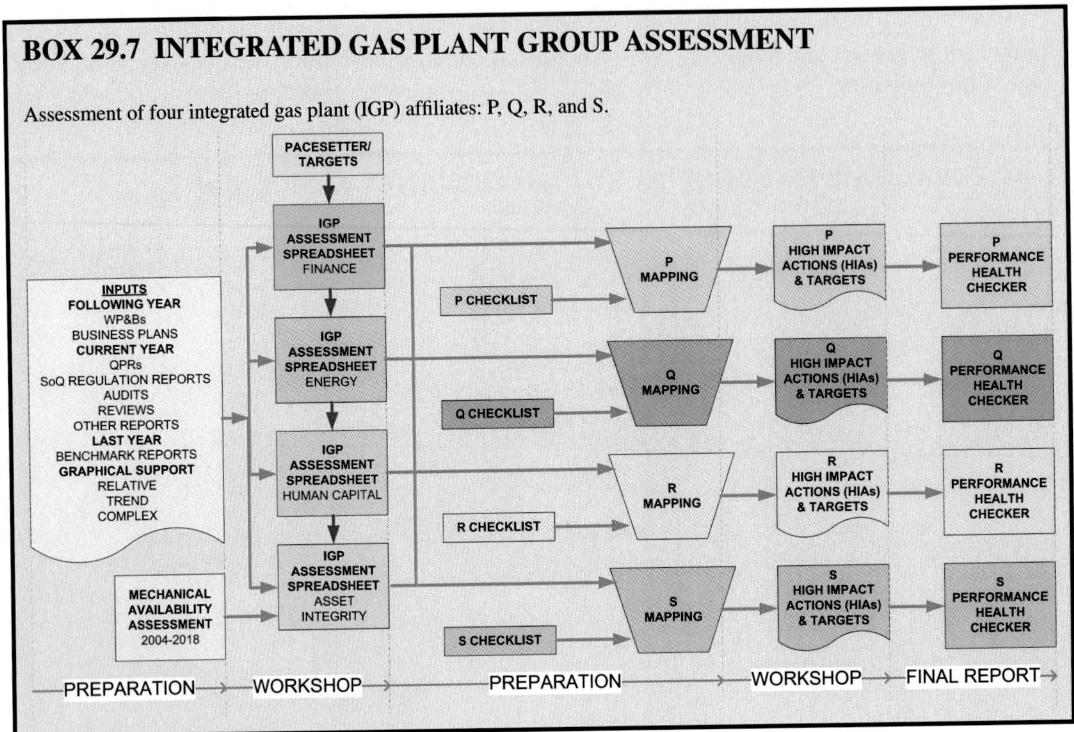

29.9 PERFORMANCE ASSESSMENT TIMELINE

Benchmarking assessments and assessments required for BP&B approval tend to be out of synch. One needs to be aware of the performance values relative to the year of performance. Box 29.8 gives an example of competitive and efficiency assessments, business risks and opportunities, and performance assessment and the relative timing.

BOX 29.8 ASSESSMENTS AND TIMING EXAMPLE

Year	YR –2	YR –1	YR 0	YR +1	YR +2
Business Unit					
Management team		Gas plant benchmark	**August: GP YR –1 BM report**	Gas plant benchmark	Gas plant benchmark
	Gas plant benchmark	**August: GP YR –2 BM report**	Gas plant benchmark	**Implement YR +1 BP&B**	
	Refinery benchmark	**August: Ref YR –2 BM report**	Refinery benchmark		Refinery benchmark

BOX 29.8 ASSESSMENTS AND TIMING EXAMPLE—cont'd

Year	YR −2	YR −1	YR 0	YR +1	YR +2
		Insurance review	**YR −1 Annual performance report**		
			YR 0 Monthly performance reports		
JV Board			**November–December: Approve BP&B for YR +1**		
Corporate					
Shareholder representative			**April: Letter of wishes to JV management**		
			August–October: Review BP&B for YR +1		
			August: Strategic risk assessment of new capital projects		
Board			**November–December: Approve BP&B for YR +1**		

Bold items used as part of YR +1 BP&B assessment and approval process.

29.10 ASSESSMENT BY CONSULTANTS

Consultants can speed up the process of assessment and decision-making on strategies and actions.

JURAN

This is discussed in Box 28.4, Juran Assessments.

SOLOMON ASSOCIATES[2]

Solomon Associates offers their proprietary assessment methodology as follows:

1. NCM[3]: performance improvement for existing plants
2. Q1Day1: maximizes return on investment on new capital investments

These are built on the foundation of Solomon Associates' Comparative Performance Benchmarking.

CELERANT[3]

Celerant endeavors to embed long-term **behavioral change** into the culture of its client organizations using a unique approach called Closework.

SHELL GLOBAL SOLUTIONS[4]

Shell Global Solutions offers the Shell Pacesetter Program outlined in Box 29.9.

BOX 29.9 SHELL PACESETTER PROGRAM

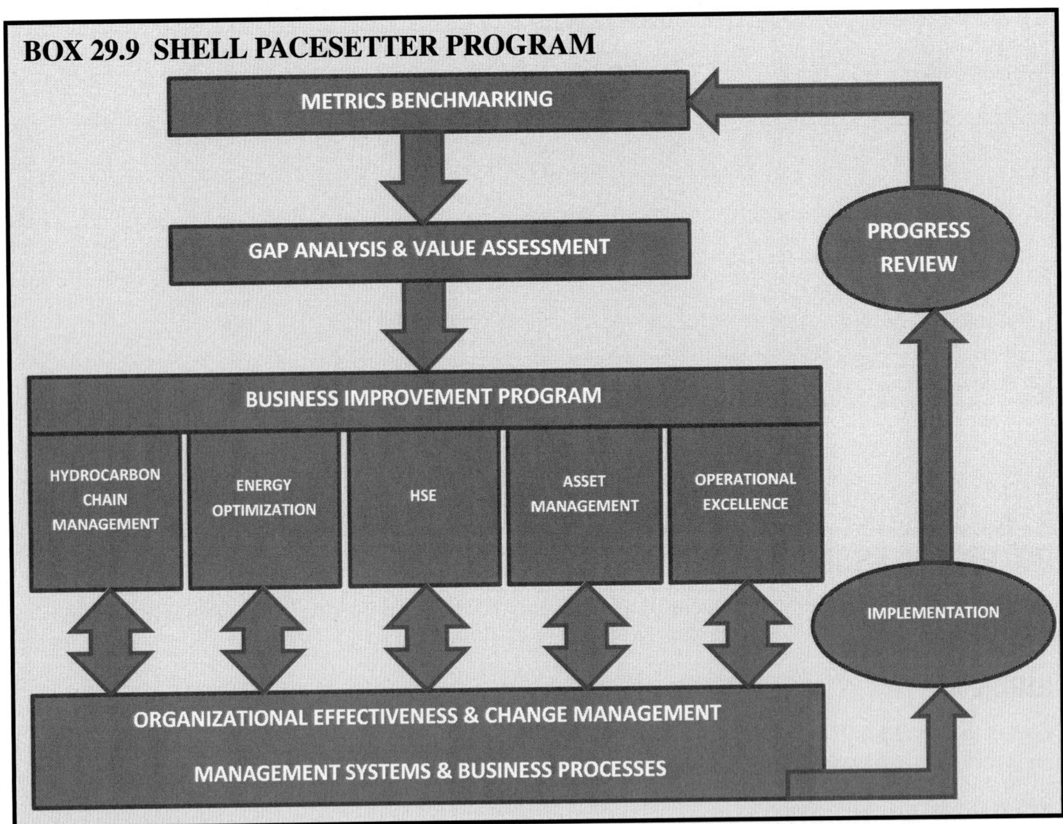

KBC[1]

For refineries, KBC offers the KBC Profit Improvement Program using their Petro-Sim refinery-wide model as described in Section 3.5, Process Simulation Models. This is outlined in Box 29.10.

BOX 29.10 KBC PROFIT IMPROVEMENT PROGRAM (PIP)

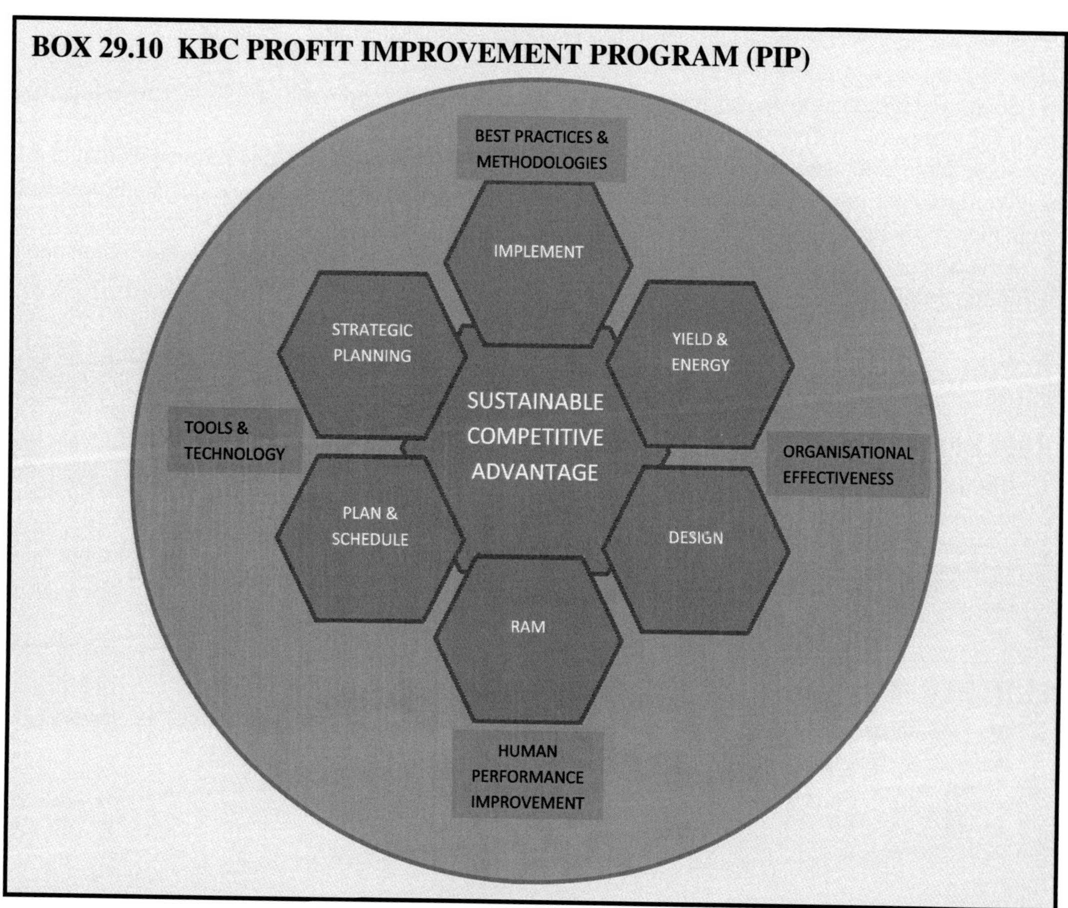

29.11 SUMMARY

Performance indicators need to be grouped into focus areas for each company. This forms the assessment template for "mapping" HIAs for the company.

Approaches to assessment could be part of the management review or directed through workshops.

Workshops focus on the few actions that will have the most impact (HIAs). Step or "breakthrough" change can be effected through the workshop route with identification of HIAs. Workshops entail "structured review techniques" using codes/standards as compliance requirements, checklists for commonly accepted practices and the experience of the participants. The assessment team needs to be multidisciplinary in order to draw on the experiences of each aspect of the business. The EFPH categories of data give focus for assessment. Risk assessment tools and techniques (using ISO 31000 process, root cause analysis software, risk assessment matrices, etc.) assess the real issues, impact, and probability.

Methods for harnessing the team's brainpower include "brainstorming" techniques, de Bono's Six Hats, and Buzan's Mind Mapping.

Shareholder budget proposal review can use this workshop approach at a higher level than that used for workshops within the business.

Some large companies and consultants have extensive sets of best practice scorecards that can be used to determine gaps. Consultants can speed up the process of assessment and decision-making on strategies and actions.

Assessment outcomes can take different forms—lists, spreadsheets, graphs or more transparent mapping. Mapping uses the 80:20 rule to show the performance indicators, gaps, and strategies required to fill the gaps.

REFERENCES

1. KBC (KBC Advanced Technologies Limited was acquired by Yokogawa Electric Corporation). www.kbcat.com.
2. Solomon Associates. https://www.solomononline.com/.
3. Celerant (acquired by Hitachi Jan 2013). http://www.hitachi.com/.
4. Shell Global Solutions (SGS). http://www.shell.com/business-customers/global-solutions.html.
5. De Bono E. *Six thinking hats.* Penguin; 2000.

FURTHER READING

1. IHS Chemical Market Advisory Service: Global Plastics & Polymers. https://www.ihs.com/products/chemical-market-plastics-polymers-global.html.

CHAPTER

STRATEGIES AND ACTIONS

30

30.1 INTRODUCTION

Assuming that the team has now completed the assessment and knows where to focus for improvement, the primary question now is:

> **How does one get management and the actioners involved to determine targets and develop strategies to achieve these targets?**

This chapter describes how to agree on targets and strategies and how to get buy-in.

30.2 QUESTIONS TO ASSESS ACTION BASED ON COMPETITIVE POSITION

The following are helpful questions that can be used to assess suitable strategies and actions relative to the company's competitive position:

- Have you reached the benchmarked "pacesetter" target?
- If the answer is "yes," then the focus is on maintaining this position. Others are continually improving, so you need to set a "stretch target" and **stay with your current strategy**.
- If the first answer is "no," then the question is: "Is the benchmarked pacesetter target within reach in the near future?"
- If the answer is "yes," you need to attempt a strategy for step improvement to achieve this target in the near future.
- If the answer is "no," you will need two targets: a near-term target without necessarily going through a step change and a long-term target that will entail a structured "change management" process. This is where a detailed strategy is required.

This is summarized in Fig. 30.1.

Performance Management for the Oil, Gas, and Process Industries. http://dx.doi.org/10.1016/B978-0-12-810446-0.00030-X

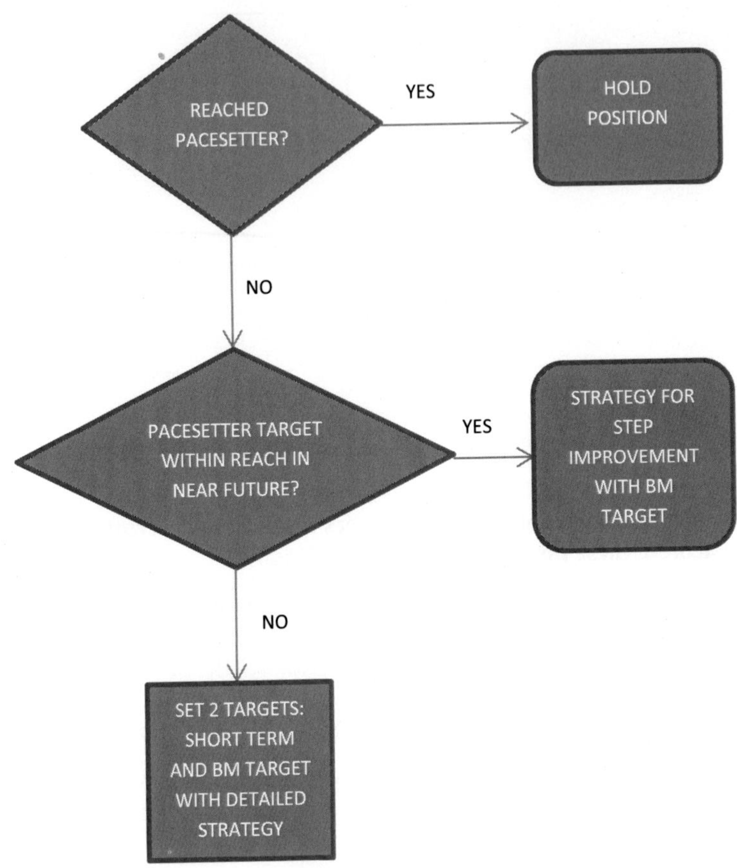

FIGURE 30.1

Action decision process.

Box 30.1 gives a case of a refinery not being able to retain "pacesetter" status.

BOX 30.1 HOLDING ON TO "PACESETTER" STATUS

Solomon Associates' definition of "pacesetter" is:

> *Achieve first or second quartile accomplishments in all the major performance areas for the entire trend period of 3 consecutive studies i.e. 6 years.*

After many years, a UK refinery eventually achieved "pacesetter" status. Management then decided to change some strategies to try to keep ahead. In subsequent benchmarking exercises, the refinery lost its "pacesetter" status.

Comment

Was this due to meddling with something that was working well?

30.3 TARGET SETTING

Short- and long-term target-setting requires buy-in from those who will "make it happen." Target-setting must be based on a thorough review of those factors that influence the indicator being addressed, as follows:

1. Consider past trends.
2. How well do others do?
3. What are the limits?
4. Understand interaction with other indicators.

The key to establishment of targets is realism. Targets need to be reasonable in view of the company's resources, expertise and partnerships. Overly ambitious targets lead to disappointment when they are not met.

Targets are to be realistic and supported by detailed strategies to ensure a reasonable chance of achievement. Targets are generally based on one or both of the following:

1. Business Plan and Budget (BP&B)—short term
2. Improvement as per a Business Improvement Plan (BIP) —long term

Achieving targets entails the three risk control mechanisms as introduced in Section 12.3, Three Ps. A recap is as follows:

1. Processes/Systems: integrated management
   - The physical Business Systems and Processes required to manage the business
2. People: organizational effectiveness
   - The motivation of people to be highly productive and open to change
3. Performance monitoring
   - The physical monitoring of performance based on facts

However, target setting has to be achieved within certain guidelines and restraints.

Some guidelines and restraints are:

- Guideline: Benchmarking "It is difficult to know what to improve if one doesn't know where one is relative to competitors."
  - Information on suitable **long-term** targets is generally obtained from benchmarking exercises.
- Restraint: risk appetite that is set by the board of directors/management committee
  - If the board of directors chooses to have a higher risk appetite for a new technology project, cost and completion time would have a higher risk of meeting planned values. Production targets may also not be achieved.

Box 33.13, Investment Risk – New Technology, gives an example of risk appetite.

Short-term targets, on the road to achieving **long-term** targets, depend on the detailed strategies that are adopted. **Short-term** targets are generally based on the previous year's performance.

Key steps are:

1. Agree on top Key Performance Indicators (KPIs) (see Chapter 16: KPI Selection Guidelines, for details).
2. Benchmark (see Part 6: Benchmarking, for details).

3. Roll down the top KPIs.
4. Set targets.
5. Agree on strategies and plans for improvement.

30.4 STRATEGIES: PLANS AND ACTIONS

Strategies could involve change management, which is discussed in detail later in Section 30.6.

SHORT- AND LONG-TERM GROUPING

These are grouped as follows:

1. BP&Bs
   a. Strategic: long term—5 to 7 years
   b. Tactical: annual
2. Performance actions
   a. Operational: daily, monthly, quarterly

   (2) is the detailed planning and implementation of **(1)**.

TYPES

The following are possible types of BIPs:

- Operational
  - Management Improvement Projects (MIPs) and Process Improvement Projects (PIPs)
    - These have a clear start and end with a **step** business process **improvement** as a result of implementation.
  - System Improvement Programs (SIPs)
    - **SIP**s are ongoing **continuous improvement** initiatives.
  - Minor capital: Health, Safety & Environment, regulatory and “replacement in kind”
- Investment
  - Major capital
    - Return on Investment: production enhancement
    - Regulatory: environmental emission reduction

Box 30.2 gives examples of reports on the progress of MIPs, PIPs, and SIPs.

BOX 30.2 EXAMPLE REFINERY BIPS

Categories

1. MIPs
2. PIPs

   MIPs and PIPs have a clear start and end with a **step** business process **improvement** as a result of implementation.
3. SIPs

   SIPs are ongoing **continuous improvement** initiatives.

Management Improvement Projects

The following are currently in progress:

KPI No.	Reg No./Cat	Title/ Description	Sponsor/ Owner/ Actioner	Target	Status	Traffic Light	Comment
8,9	5 MIP	Implement a shutdown planning system	MC/MCP/ MCL3/ITB	4Q06	80%	RED	Revised completion 2nd Q07. Reports to be completed. Project critical for 2008 Turnaround.
8,9	6 MIP	Develop inspection management and risk-based inspection systems	FM/FMI/ FMI3/ITBS5	2Q07	Software installation complete. Consultant contract in progress. (GTC 05/130/ MCO)	RED	Software contractual issues to be resolved. Consultant contract with BV extended to 4Q07.

Process Improvement Projects

The following are currently in progress:

KPI No.	Reg No./Typ	Title/ Description	Sponsor/ Owner/ Actioner	Target	Status	Traffic Light	Comment
EI	12 PIP	Balco unit effluent	TM/TMS/ TMS1	1Q06	PAL filters on order. Agitator gearbox being replaced.	RED	Revised target 2Q09.
6,7, EI	13 PIP	LPG recovery from FG system	OM/OMT/ TMS/ TMS4	4Q08	FEED in progress.	AMBER	FEED completed 1Q08. Project completion now expected 1Q09.

Continued

BOX 30.2 EXAMPLE REFINERY BIPS—cont'd

System Improvement Programs

The following are currently in progress:

KPI No.	Reg No./Typ	Title/ Description	Sponsor/ Owner/ Actioner	Targets	Status	Traffic Light	Comment
7,3	15 SIP	15-MW power generation unit operation	OM/OMU1/ MCL/MCL12/ MCE2	Operate at 12 MW	Turbine water washing complete.	RED	Outstanding problems to be resolved. Cross-functional action team required.
6	16 SIP	In-line custody transfer meters	TM/TMP/ MCE1	All custody transfer meters calibrated and in use		RED	Outstanding problems to be resolved.

IMPLEMENTATION

Measurement of progress on implementation should be automated as far as possible, so as to eliminate data transfer and calculation errors. However, verification of actual work done is essential. Implementation could entail a number of parallel strategies including MIPs, PIPs, and SIPs. Box 30.3 gives an example for flare reduction.

BOX 30.3 FLARE REDUCTION TARGET AND STRATEGIES

A list of a shareholder's wishes for inclusion in the following year's gas plant operating company's business plan is included in Box 35.2, Shareholder's "Letter of Wishes".

One of the wishes is as follows:

- Flaring percent on intake is to be reduced by 0.6%.
    - Target: 0.3% weight on intake (woi)

The gas plant operating company included the following in its business plan:

1. Implement a boil off gas (BOG) project to reliquefy BOG during liquefied natural gas (LNG) ship loading so as to prevent BOG flaring (0.3% of reduction).
2. Review LNG start-up procedure reducing cool-down time so as to reduce start-up flaring (0.2% of reduction).
3. Implement an operational flare reduction program (0.1% of reduction).

Comment:

- *Item 1 is a PIP.*
- *Item 2 is a MIP.*
- *Item 3 is a SIP.*

30.5 MECHANICAL AVAILABILITY DRILL-DOWN

Benchmarking can show that a target is achievable, but the question is: "how does one get there?" **Mechanical Availability** is one of the most important KPIs. The establishment of targets for this indicator needs to be based on a long-term detailed strategy. Previous discussion on Mechanical Availability is as follows:

- Section 17.4, Utilization and Availability
- Section 23.7, Strategic Planning and Run Length determination
- Section 29.4, Assessment Methods

Improvement in **Mechanical Availability** is based on **optimizing the full operating cycle**, which entails turnaround time and on-stream time.

Fig. 30.2 depicts the **full operating cycle**.

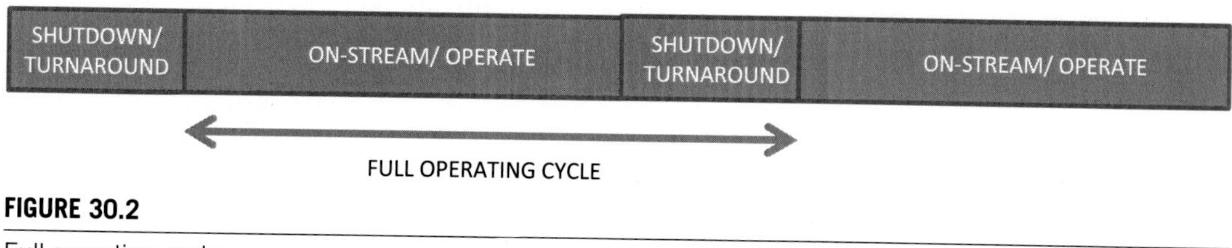

FIGURE 30.2

Full operating cycle.

The on-stream period tends to be long: often in the order of 5 years, whereas the (planned) shutdown or turnaround tends to be short and intensive—planned by the hour and taking just weeks. As per Solomon, **annualized Mechanical Availability (MA)** is generally calculated as follows:

MA = (days in current year – **TADD** – **RMDD**) * 100/days in current year

where:

TADD (Turnaround Annualized Down Days): annualized down days taken from the last reported turnaround. Annualized days are calculated by dividing the reported turnaround days by the number of equivalent years between the last two reported turnarounds.

TADD = days down × 365.25/interval between turnarounds in days.

RMDD (Routine Maintenance Down Days): the annualized down days for unplanned down days that are not accounted for in the turnaround data. The actual days for the last 2 years are averaged for a year.

So it is important to review the last on-stream period and the unplanned days in that period to identify issues to be addressed to reduce unplanned down days. Targets for **reliability** and predicted **RMDD** can then be determined based on strategies to be adopted to reduce unplanned down days.

Reliability = (total days in a year – unplanned maintenance days) * 100/total days in a year.

Then, from workshopping a predicted run length based on Section 23.7, Strategic Planning and Run Length determination, and modeling the next turnaround on the critical path software, predicted **TADD** can be determined.

Predicted **MA** can then be determined. This can then be compared to the "pacesetter" value and possibly be further adjusted, keeping in mind that strategies for improvement have to be realistic.

The above is not easy in refineries, as each type of process unit has different turnaround times, run lengths, and reliability. However, the primary process units such as crude and catalytic cracking units will dictate to the rest. The use of Solomon Associates" Profile II model to carry out "what ifs" could help in determining suitable targets.

Enhanced risk-based inspection (RBI), reliability-centered maintenance (RCM), and turnaround management are key requirements for improvement. Certification to ISO 55001 "Asset Management" would ensure continuous improvement.

"Rules of thumb" as discussed in Section 28.8, Quick Assessments, could help guide the process. Some relevant "rules of thumb" are listed next:

- Turnarounds duration
  - Crude unit: 28 days
- Turnarounds interval
  - Gas trains (turbines): 5.5 years
- Reliability
  - Liquefied natural gas (LNG) mega-train design basis: 98% reliability
    - Seven unplanned down days per year

Box 30.4 gives an example of lower-level targets rolling up to achieve long-term targets for Mechanical Availability *and* Maintenance (cost) Index.

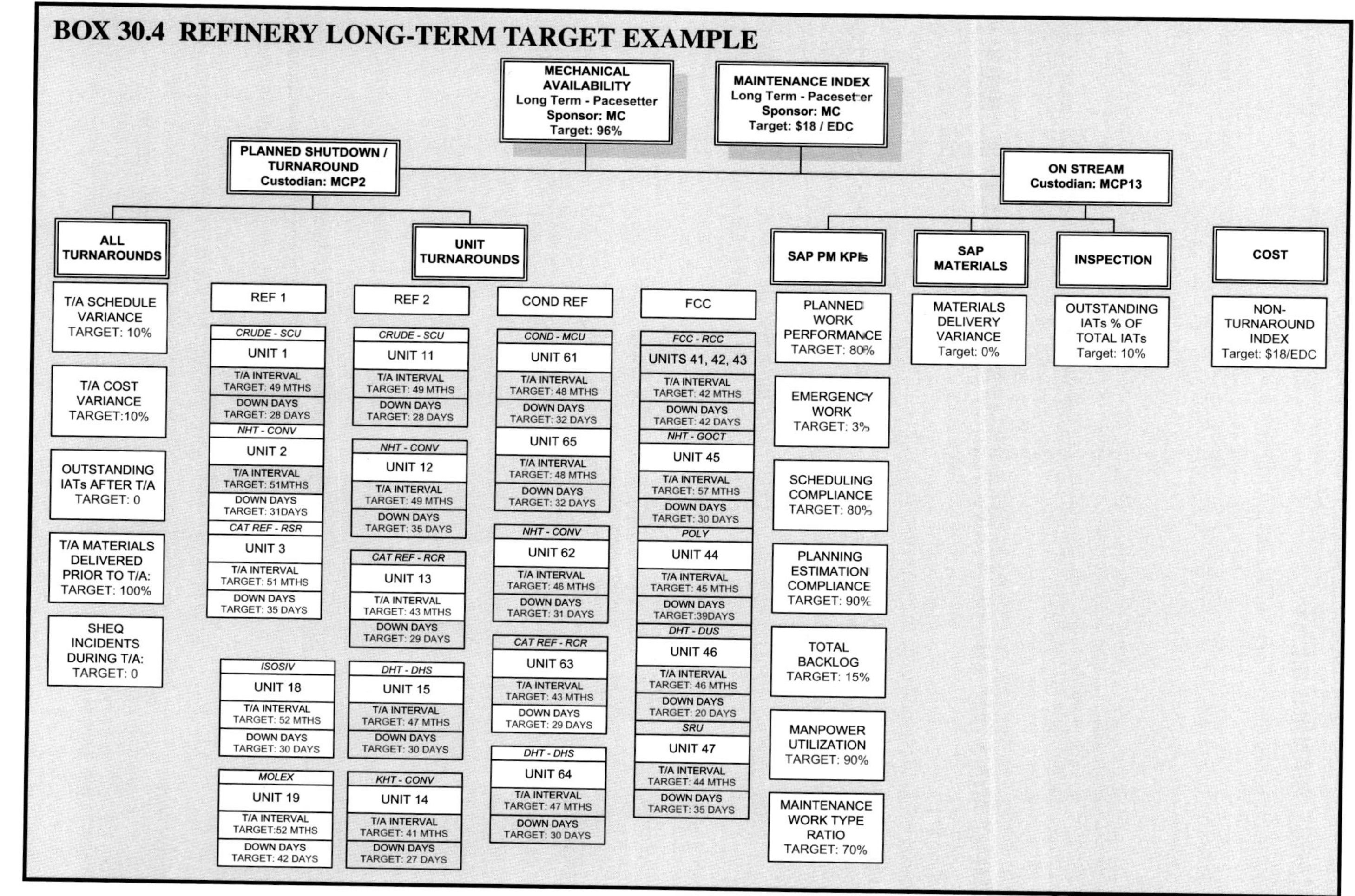
BOX 30.4 REFINERY LONG-TERM TARGET EXAMPLE
MECHANICAL AVAILABILITY Long Term - Pacesetter Sponsor: MC Target: 96%
MAINTENANCE INDEX Long Term - Pacesetter Sponsor: MC Target: $18 / EDC
PLANNED SHUTDOWN / TURNAROUND Custodian: MCP2
ON STREAM Custodian: MCP13
ALL TURNAROUNDS
T/A SCHEDULE VARIANCE TARGET: 10%
T/A COST VARIANCE TARGET:10%
OUTSTANDING IATs AFTER T/A TARGET: 0
T/A MATERIALS DELIVERED PRIOR TO T/A: TARGET: 100%
SHEQ INCIDENTS DURING T/A: TARGET: 0
UNIT TURNAROUNDS
REF 1
CRUDE - SCU
UNIT 1
T/A INTERVAL TARGET: 49 MTHS
DOWN DAYS TARGET: 28 DAYS
NHT - CONV
UNIT 2
T/A INTERVAL TARGET: 51MTHS
DOWN DAYS TARGET: 31DAYS
CAT REF - RSR
UNIT 3
T/A INTERVAL TARGET: 51 MTHS
DOWN DAYS TARGET: 35 DAYS
ISOSIV
UNIT 18
T/A INTERVAL TARGET: 52 MTHS
DOWN DAYS TARGET: 30 DAYS
MOLEX
UNIT 19
T/A INTERVAL TARGET:52 MTHS
DOWN DAYS TARGET: 42 DAYS
REF 2
CRUDE - SCU
UNIT 11
T/A INTERVAL TARGET: 49 MTHS
DOWN DAYS TARGET: 28 DAYS
NHT - CONV
UNIT 12
T/A INTERVAL TARGET: 49 MTHS
DOWN DAYS TARGET: 35 DAYS
CAT REF - RCR
UNIT 13
T/A INTERVAL TARGET: 43 MTHS
DOWN DAYS TARGET: 29 DAYS
DHT - DHS
UNIT 15
T/A INTERVAL TARGET: 47 MTHS
DOWN DAYS TARGET: 30 DAYS
KHT - CONV
UNIT 14
T/A INTERVAL TARGET: 41 MTHS
DOWN DAYS TARGET: 27 DAYS
COND REF
COND - MCU
UNIT 61
T/A INTERVAL TARGET: 48 MTHS
DOWN DAYS TARGET: 32 DAYS
UNIT 65
T/A INTERVAL TARGET: 48 MTHS
DOWN DAYS TARGET: 32 DAYS
NHT - CONV
UNIT 62
T/A INTERVAL TARGET: 46 MTHS
DOWN DAYS TARGET: 31 DAYS
CAT REF - RCR
UNIT 63
T/A INTERVAL TARGET: 43 MTHS
DOWN DAYS TARGET: 29 DAYS
DHT - DHS
UNIT 64
T/A INTERVAL TARGET: 47 MTHS
DOWN DAYS TARGET: 30 DAYS
FCC
FCC - RCC
UNITS 41, 42, 43
T/A INTERVAL TARGET: 42 MTHS
DOWN DAYS TARGET: 42 DAYS
NHT - GOCT
UNIT 45
T/A INTERVAL TARGET: 57 MTHS
DOWN DAYS TARGET: 30 DAYS
POLY
UNIT 44
T/A INTERVAL TARGET: 45 MTHS
DOWN DAYS TARGET:39DAYS
DHT - DUS
UNIT 46
T/A INTERVAL TARGET: 46 MTHS
DOWN DAYS TARGET: 20 DAYS
SRU
UNIT 47
T/A INTERVAL TARGET: 44 MTHS
DOWN DAYS TARGET: 35 DAYS
SAP PM KPIs
PLANNED WORK PERFORMANCE TARGET: 80%
EMERGENCY WORK TARGET: 3%
SCHEDULING COMPLIANCE TARGET: 80%
PLANNING ESTIMATION COMPLIANCE TARGET: 90%
TOTAL BACKLOG TARGET: 15%
MANPOWER UTILIZATION TARGET: 90%
MAINTENANCE WORK TYPE RATIO TARGET: 70%
SAP MATERIALS
MATERIALS DELIVERY VARIANCE Target: 0%
INSPECTION
OUTSTANDING IATs % OF TOTAL IATs Target: 10%
COST
NON-TURNAROUND INDEX Target: $18/EDC

30.6 CHANGE MANAGEMENT

An Economist intelligence unit survey[1] indicates the following:

> The main reason that change programs do not succeed appear to be a lack of clearly defined or achievable milestones and objectives.
>
> The second most commonly cited reason for failure is a lack of commitment from senior managers.

Change management needs to be project managed. Some pointers are described in this paragraph.

MANAGEMENT AND STAFF BUY-IN: ACKNOWLEDGMENT OF A GAP AND DECISION TO TAKE ACTION

On presentation of an identified gap, the first response is to challenge the source of the data. Thus, the derivation of the gap has to be clearly documented and based on an objective assessment of facts. Chapter 26, Benchmarking Data Verification, describes checks to ensure the recorded data are correct. Also, Chapter 36, Hydrocarbon Accounting, describes details to ensure reconciliation of hydrocarbon measurement.

Bridging a gap often entails a paradigm shift. One might think that there is no chance of improvement based on current practices. However, with benchmarking, one is made aware of others achieving a higher level of performance. This provokes the question:

"What are they doing that we are not doing?"

WHY CHANGE?

Eight major marketplace issues dictate that we must change or perish:

1. Globalization: the world is getting smaller and competition can emerge from anywhere in the world
2. Customer focus—"the customer is king": this has to be continuously reinforced within organizations
3. Product/service innovation: this is continuous and one has to be at the leading edge of innovation to keep in business
4. Profitability: this is the bottom line for shareholders to invest in your company
5. Quality, time, and cost competitiveness: continually assesses your competition and **keep ahead**
6. People and organization—the heart of your company: highly motivated people in a streamlined organization are required
7. Integration: this is necessary to blend the previous corporate cultures and Systems/Processes into one cohesive force
8. Regulations: if you don't self-regulate your industry, the government will surely step in

Every business has a need to improve its performance.

STRUCTURED APPROACH

Organizations need to manage change effectively, utilizing a structured approach to increase the probability of success.

Question

- Why and when does change implementation warrant the use of a structured approach?

Answers

- When the change is MAJOR
- When there is high COST of implementation **failure**
- When there is high RISK that certain human factors could result in implementation **failure**

A structured approach to step change is required to prevent regressing to the old way of doing things. Kotter's eight-step approach[2] is the generally accepted way to ensure this. These are:

1. Set the stage
   a. Create a sense of urgency
   b. Pull together the guiding team
2. Decide what to do
   a. Develop the change vision and strategy
3. Make it happen
   a. Communicate for understanding and buy-in
   b. Empower others to act
   c. Produce short-term wins
   d. Don't let up
4. Make it stick
   a. Create a new culture

THE CHALLENGES

- **When** to change?
- **What** to change?
- **How** to change?

When

Timing needs to be carefully determined to minimize total disruptive impact.

What

The mountain analogy best describes the process:

> To get a new improved state, one has to climb the mountain—it takes additional effort during the transition period to attain the new state. An example is learning a new skill—initially it is difficult but gets easier with practice.

Should you tackle a series of smaller mountains or one big mountain? In other words, do you effect change in stages or all at once?

How

Ten rules to guide you through change:

1. You have to believe the change is important and valuable.
2. A vision of the future is to be "seen" and understood by all and internalized.
3. Existing and potential barriers to change are to be identified and removed.
4. The entire organization is behind the strategy to achieve the vision.
5. Leaders must set examples and line managers must lead.
6. Training in a participative way is essential.
7. Measurement systems need to be established.
8. Continuous feedback to all is vital.
9. Coaching through the transition state is essential.
10. Recognition and reward systems are necessary to motivate people.

BASIC PROCESS

- Initially "awareness building" is carried out. An environment assessment and readiness assessment is completed.
- "Change" support is required throughout the transition state.
- Best practices and benchmarking are required to measure progress toward the new state, keeping the vision in mind all the time.

THE FOCUS

The focus of the change process must be on the following:

- The people
- The vision and goals
- The benefits
- The measurement of success

KEY ROLES IN THE CHANGE PROCESS

- The **stakeholders**—all the people who will be affected by the changes
- The **sponsors**—those who can authorize the changes
- The **change agents**—the key people who have the sponsors' and stakeholders' confidence to implement the change in a way that results in widespread buy-in and support—the bridge builders in the mountain analogy
- And the champion of change—the **chief executive**

The **vision and goals** need to be defined by the **chief executive** and **senior management**. The **benefits** will flow from the **goals**.

The **measures of success** need to be established to show progress toward the **goals**.

Managing change successfully can be a real competitive advantage. Project managing change provides the structure and discipline to help ensure the most complete, integrated, and effective solution possible in the shortest time.

THE FIVE STAGE PROCESS: AN INDIVIDUAL'S EXPERIENCES

An individual going through change generally experiences the following five stages of emotion:

1. Denial
2. Anger
3. Negotiation
4. Depression
5. Acceptance

Change agents need to understand these emotional stages that people go through. These are expanded as follows in respect to performance improvement:

1. Denial

An individual might:

1. question the **data** presented;
2. question the need for change and the **benefits** of change.

Data have to be:

1. factual;
2. benchmarked against similar operations.

Benefits of change:

1. have to be clear;
2. should make work easier for the individual.

2. Anger

An individual might:

1. attempt to **blame** someone or something else if the data are correct;
2. try to "**shoot** the messenger."

Blame:

1. inhibits the ability to learn what is really causing the problem;
2. prevents one from doing anything meaningful about solving the problem.

3. Bargaining

Individuals try to:

1. get the best deal out of the situation;
2. minimize their contribution.

4. Depression

One thinks that the situation is so bad that one cannot do anything about it.

5. Acceptance

An individual:

1. realizes that the change will improve the situation or solve the problem;
2. accepts the necessary changes;
3. works to make the change process easier to arrive at the improved state.

This is depicted in Fig. 30.3.

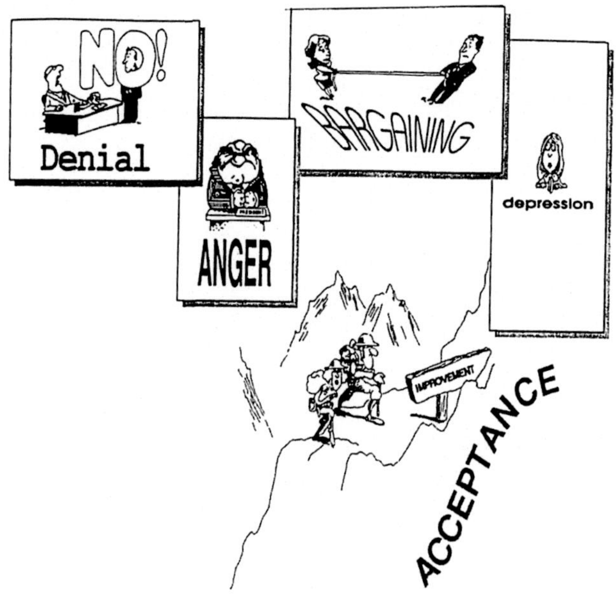

FIGURE 30.3

Five stages of emotion.

VARIOUS INDICATIONS OF RESISTANCE

These include:

1. "not invented here" syndrome, where a person thinks that nothing can be better than what he/she has done;
2. anger;
3. challenging the credentials of the data analyst;
4. attempting to "shoot the messenger";
5. questioning the validity of the data;
6. "nit-picking" minor errors in the data.

Typical character types are:

- **saboteurs**: those who actively intervene to prevent change;
- **pacifists**: those who do nothing;
- **fence sitters**: those who will wait to see which direction the process is going and then go with the crowd;
- **nikes**: those who "just do it."

MOVING FROM A BLAME CULTURE

A blame culture is where a problem is always somebody else's or attributed to an inanimate object. Rather than blaming and pointing fingers, participants in the change process need to be encouraged to sit on the same side of the table and look at the problem together. In other words, the group must be united in trying to find solutions to the problem together, rather than finding problems with each other.

Fig. 30.4 depicts the process of moving from perception to fact and then pursuing joint problem-solving activities, strategies, and action plans to achieve the required objectives.

Box 39.5, IOC Survey, identifies a significant gap between management and worker perceptions.

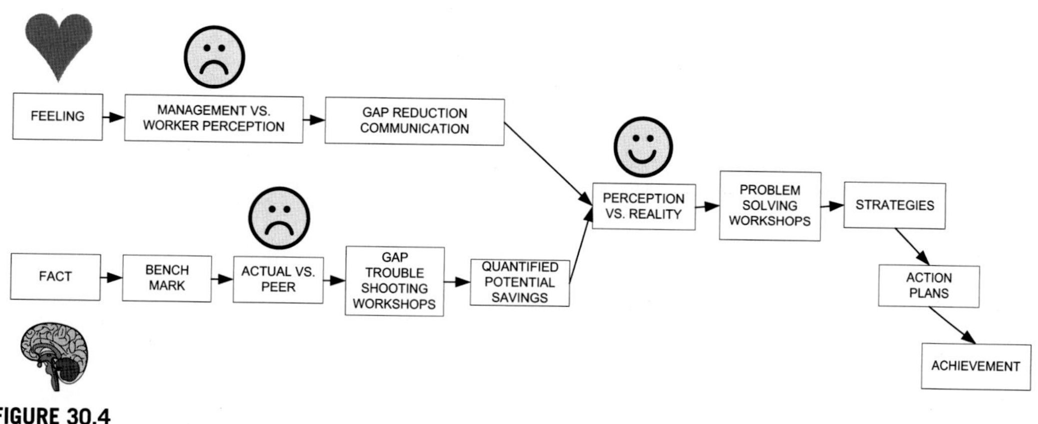

FIGURE 30.4

Moving from blame culture.

Behavior-based safety reduces the effect of the blame culture. Section 20.5, Process Safety Management (PSM) and Safety Culture, discusses this further.

LEAP-FROGGING

Advancement in technology can benefit late starters who do not have the "baggage" of the old technology.

Example

A polyethylene terephthalate production plant built in the 1970s uses a batch process for polymerization and continues to operate. (SA Nylon Spinners in South Africa)
A polyethylene terephthalate production plant built in the 1990s uses a continuous process for polymerization, thus achieving much higher production rates, at a lower cost, than the old batch technology. (Arabian Industrial Fibers Company in Saudi Arabia)

MAKING CHANGE STICK

The actioners need to agree with the strategies and targets to be achieved. This requires agreement on a short list of strategies and timing for implementation. This is to align with agreed-on short- and long-term targets for relevant KPIs.

The two main ways of making change stick are:

1. alignment with Company Vision;
2. integration with Management Systems.

Box 30.5 gives an example of change management in a power generation company.

BOX 30.5 STRATEGY WORKSHOPS AT ESB POWER GENERATION[3]

ESB Power Generation in Ireland was confronted with possible future privatization. Management decided to examine the future strategy for the company through a series of workshops involving different levels in the organization.

The process began with a 2-day top team workshop that addressed a series of questions relating to macroeconomic forces, competition, competitive advantage, and strategic options.

This was followed by two workshops with the next level of management and functional experts. These reviewed the same subject matter of the top workshops in order to establish whether they would come to similar conclusions, which they did.

Once there was buy-in at the above two levels, a series of workshops at levels varying from senior executives to supervisors in the production units were carried out to examine a change in culture and other changes needed to be carried out and the priorities for action.

Outcome

Forty years later, ESB is a vibrant international power company active in bringing power to over 120 countries.

30.7 VISUALIZATION OF TARGETS

Communication is essential to motivate everyone to achieve their contribution to the higher-level targets. Visualization helps. Various graphs are used as shown in Chapter 31, Reporting.

Box 30.6 gives an example (from before the "Internet age").

BOX 30.6 REFINERY PACESETTER TARGETS VISUALIZED

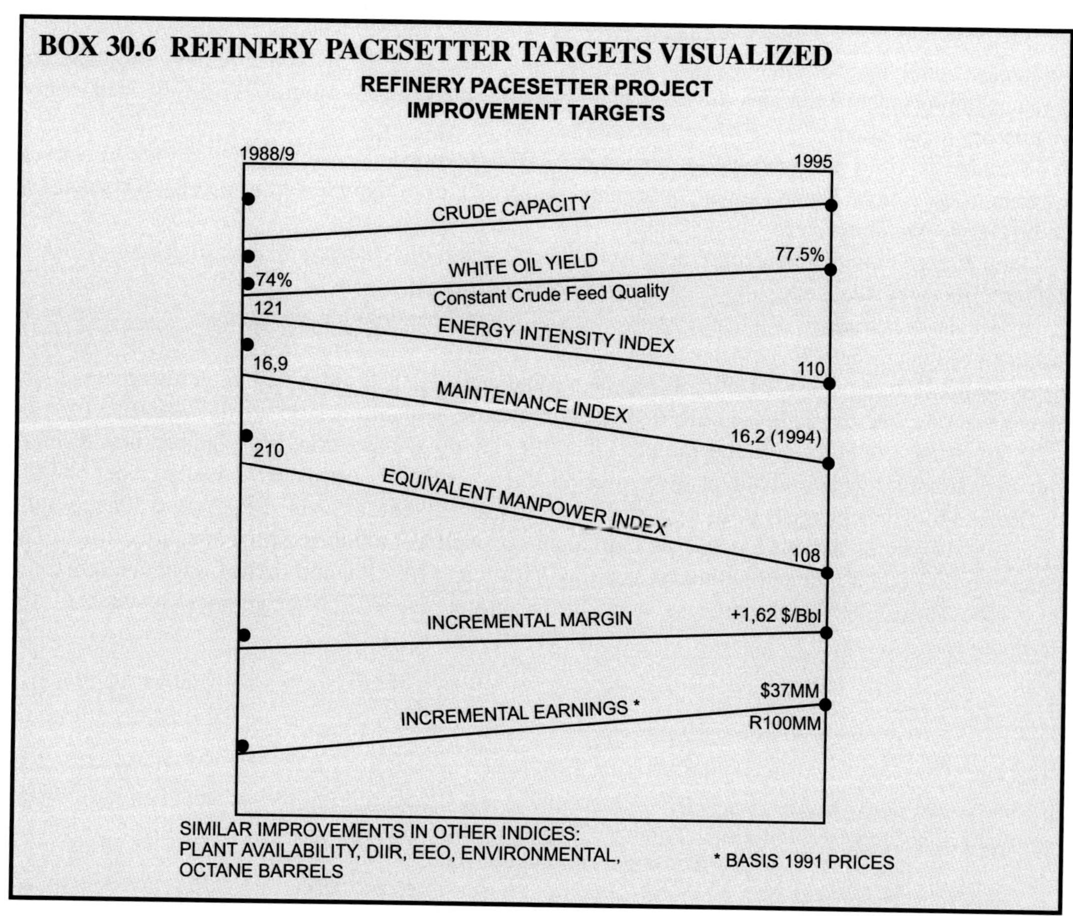

30.8 SUMMARY

Target setting needs the active participation of those who will "make it happen." Targets are either budget related or performance improvement (nonbudget) related, or they could be a combination of the two. Targets are generally categorized as short term (1 year) and long term (period to be defined for the particular target).

The interplay of three risk control mechanisms (people, process, and performance) could help to determine a target.

At the outset, the top company KPIs need to be agreed by the management team. These may, or may not, include benchmarked KPIs. Long-term targets for top KPIs may be determined by benchmarking results. High-level KPIs and benchmarked KPIs may need simplification to ensure application to the activities of the actioners. These are rolled down to where targets can be set for buy-in by the actioners. Target setting has to concur with identification of strategies for achieving targets and the time required to achieve the targets.

Plans and strategies could be grouped into two categories: operational and investment. Operational strategies could include Management Improvement Projects, Process Improvement Projects, and System Improvement Programs. Investment strategies could produce a Return on Investment or be a regulatory requirement.

Management buy-in is critical. Management has to acknowledge the gap and decide to actively champion the action. The incentive ("a burning platform"), to bridge the gap using a step change, has to be clearly communicated.

Step change entails a structured approach, usually when the change entails a high cost or risk of failure. The eight step Kotter approach is highly recommended.

The classic resistance to change stages (denial, anger, negotiation, depression, and acceptance) needs to be managed.

A blame culture, where a problem is always somebody else's or attributed to an inanimate object, may have to be overcome. Behavior-based safety programs help to break down this culture.

Step change could result in the company leapfrogging the competition. For example, new technology may simplify the production process, making a new plant more competitive than an older plant.

Embedding the change is most important so as to prevent regression to the previous state. Setting targets and agreeing actions by specific individuals, as well as monitoring progress, are critical. The change must be in alignment with the Company Vision and Mission and the actions must remain in line. The entire assessment process must be integrated into the Management Systems of the company.

REFERENCES

1. *The burning platform: how companies are managing change in a recession*. Celerant. http://www.economist-insights.com/sites/default/files/EIU_Burning%20platform%20WEB.pdf.
2. Kotter J, Rathgeber H. *Our iceberg is melting*. Macmillan; 2006.
3. Johnson, Scholes, Whittington. *Exploring corporate strategy*. Pearson; 2008.

FURTHER READING

1. De Bono E. *Six thinking hats*. Penguin; 2000.
2. Fisher R, Ury W. *Getting to yes*. Random House; 1991.
3. Ury W. *Getting past no*. Random House; 1991.
4. *Setting measures and targets that drive performance*. BSC Report; December 17, 2008. https://hbr.org/product/setting-measures-and-targets-that-drive-performance-a-balanced-scorecard-reader/an/2698-PDF-ENG.
5. Maintaining quality despite organisational change. Vermillion. http://www.powershow.com/view/ed032-MTAwZ/MAINTAINING_QUALITY_DESPITE_ORGANIZATIONAL_CHANGE_powerpoint_ppt_presentation.
6. Hubert C., Cowan M. *How to get employee buy-in for measurement*. Key activities to successfully implement performance measures. https://www.apqc.org/knowledge-base/documents/how-get-employee-buy-measurement.

CHAPTER

REPORTING

31

31.1 INTRODUCTION

Reporting must get the message across so as to facilitate effective decision-making. It must be tailored to the recipient. Basic questions are:

1. Who is asking?
2. What is the question?
3. Why is it important?
4. What are the data sources?

A report must provide an analysis of the data that is clear and meaningful to those reading it. The creator of the report needs to:

- take time to understand what information will be meaningful and beneficial to the decision-makers and key stakeholders;
- provide business–critical information that is of benefit;
- describe how the activity–level data relate to big–picture goals and strategies;
- be sensitive to recipients who may not be familiar with the data;
- provide contextual information that will explain:
  - what the measures indicate about their performance;
  - which actions were taken to arrive at the conclusions described in the report;
  - what staff can do to learn about their own process and improve business process performance.

Actioners who read the reports must be able to apply what they read to their work and be able to use the data to track personal progress toward the achievement of goals.

31.2 REPORTING FOR OPTIMAL DECISION-MAKING BY DECISION-MAKERS

ACKNOWLEDGMENT OF A GAP AND THE DECISION TO TAKE ACTION

On presentation of an identified gap, the first response is to challenge the source of the data. Consequently, the derivation of the gap has to be clearly documented and based on an objective assessment of facts.

Chapter 26, Data Verification and Chapter 29, Identification, Analysis and Evaluation of Gaps, detail the verification required to be done to ensure the recorded data are correct.

Performance Management for the Oil, Gas, and Process Industries. http://dx.doi.org/10.1016/B978-0-12-810446-0.00031-1

Bridging a gap often entails a paradigm shift. Prevailing thinking might be that there is little hope of improvement based on current practices. However, benchmarking creates an awareness of the competition achieving higher levels of performance, which provokes the question: "What are they doing that we are not doing?" Analysis of processes, compared with industry best practices, will often identify gaps.

WHERE TO GO FROM HERE?

Reporting must be focused on the target audience with specific details that enable this audience to make optimal decisions based on what has been presented.

TARGET AUDIENCE

Primary target audiences are as follows:

- Board of Directors (BoD)
- Managing Director (MD) / Chief Executive Officer (CEO)
- Leadership team
- Shareholder representatives
- Management
- Staff

REPORT GROUPING

Grouping of reports are generally as follows:

A. Business Plans/Work Plans and Budgets (WP&Bs)
  - Strategic: long term—5 to 7 years
  - Tactical: annual

B. Performance reports
  - Operational: daily, monthly, quarterly

C. Audit and Benchmarking reports
  - Benchmarking: every year for gas plants and every 2 years for refineries
  - Every year for financial and external certification audits
  - Every few years for internal audits and shareholder audits

D. Other

TYPES

Some common types of reporting are as follows:

- Tables
- Graphs
- Matrices
- Traffic light reports indicating actions are as per plan or not (green, amber, red)—see Box 30.2, Example Refinery Business Improvement Plans (BIPs).

Parameters of Traffic Light Reports

Traffic light reports could be used for monthly or quarterly performance reports showing a one pager for the top KPIs with an appropriately colored traffic light to indicate if plans are on target. They can also be used for regular reporting of improvement initiatives/strategies. Progress is categorized as follows:

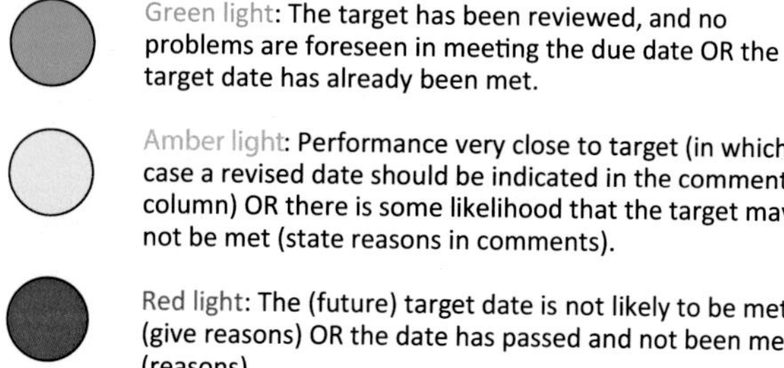

Green light: The target has been reviewed, and no problems are foreseen in meeting the due date OR the target date has already been met.
Amber light: Performance very close to target (in which case a revised date should be indicated in the comments column) OR there is some likelihood that the target may not be met (state reasons in comments).
Red light: The (future) target date is not likely to be met (give reasons) OR the date has passed and not been met (reasons)

TIMING

Reports produced generally deal with the previous interval of time and the planned future interval of time.

- Daily: operations and maintenance, production
- Weekly: operations and maintenance, production
- Monthly: operations and maintenance summary, production summary, finance
- Quarterly: operations and maintenance summary, production summary, finance versus annual plan/budget
- Annual: Annual Report, Business Plan and Budget, WP&B

FORMAT

Format could vary considerably depending on the audience.

31.3 COMMON REPORTS

A summary of common reports is shown in Table 31.1. **Bold** items are discussed in Chapter 29, Identification, Analysis and Evaluation of Gaps and Chapter 30, Strategies and Actions in detail.

31.4 THE USE OF GRAPHIC DISPLAYS

"A picture is worth a thousand words."

The value of different types of graphic displays to show deviations and trends is described here.

FINANCIAL ACCOUNTING

Top financial indicators, such as profit, should be trended and compared.

Fig. 31.1 shows the top indicators of a refinery over a period of a year (as extracted from the Solomon Associates Profile II benchmarking model). The red is the Net Cash Margin ($ profit per barrel), and the yellow is the Cash Operating Expense (COE). The target is not displayed.

Table 31.1 Summary of Common Reports

Target	Title	Content	Intent	Timing	Sources
Board of Directors (BoD)	BoD minutes of meeting	Board decisions and actions to be taken	Give direction to the company	After each BoD meeting	BoD secretary
	Business Plan and Budget (BP&B)	As required for approval of the Board prior to the budget year	For operation of the Business Unit or Function	Annual	Preparation team with management approval
	Annual Performance Report	All aspects of the performance of the business as required in the JV, O&M, and/or other agreements	For Board approval and distribution to shareholders	Annual	Preparation team with management approval
	Strategic risks	List of top risks with actions, responsibilities, and deadlines	Identify all top strategic risks and relative mitigation actions, responsibilities, and deadlines	Annual	Compiled and administered by corporate risk team
	Quarterly Performance Report	Health and Safety, Environment Integrity, Production, Human capital, Finance, Projects	Assess overall performance of the business unit(s) and/or group of companies	Quarterly	Preparation team with management approval
	Shareholder assessment of Business Units	**An evaluation of the overall performance of Business Units**	**1. Identify gaps in performance to bring to the notice of the BoD for evaluation of continued or increased investment or divestment** **2. Preparation of a shareholder's "letter of wishes" for BoD approval and submission to the business unit in anticipation of preparing the BU business plan and budget (BP&B)** **3. Assessment of the BP&B submitted to the shareholder BoD for approval**	**Annual**	**Shareholder representative team**
	Internal audit reports	**An evaluation of specific business systems**	**Provide independent assurance on the design and operation of the system of internal controls.**	**As per annual audit plan**	**Internal audit department**
Corporate/ MD/CEO	**Self-assessment of Business Unit**	**An evaluation of the overall performance of the Business Unit**	**Identify gaps in performance in line with the Corporate Governance Framework**	**Annual**	**Preparation team with management approval**
	Business Unit assurance declaration	Declaration of compliance with legal and ethical requirements and group policies, regulations, standards, manuals, frameworks, etc.	A key element of the assurance and compliance process to mitigate risk	Annual	Asset business unit head
	CoI	Conflict of Interest declaration on compliance with legal and ethical requirements of the company	Identify any potential conflicts of interest in business dealings and internal appointments	Annual	All staff to submit to CoI administrator for exception reporting to the board audit committee
	Comparative study (Benchmark)	**Comparison statistics relative to competitors in similar industries**	**Identify where we are relative to the competition**	**Annual or twice yearly (maximum of three times yearly)**	**Data compilation team and consultant**

Target	Title	Content	Intent	Timing	Sources
Business Unit management	Risk management and asset valuation report	Risk assessment and valuation of physical assets	Required for insurance of the assets	Annual	Specialist insurance assessor consultant
	Monthly performance report	Health and Safety Environment Integrity Production	Identify gaps in achievement of targets and take corrective action	Monthly	Performance monitoring team
	Statement Of Petroleum Account (SOPA)	Hydrocarbon production and royalty statistics	Hydrocarbon production and royalties due	Monthly	Hydrocarbon accounting specialist
	Financial report	Financial statements	Financial position of the business unit	Monthly	Accountant
	Production report	Production statistics	Assess plant performance for last 24 h	Daily	Production expert
	Quarterly certification compliance report	Certification requirements for reporting	Retain certification to ISO 9001, 14001, 45000, etc.	Quarterly	Performance monitoring team
	Energy Balance	Full energy balance including all hydrocarbon emissions	Emissions reporting to relevant authorities	Quarterly	Environmental expert
	Asset Integrity	Corrosion charts, progress on planned NDT work, etc.	Report on the integrity of the physical assets and preventative and corrective action being taken	Monthly	Inspection team
	Annual SMART objectives establishment	As per Personnel Management System	Alignment with corporate objectives and strategies	Annual	Each staff member
	SMART objectives progress report	As per Personnel Management System	Alignment with Corporate objectives and strategies	Biannual	Each staff member
Staff	Monthly performance report	Health and Safety, Environment Integrity Production **Plus focused performance of System(s) for which staff are responsible**	Promote ownership and accountability Ensure commitment of staff to agreed performance targets	Monthly	Performance monitoring team
	Quarterly performance report	Health and Safety, Environment Integrity Production Human capital Finance Projects	Assess overall performance of the Business Unit(s) and/or Group of Companies	Quarterly	Preparation team with management approval
	Performance appraisal	Assessment of performance	To reward staff for good performance	Annual	Management

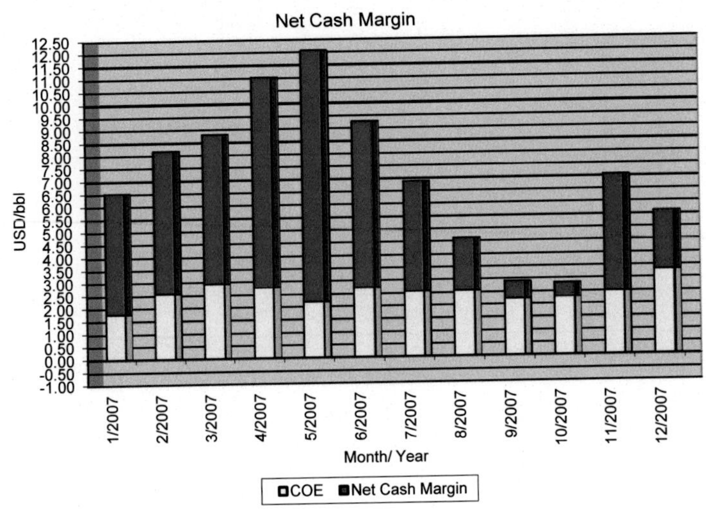

FIGURE 31.1

Refinery margins.

An example of next year's budget related to past year actual, current year predicted, and future years (at least 4 years after the budget year) is displayed in Fig. 31.2. Operating expense is split into "routine" for recurring expenses each year and "nonroutine" for turnarounds, etc. which only occur every few years. "Recovery" refers to costs that are incurred on behalf of others, such as affiliates, and will be recovered from them.

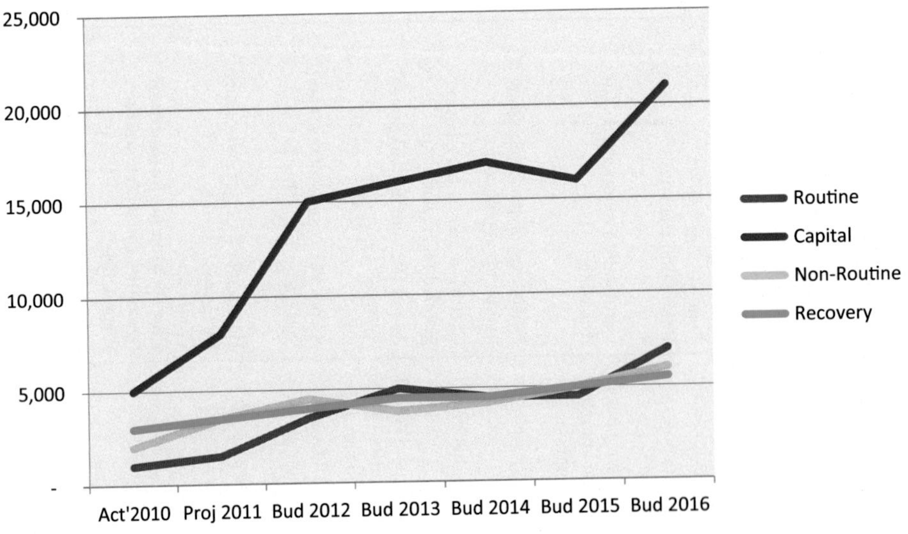

FIGURE 31.2

Budget overview.

MAINTENANCE PERFORMANCE USING TREND GRAPHS

Work orders are the basis of a Maintenance System. Fig. 31.3 shows work order completion in three levels of priority. The first graph—Priority A "Emergency Work Orders"—can also be used as a **leading KPI for asset integrity**. Too many emergency work orders indicate that the plant is not stable. The target of 3% of total work orders was determined from Oil and Gas industry norms.

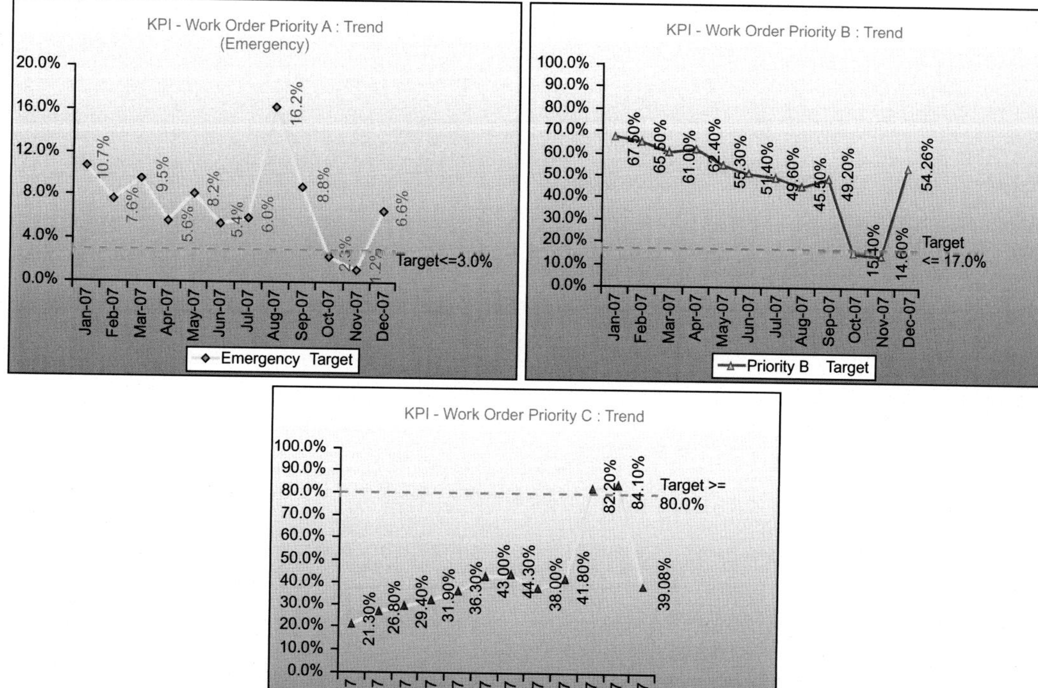

FIGURE 31.3

Maintenance work order trend graphs.

OVERALL PERFORMANCE USING A SPIDER DIAGRAM

This is useful to show top KPIs of a company against targets or pacemakers. Fig. 31.4 shows the nine top KPIs with Benchmarking outcomes for 2 study years and the targets for the last year.

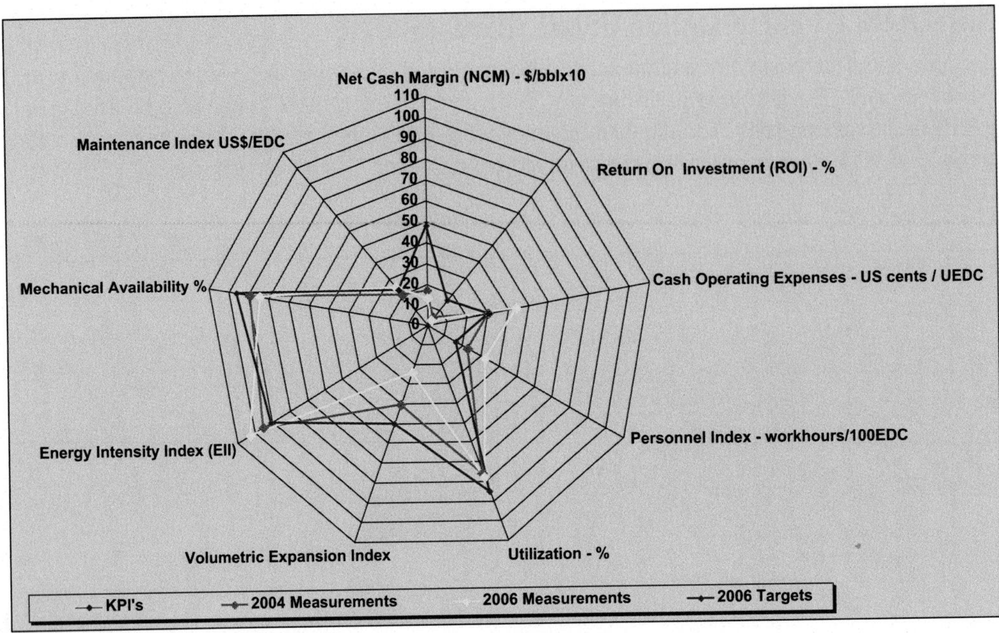

FIGURE 31.4

Refinery performance spider diagram.

TIME-RELATED TREND GRAPHS FOR BENCHMARKING

This graph shows an improvement or deterioration of an indicator (convergence and divergence). For benchmarking it can show improvement of one company relative to others, with respect to a top KPI, over an extended period of 5–6 years.

Fig. 31.5 is an example of refinery **Utilization** over time. 1Q is the average of the top quartile of refineries in the study, 4Q is the average of the bottom quartile of refineries in the study, and the red depicts actual performance of the company's refinery.

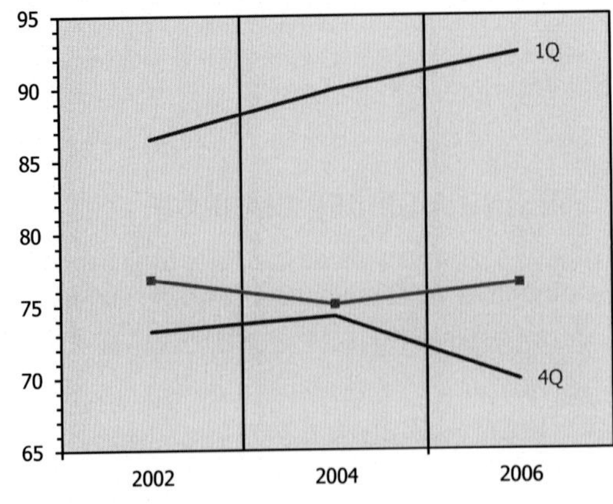

FIGURE 31.5

Refinery Utilization trend graph.

DASHBOARD

This is useful for collating critical KPIs for management to monitor on a daily, weekly, or monthly basis.

A Production, Maintenance, Safety, and Environmental example is shown in Fig. 31.6.

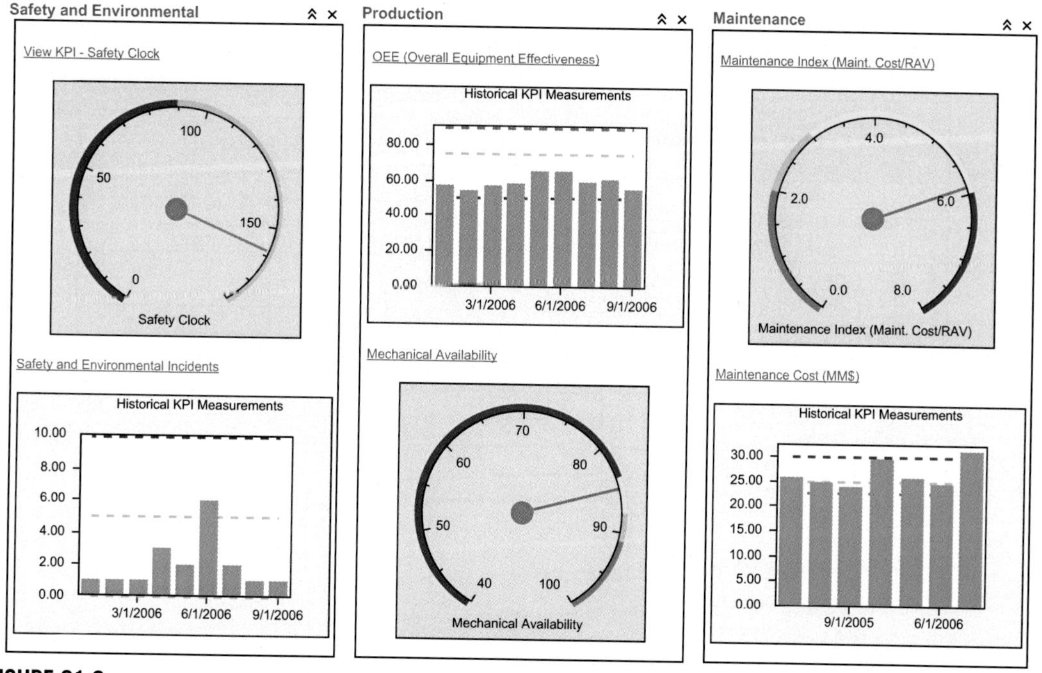

FIGURE 31.6

Production, Maintenance, Safety, and Environment dashboard.

Fig. 31.7 shows a dashboard with traffic lights showing monthly investment program performance status.

PROJECT TITLE			APPROVAL NO.		
HEALTH & SAFETY					
STATUS		*Employee hours*	*H&S incidents*	*QA NCs*	*Comments*
G	This month				
G	Cum. year to date				
G	Project total				
HUMAN RESOURCES					
STATUS		*Total*	*Vacancies*		*Comments*
G	Contractors				
G	FTE				
FINANCE					
STATUS		*Latest*	*Budget*		*Comments*
R	Total spent YTD				
R	Forecast spend for year				
A	Cum. total to date				
	Project forecast				

TOP RISKS & OPPORTUNITIES					
STATUS	*Score*	*Title*	*Risk log ID no*	*Mitigation*	*Comments*
G	25	x			
A	20	y			
R	17	z			

PROGRESS					
STATUS		*Forecast*	*Plan*		*Comments*
G	Overall				
G	Design				
A	Fabrication				
A	Installation				

Status:

- Green – progressing well
- Amber: - concerns
- Red – significant concerns

FIGURE 31.7

Investment performance status with traffic lights.

MATRIX FOR RISK REPORTING

Changes in listed risks over time are difficult to report. A risk matrix could indicate each current risk and whether the trend in control and environment for that risk has improved, stabilized or declined since the last report. It also rates each risk in terms of acceptability. Fig. 31.8 is an example that could form part of a risk report.

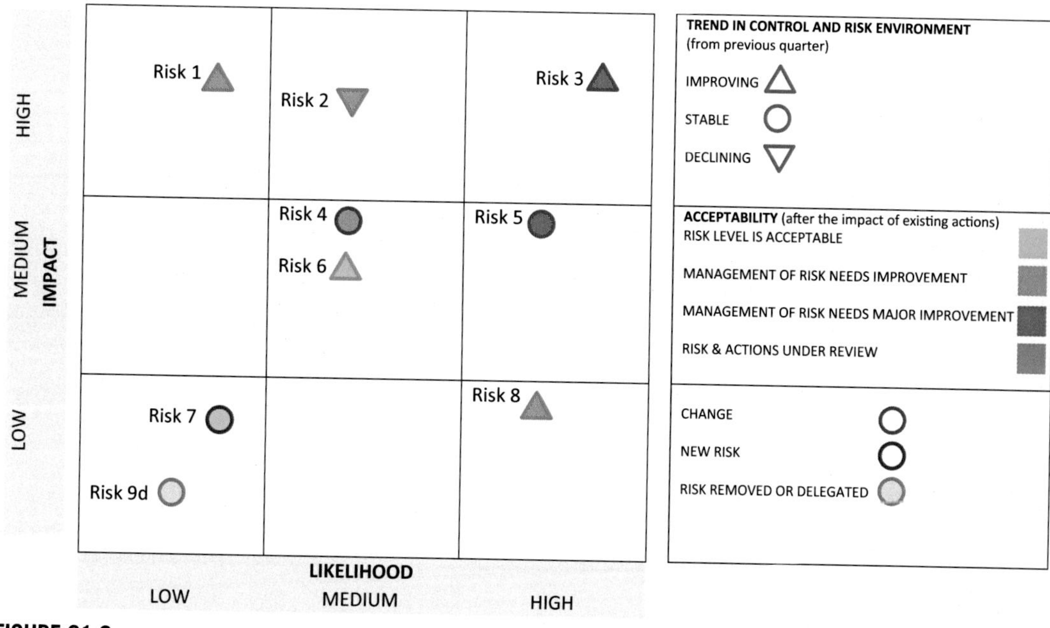

FIGURE 31.8

Risk matrix.

31.5 INTRANET-BASED REPORTING

Various types of reports can be published online. The company Intranet could display a daily dashboard (see earlier) whenever users log in. Alternatively, the Integrated Management System could be displayed on the Intranet showing all System KPIs. Box 16.2, Terminology – Refinery Utilization Cascade, is an example of cascaded reporting on the Intranet.

31.6 BP&BS AND PERFORMANCE REPORTS

Standard format and terminology, used year on year, is necessary since Business Performance Reports need to align with the Business Plans and Budget. Business Reports tend to be quarterly and operations reports tend to be monthly.

Note: Some companies use the term “Work Plan and Budget” (WP&B) in place of “Business Plan and Budget” (BP&B). These are regarded as the same in this book.

GROUP OVERVIEW

It is suggested that focus be related to critical success factors (CSFs) as follows:

1. Supply and demand
2. Profit

3. Capex
4. HSE
5. Human capital

An example of the contents of a BP&B is shown in Box 31.1.

> **BOX 31.1 BUSINESS PLAN AND BUDGET NOC EXAMPLE**
>
> **Contents**
>
> 1. Executive summary
> 2. Performance score card
> 3. Initiatives and activities
> 4. Technology plan
> 5. People requirement
> 6. "Triple plus"
>   a. Leadership
>   b. Mindset
>   c. Capabilities
> 7. Operating plan
> 8. Master budget

PERFORMANCE REPORTS

Corporate

Supply and Demand

1. Link market demands with supply (quantity and quality).
2. Identify strategic risks with clear mitigating actions for identified risks outside the company risk appetite.

Profit and Capex

3. Compare margins (in $/mm BTU for gas and $/barrel for oil) between the markets and the group.
4. Link Capex with potential increased revenue and profit.
5. Compare and identify risks and opportunities with respect to all of the group's investments.

HSE

6. Identify and compare lagging and leading HSE indicators that may affect long term production at design capacity. Clear mitigating actions for identified risks outside the company risk appetite need itemizing.

Human Capital

7. Show clear progress on nationalization (in the case of a government company) or affirmative action (in the case of balancing employment with population demographics).
8. Identify critical skill deficiencies and mitigating actions.

Business Units

For each Business Unit a standard format should be adopted for both the Company Business Plan and Company Performance Reports. With focus on the above CSFs, the following is suggested:

1. Executive summary
2. Performance score card

3. Description of Business Unit
4. Objectives of Business Unit
   a. Assumptions on which Objectives are based
5. Strategies and plans
   a. Operating
      i. Activities
      ii. Budget (Opex)
   b. Investment
      i. Activities
      ii. Budget (Capex)
6. Performance details

An example of a typical Monthly Operations Report is shown in Box 31.2.

BOX 31.2 A TYPICAL MONTHLY OPERATIONS REPORT

- Title page: monthly operations report for January 2015
- Table of contents
- Executive summary
- Consolidated financials—table
- HSE—trend graphs
- Production operations—trend graphs
- Project management—traffic light report
- Appendices

31.7 AFFILIATE PERFORMANCE REPORTS FOR THE SHAREHOLDER

The following are possible alternatives or combinations for reporting affiliate performance to the shareholder representatives:

a. Codeveloped monthly operations report
b. One pager performance report (weekly) possibly including a "dashboard"
c. Copy of management level reports
d. Access to affiliate underlying financial system (password protected)
e. Open invite to shareholder information meeting

Common requested information includes the following:

Financial

1. Profit and loss statement
2. Balance sheet
3. Cash flow report
4. Capital budget (Capex)
5. Operating budget (Opex)
6. Variance from budgets
7. Third-party audit findings

8. Annual tax reporting
9. Revenues

Operational

10. Annual strategic plan
11. Proposed business plans
12. Approved business plans
13. HSE performance
14. Production reports

Events Requiring Notice

15. Major disruption to production
16. Major HSE incident

A possible guideline for determining what to report is given in Box 31.3.

BOX 31.3 GUIDELINE FOR DETERMINING WHAT MANAGEMENT IS REQUIRED TO REPORT TO THE SHAREHOLDER[2]

1. Management shall develop, and the Board shall approve, a set of KPIs that will be tracked and reported to the Board on a regular basis.
2. Other than the above agreed KPIs, Management shall develop, and the Board shall approve, a "Register of Performance and Planning Information" to be provided to the Shareholders on a recurring basis.
3. The affiliate shall maintain a current list of other information/reporting requests from the Shareholders.
4. Shareholders shall annually review this list, and provide input on how much lead time is required for them to respond for key approvals.
5. Board and Committee members shall make any additional request for information through the Board or Committee, with the visibility and endorsement of the Board or Committee chair.
6. Substantive requests that deviate from the information register shall be coordinated by the Shareholders and channeled to the affiliate via the lead directors. The affiliate CEO shall determine if a request is substantive and that this process is required.
7. The lead directors shall be responsible for coordinating shareholder information requests and audits of the affiliate.

31.8 BUSINESS PERFORMANCE SOFTWARE

Most performance reporting entails dashboards, scorecards and spreadsheets, as discussed above. However, consolidating information into simple top reports for decision-makers is a challenge. Various tools are offered by software vendors[1] with varying effectiveness, since software that has been applied successfully to one particular industry may not be suitable for other industries. Section 3.3, Business Models, discusses this further.

31.9 SUMMARY

Reporting must get the message across so as to facilitate effective decision-making and must be tailored to the recipient. The target audience has to be clearly identified before deciding on the types and methods of reporting.

A report must provide an analysis of the data that is clear and meaningful to those reading it. In the case of reporting a gap, the derivation of the gap has to be clearly documented and based on an objective assessment of facts.

The reporting of plans and targets and measurement of performance against these plans and targets is critical for performance improvement. The use of Performance Indicators, rather than descriptive comparisons, gives a clear picture for the decision maker. The presentation of the Indicators needs to be such that, if there is a gap, the decision maker is easily able to review the action being taken in line with an agreed target, and take further action if achievement of the target is in jeopardy.

Various tools are used:

- Trend graphs can show diversion or conversion toward a target.
- Spider diagrams are useful for summarizing a number of KPIs but can be difficult to read.
- Dashboards can give current performance.
- Intranet systems facilitate drill down.
- Traffic light reports indicate probability of meeting targets.
- Bar graphs, divided into quartiles, show a clear picture of a company KPI relative to the competition.

Benchmarking reports may have scopes that are smaller than the overall Business Unit.

The current trend for Internal Audit Reports is to focus on **Management Systems** and business risks, rather than Departments.

Timing and format are important for comparison and assessment. Monthly/Quarterly Performance Reports need to be aligned with Business Plans and Budgets to clearly demonstrate deviations.

The format, content and timing of shareholder reporting for an affiliate needs to be established as part of the Joint Venture Agreement.

REFERENCES

1. The Forrester Wave. *Business Performance Solutions, Q4 2009 – comparative study of performance software*. www.forrester.com.
2. Streamlining shareholder reporting in JVs: the seven reporting practices of highly-effective JV CEOs. The joint venture exchange: Water Street Partners: issue 59 June 2013.

FURTHER READING

1. Answering the call – managing shareholder information The joint venture exchange: Water Street Partners. January 2012.
2. Demands in JVs. Gonzalez T. Dashboard design: key performance indicators and metrics. www.brightpointinc.com/dashboard-design-executive-dashboards-part-1.
3. Quinn K. *Guided Ad Hoc reporting: the most widely used business intelligence paradigm*. 2008. http://hosted-docs.ittoolbox.com/most-widely_21009.pdf.

PART

BUSINESS OVERSIGHT

OVERVIEW

WHAT IS BUSINESS OVERSIGHT?

Business oversight is the management of an asset from concept to expiry. As an investment progresses through its lifecycle, the requirements for decision-making by various parties and the roles and responsibilities for decision-making change. Specific requirements have to be enforced, at specific stages, for the investment to be a success.

THE OBJECTIVE

The objective of Part 8 is to identify key requirements, roles, and responsibilities for the complete investment cycle.

PICTORIAL OVERVIEW

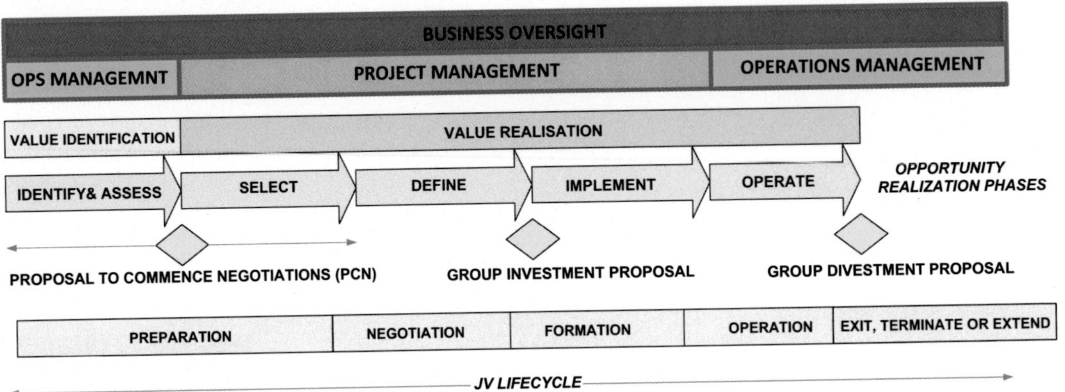

OUTLINE OF PART 8

- Primary business relationships are outlined, accentuating the value of partnerships.
- Types of companies (legal entities) are listed.
- Common agreement documentation is described with supporting examples.
- The lifecycle of an investment is detailed from "cradle to grave" with key "hold" points for review and approval in the value realization stage.
- Risk assessment throughout the lifecycle is outlined.
- The requirement for an investment policy or framework is discussed.
- An investment approval package model is outlined.
- Investment decisions and their associated risks (business, health, safety, and environment (HSE), compliance, and technology) are discussed, supported by examples.
- The process of how an investment is logged and tracked, from idea to handover, is described.
- The partnering cycle, which runs parallel to the investment cycle, is outlined.
- A discussion on divestment and decommissioning is supported by examples.
- Roles and responsibilities are described, as they change from project phase to operating phase.
- Risks related to value chain management in the operating phase are outlined, using the liquefied natural gas value chain as an example.
- Roles and responsibilities related to the type of agreement are outlined.
- Responsibilities of each role in the organization are described, noting the potential conflicts between roles, especially if an individual undertakes more than one.
- Specific duties for shareholder oversight are listed.
- Levels of management review are outlined.
- The evolution of review meetings for improved effectiveness is discussed.
- Member of the Board of Directors, management, and shareholder representative review responsibilities are outlined.
- Shareholder Business/Work Program and Budget review steps are outlined.

CHAPTER

BUSINESS RELATIONSHIPS

32

32.1 INTRODUCTION

Businesses are created by a single owner or in partnership, where each partner has complementary contributions to make. From germination of the idea to termination of the investment, relationships, roles, and responsibilities change. Restraints for decision-making also change.

Agreements are the basis of business relationships. There could be more than one agreement between business partners/operators for a particular venture.

The legal entities arising from Agreements are Companies that could consist of one or more Business Units. Those mentioned in this chapter are simplistic views of complex relationships and may not be described fully.

32.2 THE VALUE OF PARTNERSHIPS

Partnerships, in the form of one of the agreements as described in Section 32.4, are formed when the partners have complementary contributions to the venture. These could include the following:

1. Access to raw materials
2. Technology
3. Management Systems
4. Finance
5. Marketing network

Often, a National Oil Company (NOC) will have access to oil and gas reserves, whereas an International Oil Company (IOC) will have the technology to exploit the reserves and process the oil and/or gas. Other partners may have the network to market the products.

Box 32.1 gives an example of an NOC/IOC Joint Venture (JV).

BOX 32.1 JOINT VENTURE COMPANY EXAMPLE: QATAR CHEMICAL COMPANY LTD.[1]

Qatar Chemical Company Ltd. (Q-Chem) is a Private Incorporated Petrochemical Joint Venture (JV) between Mesaieed Petrochemical **Holding** Company Q.S.C. (MPHC) with 49%, Chevron Phillips Chemical International Qatar **Holdings** LLC at 49%, and Qatar Petroleum (QP) the remaining 2%. MPHC is majority owned by QP.

The Q-Chem facility is a world-class integrated petrochemical plant producing high-density and medium-density polyethylene (HDPE and MDPE), 1-hexene, and other products, using state-of-the-art technology provided by Chevron Phillips Chemical, a major producer of chemicals and plastics.

Comment

Note the use of Holding Companies. See Section 32.3.

Performance Management for the Oil, Gas, and Process Industries. http://dx.doi.org/10.1016/B978-0-12-810446-0.00032-3

32.3 TYPES OF COMPANIES

A company could be a Private Shareholding Company or a Public Shareholding Company (listed on a stock exchange).

AFFILIATE

An Affiliate is defined in this book as any company in which the parent company has a direct or indirect interest.

This includes all JVs, Subsidiaries, and Production Sharing Agreement (PSA) operators.

This excludes investments in companies in which the parent company has no rights other than the right to attend the annual general meeting as a shareholder and which the parent company holds primarily for investment purposes.

Controlled Affiliate

A Controlled Affiliate has a parent company that has either a majority of voting rights or the right to exercise a controlling influence over how the Affiliate is organized and managed.

Box 32.2 gives an example of the right to exercise a controlling influence.

> **BOX 32.2 CONTROLLED COMPANY EXAMPLE: BAHRAIN PETROLEUM COMPANY (PRE-1997)[2]**
>
> In 1981, Bahrain Petroleum Company was reconstituted as a Joint Venture (JV) owned by the Bahrain Government (60%) and Chevron (40%). As per the JV agreement, Chevron regarded it as a **controlled company,** as it had the right to exercise a controlling influence over how the company was organized and managed, even though it was a minority shareholder. For example, major investments required both shareholders' approval.
>
> In 1997, the Bahrain Government bought Chevron's shares and thus gained full control of the company.

SUBSIDIARY

A Subsidiary is a Private Shareholding Company, which is wholly owned by the parent company and is generally established as follows:

1. **Service Company** to the Asset Business Units (*for example, Inspection Services*). This is normally regarded as a **Service Business Unit**.
2. **Holding Company**: see as follows.

HOLDING COMPANIES

Holding companies do not produce goods or services, but own shares of other companies. Holding companies allow the owners to mitigate risk, and own and control a number of different companies.

JOINT VENTURE

A JV is an association of two or more participants who engage in business together, sharing profits or production and costs and liabilities, and between them, exercising control over the business through a JV business organization or structure that is distinct from the business organization of each of the participants.

Incorporated Joint Venture

An Incorporated JV is a legal entity formed to carry out the JV's business. It owns assets in its own name. Incorporated JVs include listed companies and limited liability partnerships.

Unincorporated Joint Venture

An unincorporated JV is a JV set up by contract. It does not own assets in its own name, and the assets and the operations are owned and controlled by the investors according to the terms of the agreement (contract).

Controlled Joint Venture

The parent company either has a majority of voting rights or the right to exercise a controlling influence over how the JV is organized and managed.

Noncontrolled Joint Venture

Noncontrolled JVs are those that don't comply with the previous.

Example:

A Shareholder has majority shares in an Incorporated JV through a Subsidiary Holding Company. This JV is a Controlled JV from this Shareholder's perspective, as it has majority shareholding. However, the JV is operated by an Operating Company, which has been set up by both Shareholders and which is technically controlled by the minority Shareholder through secondees with the necessary experience.

Fig. 32.1 depicts what is described in the previous example.

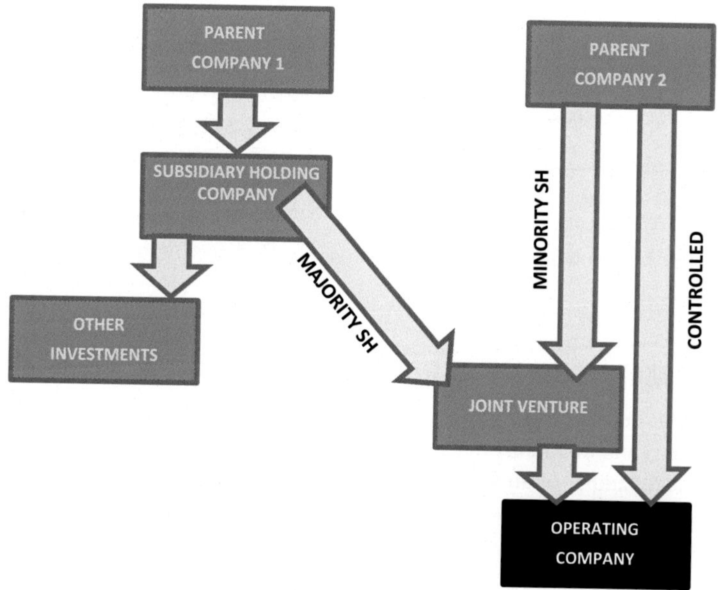

FIGURE 32.1

Joint venture example.

32.4 AGREEMENT DOCUMENTATION

As a project evolves from idea to fully operating venture, the following are typical progressive agreements:

1. Letter of Intent (LOI)
2. Memorandum of Understanding (MOU)
3. Confidentiality Agreement
4. Heads of Agreement (HOA)
5. Joint Venture Agreement (JVA)

The development of the legal and commercial aspect of a JV is briefly discussed in Section 33.11, Partnering.

Agreement documentation includes the whole of the legal and contractual documentation governing the rights and obligations of the parties involved in a JV.

Common examples include the following:

- JVAs
- Memorandum and Articles of Association (M&AofA) of a JV company
- Framework Agreements

Table 32.1 shows "**boundary**" documents of Business Units.

Table 32.1 Business Unit Boundary Documents

Directly Managed Business Unit	Subsidiary	Joint Venture (JV)	Production Sharing Agreement (PSA)	State Services
Mandate	Memorandum and Articles of Association	Agreement	Agreement	Government Decree/ Law

The relationships between the Parent Company, JV Partners, and an Operating Company are depicted in Fig. 32.2.

OPERATING AND MAINTENANCE AGREEMENT

An Operating and Maintenance (O&M) Agreement is an agreement between the Shareholders and the Company that operates and maintains the shareholder assets on behalf of the Shareholders.

Box 32.3 gives an example of an Operating Company managing a number of Assets with different JV Shareholders within an O&M Agreement.

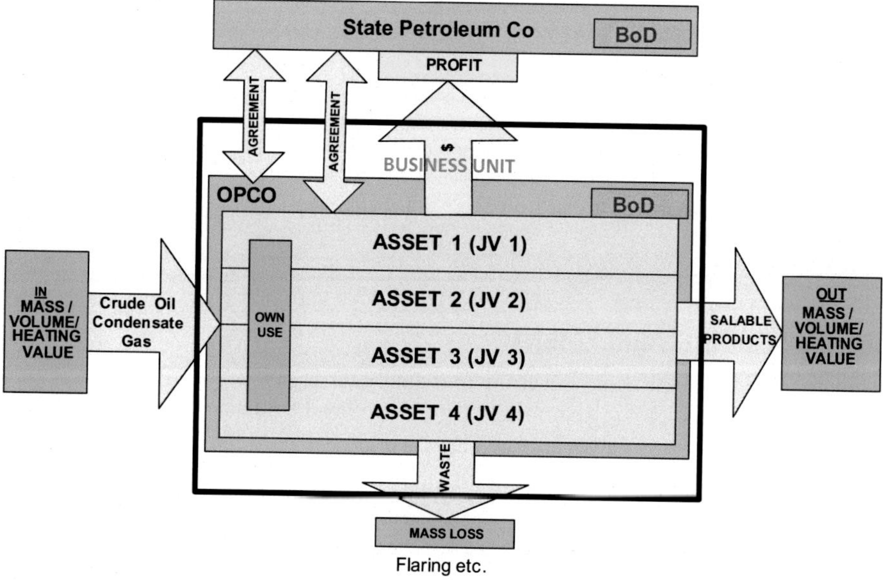

FIGURE 32.2

Joint venture relationship example.

BOX 32.3 OPERATING COMPANY EXAMPLE: RASGAS[3]

RasGas Operating Company has a number of discreet Joint Venture (JV) shareholders.

These are as follows:

I. Asset 1: RasGas1 (RL): two LNG trains
   - Shareholders: Qatar Petroleum, ExxonMobil, and other minority shareholders

II. Asset 2: RasGas 2 (RL2): three LNG trains
   - Shareholders: Qatar Petroleum, ExxonMobil, and other minority shareholders

III. Asset 3: RasGas 3 (RL3): two LNG trains
   - Shareholders: Qatar Petroleum, ExxonMobil, and other minority shareholders

Qatar Petroleum is the primary shareholder in each Asset with >50% shares. ExxonMobil is the second major shareholder in each Asset and is the technology provider for the complex. Thus a number of ExxonMobil staff are seconded to RasGas.

RasGas also operates Al Khaleej Gas and Barzan Gas, producers of flowing gas and associated by-products, on behalf of ExxonMobil (in line with Production Sharing Agreements). Res Laffan Helium is also operated by RasGas on behalf of its shareholders.

See Box 33.16, Qatar's Partnering Strategy, for more details.

SALES/SUPPLY AND PURCHASES AGREEMENT

A Supply and Purchase Agreement (SPA) is an Agreement between the Shareholders and the Marketer or Purchaser of the Salable Products.

Box 32.4 shows an example of a marketer of a State Oil Company's share of products from its various interests.

BOX 32.4 MARKETING COMPANY EXAMPLE: TASWEEQ[4]

Qatar International Petroleum Marketing Company Ltd. (Tasweeq), Q.J.S.C. is an independent state-owned company with the mandate of reliably and efficiently capturing maximum market value from the rapidly increasing exports of "Regulated Products" from the State of Qatar. Tasweeq currently delivers products to customers and markets globally.

"Regulated products" are obtained from LNG, natural gas, gas to liquids (GTL), and refinery operations in Qatar. These are liquefied petroleum gas (LPG), sulfur, field & plant condensates, and refined products such as naphtha, motor gasoline, gasoil, & jet fuel (refined products).

Tasweeq also markets crude oil and GTL entitlements from PSAs on behalf of Qatar Petroleum under a Sales and Purchases Agreement (SPA). These are termed as "Nonregulated Products."

PRODUCTION SHARING AGREEMENT

PSAs are contracts signed between a government (usually through the State Petroleum Company) and an experienced upstream technology company (the contractor). PSAs are one of the most popular of the various petroleum fiscal arrangements and are classified as in Fig. 32.3.

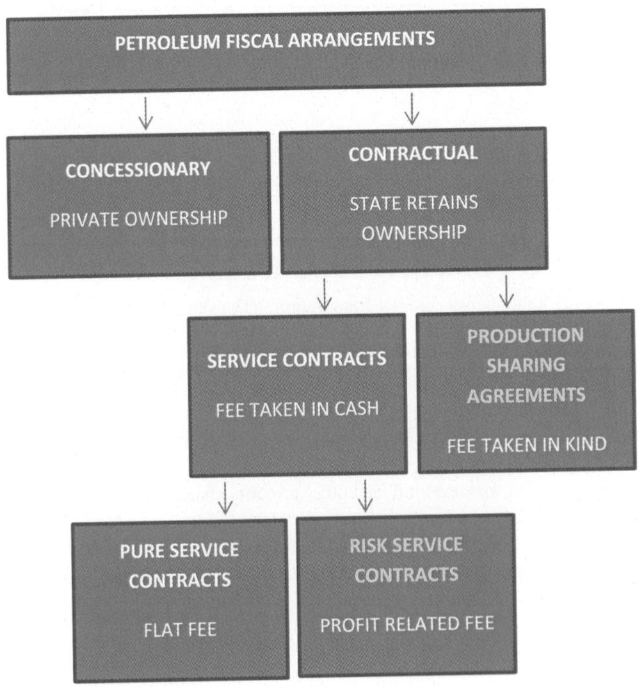

FIGURE 32.3

Classification of petroleum fiscal arrangements.

The contractor bears the risk of the initiative and explores, develops, and ultimately produces the field as required. When successful, the contractor is permitted to use the money from produced oil to recover capital and operational expenditures. The remaining money is split between the government and the contractor. PSA government fees are generally taken in kind, i.e., share of the product, which is then disposed of by the State (see Box 32.4, for an example).

PSAs can be beneficial to governments of countries that lack the expertise and/or capital to develop their resources and wish to attract foreign companies to do so. They can be very profitable agreements for the oil companies involved, but often involve considerable risk.

A PSA can either be an Exploration and Production Sharing Agreement or a Development and Production Sharing Agreement. "Exploration" is prospecting for undiscovered petroleum by the contractor. "Development" is the development of a discovered accumulation by the contractor (an accumulation is an individual body of petroleum in place).

Production Sharing is generally depicted as in Fig. 32.4.

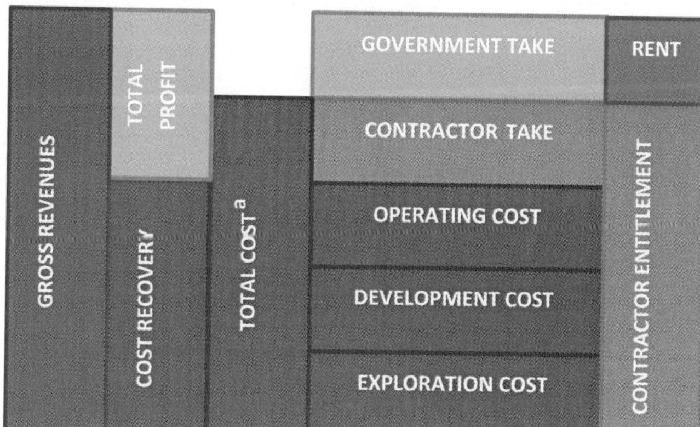

FIGURE 32.4

Allocation of revenues from production. **Note a**: total cost from the perspective of the government.

Control is exercised through a Management Committee.

A PSA may possibly be regarded as an unincorporated JV.

The Risk Services Contract,[5] a variation of the PSA, was introduced in Malaysia for marginal fields.

Box 32.5 gives an example of a Company operating under a PSA.

BOX 32.5 PRODUCTION SHARING AGREEMENT (PSA) COMPANY EXAMPLE: SHELL PEARL GAS-TO-LIQUIDS[6]

"Developed in partnership with Qatar Petroleum, Pearl Gas-to-Liquids (GTL) is the world's largest GTL plant and one of the world's largest, most complex, and challenging energy projects ever commissioned. From the origins of Shell GTL technology nearly 40 years ago, to its first commercial debut in Shell's Bintulu GTL plant in Malaysia in the early 1990s, to the creation of the world's GTL capital in Qatar today, the delivery of GTL on such a vast scale as Pearl GTL brought together almost every aspect of Shell's technical and project management capabilities."

Shell Qatar owns Pearl GTL (the asset). Pearl GTL produces liquid transport fuels and lube oils as well as various by-products from its offshore gas fields. The asset was created as a result of a Production Sharing Agreement (PSA) signed by Qatar Petroleum (on behalf of the Qatar Government) and Shell Qatar.

The annual Work Plan and Budget (WP&B) is approved by Qatar Petroleum. The Management Committee (ManCom) is a joint Shell/QP committee that oversees the work and expenditure of the venture. Revenue is divided between the two parties as per the PSA.

Shell's Gas To Liquid (GTL) technology is the driver for this project.

The relationship between the State Petroleum Company and the Contractor (Operator) as per the PSA is generally depicted as in Fig. 32.5.

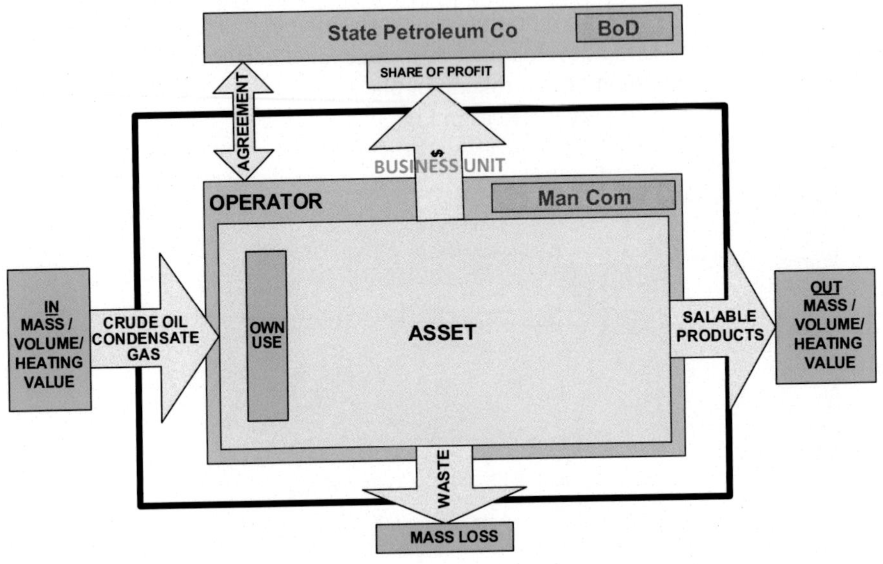

FIGURE 32.5

Production Sharing Agreement (PSA) relationship example.

Other agreements are listed in Appendix A, Glossary of Terms, under Agreements.

32.5 POLICY, REGULATION, AND OPERATION RELATED TO A NATIONAL OIL COMPANY

To promote transparency and reduce the potential for conflict of interest, oil- and gas-rich countries need to separate the policy, regulation, and operation functions when exploiting national oil and gas reserves. Separation is as follows:

1. Policy is determined by the legislative process.
2. Regulation converts policy into compliance.
3. Operation takes place within the constraints of compliance.

The separation of state "policing" activities from operation is thus required. An example is shown in Box 32.6.

An example of potential conflict of interest is discussed in Box 34.1, Role Conflict Example: National Oil Company.

BOX 32.6 EXAMPLE OF LINES OF AUTHORITY FOR AN OIL AND GAS–PRODUCING COUNTRY AND NATIONAL OIL COMPANY (NOC)

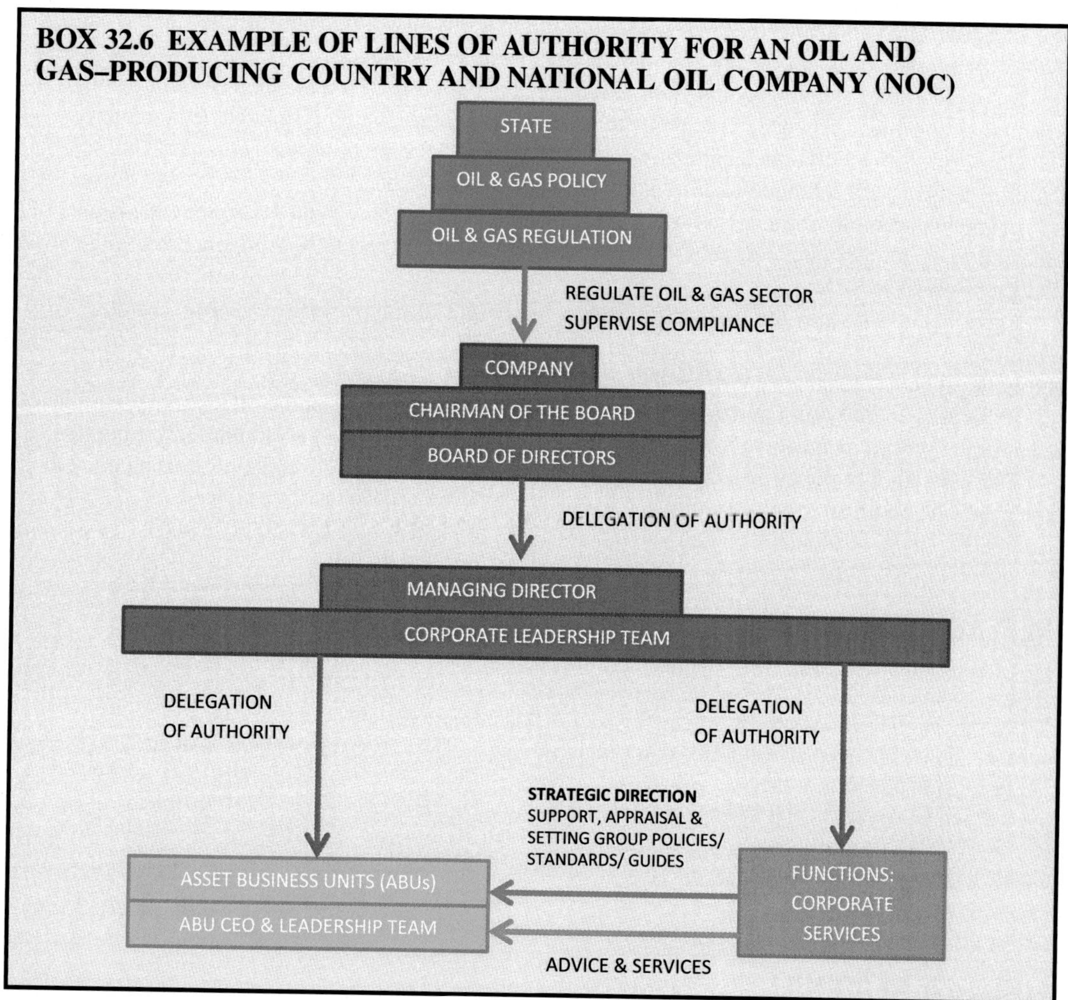

32.6 SUMMARY

Businesses are created by a single owner or in a partnership where each partner has complementary contributions to make. The business could either be a stand-alone business, a Subsidiary with no partners, or a JV.

A Subsidiary is generally a service company for a parent company, or a holding company that exists solely for the purposes of investing in a JV (an investment conduit).

In the Oil and Gas industry, a JV is often established between a NOC that controls the oil and gas reserves of the State, and an IOC, which has the technology to exploit the reserves.

In a controlled company, the shareholder has the majority of voting rights or the right to exercise a controlling influence over how the company is organized and managed.

An unincorporated company does not have a legal personality. It is a business structure with a contractual arrangement between participants to operate the venture, with no separate legal identity.

Conversely, an incorporated company is a "legal entity." It could be a listed company (public shareholding company) or a limited liability company (private company).

In another scenario, the assets of a company could be managed by a third party in what is generally referred to as an operating company. This company usually has an O&M Agreement with the shareholders, as well as SPAs.

Agreements evolve as the partnership in a new venture develops: from LoI, MoU, and HoA to a JVA.

A "legal entity" requires an M&AofA.

PSAs are generally formed between an NOC on behalf of the State, and an IOC. Usually all investment in the venture is made by the IOC, and on startup, the NOC/State gets a share of production.

The separation of policy, regulation, and operation is essential for promotion of transparency in oil- and gas-rich countries.

REFERENCES

1. QChem. http://www.qchem.com.qa/internet/Pages/default.aspx.
2. Bapco. http://www.bapco.net/en-us/about-bapco/our-history.
3. Rasgas. http://www.rasgas.com/AboutUs/AboutUs_TheCompany.html.
4. Tasweeq (Qatar International Marketing Company). https://www.tasweeq.com.qa/EN/AboutTasweeq/Pages/Tasweeq%20Mandate.aspx.
5. Risk services contracts (RSCs). http://www.thestar.com.my/business/business-news/2016/07/12/end-of-the-road-for-rscs/.
6. Shell Qatar. http://www.shell.com.qa/en_qa/projects-and-sites/pearl-gtl.html.

FURTHER READING

1. Baynham G, Bamford J. *Reconciling ownership, valuation and control JV ownership and control blueprints.* Feb 2012. The Joint Venture Exchange: Water Street Partners.
2. Johnston D. *International petroleum fiscal systems and production sharing contracts.* PennWell; 1994.

CHAPTER

THE OPPORTUNITY LIFECYCLE 33

33.1 INTRODUCTION

The performance of the investment from "cradle to grave" is critical for the success of the company. Return on Investment (ROI) is clearly a Key Performance Indicator (KPI) for an Asset Business Unit.

The basic principles for project capital investment are as follows:

- **Gates for decisions** on major expenditure **phases**
- **Requirements** for submission for approval
- **Review** authorities for aspects of proposal document
- **Approving** authority

Compliance with the following international standards is common practice:

- ISO 21500 Guidance on Project Management September 2012[1]
- ANSI 99-001-2008 Project Management Body of Knowledge (PMBOK)[2]

Major process industry companies have extensive manuals for management of their projects. Examples of IOC manuals/processes are as follows:

- Chevron Project Development and Execution Process (CHPDEP)
- Shell Opportunity Realization Manual (ORM)

For major investments, comparative analysis is also common practice.

Partnering may be desirable and the commercial/legal development of this runs parallel to the value realization phases of the investment cycle. The opportunities selected need to be maximized. This includes the following:

1. Minimizing lifecycle costs
2. Minimizing realization times
3. Maximizing production rates and volumes
4. Securing the best possible market terms
5. Securing or enlarging the license to operate.

Performance Management for the Oil, Gas, and Process Industries. http://dx.doi.org/10.1016/B978-0-12-810446-0.00033-5

33.2 PORTFOLIO MANAGEMENT

Project Management Body Of Knowledge quote:

> Portfolio management refers to the centralized management of one or more portfolios, which includes identifying, prioritizing, authorizing, managing, and controlling projects, programs and other related work, ***to achieve specific strategic business objectives***. Portfolio management focuses on ensuring that projects and programs are reviewed to prioritize resource allocation, and that ***management of the portfolio is consistent with, and aligned to, organizational strategies***.

The basic screening process is shown in Fig. 33.1.

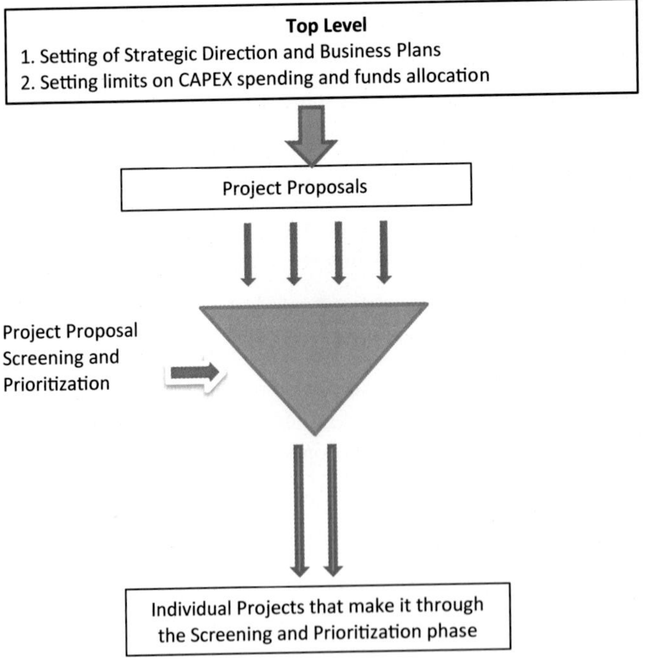

FIGURE 33.1

Basic capital project screening.

33.3 THE INVESTMENT CYCLE

The investment cycle is divided into two: value identification and value realization. The investment cycle is described as follows:

VALUE IDENTIFICATION

Identify

This starts with initiation of the opportunity. Ideas are generated. Alignment with business strategy and objectives is verified. Potential value is identified, and the initial decision to fund is taken. This is generally referred to as the **Project Initiation Note (PIN)**.

VALUE REALIZATION

Assess

This phase demonstrates the feasibility of the various options. Risks are assessed for different options and relevant realizations/outcomes. The options with their relative pros and cons are taken forward to selection after fund approval. This is generally referred to as the **Initial Investment Decision (IID)**.

Select

This phase is sometimes referred to as the pre–Front End Engineering and Design phase (FEED) or feasibility phase. The best concept solution for delivering value from the opportunity is selected. The reasons for alternative choices not being selected need to be explained. If a business partner is to be selected, this then leads to a mandate to negotiate. The gate at the end of this phase is normally referred to as the **Preliminary Investment Decision (PID)**.

Define

This is normally referred to as the FEED. The technical scope, cost and schedule are developed and the commercial approach [joint venture (JV), joint operating agreement, or other] will be decided. These are put forward for gate approval as the **Final Investment Decision (FID)**.

Implement (Execute)

This is where the physical asset is built and is often referred to as the Engineer, Procure, and Construct (EPC) phase. The ideal outcome is a project completed on budget and on schedule and able to produce at the required performance level. A prestartup audit normally takes place for major projects.

Operate

The asset is operated to maximize the return to shareholders. A post project audit/review normally takes place about 1 year after startup to check the asset performance against the final investment proposal.

Divest/Decommission

At the end of the operating life, the decision is made as to divest or decommission. The investment cycle is depicted in Fig. 33.2, where performance is monitored throughout the cycle.

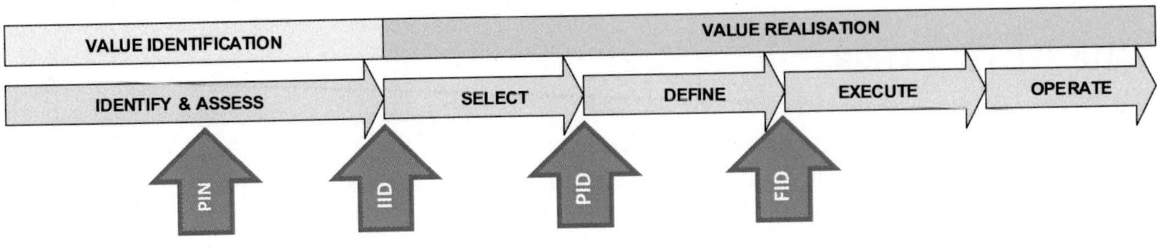

FIGURE 33.2

Investment lifecycle.

Phased approval is essential as the scope of the project is more clearly defined and the justification could change. This approach also minimizes the risk of a major capital outlay at the start of the development.

Box 33.1 emphasizes the value of phased approval where a project was terminated.

> **BOX 33.1 THE VALUE OF PHASED APPROVAL—TERMINATION OF A PROJECT**
>
> A Middle Eastern National Oil Company (NOC) decided to build a new refinery in their home country. The project proceeded through the Initial Investment Decision (IID) gate and Preliminary Investment Decision (PID) gate. However, when it came to the Final Investment Decision (FID) gate approval, it was realized that the supply and quantity from the state's oil fields could not be guaranteed. The project was abruptly shelved.
>
> **Comment**
>
> The project should never have got to FID submission as the source and amount of crude should have been firmed up long before the PID gate.
>
> Inadequate gate approval processes were to blame for the project not being stopped at an earlier stage.

Box 33.2 emphasizes the value of phased approval where a project was reassessed in the FEED phase.

> **BOX 33.2 THE VALUE OF PHASED APPROVAL: REEVALUATION OF A PROJECT IN FEED PHASE**
>
> A Middle Eastern refinery decided to upgrade the quality of diesel produced, so as to comply with the latest euro standards.
>
> The project proceeded through the IID and PID gates and, at the end of the FEED phase, the project was reassessed and found not to be viable with the selected licensor. The project team decided to redo the FEED with a different licensor and the project was finally approved at the FID gate for Engineering, Procurement and Construction to proceed.
>
> **Comment**
>
> The first FEED cycle cost the company US$8 million, but this was still considered preferable to being trapped with a low ROI for the rest of the life of the plant, with an unsuitable licensor.

33.4 INVESTMENT RISK ASSESSMENT

BUSINESS RISKS

Business risk is assessed by using Independent Reviews (IRs). IRs should take place before each investment approval gate—IID, PID, and FID, as well as 1 year after startup. It should be undertaken by experienced project and line managers, who are not involved with the project and can, therefore, offer unbiased feedback.

The reviews, before each approval gate, give the decision-makers confidence that the project team has done everything necessary to make an appropriate decision. This includes identification of all significant risks and opportunities that can be managed and that work has been completed to the necessary quality so that the project is not unnecessarily exposed.

The review after 1 year of operation (post project review) identifies lessons learned and compares the actual performance of the assets against that in the project proposal.

PROJECT RISKS

Project risk management should be in line with the PMBOK Chapter 11, Project Risk Management.[2]

Project risks are managed in parallel to the business risks for the duration of the project. At handover to the operator, the residual risks from the project are integrated into the business risk model for monitoring and review by the operator. Fig. 33.3 depicts the full risk process.

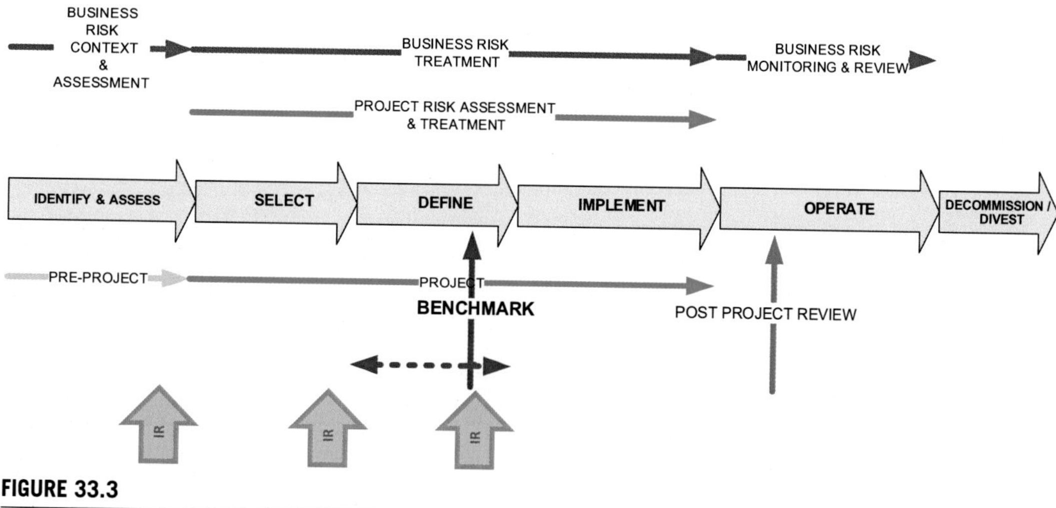

FIGURE 33.3

Full investment risk process.

33.5 PROJECT RISK MITIGATION

RISK REGISTER

Project risks need to be tabulated in a risk register. The project risk register lists risks and the agreed mitigation for each risk. Tools such as "Easyrisk" are useful for project risk management.

Example:

In Qatar, the Shell Pearl GTL project team successfully used Easyrisk.[3] *This was Shell's biggest project at the time.*

RISK MITIGATION PROCESS

Fig. 33.4 depicts an example of a State Oil Company project risk mitigation process.

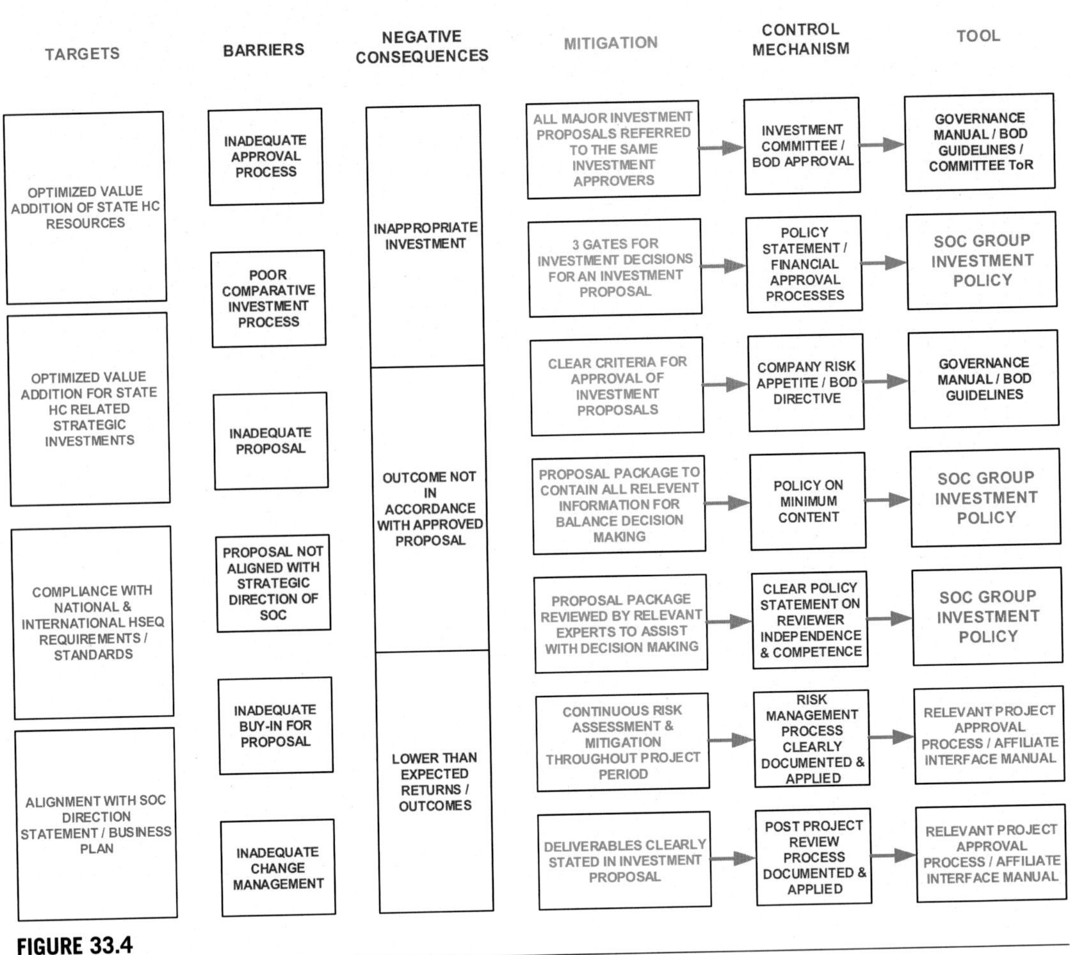

FIGURE 33.4

Risk mitigation process example—SOC.

BENCHMARKING

For projects greater than a certain value, major project developers carry out a benchmarking exercise. This entails comparing the project to similar previous projects for cost and time of implementation. Benchmarking for large projects is discussed in Section 24.10, Project Benchmarking.

SPECIFICATIONS

The Basic Engineering Design Specification (BEDS) is a key document in the Project Proposal Package and EPC Bid Package. Two key elements of potential major risk to the project are discussed as follows:

1. Analysis of raw materials
2. Analysis of ground conditions for construction

Analysis of the Raw Materials

Analysis of the raw material expected to be processed is a key element in determining product output. Box 33.3 gives an example of what is required before going out to bid for a new process plant so as to mitigate the risk of noncompliance with a performance guarantee.

> **BOX 33.3 INVESTMENT RISK MITIGATION—DESIGN SPECIFICATION CONFIRMATION**
>
> When a new cement plant was designed for Cape Portland Cement Company in South Africa, tons of limestone from the quarry was shipped to Germany for analysis and processing. This determined the type of processing required, the expected throughput, and product quality. This was before inclusion of the raw materials specification in the tender documents for the new cement kiln train.
>
> The tender included the requirement for a performance guarantee based on the raw material specification. On startup, the new kiln train met the performance guarantee for throughput and quality.
>
> **Comment**
>
> To ensure enforcement of a performance guarantee, a good analysis of the raw material to be processed is required to be included with the design specification.

In the case of oil and gas feedstock, the sulfur content influences the materials of construction and process plant design. Box 33.6 gives an example where the sulfur content of the feedstock could have been a factor in inadequate design.

Analysis of Ground Conditions

Full analysis of ground conditions is essential. This includes physical analysis by taking core samples at site, as well as analysis of seismic activity, especially on a coast with a risk of both earthquakes and tsunamis. Design, construction, and operation could be seriously compromised if these risks are not addressed and mitigated, preferably in the design phase. If necessary, an earthquake resistant design philosophy has to be adopted (see also Section 12.7: Black Swan for a discussion on tsunamis).

Box 33.4 gives an example of project delay caused by ground conditions.

> **BOX 33.4 INVESTMENT RISK MITIGATION—GROUND CONDITIONS**
>
> A new cement terminal was planned for Cape Town, South Africa. Soil analysis over the entire site was undertaken before commencement of design, resulting in 180 piles being specified for the 60-m-high cement storage silo.
>
> Right from the start, the project was delayed when it was found that the pile-driving machines could not penetrate a layer of clay that was close to the surface at the precise location where the silo was to be built. Tons of clay had to be removed and replaced with sand for the piles to drive through.
>
> **Consequence**
>
> This caused a 2-week delay in the project schedule. The soil analysis should rather have focused on the exact location of the silo and not spread across the whole area.

POOR RISK MITIGATION

There are also cases where risks are listed in the risk register but the required risk mitigation is not carried out. Poor risk mitigation is demonstrated in Box 33.5, which gives an account of inadequate preparation of the plant before startup.

BOX 33.5 POOR PROJECT RISK MITIGATION—PETROCHEMICAL PLANT

A new petrochemical plant was started up without proper flushing and passivation of the cooling water system. This resulted in excessive corrosion in the heat exchangers, thus reducing the life of the exchanger bundles.

Further to this, the stainless steel product lines were not adequately cleaned before initial startup, resulting in rejection of the finished product. To attain the required standard of product quality, the plant had to be shut down and the process lines recleaned and flushed before starting up again. This took a number of weeks.

Comment

Experience in passivation of cooling systems is a "must" for project commissioning staff. Their contractors also need to have a proven track record for this sort of work.

In addition, the preparation of other piping systems, especially product lines, is as critical.

Some risks are not anticipated and only come to light during or after startup. These need to be registered in the project close-out report under "lessons learned" to ensure that the same mistakes are not encountered in future projects.

Box 33.6 gives an example of attempting to determine the cause of failure after startup, with alternative possible causes being poor raw materials specification and poor quality control during manufacture.

BOX 33.6 PROJECT RISK ROOT CAUSE ANALYSIS (RCA) – CAT CRACKER WOES

A major refinery expansion project entailed building a Fluidic Catalytic Cracking Unit (FCCU).

During initial operation after start-up, spalling started to occur in the reactor vessel. The licensor and contractor were called in to identify the problem and repair the damage.

Two experts, one each from the licensor and the vessel manufacturer, presented theories for the failure.

Theory 1

The vessel had been hydro-tested in the factory with the lining in place, and then put on the quay ready for shipping where it had lain for an extended period. It was surmised that residual water from the hydro-test had penetrated the lining and started to corrode the vessel, thus resulting in the separation after start-up.

Theory 2

The high sulfur content in the feedstock (higher than the design value) had caused the spalling.

Outcome

The result was that the owner paid for a new (replacement in kind) reactor vessel which was installed in the first turnaround 3 years after startup. Unplanned downtime to repair the regenerator in the first 3 years of operation was a major contributor for the refinery being in the bottom quartile of the Solomon Benchmarking study.

Comment

Attention to detail with respect to required standards during manufacture and pre-delivery is essential to ensure that this sort of thing does not happen. See also Section 18.6, Third Party Compliance Checks.

Furthermore, it is vital that the design specification be based on an analysis of samples of raw materials that the plant is expected to process. See item 1, Analysis of Raw Materials section.

BLACK SWANS: "UNKNOWN–UNKNOWNS"

Project cost and schedule contingency is included to cater for "known–unknowns." However, totally unforeseen events may also occur. These are "unknown–unknowns."

"Unknown–unknowns" are discussed in Section 12.7, Black Swan.

And Box 33.7 gives an example of an "unknown–unknown," where project time was recovered with a "plan B."

> **BOX 33.7 PROJECT RISK—"UNKNOWN–UNKNOWN"**
>
> An Engineer, Procure, and Construct contract was awarded for a new cement plant in South Africa. The plant was to be built in France, using the French government's tax initiatives for heavy industry, and then shipped to South Africa.
>
> The kiln train and rotary ball-mill were shipped in several shipments as deck cargo. However, one ship encountered a heavy storm in the Bay of Biscay and subsequently lost its deck cargo, which included the large rotary ball-mill.
>
> The first problem was the cost, since the insurance of the shipments did not fully cover the loss of the single ball-mill.
>
> The second problem was that the ball-mill was on the critical path of the project and was desperately needed get the project back on track.
>
> Luckily, a manufacturer in South Africa was able to produce a new ball-mill on a fast-track basis and then ship it to site on a large low-bed road transporter.
>
> The project team eventually got the project back on track but the incident consumed a large part of the project contingency, being the difference between the insured value and the final cost of manufacture of the second ball-mill. Fortunately, the contingency was sufficient to cover these unexpected costs.
>
> **Comment**
>
> Normal practice for "unknown–unknowns" is to use "management reserve" by applying for additional funding.

33.6 INVESTMENT FRAMEWORK AND POLICY

An investment framework and policy are essential to establish policy, guidelines, and processes for investment. Fig. 33.5 shows an example of a State Oil Company investment framework and policy.

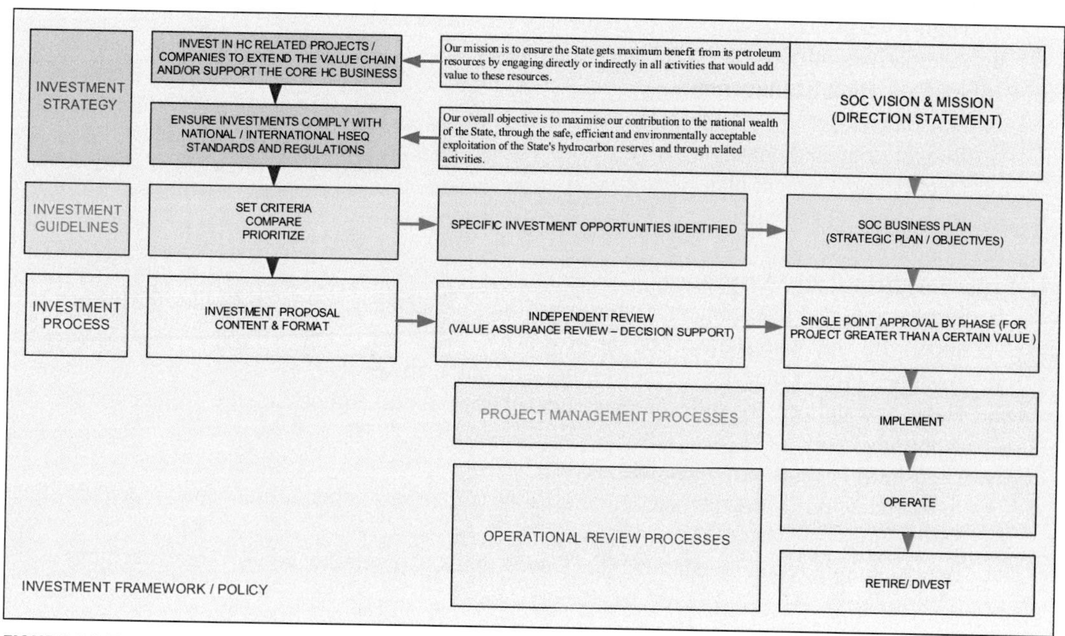

FIGURE 33.5

Company investment framework and policy.

Note that the value-addition entails supporting the Core Systems and extending the value chain and **not** diversifying the business.

Box 33.8 gives an example of the need for an investment policy as described in Fig. 33.5.

> **BOX 33.8 THE VALUE OF AN INVESTMENT POLICY**
>
> A Middle Eastern NOC wished to build a refinery in South America. The project was driven by the project manager from conception with no formal IID process or investment Policy in place. The project was canceled at the PID gate after the company investment policy had been formalized.
>
> The project did not comply with the company investment policy of extending the supply chain, as the crude would have to have been sourced on the international market.
>
> **Discussion**
>
> This example shows the need for an investment policy, as well as the basic requirement that a project manager not be appointed until after the PID gate. He should be a team member but certainly not the key driver of the project before PID. This is in line with the requirements of the PMBOK project management standard (see PMBOK Section 3.3: Initiating Process Group).[2]

33.7 APPROVAL PACKAGE

Each investment approval gate (IID, PID, and FID) requires the same format for the investment package. The items listed next are suggested with increased definition as the project develops to FID.

1. **Project Definition—Business Case**
   i. Project purpose and justification
   ii. Measurable project objectives (deliverables) and related success criteria
   iii. High-level (business) requirements (completion date, etc.)
2. **Technical Proposal**
   i. High-level project description (statement of requirements)
   ii. Alternative solutions (at end of selection phase)
3. **Project Execution Management**
   i. Schedule
   ii. Budget (total and phase)
   iii. Quality management plan
   iv. Risk management plan
   v. Contracting strategy
4. **Project Approval and Assurance**
   i. Project approval requirements (what constitutes project success, who decides the project is successful, and who signs off on the project)
   ii. Assigned project manager, responsibility, and authority level
   iii. Name and authority of sponsor, owner, or other person(s) authorizing the project and budget proposal
   iv. High-level business and technical risks
   v. Independent project review recommendations (for capital expenditure, equal to or exceeding $50 million)
   vi. Benchmarking recommendations (for O&G capital expenditure, equal to or exceeding $500 million)

Fig. 33.6 shows a model that could be applied for this purpose for a JV project.

Project Business Case Category approval	
Project Objectives and Statement Of Requirements (SOR)	Discipline approval
Project fit with corporate strategy (big picture) and approved development plans	Discipline approval
Project development options and any project risks and uncertainties	Discipline approval
Project benefits and economics – (NPV, Payout, IRR, ROI)	Discipline approval
Project interface with other developments	Discipline approval

Technical Proposal & PlanCategory approval	
Project Scope Of works (SOW) and any outstanding options	Discipline approval
Interface with other facilities	Discipline approval
Operations aspect – operating philosophy, operations plan	Discipline approval
Engineering status – completed studies and key deliverables	Discipline approval
Engineering risks	Discipline approval

Project Execution Management & PlanCategory approval	
Project Execution Plan	Discipline approval
Quality Plans & Review schedule	Discipline approval
HSE Plans	Discipline approval
Project Control Schedule & Cost for complete project & for each generic phase	Discipline approval
Project Execution Risks and Risk Management Plan	Discipline approval
Statutory approval plans and status	Discipline approval
Contracting Strategy	Discipline approval
Organisation Plan	Discipline approval
Interface management plan	Discipline approval

Agreement Plan Category approval	
Licensors	Discipline approval
LoI / HoA / MoU / HoA / JVA	Discipline approval
EPC	Discipline approval

Funding Plan Category approval	
Internal	Discipline approval
External	Discipline approval

Project Approval & Assurance Plan Category approval	
Project Milestone Review report	Discipline approval
Need for Benchmarking report (typically projects > 2 billion) been addressed	Discipline approval
Project Risk Register & Risk Mitigation Plan	Discipline approval
List of all Assurance Reviews to be carried out (including IA and benchmarking)	Discipline approval

PROJECT PROPOSAL PROPOSER & APPROVERS

FIGURE 33.6

Model investment proposal package for JV project.

33.8 INVESTMENT DECISIONS

Further to Section 21.5, Capital Expenditure/Investment, the decision to invest more initially, to ensure a safer plant or lower operating cost, is a difficult one. Box 33.9 gives an example of higher investment based on inherent safety.

Box 33.10 gives an example of higher investment based on long-term economic factors.

BOX 33.9 GOOD INVESTMENT DECISION—INHERENT SAFETY

The French pressurized water reactor (PWR) design was used for Koeberg Nuclear Power Station situated outside of Cape Town in South Africa.

The decision was made to put several safety measures in place, even though this involved extra cost. For example, the reactors' control rods shut the plant down by gravity when released and therefore no power is required for the operation. The reactor buildings are mounted on large rubber seismic cushions, which are designed to withstand the force equal to the direct impact of a Boeing 747.

In addition, the location was carefully evaluated. Koeberg is situated on the coast 50 km north of Cape Town. Cooling water comes from the cold Benguela current, which sweeps up the coast and away from the city. Although highly unlikely, if cooling water contamination reaches the sea, this contaminated water would be swept away from densely populated areas.

Comment

TEPCO Nuclear Power Stations in Japan, when built, had the known possibility of tsunamis and earthquakes. Nonetheless, design precautions were inadequate and could not stop the meltdowns of 2011.

BOX 33.10 GOOD INVESTMENT DECISION—LOWER OPERATING COST

The Cape Portland Cement (CPC) 2000-ton per day distribution terminal in Cape Town was designed for minimum power consumption. Cement is offloaded from rail-tankers and lifted 60 m into two concentric silos by means of a belt-bucket elevator. Using gravity, the cement then flows from the silos into road-tankers and bagging machines.

At the start of the project, the economics of extra investment in a belt-bucket elevator with a low electrical operating cost as opposed to a cheaper low-density pneumatic conveyor with a high electrical operating cost was carried out. The difference in electricity charges over the life of the project clearly favored the extra initial investment in a belt-bucket elevator.

The conventional separate conveyors and loading hoppers above the truck loading bays and bag packing machines were eliminated by having these bays and machines under the silos. The configuration is shown here.

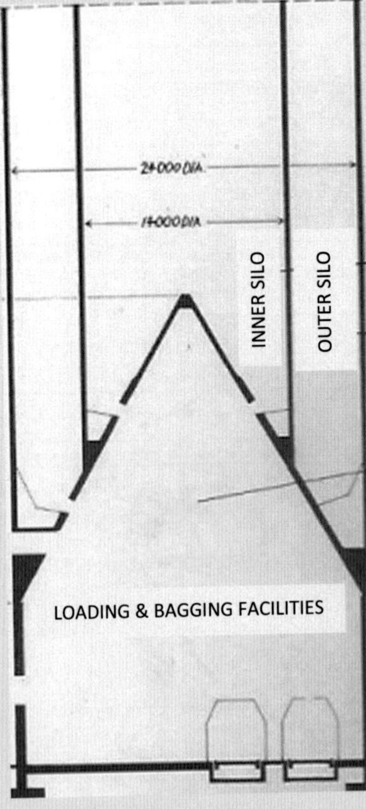

Comment

With major increases in the cost of electricity in South Africa, the decisions made at the project design phase have resulted in a major saving in operating costs.

Cutting capital investment costs to get a project approved may have long-term detrimental effects. Box 33.11 discusses the alternatives for FCCU flue gas scrubbing.

BOX 33.11 TECHNOLOGY INVESTMENT CHOICE—FLUE GAS SCRUBBING.

Box 12.6, Tripod: Tank Farm Flooding, shows an RCA for a problem raised by the use of wet gas scrubbing. The investment choices were “wet gas scrubbing” or “dry gas scrubbing.”

Wet Gas Scrubbing

- High efficiency
- **Higher operating costs**
- **Lower investment costs**
- Few users for FCCU flue gas scrubbing

Dry Gas Scrubbing

- High efficiency
- **Low operating costs**
- **High investment costs**
- Many users in the process industry including FCCUs

Comment

The final choice of wet gas scrubbing might not have been a good choice due to the higher operating costs. In addition, the choice of new or alternative technology that is not currently common in a specific industry always involves higher risk.

Sometimes poor investment decisions come to light only after startup. Box 33.12 discusses a decision that looked good from an ROI perspective but failed to assess the environmental implications.

BOX 33.12 POOR INVESTMENT DECISION—TABLE BAY POWER STATION

In the early 1970s, the Cape Town City Council decided to convert the boilers of its aging 200-MW power station from coal to oil firing. The justification was the supply of cheap fuel oil from the local “topping” refinery and so was largely driven by economic motives and the environmental impact was not considered.

The power station was on the edge of the city with high-rise buildings on one side. After conversion, it was found that the high sulfur fuel resulted in acid “smut” being emitted from a chimney during startup of a boiler. This caused damage to the paintwork of nearby vehicles, and complaints of women’s stockings dissolving were also reported! In addition, during certain wind conditions, the fumes from the stacks entered the air-conditioning of nearby buildings, resulting in complaints from the occupants.

Various measures were undertaken to reduce emissions and to keep the power station running.

Meanwhile, the refinery decided to install a catalytic cracking unit to “crack” the fuel oil and get better value for the resulting products. Thus, the availability and cost of fuel oil to the power station became such that it was impossible to continue running the power station. After a few years, it was demolished and the site became prime real estate in the expanding city.

Comment

The changing environment in which a process plant operates puts pressure on the owners to continue to mitigate emissions. Changes in technology and resulting different emissions call for an environmental impact assessment before proceeding with a project.

The long-term secure supply of raw materials, at a reasonable price, also needs to be assessed from a risk perspective.

Continued

BOX 33.12 POOR INVESTMENT DECISION—TABLE BAY POWER STATION—cont'd

Table Bay Power Station (TBPS) in Its Heyday

Photographer unknown.

As discussed in Section 11.5, Risk Appetite, the company's "risk appetite" determines the level of risk the company is willing to take. New technology generally entails a higher-than-normal business risk. Box 33.13 discusses a fortunate positive outcome from a risky investment in new technology.

BOX 33.13 INVESTMENT RISK—NEW TECHNOLOGY

Having major gas reserves, an NOC decided to invest in a gas-to-liquid plant that involved new technology. It signed a JV agreement with a partner who had previously designed and operated similar plants.

The Processes/Systems, People, and Performance management (3P) aspects were all well established based on those of the venture technology partner. However, new technology involved in the process was based on a small-scale pilot plant and had not been tested in a full-scale commercial plant.

ROI was based on a specific throughput (design capacity) and product "slate" price. After major project cost and time overruns, the plant finally started up but could achieve only half the design capacity. Nevertheless, the shareholders were delighted as, by a happy coincidence, the product "slate" price had doubled on the market.

The plant eventually achieved design capacity, after more than 1 year of production.

Comment

Shareholders were aware of the risk of the new technology but were still willing to take that risk.

33.9 SUSTAINABLE DEVELOPMENT PROJECTS

In Section 21.8, Carbon Credits, UNFCCC and the Kyoto Protocol are introduced with three market based mechanisms for emission reduction: JI, Clean Development Mechanism (CDM), and IET.

The core of the two-prong objective of CDM is that CDM project activities must contribute to **sustainable development** in the host developing country.

A formal ranking process covering all Sustainable Development (SD) projects is suggested. The ranking process enables SD projects to be prioritized to guide implementation. Box 33.14 gives a sample ranking matrix.

BOX 33.14 SUSTAINABLE DEVELOPMENT PROJECT RANKING MATRIX

The criteria span **economic, environmental, and social pillars** of a sustainable development project. Each pillar is subdivided into several criteria (elements) with allotted weightings. Additionally, each criterion is provided with a grading so as to determine the degree to which it contributes to sustainable development requirements. The score for each criterion is provided by the product of the weighting and the grade. For a number of sustainable developments projects, the sum of scores of the criteria is used for ranking the projects.

	Criterion	Allocated Weight (W)	Grading (G) L	M	H	Score W*G	Range/notes
1.0	**Economic**	**0.40**	1	3	6		
1.1	Profitability Index	0.15					L ? 1; 1 < M ? 1.2 H > 1.2
1.2	Technology Transfer	0.10					L = Commercially Available Technology M=Modern Technology H= Advanced Technolgy
1.3	Energy Savings	0.05					L <10 %, 10% < M ?15%, H> 15% TOE/year of BAU
1.4	Water Savings	0.05					L <10 %, 10% < M ?15%, H> 15% TW/year of BAU
1.5	Adding Value to Resources	0.05					L = No value added to resources M = Value addition to resources H = Secondary value addition
2.0	**Environmental**	**0.40**	1	3	6		
2.1	Reduction of Greenhouse Gas Emissions	0.12					L<5 %, 5% < M?10%, H> 10% Tons/year of BAU emissions/effluents/waste
2.2	Satisfies Air Quality Standards	0.06					L = Comply with MoE Legislations M = Comply with WHO Guidelines H = Better than WHO guidelines
2.3	Satisfies Water Quality Standards	0.06					L = Comply with MoE Legislation M = Comply with EU Legislation H = Better than EU Legislation
2.4	Satisfies Soil Quality Standards	0.06					L = Comply with MoE Legislations M = Comply with Canada Guidelines H = Better than Canada's Guidelines
2.6	Depletion or Renewal of Natural Resources (e.g. oil, gas, flora, etc)	0.10					L = Depletion of natural resources inventory M = No change in natural resources inventory H = Renewal of natural resources inventory
3.0	**Social**	**0.20**	1	3	6		
3.1	Job Creation	0.10					L = job losses/no job created M = At least one job created H = More than one job created
3.2	Increase in Social Amenities	0.05					L = No social amenity provided M = One social amenity provided H = More social amenities provided
3.3	Improvement in Quality of Life	0.05					L = No improvement or decrease in quality of life M = Marginal improvement in quality of life H = Significant improvement in quality of life
	Total	**1.00**					

33.10 INTEGRATION WITH THE BUSINESS PLANNING CYCLE

All proposed projects that have passed the PIN phase are deposited in the "Portfolio of Opportunities" register. Those that go forward to the selection phase, thus incurring expenditure, are registered in the "Capex register." All projects in this register are then taken forward into the Business Plan and Budget and entered in the company asset register. The business plan and budget is normally at least a 5-year projection (see Fig. 33.7).

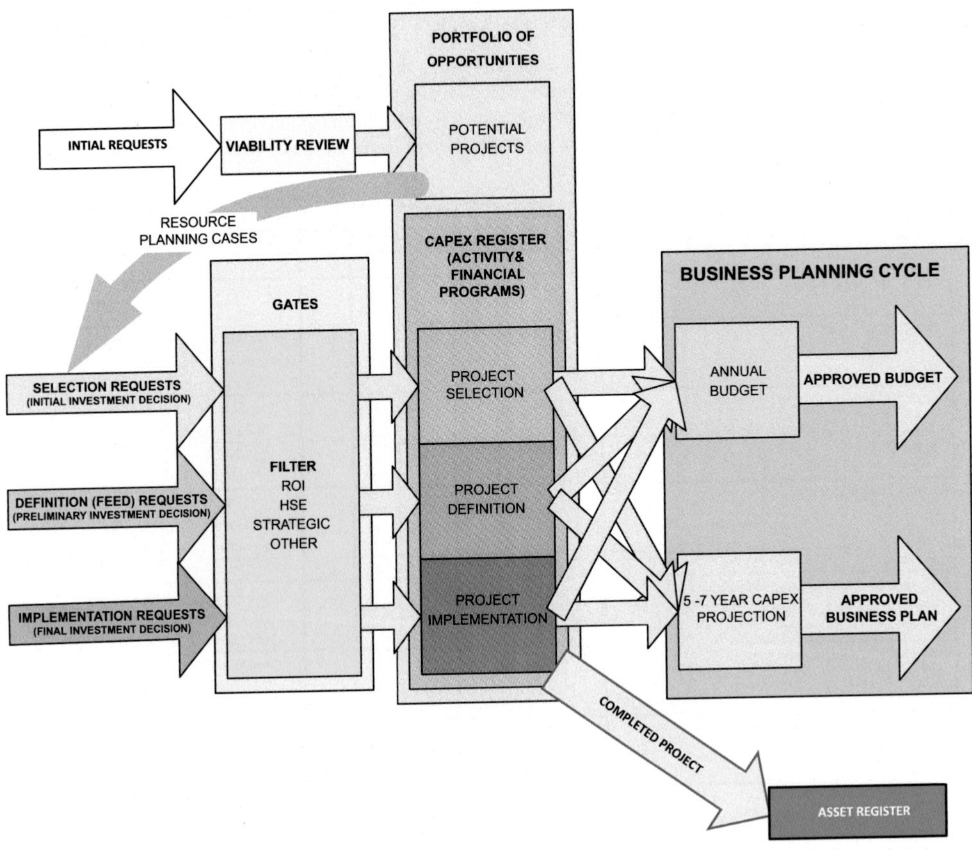

FIGURE 33.7

Capital project integration with the business planning cycle.

33.11 PARTNERING

PHASES

If a partner is sought, the process is generally as follows:

Preparation

A partner is normally sought if they can bring something significant to the table—technology, finance, supply chain, etc.

As with any other business opportunity, the early phases of the opportunity realization process should provide the most potential to optimize value in the JV structure and governance arrangements. Once the JV documentation and other relevant agreements are concluded, there will be limited opportunities to change the way in which the JV is governed and managed. It is therefore vital to identify the key governance and management issues early in the preparation phase to maximize value over the whole lifecycle of the JV. Where possible, appropriate enabling clauses for subsequent phases in the JV documentation should be included at the start of the JV.

Due diligence on the proposed JV and the other participants should be conducted during this phase to:

a. evaluate governance and management risks;
b. consider how the **foundation** components of the **governance framework** are reflected (see Section 8.5: The Model Governance Framework document: Discussion of contents);
c. assess the reputational risk, compliance history, and financial capabilities of the prospective participants;
d. assess the prospective JV management team if prospective participants are already identified.

Due diligence can continue into the negotiation phase but must be completed before making a binding commitment to form a JV.

During this phase, a **Proposal to Commence Negotiations (PCN)** would be prepared. A **PCN** must be approved before the start of any negotiations or other material engagement with a third party on a potential new or supplementary business opportunity. A mandate (opening position, target, walk away) to agree the terms that will apply to a potential commitment should be based on an approved **PCN**. The risks (e.g., financial, commercial, political, or other) to which the JV will be exposed should be highlighted in this document.

Negotiation

As the potential relationship is long term, the strategic fit has to be right.

During the negotiation phase, all key governance and management aspects of the JV are agreed as part of the overall commercial deal. The negotiations cover both the arrangements between participants and arrangements with the parent company as service provider. In these negotiations, it is therefore important for negotiators to consider in what capacity they operate (participant or prospective service provider) and to focus on delivery of an outcome that maximizes value for the parent company as a whole.

Formation

The partnership agreement acceptable to all participants is finalized and signed. This phase concentrates on the activities required to prepare for the startup of the JV. JV setup activities can be time consuming because of the multiple parties involved and therefore should be planned well in advance of the **legal agreements** becoming effective. JV risks can best be mitigated when there is sufficient time for project teams to identify and assess the risks and develop mitigation measures.

This is also the start of the application of the JV governance processes following agreement on these in the negotiation phase. Individuals who have been appointed by the JV participants would start to put in place key foundations and organization activity components of the JV control framework, although handover would not yet have been formally transferred to the JV management team.

During the formation phase, the participants of the JV may still make decisions on the JV's control framework, to the extent that it was not finalized during the negotiation phase. *For example, more details on the structure, tax, planning processes, delegation of authorities, assurance mechanisms, and financial and project progress reporting may still be negotiated at this phase.*

The formation phase ends with the formal handover of responsibilities from the opportunity realization team to the JV governance and management teams.

Operation

The ongoing shareholder rights are enforced as per the partnership agreement.

During the operations phase, the JV is run in line with the JV documentation. This phase covers the development, construction, and operation phases of an asset. During this phase, the governance by the parent company and management of activities by the JV are carried out through the relevant governance, management, and service provider roles described in Chapter 34, Roles and Responsibilities.

The operations phase is typically the longest phase in the JV lifecycle. During this time, the objectives of the JV and individual participant may diverge. If this leads to a restructuring of the JV, then a new lifecycle will start.

Extension, Exit, or Termination

Extension, exit or termination of a JV are investment or divestment activities and should follow a separate opportunity realization phase.

Extension

If the JV is set up for a defined duration, it may be necessary to negotiate an extension if the participants wish to continue their involvement in the JV. The extension of the JV's activities may be dependent on the extension of key licenses that support the activities of the JV. In those cases, the license extensions drive the extension of the JV itself and changes in the license conditions may result in a requirement to change the JV management.

Exit/Divestment

The parent company's involvement in a JV may end as a result of decision to exit (i.e., divest) or because of a decision by the participants to terminate the activities of the JV. An exit or a divestment may be triggered by the realization that the parent company's interest in the JV is worth more to others than to the parent company (value maximization) and/or realization that the JV no longer fits the strategy of the parent company. Postexit considerations may apply when the parent company exits a JV that continues to operate.

Termination

Termination of the activities of a JV usually occurs when its profitability is insufficient to continue its operations. Termination can take many years and during this time applicable regulations need to be complied with and posttermination liabilities, such as those related to staff pensions, need to be managed.

The parent company needs to maintain a nominated asset oversight team leader until any postexit or posttermination liabilities have been reduced to a level where they no longer present material risks. Fig. 33.8 depicts the parallel phasing.

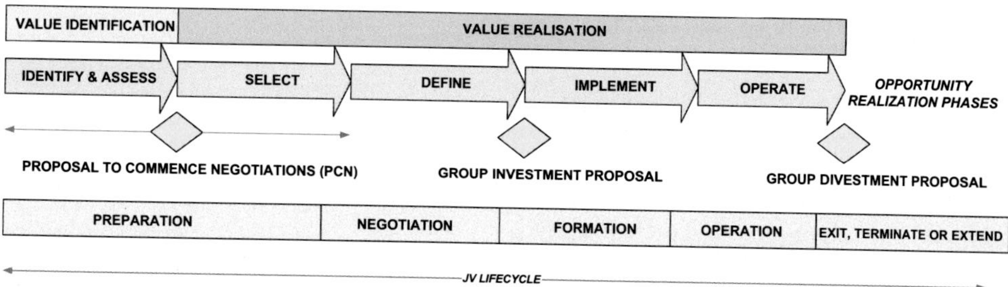

FIGURE 33.8

Parallel commercial and value realization phases of an investment.

STRATEGY

Partnering strategy is based on leverage. The question is "Will a particular partnership bring greater profit to the parent company?" Thus, partners need to have complementary contributions to the JV.

The perspectives of risk from the point of view of each partner and the JV could be different. Box 33.15 shows an example of the risk matrices of a parent company and the JV.

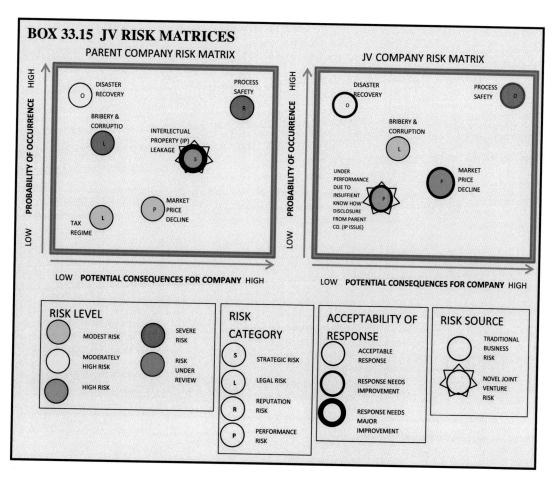

BOX 33.15 JV RISK MATRICES

PARENT COMPANY RISK MATRIX

PROBABILITY OF OCCURRENCE (LOW – HIGH)

DISASTER RECOVERY (O)
PROCESS SAFETY (R)
BRIBERY & CORRUPTIO (L)
INTERLECTUAL PROPERTY (IP) LEAKAGE (S)
MARKET PRICE DECLINE (P)
TAX REGIME (L)

LOW POTENTIAL CONSEQUENCES FOR COMPANY HIGH

JV COMPANY RISK MATRIX

PROBABILITY OF OCCURRENCE (LOW – HIGH)

DISASTER RECOVERY (O)
PROCESS SAFETY (O)
BRIBERY & CORRUPTION (L)
UNDER PERFORMANCE DUE TO INSUFFIENT KNOW HOW DISCLOSURE FROM PARENT CO. (IP ISSUE) (P)
MARKET PRICE DECLINE (P)

LOW POTENTIAL CONSEQUENCES FOR COMPANY HIGH

RISK LEVEL

- MODEST RISK
- MODERATELY HIGH RISK
- HIGH RISK
- SEVERE RISK
- RISK UNDER REVIEW

RISK CATEGORY

- S STRATEGIC RISK
- L LEGAL RISK
- R REPUTATION RISK
- P PERFORMANCE RISK

ACCEPTABILITY OF RESPONSE

- ACCEPTABLE RESPONSE
- RESPONSE NEEDS IMPROVEMENT
- RESPONSE NEEDS MAJOR IMPROVEMENT

RISK SOURCE

- TRADITIONAL BUSINESS RISK
- NOVEL JOINT VENTURE RISK

The objectives of each partner could be different and even contradicting.

Example 1:

An NOC objective may be to maximize the utilization of energy from the gas wells under its jurisdiction and thus ***minimize flaring****.*
An IOC partner objective is to maximize profit, which may entail ***increased flaring*** *of low-grade gas to increase throughput of high-value LNG.*

Example 2:

An NOC and IOC were investigating the construction of a petrochemical complex as a potential JV. The NOC perspective was that the ROI was insufficient to continue to the FEED phase. The IOC perspective was that the project was viable, even with the low ROI, as it controlled the supply chain network for further value addition of the products.

Box 33.16 discusses a national partnering strategy that maximizes the value addition for the state.

BOX 33.16 QATAR'S PARTNERING STRATEGY

Qatar has the largest nonassociated gas deposit in the world. The question was how to maximize exploitation of this reserve. The intent was to stretch the value chain.

Qatar Petroleum (QP), the NOC, thus partnered with the following major players:

1. Exxon-Mobil
2. Shell
3. Sasol
4. Total
5. Chevron-Philips
6. Yara/Hydro

Exxon-Mobil

Exxon-Mobil exploited the technology for large-scale production of LNG. They dramatically changed the approach to LNG production and shipping as follows:

1. Eliminated manned offshore platforms by bringing the wet gas onshore, before dehydrating. This reduced the investment as well as the operating costs associated with manned offshore platforms with gas dehydration facilities.
2. Increased gas compression power using the biggest gas turbines (100-MW GE Frame 9E) and compressors (Nouvo-Pignone) in the world.
3. Increased refrigeration capacity by using Air Products AP-X technology.
4. Extracted helium using Air Products and Chemicals Inc. (APCI) technology.
5. Increased LNG storage and tanker capacity by using membrane technology.
6. Changed the tanker propulsion from steam turbine, at 30% efficiency, to slow-speed diesel at 50% efficiency
7. Changed boil-off gas (BOG) management. LNG boils continuously due to heat transfer through the tank insulation. Typical boil-off rate for a modern LNG tanker is 0.15% of the tank inventory per day. Previously the BOG on the journey was used to supplement the fuel for the steam boilers feeding the steam turbines. The BOG generated on the journey is now reliquefied through a skid-mounted reliquefaction unit in the cargo machinery room. (In the future, the BOG may be fed straight to the low-speed diesel.) Also, a large amount of BOG is created when loading the tanker. The BOG, generated when loading, is now recycled to the gas plant.

Each LNG mega-train in Qatar now has the design capacity of 7.8 million tons/annum (MTPA). LNG ships have increased from a maximum of 138,000–216,000 m^3 for the Q-Flex Membrane LNG Ship and 266,000 m^3 for the Q-Max Membrane LNG Ship.

BOX 33.16 QATAR'S PARTNERING STRATEGY—cont'd

Exxon-Mobil is the prime partner with QP in Rasgas.

Shell

Shell had developed gas-to-liquid (GTL) technology and had a full-scale production plant running in Bintulu, Malaysia. They built the largest GTL plant in the world, Pearl GTL, through a production sharing agreement (PSA) with QP. Also, Shell's vast experience in petrochemicals resulted in them partnering with QP in a major petrochemical complex.

Sasol

South African Synthetic Oil (SASOL) had adopted coal-to-gas technology from Germany in the 1950s and built the biggest coal-to-liquid plants (Sasol 2 and 3) and petrochemical plant (Coalplex) in the world. They used the technology to develop GTL from offshore fields at Mossel Bay in South Africa. Part of the technology was revised, and a pilot plant was built to demonstrate the new technology. QP decided to partner with SASOL in building the first full-scale plant using this new technology in Qatar–Oryx GTL.

Total

Total's experience in the gas industry, in Algeria and elsewhere, was exploited in the establishment of a PSA with QP and others, in the form of Dolphin Energy, to produce gas for export to the United Arab Emirates (UAE) through an undersea pipeline.

Chevron-Philips

Chevron-Philip's extensive experience in the production and marketing of low-density polyethylene (LDPE) resulted in the Q-Chem joint venture with QP.

Yara/Hydro

Yara, the biggest fertilizer company in the world, partnered with QP to build the biggest fertilizer plant in the world—Qafco.

Hydro, one of the biggest aluminum manufacturers in the world, established Qatalum in partnership with QP (this is, by some, referred to as gas to solids). Gas turbines generate the electricity (1350MW) for the electric arc furnaces, producing aluminum ingots.

Yara and Hydro derived from the same company, Norsk Hydro, but are now separate businesses.

33.12 EXIT/DIVESTMENT OR TERMINATION/DECOMMISSIONING DECISION-MAKING

The questions are as follows:

a) When do you divest or dispose of your interests in the venture?

or

b) When do you terminate and decommission the assets?

The answer to the first question would be related to the marketability of the asset. When the time is reached when no one is interested in the asset, it needs to be decommissioned.

Alternative options for the real estate need to be considered if the plant cannot be disposed of as a going business. Some convert refineries to terminals as the infrastructure is already there. Others sell off the site or develop the site themselves for urban or industrial development.

Decommissioning is generally decided based on either, or both, of the following:

a) Economic viability

b) Asset integrity

The economic viability is discussed further in Chapter 21, Finance, and the asset integrity decisions are discussed in Chapter 22, Physical Asset Performance Management.

Box 33.17 gives examples of decommissioning decisions.

BOX 33.17 DIVESTMENT AND DECOMMISSIONING DECISIONS

Chevron Sydney Refinery

In 2014, Chevron decided to close their Sydney Australia Refinery and convert it to an oil import terminal.

UK Refineries

Over the years, most UK refineries have been closed on economic grounds. Milford Haven Refinery was converted to an LNG import terminal.

South African Refineries

All South African refineries are supported economically by the government fuels pricing structure and, thus, remain in operation. The oldest refinery was built in the 1950s.

Bahrain Refinery

The Bahrain Refinery continues to operate with its oldest operating process unit having been built in 1945. It is a strategic decision, by the Bahrain government, to continue operating the refinery. Unit upgrades are continuously carried out to maintain or enhance design throughput and/or product quality. Asset integrity management is of the highest standard.

Comment

Different circumstances lead to different decisions.

33.13 SUMMARY

The performance of the investment from "cradle to grave" is critical for the success of the company.

Portfolio management refers to the centralized management of one or more portfolios, which includes identifying, prioritizing, authorizing, managing, and controlling projects, programs, and other related work, to achieve specific strategic business objectives.

For major investments, there are clear phases with "hold points" or "gates" for approval before continuation to the next phase. The investment must take place within a "framework" containing an investment policy/strategy, guidelines, and process.

The process generally starts with a request to initiate the project—a PIN. Once the request has been assessed, an initial investment decision is requested. The feasibility of the investment is carried out and, if the opportunity is viable, a preliminary investment decision is requested. The project is then defined in the FEED phase and, at the end of this phase, a final investment decision is requested. The assets are constructed and, at startup, handed to the operator who operates and maintains the assets until termination of the investment.

Before each approval "gate," an independent review of the proposal should take place. In addition, until the handover, project risk management is carried out in accordance with the Project Management Body Of Knowledge. In parallel, a strategic investment risk assessment takes place. After the handover, residual risks in the project risk register are transferred to the operating company's enterprise risk management register.

At each project approval gate, the approval package must contain all relevant information for the decision-makers, including the business case, the technical proposal, how it is going to be constructed, funded, and partnered, as well as the approval and assurance plans.

Every investment proposal must be incorporated into the business planning cycle, initially in the "portfolio of opportunities" and then, if viable, in the "capex register." On startup, the investment is transferred to the company's asset register.

The partnering process runs parallel to the investment phases. The steps are: preparation, negotiation, formation, operation, and exit.

The decommissioning decision is generally based on either economic and/or integrity grounds. However, divestment is based solely on economic grounds.

REFERENCES

1. ISO 21500:2012. *Guidance on project management.*
2. ANSI/PMI 99-001-2013. *A guide to the project management body of knowledge (PMBOK).* 5th ed.
3. Risk management—Synergi Life (successor to Easyrisk). https://www.dnvgl.com/services/risk-management-synergi-life-1251?gclid=Cj0KEQjwztG8BRCJgseTvZLctr8BEiQAA_kBD8yICfL3Ifh6DIejAWSNRvxVxJ22t3SVx66uyI3yBBwaApyn8P8HAQ.

FURTHER READING

1. EV essentials: fast, affordable earned value management: Deltek. www.deltek.comhttps://education.deltek.com/pdf/ev/ds_ev_essentials.pdf.
2. Open Plan website. www.deltek.com/openplan.
3. AACE International recommended practice No. 17R-97 cost estimate classification system TCM framework: 7.3 – cost estimating and budgeting.
4. *Joint venture risk – and how to manage it the joint venture exchange.* (Issue 55). Water Street Partners; February 2013.
5. Whittaker R. *Project management in the process industries.* John Wiley; 1995.

CHAPTER

ROLES AND RESPONSIBILITIES 34

34.1 INTRODUCTION

This chapter discusses the roles and responsibilities throughout the investment lifecycle.

As the "value realization" moves from the project phases to operation, the oversight changes from the sponsor and project manager to the operator. Key objectives and Key Performance Indicators need to be kept in focus. Fig. 34.1 depicts the lifecycle with the changing oversight.

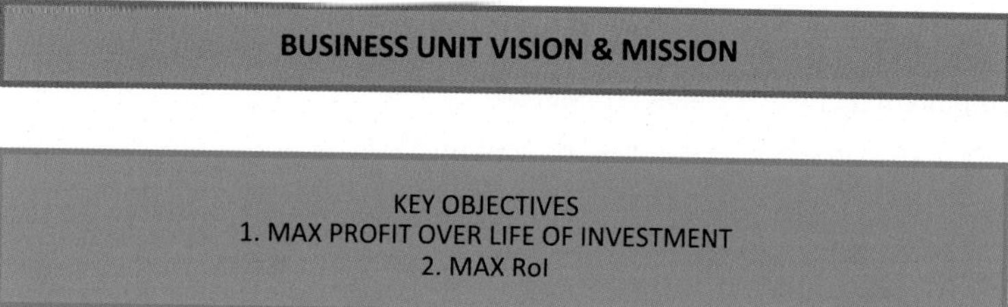

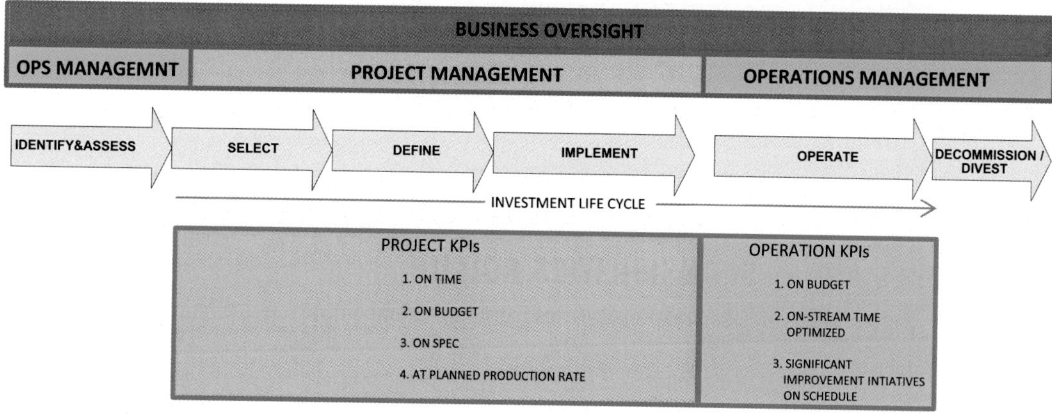

FIGURE 34.1

Investment lifecycle and oversight.

Performance Management for the Oil, Gas, and Process Industries. http://dx.doi.org/10.1016/B978-0-12-810446-0.00034-7

34.2 VALUE CHAIN OVERSIGHT AND THE CHANGING RISK PROFILE

Once an investment is operational, the oversight extends either along the entire value chain or just in those components in which the shareholder has interests. An understanding of the whole value chain is needed to minimize the risks in getting product to the customer.

To minimize risks and maximize opportunities, it is critical that this complex **value chain** is managed on an integrated basis across all Core Systems. Shareholder interests have to be governed in each of the Core Systems as well as across the **value chain**, given the integrated nature of the business.

For example, the **liquefied natural gas (LNG) value at risk** is distributed approximately as follows:

1. **10%–15% Produce**: all well and offshore surface operations, and pipelines to shore
2. **25%–35% Extract royalties**: the physical metering, review of quantity and quality, and checking and approval of royalties and taxes due to the **government** as per the relevant production agreements and marketing destinations
3. **10%–15% Process (Midstream)**: all onshore processing and loading onto LNG tankers
4. **20%–45% Market**: global marketing of LNG
5. **10%–15% Ship (and Terminals)**: bulk LNG tankers and terminals including terminals where the parent company may have a financial interest

The generic value chain is as shown in Fig. 34.2.

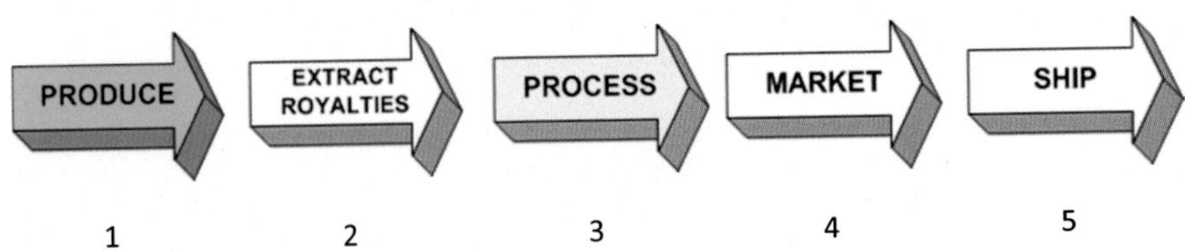

FIGURE 34.2

Generic value chain.

34.3 ROLES AND RESPONSIBILITIES ROLLUP

The following are typical roles and responsibilities from individual employee to Chairman of the Board.

1. LINE MANAGEMENT

Business Unit/Affiliate

Individual Staff Member

- Carry out duties in line with the values of the company so as to achieve agreed individual **S**imple **M**easurable **A**ligned **R**epeatable **T**imely (SMART) objectives. SMART objectives need to align with the company strategy and objectives.

Team

- Carry out duties in line with the values of the company and company strategy and objectives.

Process Owner

- Refine Processes.
- Agree on Process performance targets.
- Monitor Process performance.
- Ensure Process adoption.

System Owner

- Provide System direction by developing System vision, strategy, and objectives in line with the company strategies and objectives and certification requirements.
- Develop and implement System improvement initiatives.
- Define the Processes.
- Agree on System performance targets.
- Monitor System performance.
- Develop and manage policies and procedures related to the System.
- Ensure System adoption.

System Sponsor

- Act as the "champion" of the System: normally a senior member of staff—a department manager or higher.

Management (Certification) Representative

- Fulfill the requirements of management representative as per the relevant ISO standard(s) to which the Business Unit is certified.

Department Manager

- Maintain accountabilities and responsibilities in line with the departmental mandate and corporate direction statement.

Business Unit Management Team

- Maintain accountabilities and responsibilities in line with related agreements and corporate direction statement.

Business Unit Managing Director (MD)

- Manage the Business Unit in line with related agreements and corporate direction statement.

Advisory Committee Member

- Advise on technical and commercial aspects of the Business Unit.

Board of Directors (BoD) Member/Management Committee (ManCom) Member

- Carry out duties of Board/Committee member in line with related agreements and good governance principles.

Corporate

Shareholder Representatives

- The parent company proxy (see Section 34.5).

Asset Oversight Team (AOT)

- Advise shareholder representative of the performance of the Business Unit.

Group MD

- Manage the group of companies.

Group Chairman

- Manage the BoD.

2. PROJECTS

Decision Executive (Line Manager)

- Single point accountable for the opportunity/project.
- Chair Decision Review Committee (DRC).

Decision Review Committee

- Chaired by line manager (decision executive).
- Members to have complementary skills to decision executive.

Business Opportunity Manager (BOM)—Project Sponsor

- Articulates and protects the business case.
- Responsible for project initiation note, proof of commercial, and financial feasibility.
- Writes the group investment proposal.

Project Manager

- Responsible for delivering the technical elements of the opportunity and running the technical team.
- Manages the project from initial approval of the investment proposal **[initial investment decision (IID)]** to handover to operator of the asset.

Operations Manager

- Represents the future owner of the opportunity and is responsible for the long-term operation of the asset to generate value.

The above is summarized in Fig. 34.3.

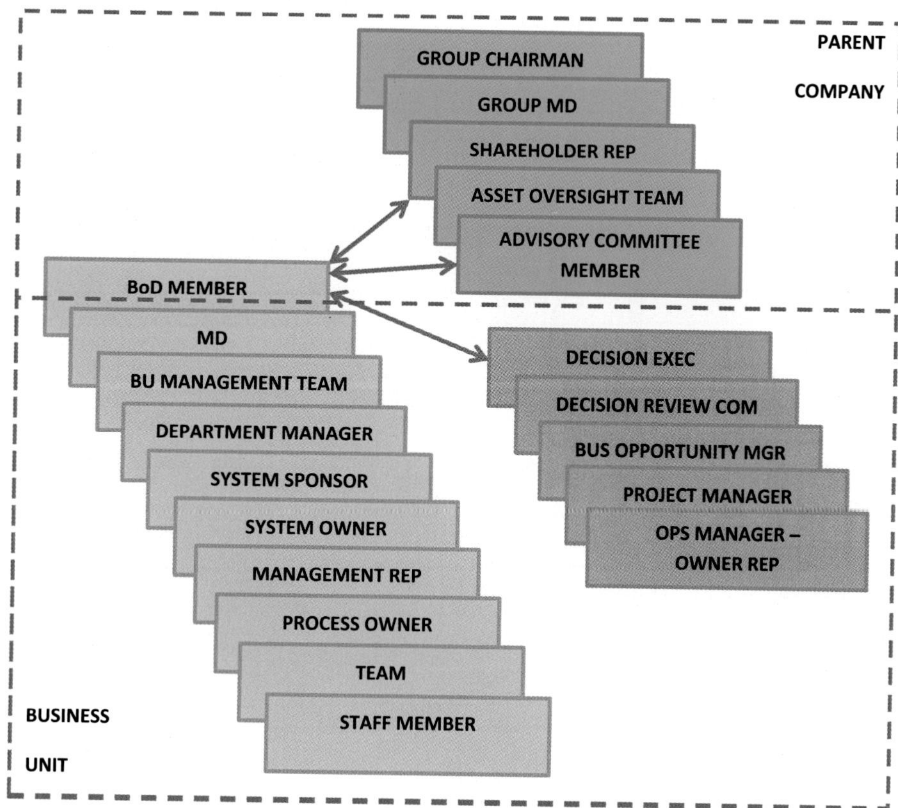

FIGURE 34.3

Roles and responsibilities summary.

34.4 ROLES AND RESPONSIBILITIES RELATED TO TYPE OF AGREEMENT

Two basic types of agreements are discussed in this section.

INCORPORATED JOINT VENTURE (JV)

Parent company roles and responsibilities for a JV are generally as follows:

1. Corporate BoD
2. Shareholder representative
3. Asset oversight team
4. Asset Board member
5. Secondee
6. Service provider
7. Regulator (as required)

These are discussed in detail in Section 34.5.
This is depicted for a State Oil Company in Fig. 34.4.

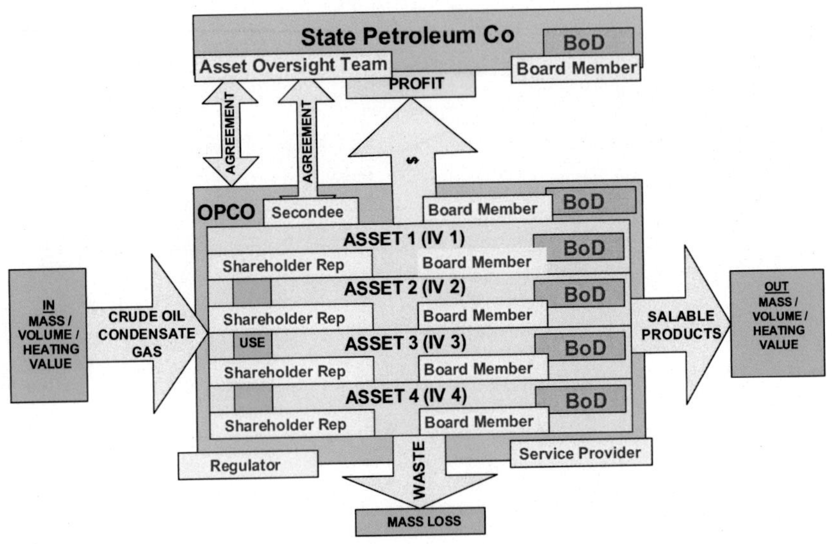

FIGURE 34.4

JV and Opco roles and responsibilities.

PRODUCTION SHARING AGREEMENTS (PSAS) AND UNINCORPORATED JVS

Parent company roles and responsibilities for PSAs and unincorporated JVs are generally the following:

1. Corporate BoD
2. Oversight team
3. ManCom member
4. Technical committee member
5. Tender committee member
6. Secondee
7. Service provider
8. Regulator (as required)

PSAs' roles and responsibilities are depicted as shown in Fig. 34.5.

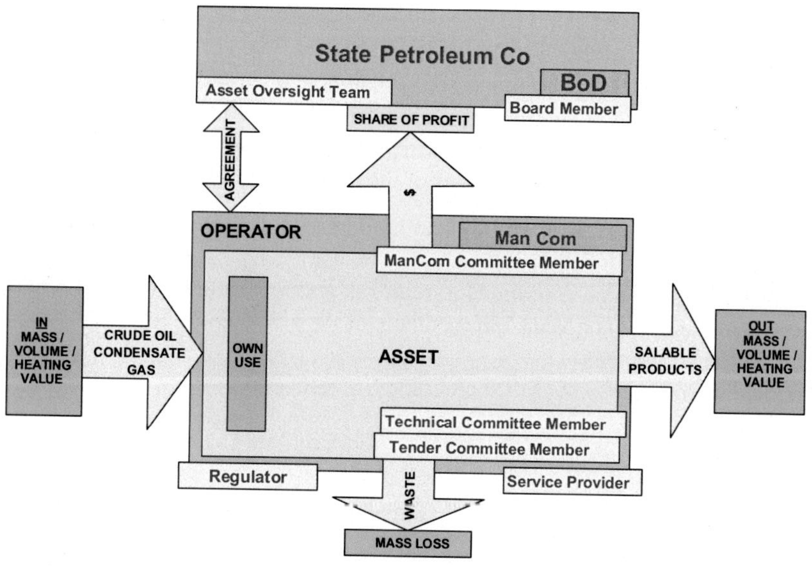

FIGURE 34.5

PSA roles and responsibilities.

34.5 AFFILIATE ROLES AND RESPONSIBILITIES—DETAILS

Focus is on the Business Unit management team and the shareholder representative. The following refers to the legal entity, being the affiliate, which consists of one or more Business Units.

AFFILIATE MANAGEMENT

These are the general categories:

1. Affiliate governance roles
   a. Shareholder representative
   b. Asset Oversight Team
2. Affiliate management roles
   a. Director/Board member or ManCom member
   b. Management team
   c. Advisory committee member (technical, tender, etc.)
   d. Operator
3. Other roles
   a. Secondee
   b. Service provider
   c. Regulator

These are discussed in detail as follows.

1. Affiliate Governance Roles

a. Shareholder Representative

A **proxy** is a corporate authority given to a person (shareholder representative) appointed to represent the parent company ownership in an affiliate, such as in a shareholders' or participants' meeting. The AOT is responsible for providing the shareholder representative with support in his or her duties.

A meeting of the shareholder representatives of the affiliate will typically:

1. have power over all matters reserved for shareholders by law and/or in the affiliate documentation;
2. have the power to recommend appointment, removal, and replacement of the parent company Directors (Board members) or ManCom members of the affiliate;
3. monitor affiliate management;
4. use appropriate methods to look after shareholder' investments;
5. appoint external auditors.

b. Asset Oversight Team Member

The AOT is internally appointed by the parent company to govern the parent company's investment in the affiliate. This is an advisory role, which will be discussed in detail in Section 34.8.

2. Affiliate Management Roles

a. Board of Directors

The parent company Directors (Board members) of incorporated JVs or subsidiaries are nominated for appointment by the parent company. Directors (Board members) should:

- be familiar with all relevant agreements;
- be familiar with applicability of the parent company governance requirements;
- take legal advice and training on the legal aspects of their role (see Section 9.4: The Board: The "Fulcrum" of Business Performance);
- be aware of the need to manage conflicting interests.

b. Management Committees

In an unincorporated JV or PSA relationship, the ManCom:

- monitors that the operator operates in compliance with the agreement documentation;
- approves the annual Work Programs and Budgets;
- monitors performance of the operator.

The ManCom includes representatives from each participant.

c. Advisory Committees

Advisory committees (e.g., technical and major project tender committees) may be established to provide support and recommendations to the affiliate Board or ManCom.

d. Operator

The operator runs the operations of the affiliate and is responsible for the implementation of the affiliate strategy, performance of the affiliate in compliance with relevant laws and regulations, internal affiliate standards and shareholders and Board decisions.

Examples of operators are Rasgas Co Ltd. and Qatargas Operating Co.

3. Other Roles

a. Secondees

Secondees are the parent company employees who have been assigned to an affiliate under a service agreement to technical, managerial, financial, or commercial positions. During their assignment, they work solely under the direction of the affiliate management in the interests of the affiliate. They must respect relevant agreements and applicable laws and be aware of restrictions on how the parent company and the affiliate can interact.

b. Service Providers

The parent company employees (who are not seconded to the affiliate) may have a service provider role in the context of arms-length service agreements that the affiliate has entered into with the parent company. This is the case when the parent company is technical advisor or provides other services to the affiliate. In these cases, employees must abide by the agreement that the parent company has with the affiliate, which may include restrictions on the transfer of information between the parent company and the affiliate. This is normally covered by a TSA.

c. Regulator

The State Oil Company may undertake the role of regulator on behalf of the State. This includes the following:

1. Allocation of exploration blocks
2. Reservoir oversight
3. Royalty metering oversight
4. GHG emissions

34.6 POTENTIAL CONFLICTS BETWEEN ROLES

There should be checks and balances that ensure that no individual has levels of authority or influence that could pose a significant risk to the achievement of the parent company's or the affiliate's business objectives.

In most circumstances, individuals should not have both a governance and management role in relation to the same affiliate, because this may give rise to a real or perceived conflict of interest. Specifically, individuals should generally not be appointed as both shareholder representative and Board director of the same affiliate. Any exceptions should be explicitly justified by the governing Business Unit or Function.

The severity of the risk associated with "double-hatting" depends on the circumstances.

Sometimes a National Oil Company has both a shareholder and a regulator role. Box 34.1 describes the potential conflicts between roles.

BOX 34.1 ROLE CONFLICT EXAMPLE—NATIONAL OIL COMPANY (NOC)

A NOC had the dual role of regulator and shareholder. A parent company specialist, as a service provider, wished to advise the affiliate on mass balance. The head of production planning in the affiliate queried the specialist's role—"advisor or regulator?"

The complication arose as both the parent company specialist and the state metering regulator, including royalty metering, were in the same department of the parent company. The metering regulator was responsible for enforcing the state regulations on both royalty and fiscal metering.

The situation was resolved by the specialist and metering regulator clearly stating their roles, and keeping their communication with the affiliate separate.

34.7 THE VALUE OF SECONDEES[1]

DEFINITION

A secondee is a legal employee of one company who has been temporarily loaned to another organization, which could be a joint venture in which the company holds an interest, to perform a specific role within that other organization. Typically, a period of secondment will last for 1–5 years.

Secondee relationships are depicted in Fig. 34.6.

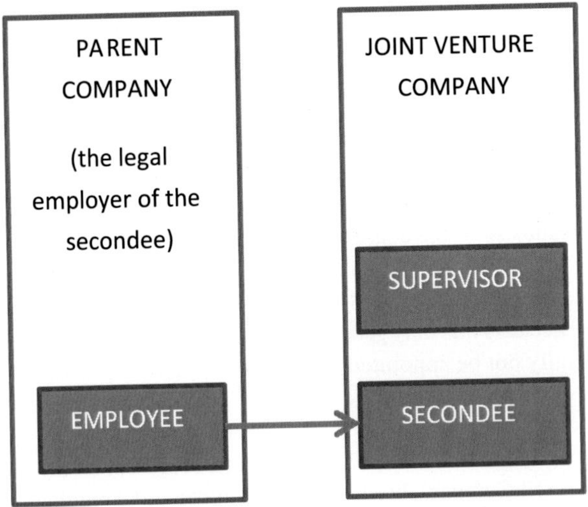

FIGURE 34.6

Secondee relationships.

VALUE

Secondees:

1. allow the parent company to quickly place trusted talent into a key asset or new business;
2. help ensure that an affiliate puts processes in place (e.g., budgeting, financial management, health, and safety) that are consistent with the parent company's standards;
3. serve as a key means for the JV to access resources from the parent company—as well as help the parent company access skills from the venture.

If the affiliate has more than a few secondees, the affiliate Board should put in place a "secondee strategy" that includes the overall target number of secondees phased over time, the relative balance of secondees between the parent companies, guidance on the types of positions where secondees are most valuable, and other items.

Secondees could sometimes result in negative value to the shareholder. An example that had serious implications for the joint venture is discussed in Box 34.2.

> **BOX 34.2 TNK-BP JV SECONDEES**
>
> A JV was formed in 2003 when BP acquired certain Russian assets and paid a group of Russian oligarchs $8 billion for a 50% interest in what then became TNK-BP.
>
> In 2008, Russian shareholders in the venture sued BP for illegally employing too many secondees, costing the JV $100 million per year. A Russian court ruled that this cost constituted a special dividend for BP at the expense of the other shareholders.
>
> BP's partners also alleged that BP was using secondees to create a parallel management structure inside TNK-BP, with no accountability to the venture. BP was forced to remove 148 seconded technical specialists, creating a situation that interfered with its ability to deliver important technical support to the venture.

34.8 SHAREHOLDER OVERSIGHT FOCUS AREAS[2]

Shareholder oversight could be grouped as follows:

- Governance
- Strategy
- Operations
- Assurance
- Business development
- Best practice sharing
- Overall assessment

Box 34.3 lists some pointers for the shareholder Asset Oversight Team, parts of which could be useful as an internal assessment by the affiliate management team.

BOX 34.3 SHAREHOLDER OVERSIGHT FOCUS AREAS

Governance

1. Establish and document focal points to act as a conduit for various parent company entities that have an interface with the affiliate.
2. Manage relationships with key affiliate stakeholders—other participants, etc.
3. Identify, drive, and monitor performance improvement initiatives that are in the best interests of the parent company.
4. Set priorities to protect, optimize, and grow the value for the parent company group.

Strategy

1. Review affiliate's strategy, direction, and annual plan (Business Plan and Budget).
2. Evaluate business performance appraisals carried out by the affiliate.
3. Provide owner perspective on market dynamics such as emerging customer needs, competitor offerings, regulatory environment, etc.
4. Review/react to affiliate investment roadmap/capex plan.
5. Review/react/secure internal resources to help strengthen business cases/underlying financials within the investment roadmap.
6. Prepare own affiliate Board members/ManCom members for strategy discussions at affiliate Board/ManCom meetings.
7. Present/clarify affiliate strategy and plan to internal parent company planning meetings.
8. Provide affiliate with core economic assumptions (e.g., long-term market price forecasts, other macroeconomic assumptions), customer needs, competitor offerings, regulatory environment, etc.
9. Monitor alignment of the parent company and affiliate's strategy and ensure that any misalignment is managed.
10. Cosponsor individual initiatives.

Operations

1. Codevelop (with affiliate management) a quarterly scorecard.
2. Review "hot topics" list—areas that could impact near-term operational performance that are not captured on the scorecard.
3. Provide affiliate management with direct feedback on improvement.
4. Review affiliate performance against targets/scorecard (including mass balance/margin analysis/product yields, energy efficiency, opex, utilization, major unplanned events).
5. Finance: verify optimum application of operating expense.
6. Production: verify sustainable production.
7. Asset management: verify asset integrity for sustainable production.
8. Personnel: verify required core competencies.

Assurance

1. Provide assurance to the parent company top management on adequacy of affiliate's system of internal control and risk management.
2. Provide assurance on compliance with the parent company Corporate Business Plan and the parent company's HSE policy.
3. Review the parent company IAD JV/PSA audits and advise Internal Audit of operational concerns.
4. Review observations from studies (including Benchmarking), reviews, etc.

Business Development

1. Review and endorse major business development investment proposals in advance of Board decision.
2. Ensure individual major investments are reviewed independently before submission for owner Board approval.
3. Ensure investment proposals are categorized such as Return on Investment, regulatory requirement (HSE, etc.), operational necessity/enhancement.
4. Embed staged approval for major investments—initial, prior to FEED, prior to EPC.
5. Coordinate global marketing strategy execution across affiliates.
6. Negotiate new agreements and amendments to agreements.

Best Practice Sharing

1. Establish networking and ideas forums (including turnarounds, major rotating equipment/asset integrity, etc.).
2. Coordinate common cost improvement initiatives—Benchmarking, common facilities, shipping, etc.

Overall Assessment

1. Complete **affiliate governance health checker** annually (see Appendix D).

Lack of adequate shareholder oversight could have serious consequences for the parent company. Box 34.4 describes a nightmare for the parent company.

BOX 34.4 INADEQUATE SHAREHOLDER OVERSIGHT RESULTING IN A MAJOR REPUTATIONAL RISK

New Zealand's Fonterra discovered that reputational risk takes on unusual features in a JV. A company is potentially exposed to damage not only from its own missteps but also from those of its **partner, over which the company may have almost no visibility into or control**.

Fonterra, the world's fourth largest dairy company and New Zealand's biggest business, confronted a PR nightmare a few years ago when it was discovered that corruption, mismanagement, and lack of adequate controls in a minority-owned Chinese baby formula JV had led to widespread product contamination. A poorly controlled supply meant that JV suppliers were falsifying protein levels, a key quality product indicator. This led to the inadvertent melamine contamination of large batches of baby formula, resulting in six deaths and the hospitalization of more than 300,000.

The CEO of the JV was sentenced to life in prison, the JV was disbanded, and the CEO of Fonterra, Andrew Ferrier, came under severe pressure to resign, in part for **committing the company to a deal that gave the company severely limited access to and influence over operations**. According to Ferrier: "The biggest single lesson that Fonterra has learnt is: **We didn't have enough say in the management of the business.**"

Water Street Partners[3] assists companies with ongoing advice on JV management.

34.9 SUMMARY

As the value realization moves from the project phases to operation, the oversight changes from the sponsor and project manager, to the operator.

A thorough understanding of the risks in getting the product to the customer, as roles and responsibilities change along the value chain, is required.

Roles and responsibilities vary according to the types of business relationships.

Roles and responsibilities are named at both shareholder and company levels. The shareholders' representatives and their asset oversight teams have a responsibility to safeguard the shareholders' interests in line with all agreements with the company. In proportion to the extent of shareholding, the shareholding company can appoint individuals to the operating company's Board of Directors as well as to various leadership positions in the company. Others could be seconded to the company for a limited period by the shareholders.

A National Oil Company may have been appointed as a "regulator for the state." This would include allocation of exploration blocks, reservoir oversight, royalty meter oversight and greenhouse gas emission monitoring.

Conflict of interest issues may occur and checks and balances need to be in place to manage these. For example, there could be a conflict of interest between allegiance to the shareholder company and to the JV.

In addition to undertaking an overall assessment of the company, shareholder representative roles and responsibilities can be categorized into governance, strategy, operations, assurance, business development, and best practice sharing. The company leadership team should ideally follow a similar format.

REFERENCES

1. JV secondee guidelines – getting a grip on secondees. The Joint Venture Exchange: Water Street Partners; November 2011.
2. Bamford J, Snyder A. Strategy under scrutiny: how a strategy committee might actually help increase speed and drive alignment among ever-watchful JV shareholders. The Joint Venture Exchange: Water Street Partners; March 2012.
3. Water Street Partners: A Joint Venture and Alliance Advisory Firm. https://www.waterstreetpartners.net/.

CHAPTER

MANAGEMENT REVIEW 35

35.1 INTRODUCTION

This chapter covers the advisors and decision-makers. The intent of "management review" is to give direction to the company and initiate any required corrective/improvement action. Assessment is the prelude to review. The outcomes from assessments are presented for review and decision-making. Assessment and review are depicted in Fig. 35.1.

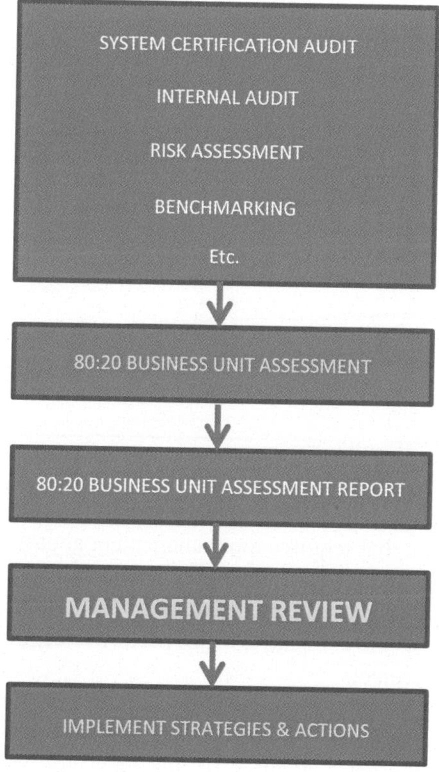

FIGURE 35.1

Assessment and management review relationships.

Performance Management for the Oil, Gas, and Process Industries. http://dx.doi.org/10.1016/B978-0-12-810446-0.00035-9

Fig. 35.2 shows assessment and review superimposed on the "After" case in Fig. 8.1, Before and after applying a Governance Framework and Systems Approach.

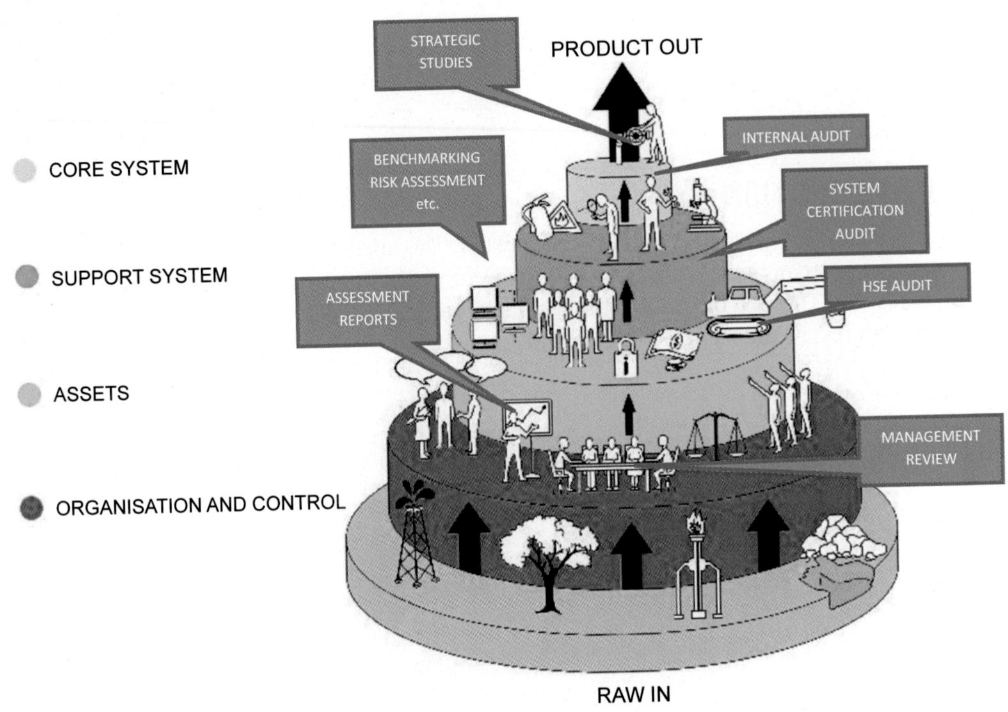

FIGURE 35.2

The "after" case with assessments and reviews in place.

35.2 SELLING A STRATEGIC ISSUE

Part 7, Assessment and Reporting covers the processes required for assessment, change management, and reporting of a strategic issue that requires top management approval.

"Selling" a strategic issue to obtain a management decision requires careful planning and strategizing to ensure approval. The strategic issue could entail proposals for dealing with a high risk to the company (i.e., outside the "risk appetite" of the company), major performance improvements, or even approval to continue to the next phase of a project.

Various proposals compete for top management's attention and managers therefore have to "sell" their particular idea to top management and other important stakeholders. They cannot assume that their strategic issue will get automatic attention or that they will necessarily win support, however important the issue might be to them in particular. Managers need to consider at least four aspects when seeking attention and support for their proposals.

1. STRATEGIC ISSUE PACKAGING

Care should be taken with how strategic topics are packaged or framed. Clearly, the importance of the strategic issue needs to be underlined, particularly by linking it to the direction statement, critical strategic goals, or performance metrics for the company. The presentation of the proposal should be consistent with the cultural norms of the company, but generally clarity and succinctness win over complexity and length. It can easily be parked as too difficult to address, if no ways forward are offered at the same time. Appropriate reporting tools, including graphics, need to be included in the package (see Chapter 31: Reporting).

Project proposal packaging is discussed in Section 33.7, Approval Package.

2. FORMAL OR INFORMAL CHANNELS

Managers need to balance formal and informal channels of influence. Formal channels are split among corporate, line, and staff. At the corporate level, they include the annual business reviews and the annual strategy retreats of the executive team. The line channel involves the regular line interaction of operational managers and the Chief Executive Officer (CEO) and other executive directors. There are also various reporting systems for staff functions, including finance, human resources, and strategic planning. Formal channels need to be two-way lines of communication.

Informal channels can be very important and often decisive in some organizational cultures. Informal channels might include ad-hoc conversations with influential managers.

3. SELL ALONE OR IN COALITIONS

Managers should consider whether to press their issue on their own or to assemble a coalition of supporters, preferably influential ones. A coalition adds credibility and weight to the issue. The ability to gather a coalition of supporters can be a good test of the issue's validity: if other managers are not persuaded, then the CEO is unlikely to be persuaded either. However, enlisting supporters may involve compromises or reciprocal support of other issues.

4. TIMING

Managers should also time their issue-selling carefully. For example, a short-term performance crisis or the period before the handover to a new top management team, are not good times to press long-term strategic issues.

35.3 LEVELS OF REVIEW

Various levels of review take place. Typically, they are as follows:

Team

1. System/departmental
2. Certification

Operational

3. Business Unit leadership: Executive Committee (ExCom)

Strategic

4. Business Unit Board of Directors (BoD)/Management Committee (ManCom)
5. Shareholder representative/Asset Oversight Team (AOT)
6. Group leadership—BoD
7. Independent: internal audit/external audit
8. Major investment project.

These are discussed in more detail as follows.

1. SYSTEM/DEPARTMENTAL

This is referred to as a peer review. ISO-certified internal auditors review the Business Systems/departments within the Business Unit.

2. CERTIFICATION

This is required for annual renewal of ISO certification. An internal review of the complete set of Business Unit Management Systems is carried out before the certification authority carries out its review.

3. BUSINESS UNIT LEADERSHIP (EXECUTIVE COMMITTEE)

This normally takes place as a weekly/monthly/quarterly performance review meeting led by the CEO.

4. BUSINESS UNIT BOARD OF DIRECTORS/MANAGEMENT COMMITTEE

This normally takes place as a quarterly performance review meeting led by the Chairman of the Board.

5. SHAREHOLDER REPRESENTATIVE/ASSET OVERSIGHT TEAM

The shareholder representative reviews the performance reports from the Business Units. These could be daily/weekly/monthly and quarterly. They also undertake an annual review of the affiliate Business Plan and Budget (BP&B).

6. GROUP LEADERSHIP—BOARD OF DIRECTORS

This is normally based on a consolidated report from all affiliates and is usually at group BoD level led by the group Chairman. Review and approval of the annual BP&B is also required of the BoD.

7. INDEPENDENT: INTERNAL AUDIT/EXTERNAL AUDIT

These are referred to as audits. Internal audits focus on the health of Business Systems and high-level risks to the business, whereas external audits focus on the financial health of the company. Internal auditors are normally Certified Internal Auditors (CIAs), whereas the external auditors are more often Chartered Accountants (CAs).

8. MAJOR INVESTMENT PROJECT

Independent Review (IR) takes place prior to each approval gate. This is discussed in Section 33.4, Investment Risk Assessment. The Decision Review Committee (DRC) then takes the decision whether to proceed to the next phase of the project.

35.4 REVIEW MEETING EVOLUTION

Review meetings need to be structured for maximum appraisal of strategic issues and their implications. Traditionally, performance was presented at the review meeting as per an agreed-on agenda but without prior circulation of the performance statistics, which would give participants time to review and discuss items with interested parties. There was often no time left in the meetings to "look ahead."

Current practice is for performance statistics to be circulated at least a week in advance of the meeting to allow participants to confer with reviewers, actioners, and other interested parties. He or she is then fully prepared to discuss and resolve problem areas in the meeting and have time to discuss strategic issues related to the future. The key is that reports be circulated timeously so that participants have time to study the data and discuss with actioners prior to the review meeting.

Box 35.1 gives an example of how a refinery triannual management review meeting evolved into one where review of strategic issues and their implications occupied most of the meeting time.

BOX 35.1 EVOLUTION OF A MANAGEMENT REVIEW MEETING

PAST	PRESENT (EVENT DRIVEN LEARNING)	FUTURE (CONTINUOUS LEARNING)	
TRI-ANNUAL MANAGEMENT REVIEW MEETING (MRM)	TRI-ANNUAL PERFORMANCE REVIEW MEETING (PRM)	BETWEEN THE MEETINGS	TRI-ANNUAL PERFORMANCE REVIEW MEETING (PRM)
REVIEW STRATEGIC ISSUES (0%)	REVIEW STRATEGIC ISSUES (10%)	PROVIDE INPUT TO STRATEGIC ISSUES UNDER DISCUSSION	REVIEW STRATEGIC ISSUES (60%)
DISCUSS IMPLICATIONS (0%)	DISCUSS IMPLICATIONS (40%)	DIALOGUE ABOUT PERFORMANCE EXPLAIN DIFFERENCES, SUGGEST SOLUTIONS, IDENTIFY STRATEGIC ISSUES FOR DISCUSSION AT NEXT MEETING	DISCUSS IMPLICATIONS (30%)
REVIEW PERFORMANCE (100%)	REVIEW PERFORMANCE (50%)	REVIEW PERFORMANCE DATA	REVIEW PERFORMANCE (10%)

MEETING MANAGEMENT: TRAINING EXAMPLE[1]

John Cleese's amusing training video *Meetings, Bloody Meetings* from the 1970s is still valid, giving an excellent guide for shorter and more productive meetings. The key points are as folows:

1. Plan
   - Be clear in your mind of the precise objectives of the meeting.
   - Be clear why you need it and list subjects to be covered in the meeting.
2. Inform
   - Make sure everyone knows exactly what is being discussed, why, and what you want from the discussion.
   - Anticipate what information is needed.
   - Make sure participants are present.
3. Prepare
   - Prepare the logical sequence of items to be discussed.
   - Prepare time allocation of each item based on its importance and not its urgency.
4. Structure and control
   - First present evidence, then interpretation, and then action.
   - Stop people from jumping ahead or going back over old ground.
5. Summarize and record
   - Summarize all decisions and immediately record with the name of person responsible for any action.

REFERENCE: ROBERT'S RULES OF ORDER[2]

A good reference for management of meetings is *Robert's Rules of Order,* which is a guide for conducting meetings and making decisions as a group. The purpose of the book is "to enable assemblies of any size, with due regard for every member's opinion, to arrive at the general will on the maximum number of questions of varying complexity in a minimum amount of time and under all kinds of internal climate ranging from total harmony to hardened or impassioned division of opinion."

It is recognized as "the most widely used reference for meeting procedure and business rules in the English-speaking world."

35.5 MANAGEMENT REVIEW RESPONSIBILITIES

BOARD OF DIRECTORS (BUSINESS UNIT AND GROUP)

The Board is collectively responsible for the success of the company. Its role is to provide leadership within a framework of prudent and effective controls, to consider strategy and approve strategic aims and business principles, to review management performance against those aims, to set values and standards and to ensure that the company meets external requirements and its obligations to shareholders/participants and other stakeholders.

The Board's role generally encompasses the following:

a. Strategic guidance and management
b. Corporate governance
c. Performance monitoring and **review**

d. Risk and audit **review**
e. Conflict of interest (COI) **review**
f. Corporate financial statements integrity **review**
g. Annual BP&B **review** and approval
h. Large investment **review** and approval (phased)

Section 9.4, The Board: the fulcrum of business performance discusses the BoD in more detail.

THE BOARD AUDIT AND RISK COMMITTEE

The board audit and risk committee reviews all audits and high level risks of the company and makes recommendations to the BoD.

MANAGEMENT COMMITTEE

A ManCom is a joint committee set up as a result of a Production Sharing Agreement or an unincorporated JV. The ManCom is collectively responsible for the success of the venture. Its role is to provide leadership within a framework of prudent and effective controls, to consider strategy and approve strategic aims and business principles.

The ManCom role generally encompasses the following:

a. Strategic guidance and management
b. Performance monitoring
c. Financial statements review and approval—Work Plan and Budget (WP&B)
d. Large investment approvals

Box 32.5, Production Sharing Agreement (PSA) Company Example: Shell Pearl Gas-to-Liquids, gives an example.

LEADERSHIP TEAM (EXECUTIVE COMMITTEE)

The leadership teams normally review performance and present Quarterly Performance Reports (QPRs) to Directors and participants/shareholders.

ADVISORY COMMITTEES (PSAS AND UNINCORPORATED JV)

Technical committees review major technical proposals such as those for energy efficiency improvements and propose for ManCom approval. Tender committees review bids for services and goods and recommend the award of contracts to management.

PARTICIPANT/SHAREHOLDER REPRESENTATIVES

The participant/shareholder and AOT team are required to review the annual BP&B. They should also complete an "affiliate operational performance health checker" annually after going through the **assessment** process described in Part 7, Assessment and Reporting. A sample "affiliate operational performance health checker" is shown in Appendix D.

Overall responsibilities are identified in Chapter 34, Roles and Responsibilities.

35.6 SHAREHOLDER REVIEW[3]

Section 31.7, Affiliate Performance Reports for the Shareholder, gives alternatives for reporting to the shareholder. This information is the basis for shareholder reviews. The business agreement (JV and/or maintenance and operation) needs to stipulate what is required to be reported for the shareholder to undertake an adequate review of the Business Unit.

To ensure the reporting of required information, a schedule could be attached to the business agreement, including:

a. a register of information that will be provided to the shareholders, including frequency of reporting;
b. the contents and format of the reports;
c. agreeing that the JV will adopt the same financial system (e.g., SAP) as the shareholders.

Specifically, the affiliate BP&B/WP&B is required to comply with the business agreements with its shareholders. Quarterly performance is measured against the targets set by the BP&B/WP&B, derived from the requirements of the business agreements, as well as targets derived from benchmarking exercises. This is depicted as shown in Fig. 35.3.

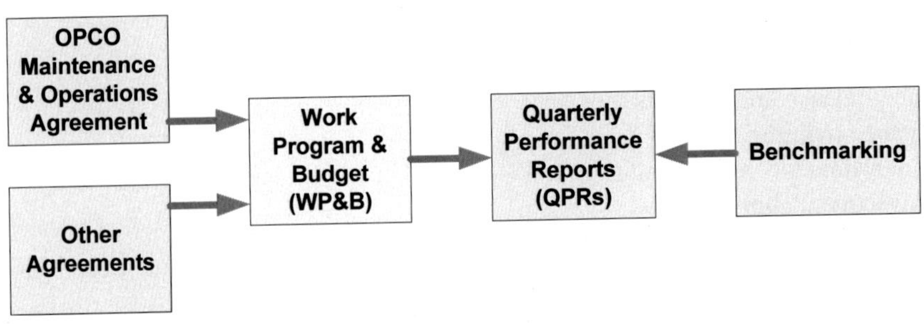

FIGURE 35.3

Business/Work Program and Budget review.

As per the ***assessment*** process described in Part 7, Assessment and Reporting, review focus areas, for an oil and gas or petrochemical plant, typically are as follows:

1. Strategy
2. HSSE
3. Hydrocarbon management
   a. Royalty accounting
      i. Mismeasurement
   b. Mass balance
      i. Unaccounted losses
4. **Production and downtime**
   a. **Salable products**
5. **Finance**
   a. **Opex**
      i. **Comparison**

6. Physical assets
7. Human assets
8. Subsurface/raw materials
9. Shipping/customer

KEY STEPS FOR BUSINESS PLAN/WORK PROGRAM AND BUDGET REVIEW

These steps follow from the assessment process described in Part 7, Assessment and Reporting.

Key steps are as follows:

1. Ensure required **sources of information** are available.
2. Use **checklist** and **rules of thumb** for identifying areas of interest.
3. Look in WP&B/BP&B for sources of **top KPIs.**
    a. List "not reported."
4. Identify **comparative KPIs.**
    a. Use rules of thumb.
    b. Comment on comparisons.
5. Request **further information** from affiliate/line management.
6. Review affiliate/line management feedback.
7. Compile **reports** and prepare presentation for management.

After carrying out an overall assessment of the company, but prior to the formulation of the annual Business Plan and Budget (i.e., at the start of the planning cycle for the following year), the shareholder representative may send a "**letter of wishes**" for the BoD to consider when formulating the following year's BP&B. This is normally in line with an operating and maintenance (O&M) agreement with the operating company. Box 35.2 gives an example of the contents of such a letter.

BOX 35.2 SHAREHOLDER'S "LETTER OF WISHES"

List of a shareholder's wishes for consideration when preparing the following year's gas plant operating company's Business Plan and Budget.

Safety, Health, and Environment

1. Flaring percent on intake is to be reduced by 0.7%
    - Target: 0.3% woi
2. Energy use on intake is to be reduced by 0.5%
    - Target: 8.1% woi

Reliability and Delivery

3. Reliability is to be increased by 1%
    - Target: 97%

Cost Optimization

4. Fixed cost $/ton of salable product is to be reduced by 10%
    - Target: $12/ton of salable product

People

5. Staffing is to be reduced by 200
    - Target: 2000

35.7 INDEPENDENT REVIEW: R/P RATIO[4]

The reservoir production ratio is one of the most important indicators for an oil and gas company that undertakes exploration and development.

This is the ratio between what can be extracted from the reservoir and what can be produced in a year. The ratio gives the number of years of production that the company has available.

Example:

BP recently published the R/P for its North Sea Operations as 10 years. By contrast, Qatar (Petroleum) has in excess of 100 years.

Critical independent review of the basis for this published indicator is generally required.

Box 7.9, The Leaders in Corporate Governance! discusses the misreporting of the R/P ratio in an annual report.

35.8 SUMMARY

Review responsibilities roll up from System and Process owners, department managers, leadership teams, and Board committees to the BoD and shareholders. Process owners maintain and improve their Processes in line with agreed targets. System owners review the health of their Systems and monitor performance in line with agreed targets. Peer auditing takes place at this level in line with ISO certification requirements. Departmental managers manage the Systems within their departments and take corrective action when and where necessary to stay on track to meet agreed targets. The leadership team reviews and reports business performance on an ongoing basis, and takes the necessary action to stay on track. It also undertakes a formal internal review at least once per year as required for ISO certification.

The internal audit department undertakes audits of selected Systems, as set out in the company annual audit plan approved by the Board audit committee. External certification audits take place annually for certification revalidation. The Board audit committee reviews audits and major risks and makes recommendations to the BoD. The BoD's roles include strategic guidance, governance issues, performance monitoring and review, approval of large investments, review of risk and audit, conflict of interest issues and the integrity of financial statements.

The shareholder representatives review the annual Business Plan and Budget and monitor the company performance, preferably on at least a quarterly basis. The level and depth of performance monitoring depends on the relationship with the company. After carrying out an overall assessment of the company, but prior to the formulation of the next year's Business Plan and Budget, the shareholder representative may send a "letter of wishes" for the Board to consider when formulating the following year's BP&B.

The key to effective review meetings is that reports be circulated timeously so that participants have time to study the data and discuss with actioners prior to the review meeting.

REFERENCES

1. John Cleese's training video 'Meetings bloody meetings'. https://www.youtube.com/watch?v=vE7jfQt2ic4.
2. Robert's Rules of Order. http://www.robertsrules.com/.

3. *Answering the call – managing shareholder information the joint venture exchange*. Water Street Partners; January 2012.
4. Olsen GT. *Reserves overstatements: history, enforcement, identification, and implications of new SEC disclosure requirements*. Texas A&M University; May 2010.

FURTHER READING

1. Johnson G, Scholes K, Whittington R. *Exploring corporate strategy*. Pearson; 2008.
2. BP Energy Outlook 2016 Edition. https://www.bp.com/content/dam/bp/pdf/energy-economics/energy-outlook-2016/bp-energy-outlook-2016.pdf .

PART

OIL AND GAS ISSUES

9

OVERVIEW

WHAT ARE THE MAIN ISSUES FOR THE OIL AND GAS INDUSTRY?

Energy management is the primary focus. Balancing inputs and outputs ensures accounting for hydrocarbons sold, used, and lost. Losses to the environment have to be accounted for and mitigated.

THE OBJECTIVE

The objective of Part 9 is to outline the "big issues" specific to the oil and gas industry that need constant attention in the quest for continuous performance improvement.

PICTORIAL OVERVIEW

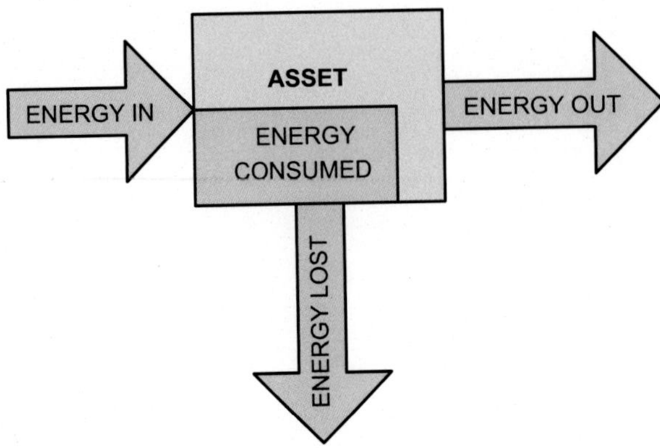

OUTLINE OF PART 9

- The hydrocarbon accounting process is described: from measurement through reconciliation to allocation.
- Accuracy is detailed for acceptance of a mass and energy balance.
- Refining issues, focusing on energy efficiency and reliability, are summarized.
- Gas plant issues are summarized, focusing on loss of primary containment, greenhouse gas emissions, and reliability.
- The implementation of a process safety program is discussed.
- The gas-to-liquid process is outlined.

CHAPTER

HYDROCARBON ACCOUNTING 36

36.1 INTRODUCTION

Hydrocarbon accounting lies at the heart of the oil and gas business and is closely aligned with financial accounting. The process is summarized in Fig. 36.1.

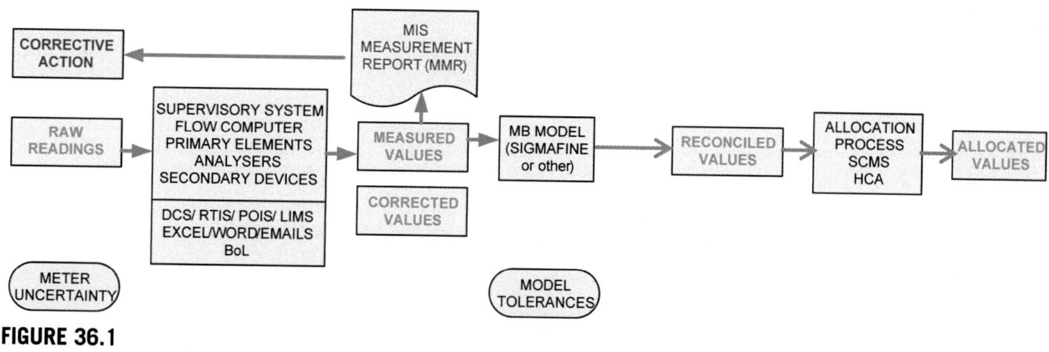

FIGURE 36.1

Hydrocarbon accounting process.

Most discussion in this chapter centers round gas plants. Refineries are comparatively much simpler.

36.2 MASS BALANCE

REFINERY

A fuels refinery will always have a volume gain and mass loss. Fig. 36.2 depicts this.

Performance Management for the Oil, Gas, and Process Industries. http://dx.doi.org/10.1016/B978-0-12-810446-0.00036-0

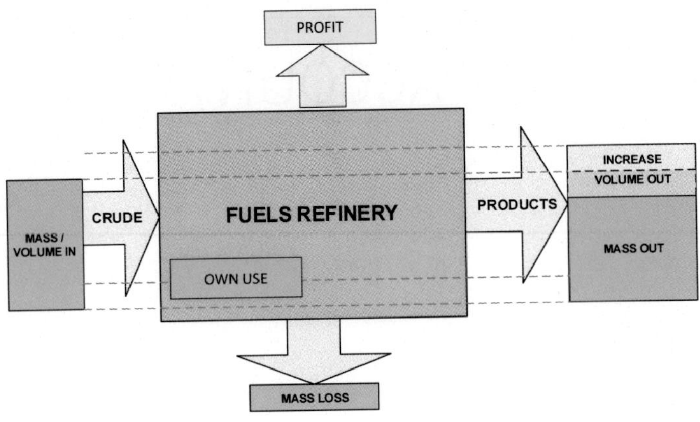

FIGURE 36.2

Fuels refinery mass balance overview.

Crude and transport fuels are generally priced in $/barrel and other products in $/ton.

GAS PLANTS

A gas plant consumes (for own use and losses) between 3% (for flowing gas) and 8% [for liquefied natural gas (LNG)] of the intake. Fig. 36.3 depicts this.

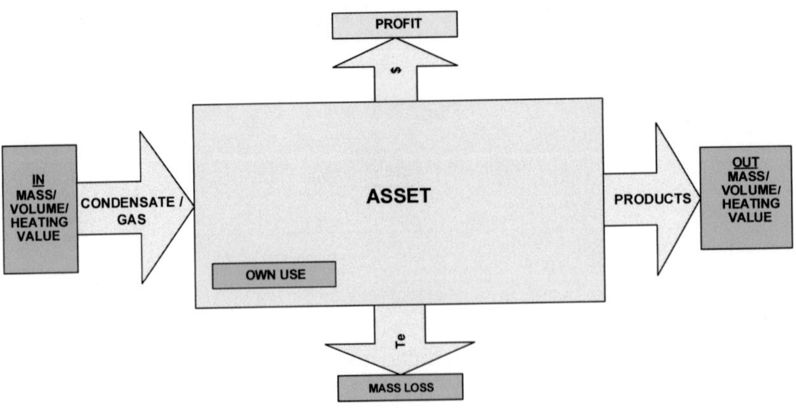

FIGURE 36.3

Gas plant mass balance overview.

Mass balance is vital for hydrocarbon accounting and royalty payment, whereas energy balance is essential for marketing, where the product is priced in $/million BTUs.

36.3 MEASUREMENT SYSTEM CLASSIFICATION

A measurement system is an assembly of primary elements, secondary devices, analyzers, and flow computers that serve to measure the quality and quantity of a raw material or product. Primary measurement system classifications are as follows:

ROYALTY

"Royalty" measurement refers to a measurement process in a dedicated measurement station (or an element of such a system) that is used to determine hydrocarbon net value from which government royalties are paid.

Some LNG joint ventures (JVs) pay royalties based on LNG sales to particular destinations back calculated to the gas intake royalty measurement system, minus the plant condensate.

FISCAL/CUSTODY TRANSFER

"Fiscal/custody transfer" measurement refers to a measurement process in a dedicated measurement station, which is associated with the transfer of ownership of, or financial transactions involving, the measured product; but which is not subject to the payment of government royalties.

ALLOCATION

"Allocation" measurement refers to a measurement process, via which measured quantities of hydrocarbons are attributed to different partners and/or sources.

36.4 MEASUREMENT SYSTEM DETAILS

The complete measurement system consists of primary elements, analyzers, secondary devices, and the flow computer. Fig. 36.4 depicts the complete measurement system.

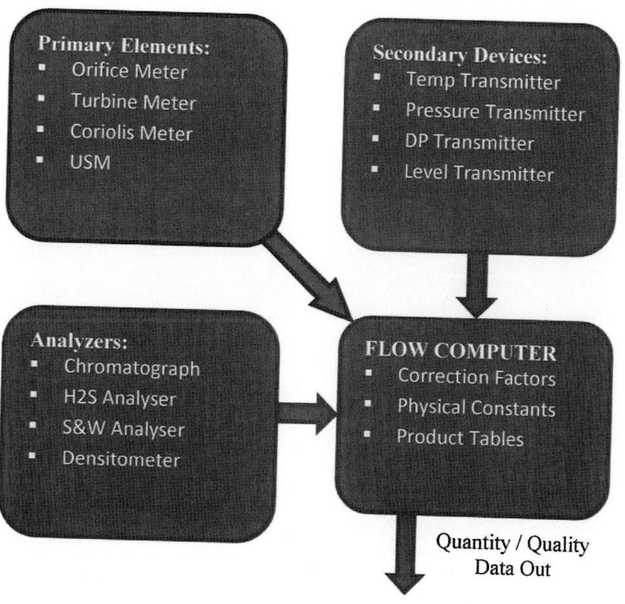

FIGURE 36.4

Complete measurement system.

MEASUREMENT SYSTEM UNCERTAINTY

When a **measurement** of quantity is made, the result is not the actual **true value** of the quantity but only an **estimate of the value**. The uncertainty of a measurement is the size of the margin of doubt. It is evaluation of the quality of the measurement. To fully express the result of a measurement, three values are required:

1. The **measured value.** This is simply the figure indicated on the measuring instrument
2. The **uncertainty** of the measurement. This is the margin or interval around the indicated value inside which you would expect the true value to lie with a given confidence level
3. The **level of confidence** attached to the uncertainty. This is a measure of the likelihood that the true value of a measurement lies in the defined uncertainty interval. In industry, the confidence level is usually set at 95%.

Example

The measured value from a flowmeter is 10 m³/hr. It has been determined by analyzing the measurement system that the uncertainty, at 95% confidence, is 3%. The result of this measurement should be expressed as: 10.0 ± 0.3 at 95% confidence. That is, we are 95% confident that the true value of this measurement lies between 9.7 and 10.3 m³/hr.

Uncertainty must not be confused with error. Uncertainty is the margin of doubt associated with a measurement. Error is the difference between the measured value and the true value.

Example of error

An oil field produces 10,000 barrels per day and the cost is $100 per barrel. The flow meter overreads by 1%, resulting in losing $20,000 per day.

Accuracy and repeatability relationships are depicted in Fig. 36.5.

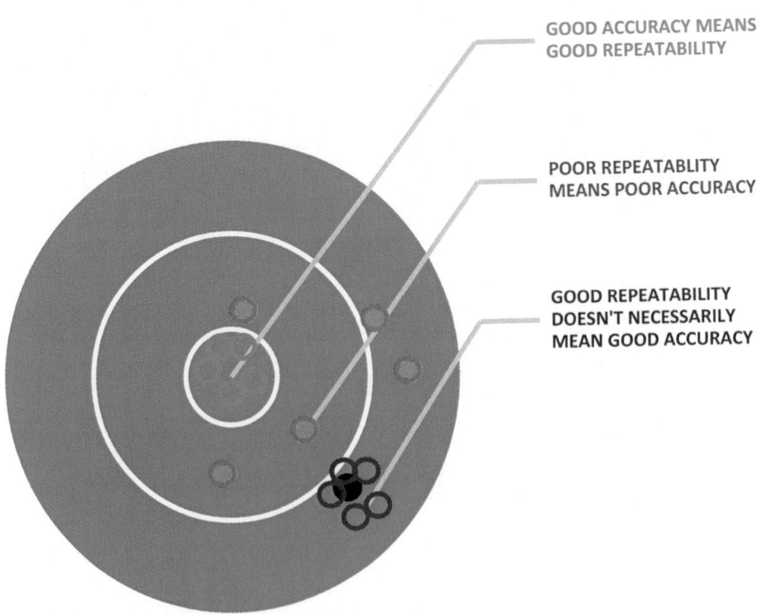

FIGURE 36.5

Accuracy and repeatability relationships.

Based on industry good practice for royalty and fiscal/custody transfer measurement systems, the overall system uncertainty in mass or standard volumes reported should not exceed ±1.0% for gas, or ±0.25% for liquid, at the 95% confidence level (coverage factor k = 2). The maximum uncertainty limits for other system classifications are shown in Table 36.1 for mass or standard volume and Table 36.2 for energy.

Table 36.1 Limits on Total System Uncertainty for Mass or Standard Volume

Measurement System Classification	Total System Uncertainty (%)	
	Liquid	Gas
Royalty	0.25	1.0
Fiscal/custody transfer	0.25	1.0
Allocation	3.0	10.0
Flare gas	N/A	7.5
Well test	10.0	10.0

Table 36.2 Limits on Total System Uncertainty for Energy

Measurement System Classification	Total System Uncertainty (%)	
	LPG/LNG	Gas
Royalty	0.8	1.0
Fiscal/custody transfer	0.8	1.0
Allocation	3.0	10.0
Flare gas	N/A	7.5

METER UNCERTAINTY

Specific meter uncertainties are calculated by specialist consultants, such as Kelton Engineering,[1] using standard industry practices. Common sources of uncertainty to be accommodated in the preceding calculations include the following:

- Measuring instrument
- Effect of the environment
- Operator skill
- Process of taking the measurement
- Variation in the measured quantity

Thresholds are set for mass balance based on system and meter uncertainty. Box 36.1 gives an example of the use of "tram-lining" using these thresholds.

BOX 36.1 MASS BALANCE: STEP CHANGE IN A TREND GRAPH

Trend graphs are used to monitor processes so as to keep the process limits within tram-lines. It can also identify a particular action taken at a specific time.

A monthly mass balance was undertaken for a particular gas plant using the model mentioned in this chapter.

The mass balance exceeded the acceptable variation limit of 0.6% in March and April 2010. A reference gas was used to recalibrate a Chromatograph in April 2010 correcting the mass balance but balance ran outside the lower limit in June and August 2010. It was discovered that the reference gas was incorrect. Recalibration with the correct reference gas brought the balance within limits.

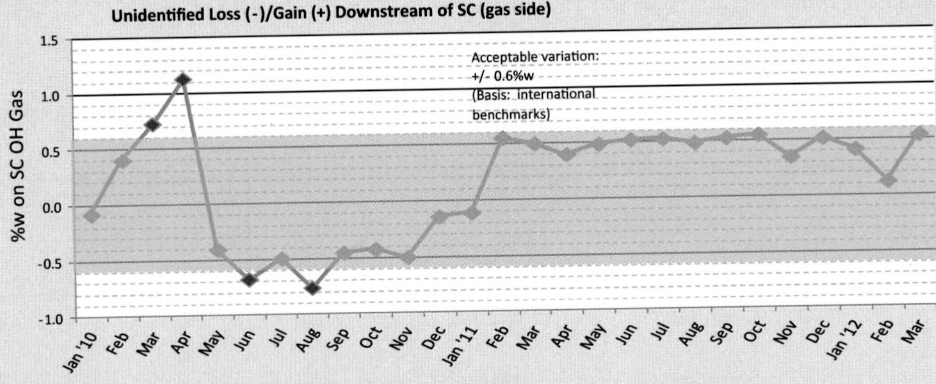

Comment

This case study shows the power of using a mass balance model to cross-check the measurement and reconciliation process and identify where a problem might be.

36.5 IDENTIFYING INPUTS AND OUTPUTS

REFINERIES

Refineries are straightforward with the following:

1. Inputs
   a. Crude
   b. Condensate
   c. Fuel gas
2. Production
   a. Fuel gas
   b. Fuel oil and/or coke
3. Losses
   a. Flaring
   b. Fugitive emissions
4. Products
   a. Transport fuels
   b. LPG
   c. Petro-chemical naphtha (PCN), coke, and/or asphalt

Refineries are generally depicted as shown in Fig. 37.1, in Chapter 37, Fuel Oil Refineries.

GAS PLANTS

Gas plant raw materials and products are generally depicted as shown in Fig. 38.1 in Chapter 38, Gas Plants. Further definition is as follows.

Well Head Gas

This is gas coming out of the well. This is estimated based on the well test, and the official figures are back-calculated from the royalty gas and field condensate measurement system.

Royalty Gas

The royalty gas measurement system is normally located at the outlet of the slug catcher.

Intermediate and Waste Products for Gas Plants

Process

GTL (Shell)

1. Lean gas from feed gas preparation (FGP) to Shell gasification process (SGP)
2. Syngas from SGP to heavy paraffin synthesis (HPS)
3. Wax from HPS to liquid processing unit (LPU) split between hydrogenation unit (HGU) and heavy paraffin conversion (HPC)
4. Light detergent feedstock (LDF) from between HGU to storage
5. Syncrude from HPC to synthetic crude distiller (SCD)
6. Base oil feedstock from SCD to high vacuum unit (HVU) and catalytic dewaxing (CDW).

Other

1. Stabilized condensate from condensate stabilization unit to condensate treatment unit
2. Off gas from condensate stabilization to reception facilities
3. Sweet gas from acid gas recovery (AGR) to demethanizer
4. Sour gas—H_2S and CO_2
5. Acid gas from AGR to acid gas enrichment/sulfur recovery
6. Lean gas from demethanizer to LNG or sales gas
7. Disulfide oil (DSO) from caustic regeneration and condensate treatment

Fuel Gas

Fuel gas makeup, as per the energy efficiency model described in Section 3.4, Mass and Energy Balance Models, is as follows:

1. Lean gas
2. Fuel from feed (FFF)
   a. Lean
   b. Rich
3. AGR flash gas
4. Regen gas
5. End flash gas (EFG) from LNG liquefaction
6. Sour gas (H_2S only) to sulfur recovery unit (SRU)
7. Boil-off gas (BOG)—may be excluded from fuel gas pool
   a. Tankage
   b. Loading—jetty BOG (JBOG)
8. Selexol/sulfinol treatment flash gas

Flare Gas

Flare gas consists of the following:

1. Purge/sweep fuel gas
2. Acid gas—process flaring
3. Passing valves
4. Catalyst regenerator off-gas (GTL)

Royalties Depending on the agreement, royalties could be imposed on flare gas in certain of the following categories:

1. Onshore for SHE reasons
2. Unplanned shutdown and restart of facilities
3. In connection with equipment breakdown and repair

Salable Products for Gas Plants

The assumption is that salable products pass from a JV to a terminal operator (TO) that ships the products on behalf of all ventures in the petrochemical complex.

Field Condensate

Field condensate is normally allocated at the **royalty** measurement system into TO but is based on the official actual sales by every single venture using a fiscal/custody transfer measurement system.

Plant Condensate

Plant condensate is normally measured at the **royalty** measurement system into TO.

LNG

Rundown Fiscal/custody transfer measurement system takes place at the point entering the LNG tankage.

Loading Fiscal/custody transfer measurement process takes place by tank gauging onboard the ship. BOG measurement should account for the difference between rundown and loading.

The process for measurement should be in accordance with the International Group of LNG Importers (GIIGNL), which has evolved into **ISO 10,976:2012 Procedure for Measurement (Volume, Density and Gross Calorific Value) of LNG aboard LNG Carriers**. Uncertainty could be up to ±0.76% (see also Section 36.4). *It is estimated that a 1% error could have a financial implication of US$500,000.*

Ethane

Fiscal/custody transfer measurement process could take place at the inlet to petrochemical complexes if the pipelines are owned by the gas plant (see Section 2.4: Boundaries and Content of a Business Unit). Otherwise, the metering should be at the site boundary fence of the gas plant.

Propane

Royalty measurement process normally takes place at the inlet to the TO, but it is allocated based on the actual sales. The sale is a fiscal/custody transfer measurement process and it takes place at loading. In certain cases, and based on contractual terms, the royalty measurement process takes place at the actual sales.

Butane

Royalty measurement process normally takes place at the inlet to the TO, but it is allocated based on the actual sales. The sale is a fiscal/custody transfer measurement process, it takes place at loading. In certain cases, and based on contractual terms, the royalty measurement process takes place at the actual sales.

Helium

Fiscal/custody transfer measurement process takes place at the inlet to helium refinery. Fiscal/custody transfer (sales) takes place at loading.

Sales (Piped/Flowing) Gas

Fiscal/custody transfer measurement process (sales) takes place at the measurement system managed at the boundary of the JV/operator.

Sulfur

Fiscal/custody transfer measurement process normally takes place at the inlet to the common sulfur plant of the petrochemical complex (part of terminal operations). Fiscal/custody transfer (sales) measurement process takes place at loading.

GTL Products

Fiscal/custody transfer measurement process takes place at the inlet to the TO. Fiscal/custody transfer (sales) measurement process takes place at loading.

36.6 THE MEASUREMENT, RECONCILIATION, AND ALLOCATION PROCESS REGULATION

Regulation is normally in accordance with a government document.[2,3]

METER CALIBRATION

Meters are generally classified as follows for calibration:

1. Fiscal and key allocation meters and analyzers
2. Allocation meters and analyzers
3. Mass balance meters and analyzers
4. Process meters and analyzers

Meter classification and calibration are generally as shown in Fig. 36.6.

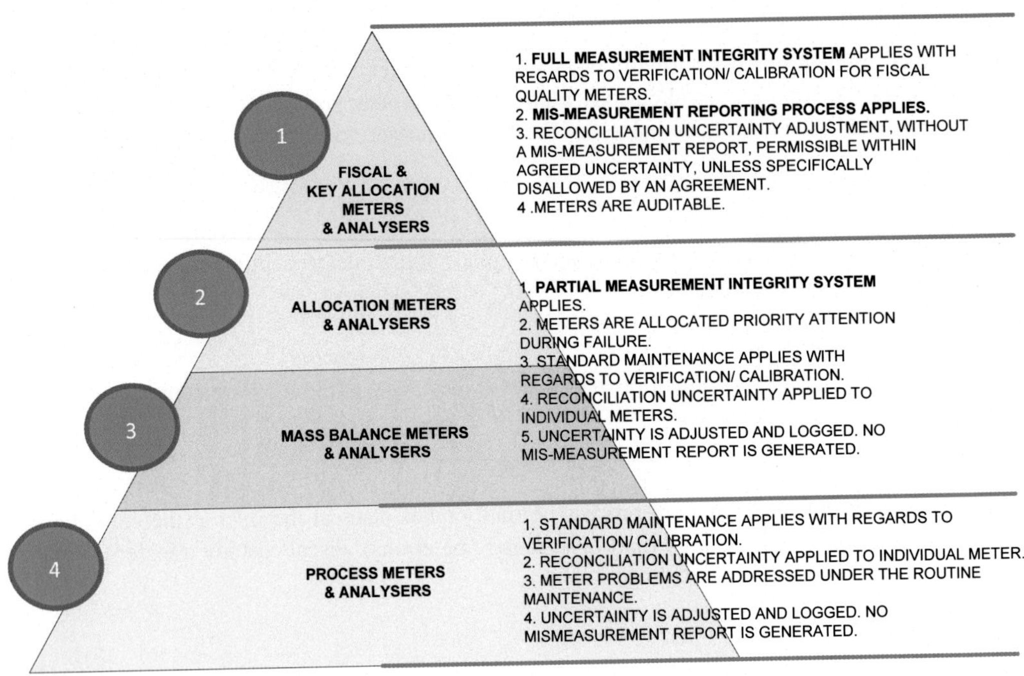

FIGURE 36.6

Meter classification and calibration.

PROCESS

The generic process of measurement, reconciliation, and allocation is shown in Fig. 36.7.

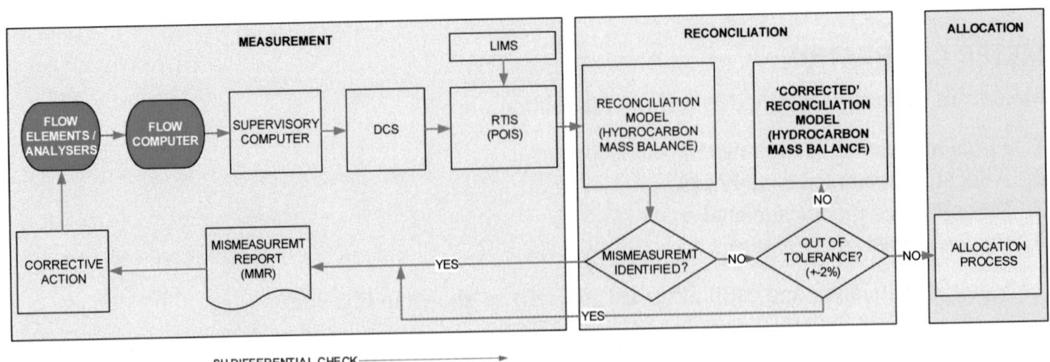

FIGURE 36.7

Generic measurement, reconciliation, and allocation process.

Details are explained as follows.

Measurement

Measurement is undertaken in primary flow elements, analyzers, and secondary temperature, pressure, and level devices. The flow computer corrects the readings and converts them to readable values. The data are then transferred through the supervisory computer and Distributed Control System (DCS) to the Real-Time Information System (RTIS). Analyzed values are also dumped from the Laboratory Information Management System (LIMS) into RTIS. All required data for reconciliation are then transferred to the reconciliation model.

Reconciliation

Reconciliation is undertaken using a computer model for complex systems.

The theory for behavior of the model is explained in the following example.

Process

The model is populated with the uncertainty value (bracketed) for each flow meter.

A node is a point in a network where all incoming and outgoing flows are compared.

The computer calculates the difference at the node and adjusts the values of each in and out measurement system within the uncertainty values so as to obtain complete mass balance at the node ("Optimum" correction is done by "root sum squared" calculation).

In this case, the inlet measurement system and the top outlet measurement system are adjusted to the maximum uncertainties. However, the bottom outlet measurement system exceeds the uncertainty that identifies a potential measurement system malfunction.

Fig. 36.8 gives the values as inputted in **green** (before reconciliation) and the calculated values (after reconciliation) as **red**.

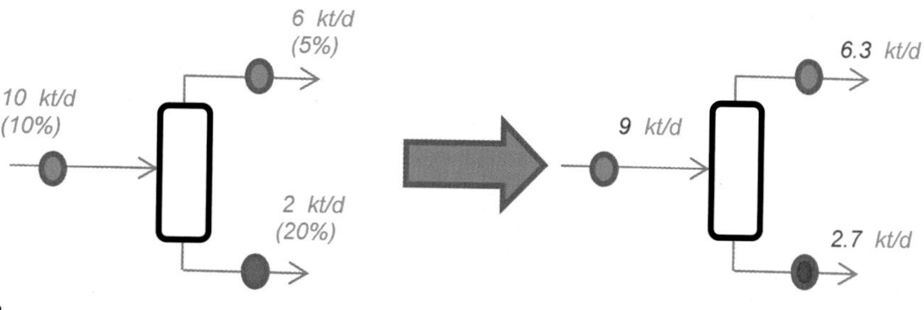

FIGURE 36.8

Reconciliation process illustration.

Accepted practice is to set the uncertainty value to zero in the model for the royalty and fiscal/custody measurement systems. The mass balance of the model should generally close with a tolerance of ±2% at the end of each month.

Computerized reconciliation is used to:

1. identify potential measurement system malfunctioning (if correction for a meter > measurement system uncertainty);
2. perform online validation of meter consistency.

Allocation

The allocation process varies in complexity. An example of complex allocation is where royalties are allocated based on the destination of LNG.

36.7 MODELS

MASS BALANCE

Mass balance should be used as a check on the measurement and reconciliation process as well as the energy balance process. The data input blocks of a typical gas plant model is depicted in Fig. 36.9.

		Feedstocks, Rundowns, Own Use, Known Losses	Loaded on ship + Delta stocks
Offshore	In:	WH Gas	
	Out:	SC Feed	
Slugcatcher	In:	SC Feed	
	Out:	SC OH Gas, Stab. FC	
Plant	In:	SC OH Gas	
	Out:	NGL, LNG, Own Fuel cons., Flaring	
Storage & Loading	In:	NGL, LPG, LNG	
	Out:		NGL, LPG, LNG

Note: List of streams is not exhaustive

FIGURE 36.9

Gas plant model: data input block.

A typical gas plant model flowchart is depicted in Fig. 36.10.

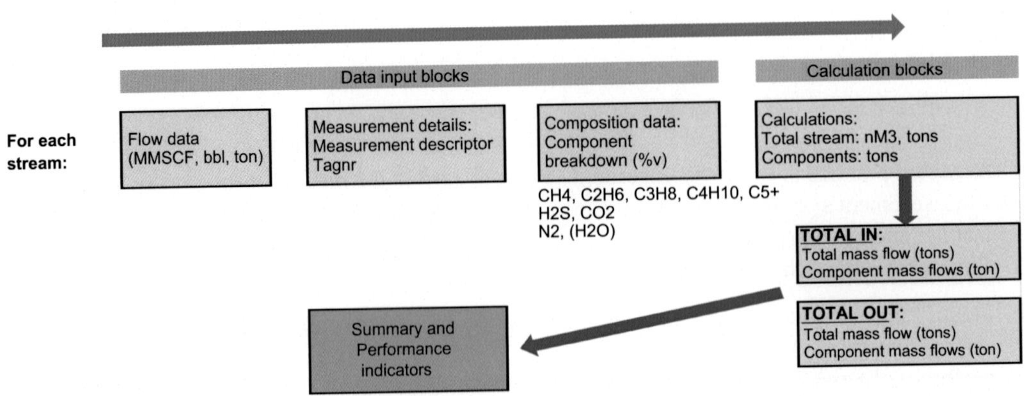

FIGURE 36.10

Typical gas plant model flowchart.

ENERGY BALANCE

Energy balance should be used as a check on the mass balance process.

A typical gas plant model is depicted in Fig. 36.11.

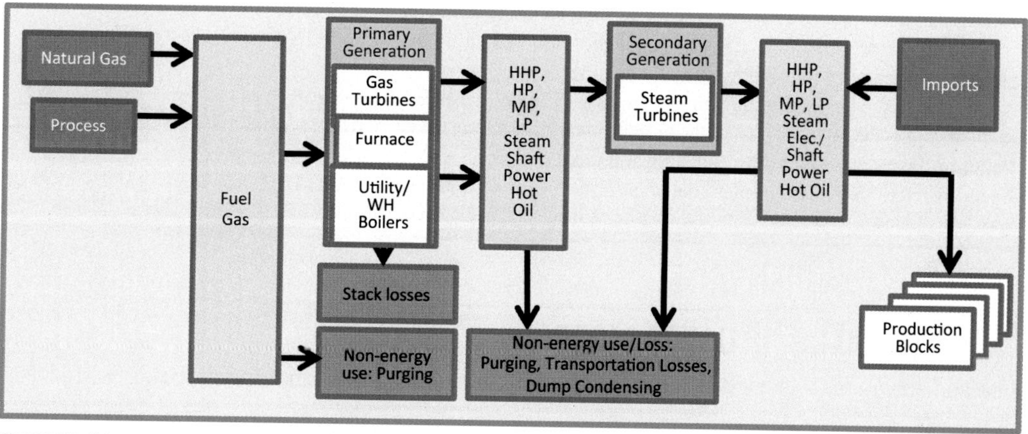

FIGURE 36.11

Gas plant model.

The use of equivalent energy is required to compare the value of imported and internally generated energy. Section17.7, Energy discusses this.

36.8 BALANCING PROBLEMS

Balancing should be within certain limits determined by internal procedures based on measurement system uncertainty, as described in Section 36.4. However, industry norms for internal energy consumption give an indication of efficiency of conversion. Table 36.3 gives typical consumption of energy for gas plants and refineries.

Table 36.3 Gas Plant and Refineries: Energy Consumption as % Weight on Intake

	Total WH Flow per Train/Stream (MMscf/d)	Total Energy Consumption % woi as Reference Gas
LNG (medium size train)	790	7.0
LNG (big size train)	1470	8.3
Flowing (sales) Gas	1300	3.4
GTL (initial data)	770	25
Hydro-skimming oil refinery	N.A.	4.0

Box 36.2 identifies some problems when carrying out balancing.

BOX 36.2 MASS AND ENERGY MODEL BALANCING PROBLEMS—EXAMPLES

Gas-to-Liquid (GTL) Plant: Water and Oxygen as Feedstock

On carrying out a mass balance of a GTL complex, it was found that water and oxygen, as key components of the chemical reaction, were required as inputs, even though they are not hydrocarbons.

Fig. 38.6, Basic gas to liquid (GTL) process—Shell technology shows the blocks for mass and energy balance.

LNG Plant: Incorrect Selection of Inputs and Outputs

A particular LNG complex used Sigmafine for mass balance reconciliation. Nonhydrocarbon inputs were used, which gave spurious outputs that were adjusted by the model. However, they were not out of range, so no corrective action was taken.

Refinery: Mass and Energy Do Not Both Balance

A refinery "slops" meter was operating out of range, but the mass balance was in limits. The Profile II mass and energy model gave "out of limits" values for Volumetric Expansion Index (VEI) and Energy Intensity Index (EII) (see also Section 36.10).

36.9 THE VALUE OF USING MODELS

Mass and energy balance models are introduced in Section 3.4, Mass and Energy Balance Models.

COMPONENT IMBALANCE

A mass balance model that includes component balance can detect poor recovery of high-value products. Box 36.3 gives an example.

BOX 36.3 MASS COMPONENT BALANCE IN A GAS PLANT

Using a mass balance model, it was determined that separation of butane and propane was worse than design. Also there was excess C5+ in condensate. However, sulfur recovery was better than design.

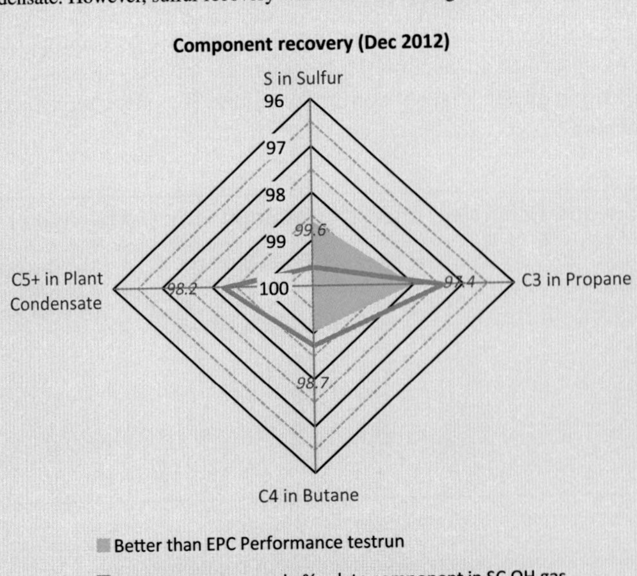

ENERGY EFFICIENCY

Energy efficiency gaps can be detected by using an energy model. An example of using an energy model for identifying wasted energy is illustrated in Box 36.4.

BOX 36.4 ENERGY EFFICIENCY: GAS PLANT WASTE HEAT BOILER (WHB)

A WHB problem was identified using an energy model as reported here.

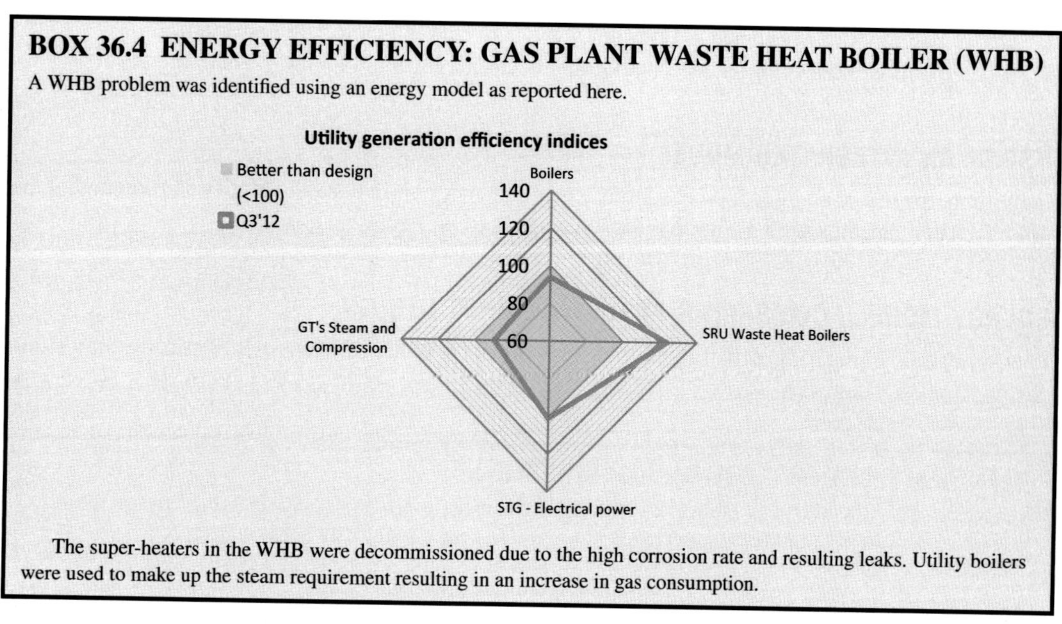

The super-heaters in the WHB were decommissioned due to the high corrosion rate and resulting leaks. Utility boilers were used to make up the steam requirement resulting in an increase in gas consumption.

Box 36.5 illustrates the case of poor energy utilization using an alternative amine treatment.

BOX 36.5 ENERGY EFFICIENCY: GAS PLANT GAS TREATMENT

Two trains using different amines for gas treatment were compared. The one train had the amine prescribed by design and the other had amine that was available at the time of loading, as the design amine was not available.

Using an energy model, an amine treatment problem was identified in the gas treatment block of the second train. This problem is shown here.

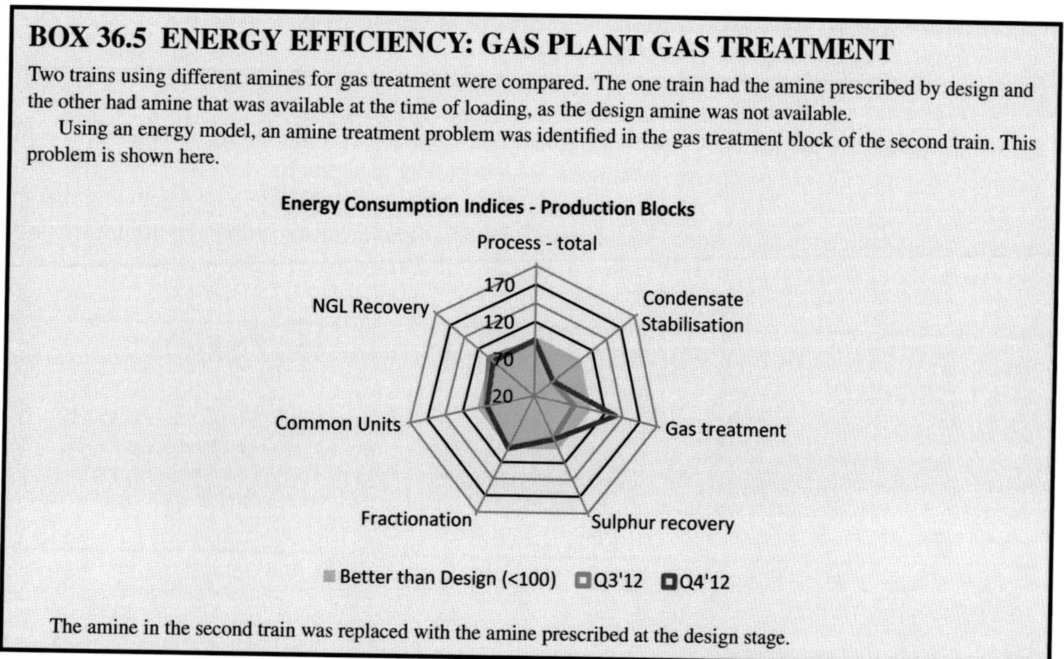

The amine in the second train was replaced with the amine prescribed at the design stage.

36.10 THE LIMITS OF USING MODELS

Mass balance models have clear limits based on system and meter uncertainty.

Two scenarios follow:

1. When the model **does** balance, it does not always give an indication of a problem.
2. When the model **does not** balance within the set limits, it cannot identify where the problem is.

USE OF AN INTEGRATED MODEL

In refining, the use of an integrated model for mass and energy is needed to identify a problem that may **not** have been detected with a simple mass balance. Box 36.2 gives an example.

ENERGY MODELS COMPARING ACTUAL AGAINST DESIGN

Energy models comparing actual against design do not identify improvements outside the original design. Box 36.6 describes a case.

> **BOX 36.6 ENERGY MODEL LIMITATIONS**
>
> The energy model described in this chapter was used for joint venture companies consisting of one or more gas trains. A number of joint venture companies were operated by an operating company within the same complex.
>
> Each JV was built to generate its own power without linkage to other JVs in the complex or to the national grid. The model was used to compare actual energy consumption against design as well as against best operating practice.
>
> Variances against design and best operating practices have been addressed.
>
> When designing each JV the generation capacity was designed to be just sufficient for the JVs own needs. Thus, a number of gas turbines (GTs) were "open cycle." There has been no financial incentive for a JV to conserve energy by installing more heat recovery steam generators (HRSGs) and steam turbines.
>
> **There is now a state initiative to conserve energy.** It is estimated that up to 300 MW can be generated from the exhausts of the "open cycle" GTs. What needs to be put in place is an incentive for the JVs to invest in HRSGs and steam turbines and to be able to sell electricity into the national grid.

In some countries, the national electricity generation company is a monopoly and, as such, is not interested in importing excess energy from those who normally generate for their own use. They tend to lock in major consumers on a "take or pay" agreement. Many countries are now legislating that the "power grid" is obliged to take excess power from consumers and are also setting a premium for those who exceed the agreed maximum MW demand. Box 36.7 discusses a case study on the subject.

> **BOX 36.7 TAKE OR PAY POLICY OF THE NATIONAL ELECTRICITY DISTRIBUTION COMPANY (NEDC)**
>
> The energy model described in this chapter was used for a gas plant exporting piped gas to other countries. To ensure a reliable supply of flowing gas to its customers, the gas plant was designed to have sufficient generating capacity so as not to depend on the NEDC national grid. Thus, operating philosophy was to draw a certain percentage of its power needs from the NEDC national grid and use the NEDC national grid as backup if its own power-generating capacity was jeopardized.

BOX 36.7 TAKE OR PAY POLICY OF THE NATIONAL ELECTRICITY DISTRIBUTION COMPANY (NEDC)—cont'd

All gas turbines are combined cycle. The model was used to compare actual energy consumption against design as well as against best operating practice.

Variances against design and best operating practices have been addressed. One recommendation has been to replace the type of amine in one of its trains to save energy (see Box 36.5).

The NEDC national grid had a take or pay policy and thus there was no incentive for companies to consume less than the agreed MWh in a month.

At the time of writing, there is a state initiative to conserve energy. The NEDC national grid has to change its policy to a "maximum demand" policy where there is a premium for exceeding an agreed MW value.

36.11 SUMMARY

Hydrocarbon accounting is at the heart of the oil and gas business and is closely aligned with financial accounting.

The basic process is measurement, reconciliation, and allocation.

Measurement system classification is, in order of importance, royalty measurement at the top, then fiscal/custody transfer measurement, and, finally, allocation measurement.

Mass balance for refineries must always be compared with volume balance with a decrease in mass and increase in volume. The pricing is based on $ per barrel or ton.

Mass balance for gas plants is required for royalty payments, but energy balance is required for marketing where pricing is in $ per million BTUs.

Measurement takes place in primary devices (flowmeters), analyzers, and secondary devices (temperature, pressure, and level) and is fed into the flow computer. The data pass through the Distributed Control System to the Real Time Information System (RTIS). Data from Laboratory Information Management System are also sent to the RTIS. The data are then downloaded to the reconciliation software.

Reconciliation for refineries is usually straightforward. However, gas plants use a computer model to manipulate the measurements within limits to obtain a balance. Allocation is normally straightforward but could become complex, as in some liquefied natural gas companies.

Mass and energy balance models are used as a cross-check on the reconciliation process.

REFERENCES

1. Measurement Consultancy. Kelton. http://www.kelton.co.uk/.
2. National Measurement System Good Practice Guide. Flow Measurement Uncertainty and Data Reconciliation. www.tuvnel.com; http://www.tuvnel.com/_x90lbm/Flow_Measurement_Uncertainty_and_Data_Reconciliation.pdf.
3. National Measurement System Guidance Documents. http://www.tuvnel.com/assets/content_images/GD%20 Energy.pdf.

FURTHER READING

1. Petroleum Resources Management System – Society of Petroleum Engineers (SPE), American Association of Petroleum Geologists (AAPG), World Petroleum Council (WPC), Society of Petroleum Evaluation Engineers (SPEE). http://www.spe.org/industry/docs/Petroleum_Resources_Management_System_2007.pdf.
2. Buried treasure: history – Middle East reservoir review. http://www.slb.com/~/media/Files/resources/mearr/num1/buried_treasure.pdf.
3. Gas-Gold for the future: history – middle East reservoir review. http://www.slb.com/~/media/Files/resources/mearr/num1/gold_future.pdf.
4. Sigmafine from OSI. www.osisoft.com, Mass balance model.
5. ISO 10976:2012. *Procedure for measurement (volume, density and gross calorific value) of LNG aboard LNG carriers.*
6. The International Group of LNG Importers. www.giignl.org.

CHAPTER

FUEL OIL REFINERIES

37

37.1 INTRODUCTION

Refinery-specific performance issues are discussed in this chapter.

Fuel oil refineries are generally depicted as shown in Fig. 37.1.

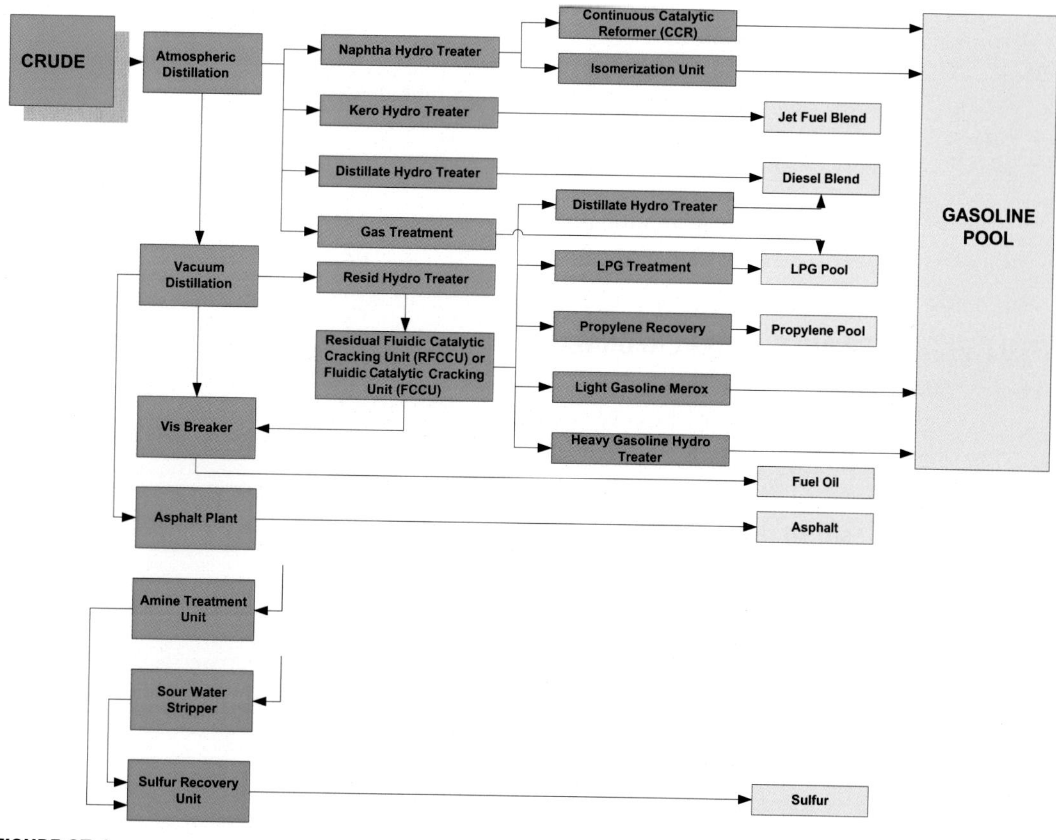

FIGURE 37.1

Refinery block diagram.

Performance Management for the Oil, Gas, and Process Industries. http://dx.doi.org/10.1016/B978-0-12-810446-0.00037-2

Deciding what specific action is to be taken is often difficult, as there may be so many issues at hand. The assessment processes described in Part 7, Assessment and Reporting, are guides. When the problems come down to process operations issues, assistance may be found in Norm Lieberman's book *Troubleshooting Process Operations*.[1]

37.2 REFINERY-SPECIFIC PERFORMANCE ISSUES COVERED IN PREVIOUS CHAPTERS

Figures and boxes giving examples related to refining are outlined as follows:

PART 1: SYSTEMS

Chapter 3: Models

Figure 3.2: Refinery mass balance model.

Chapter 4: Systems Requirements

Box 4.2: Core Management System Intranet Front Page
- Box 4.3: Core Management System With Underlying Computer System—Refinery
- Box 4.5: Refinery Integrated Management System
- Box 4.7: Refinery Integrated Management System Guideline

Chapter 6: Business Process Management

Box 6.4: Bharat Petroleum SAP Implementation

PART 2: GOVERNANCE AND PERFORMANCE

Chapter 10: Alignment

Box 10.4: Refinery Business Unit Objectives Example and Explanation
- Box 10.5: Sample Commitment Statement and Explanation

PART 3: RISK AND PERFORMANCE

Chapter 12: Risk Control Mechanisms

Box 12.5: Cause and Effect Diagram: Amine System Corrosion
- Box 12.9: Event Potential Matrix Refinery Examples

PART 4: PERFORMANCE INDICATOR SELECTION

Chapter 16: Key Performance Indicator Selection Guidelines

Figure 16.3: Refinery Key Performance Indicators categorization example.
- Box 16.2: Terminology—Refinery Utilization Cascade
- Box 16.3: Refinery Mechanical Availability Relationships

Chapter 17: Relationships

Box 17.1: Mass, Volume, and Heating Value: Refinery
Box 17.4: Refinery Availability
Box 17.7: Cross Linkage—Refinery Example

PART 5: ASSET PERFORMANCE MANAGEMENT

Chapter 19: Information

Box 19.4: Example of Refinery IT Integration

Chapter 21: Finance

Box 21.4: Refining Margins Example
Box 21.5: Reliance Refinery—Profits and Investment

PART 6: BENCHMARKING

Chapter 25: Common Denominators

Box 25.1: Equivalent Distillation Capacity
Box 25.2: Comparison Using Equivalent Distillation Capacity as Denominator

Chapter 27: Using Benchmarking Results

Figure 27.2: Fuels refinery benchmarking envelope
Box 27.4: Time-Based Line Graph—Refinery Utilization
Box 27.6: Refinery Benchmarking Summary
Box 27.7: Reliance Refinery Benchmarking Performance

PART 7: ASSESSMENT

Chapter 29: ID, Analysis, and Evaluation of Gaps

Box 29.6: Mapping—Refinery Performance Assessment

Chapter 30: Strategies and Actions

Box 30.1: Holding on to "Pacesetter" Status
Box 30.2: Example Refinery Business Improvement Plans
Box 30.4: Refinery Long-Term Target Example
Box 30.6: Refinery Pacesetter Targets Visualized

Chapter 31: Reporting

Figure 31.1: Refinery margins
Figure 31.4: Refinery performance spider diagram
Figure 31.5: Refinery utilization trend graph

PART 8: OVERSIGHT

Chapter 33: The Opportunity Lifecycle

Box 33.2: The Value of Phased Approval—Reevaluation of a Project in FEED Phase

Box 33.6: Project Risk Root Cause Analysis—Cat Cracker Woes

Box 33.11: Technology Investment Choice—Flue Gas Scrubbing

37.3 REFINERY PERFORMANCE INITIATIVES

Some performance initiatives are listed next.

ENERGY CONSERVATION[2]

Energy accounts for 45–55% of total cash operating expenses. It is, therefore, a good place to start for performance improvement. The following are some performance improvement examples:

- Modifications to heat recovery systems
  - Waste heat recovery
  - Rearrangement of heat exchanger network (pinch technology)
- Improvement in operation
  - Reduction in excess combustion air
  - Application of advanced process control
  - Run fired heater workshops
  - Control hydrogen balance
- Reduction in electric power consumption
  - Power recover turbine
  - Flue gas expander for FCC unit
  - Switching pump operation from motor to steam turbine driver
- Reduction of heat radiation loss
  - Insulation of bare pipe upstream of steam traps
  - Replace steam traps with proper type
  - Apply optimum thickness insulation on piping and vessels
  - Insulate heat exchanger flanges

New technologies that can help are as follows:

- High-efficiency burners
- Plate heat exchangers
- Cogeneration of power
- Gas recovery systems for flare and hydrogen

Valero Refinery undertook an energy assessment. This is outlined in Box 37.1.

BOX 37.1 VALERO REFINERY ENERGY ASSESSMENT[3]

Houston Refinery Uses Plant-Wide Assessment to Develop an Energy Optimization and Management System

Summary

Valero Energy Corporation recently undertook a plant-wide energy assessment at its refinery in Houston, Texas. The assessment consisted of an energy systems review to identify the primary natural gas and refinery fuel gas users, electricity- and steam-producing equipment, and cooling water systems and to develop an energy optimization and management system. It also addressed ways to reduce water use and environmental emissions. If all the projects identified during the study were implemented, the assessment team estimated that total annual energy savings at the Houston refinery would be about 1.3 million MMBtu (fuel) and more than 5 million kWh (electricity). Total annual cost savings would be about $5 million.

Energy Optimization and Management System Development

Assessment data and energy system information were gathered to develop an **Aspen Utilities 1 computer model** of the primary refinery processes and the energy production and distribution systems. The model can be used to determine the most efficient loading of individual pieces of equipment.

PROCESS IMPROVEMENT

Simulation

Simulation demonstrates what the outcome would look like when certain inputs are applied to a model. Favorable outputs can then be applied to the real situation. Two common models are as follows:

1. Process: Petro-SIM[4] by KBC
   - This can simulate the gross margin of the refinery.
2. Business Unit: Profile II[5] by Solomon Associates
   - This can, for example, simulate the impact of improvement of a particular KPI on other KPIs.

Process Field Data Collection and Monitoring

Honeywell's IntelaTrac Process Knowledge System (PKS)[6] uses mobile technology to deliver reliable field data faster to the right people, resulting in lower costs, better return on plant assets, and improved profits.

MAINTENANCE IMPROVEMENT

Tank and Pipeline Cleaning

Process for the Thermo-Chemical Cleaning of Storage Tanks—US Patent: 5,580,391 A[7]

This process could solve various problems related to thermo-sensitive materials such as paraffin, hydrates, and asphaltenes. Paraffin deposits occur in certain areas of the petroleum industry, such as production sub-sea flow lines, offshore and on shore pipelines, crude oil storage tanks, mud separators, and other critical areas of the production, refining, and storage of crude oil and its products.

The thermo-chemical process could be used for the cleaning of storage tanks that have been fouled by sludge that precipitates out of crude oil or petroleum-related products. This process fluidizes the sludge as a result of heat generated by the interaction of organic salts and solvents. Nitrogen gas is given off during the process, which further agitates the solution. The resulting petrochemical compound can then be separated and introduced back into the refinery stream.

The main benefits of the technology are as follows:

- Environmentally friendly with virtually complete oil recovery
- Oil production enhancement
- Work performed in situ
- Reduction of maintenance time and low follow-up costs
- No heavy equipment or civil construction required
- No restrictions on pipeline length or placement.

SALABLE PRODUCT QUALITY AND QUANTITY

In-Line Blending

The use of in-line analyzers and blending of the final product has become common. Box 37.2 outlines a tank reduction program as a result of in-line blending.

> **BOX 37.2 BAHRAIN PETROLEUM COMPANY (BAPCO) TANK REDUCTION PROGRAM**
>
> The Bahrain Refinery has two tank farms. In the 1990s, Bapco embarked on a tank reduction program since at that time tank maintenance was a significant component of refinery operating costs.
>
> At the start of the program, there were some tanks, including a few that were riveted, dating back to the 1940s. Blending was done between tanks. The blend tank was then sampled and a laboratory certificate was required before the shipping of product.
>
> In-line analysis and blending was introduced, resulting in the scrapping of a significant number of tanks. In parallel, a project was carried out to reduce the "heel" in tanks. (The heel is the remaining product in the tank after pumping out as much of the product as possible.)
>
> **Outcome**
>
> The project resulted in a major reduction in tank maintenance and the related costs. Also, product quality was found to be more consistent.

Accurate Fiscal Metering

Sometimes, in the project design stage, inadequate attention is given to fiscal metering at key places in the plant. Tank gauging in the crude and export tanks should be used only as a backup and as a cross-check on skid-mounted fiscal metering units for crude into the refinery, and finished products out of the refinery. An example of poor hydrocarbon accounting is given in Box 37.3.

> **BOX 37.3 FISCAL METERING**
>
> After a major upgrade, the new refinery export metering units were never commissioned since the meter "prover" unit could not be attached to the meter for calibration. Fiscal measurement was carried out based on ship ullage and cross-checked against tank gauging.
>
> This resulted in a recurring Nonconformance Report (NCR), as the inaccuracy of measurement would always be biased toward giving the customer more product.
>
> At the time of writing the issue had not been resolved.
>
> The loss to the company is unknown as the accuracy of mass balance could not identify this issue.

WASTE DISPOSAL AND RECYCLING

Platinum Recovery

Contracts need to be established for the removal and banking of platinum from spent catalyst. The platinum is then reused in fresh catalyst for the company by the catalyst manufacturer. Excess platinum could be sold. An example of a company earning a healthy windfall is given in Box 37.4.

> **BOX 37.4 PLATINUM WINDFALL**
>
> Over many years, Qatar Petroleum Refinery stockpiled spent catalyst until an improvement project was initiated to remove it.
>
> Some catalyst was not recyclable and so was disposed of, while other catalyst was regenerated, and catalyst containing platinum was sent to a specialist recovery plant for removal of platinum.
>
> The recovered platinum was used in fresh catalyst supplied to the company, but all excess platinum was sold, giving the company an unexpected bonus.

Disposal of Toxic Sludge

Common practice is to establish an agreement with a local cement plant to incinerate toxic sludge in the cement rotary kiln. The heating value of the sludge may offset the cost of installing and maintaining the sludge injection system, resulting in a zero value agreement.

37.4 REFINERY PROFITABILITY

The refining **(gross) margin** is the difference between the wholesale value of the oil products a refinery produces and the value of the crude oil from which they were refined.

Hydro-skimming is one of the simplest types of refinery used in the oil industry. A hydro-skimming refinery is defined as a refinery equipped with atmospheric distillation, naphtha reforming and necessary treating processes. A hydro-skimming refinery is, therefore, more complex than a topping refinery (which just separates the crude into its constituent petroleum products by distillation, known as atmospheric distillation, and produces naphtha but no gasoline) and it produces gasoline. However, a hydro-skimming refinery also produces a surplus of fuel oil, which has a relatively unattractive price and demand.

Most refineries, therefore, add vacuum distillation and catalytic cracking, which adds one more level of complexity by reducing fuel oil by conversion to light distillates and middle distillates. A coking refinery adds further complexity to the cracking refinery by high conversion of fuel oil into distillates and petroleum coke.

Catalytic cracking, coking, and other such conversion units are referred to as secondary processing units.

Box 37.5 shows the relative profitability of “hydro-skimming” refineries versus “hydro-cracking” refineries.

BOX 37.5 "HYDRO-SKIMMING" VERSUS "HYDRO-CRACKING" REFINERIES

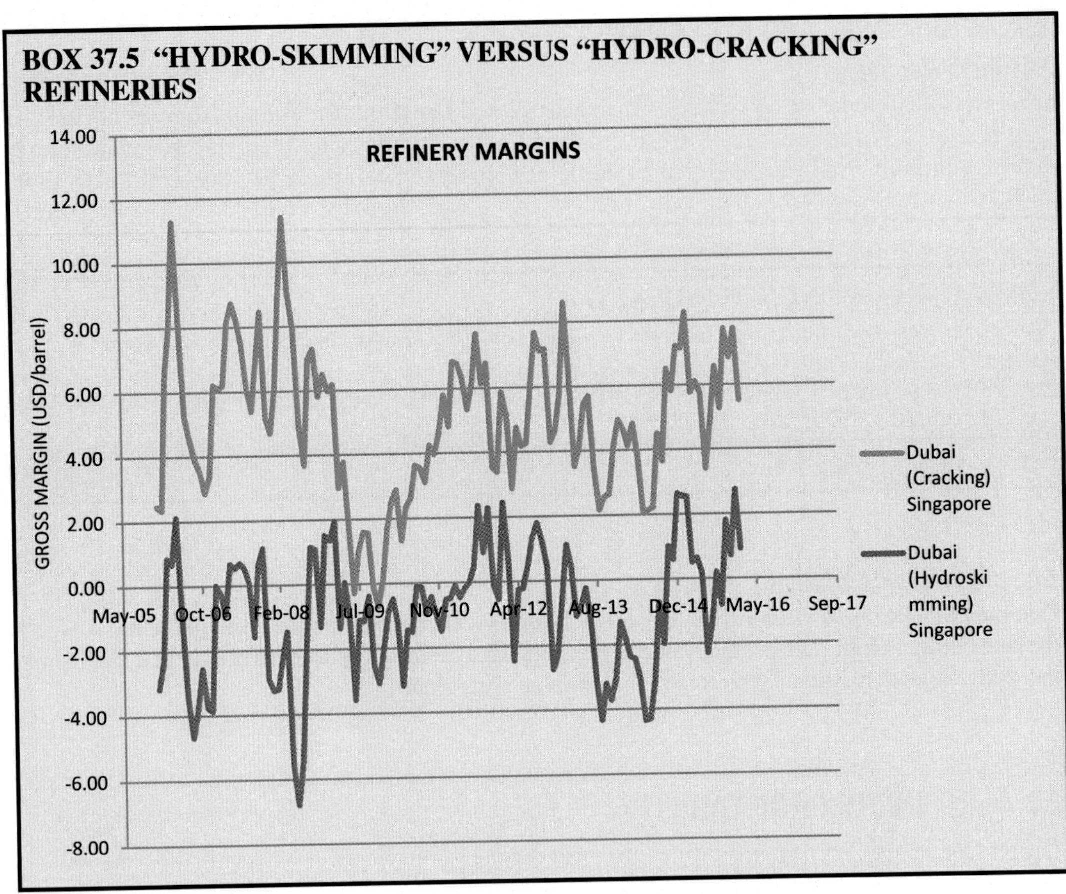

37.5 REFINERY TOP KEY PERFORMANCE INDICATORS

The top indicators are selected based on those top indicators that can be compared worldwide. These are the top nine Solomon indicators that are used for trending over the years. Solomon is able to compare more than 350 refineries worldwide and trend the top indicators over 6 years. However, some indicators are Solomon proprietary indicators based on Equivalent Distillation Capacity (EDC). EDC is based on unit capacities and process complexity so that "apples" can be compared with "apples." Chapter 25, Common Denominators, details the derivation of EDC.

In addition to these Solomon indicators, an additional three have been added to comply with the requirements of ISO 9001, ISO 14,001, and ISO 45,001, giving a total of 12 top indicators.

The top 12 are described as follows:

1. Net Cash Margin (NCM)
   NCM is the profit (loss) after deducting the raw material and conversion costs (operating expenses and plant depreciation) from the average selling price of the products. Zero $ per barrel is the financial break-even point.
2. Return on Investment (ROI)

The investment is defined as the replacement value of the physical assets and materials held in inventory. This should result in a profit. The percentage gain on the investment is the ROI.

3. Cash Operating Expense (Cash OPEX)
   This entails all funds required to operate a refinery, including shutdown expenses, but excluding cash invested in improving the physical plant and equipment. This value is divided by the utilized Equivalent Distillation Capacity to obtain a cost per equivalent barrel.
4. Personnel Index
   This is total number of work hours in a year by all direct and indirect staff divided by 100 and multiplied by EDC.
5. Utilization
   Utilization is actual production divided by design production capacity indicated as a percentage.
6. Volumetric Expansion Index (VEI)
   Liquid hydrocarbon raw materials are "boiled" to produce lighter high-value products. The VEI is determined using a complex set of formulae based on type of unit, feed, and utilized unit capacity to compare output and input volumes.
   All crude oil refineries must have a positive VEI; see Section 36.2, Mass Balance for details.
7. Energy Intensity Index
   Total actual energy consumed divided by a "typical" energy consumption for this process unit and technology type to process the same capacity.
8. Mechanical Availability
   This is expressed as the total days in the year minus days that the unit is out of service for maintenance in a year, including annualized turnaround days, divided by the total days in the year expressed as a percentage.
9. Maintenance Index
   Total cost of maintenance (direct costs of labor and materials as well as indirect management and support costs) divided by the EDC.
10. Carbon Emission Index
    Actual greenhouse gases based on reported energy consumption data by fuel type, divided by standard greenhouse gas emissions.
11. Lost Time Index
    This is calculated a lost time per million man-hours worked.
12. Customer Satisfaction Index
    This is an index based on a standard customer satisfaction survey form.

37.6 SUMMARY

Refineries are complex process plants. Homing in on the few strategies that will make a big difference to performance entails a lot of analysis and discussion with interested parties. There are many strategies to choose from, but the trick is to choose only a few. Once these few strategies are decided, the change process needs to be structured to ensure a step improvement to the next level of performance. Refinery-specific performance issues covered in previous chapters are listed for review. Some performance initiatives include energy conservation, process improvement, maintenance improvement, product quality and quantity, and waste disposal. Top indicators are selected based on those top indicators that can be compared worldwide.

REFERENCES

1. Lieberman N. *Troubleshooting process operations*. PennWell; 1991.
2. Cosmo Refinery energy initiatives. http://ceh.cosmo-oil.co.jp/eng/csr/sustain/pdf/2002/02p31.pdf.
3. Refinery-wide energy optimization model. Valero Refinery energy assessment. http://texasiof.ceer.utexas.edu/texasshowcase/pdfs/casestudies/cs_valero.pdf.
4. Petro-SIM from KBC. www.kbcat.com.
5. Profile II from Solomon Associates. https://www.solomononline.com/.
6. Honeywell's IntelaTrac Process Knowledge System (PKS). http://www.automation.com/automation-news/industry/honeywell-introduces-mobile-pks-to-expand-process-knowledge-management-through-wireless-technology .
7. Thermo chemical cleaning. https://www.onepetro.org/conference-paper/SPE-105765-MS and http://www.gsosweden.com/industries/oil-and-gas/.

CHAPTER

GAS PLANTS

38

38.1 INTRODUCTION

Gas plant–specific performance issues are discussed in this chapter. Gas plants are generally depicted as shown in Fig. 38.1.

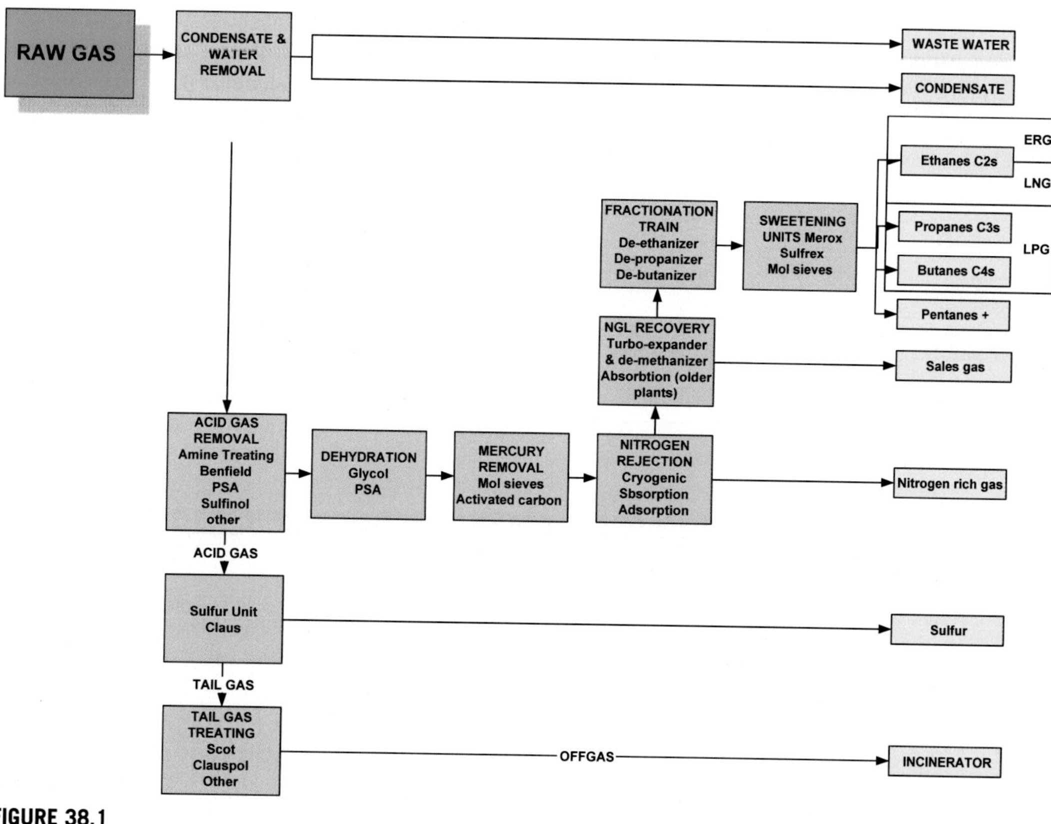

FIGURE 38.1

Gas plant block diagram.

Performance Management for the Oil, Gas, and Process Industries. http://dx.doi.org/10.1016/B978-0-12-810446-0.00038-4

Subtle differences in designation of names are explained as follows:

NATURAL GAS LIQUIDS

NGL is an acronym for natural gas liquids—the liquid hydrocarbons normally associated with "natural" gas, which is composed mainly of methane, ethane, and "heavies" like propane and butanes. Depending on the gas pressure, the heavies will condense and constitute the liquid, or "wet," portion of the natural gas. NGLs are much more valuable as raw material for further processing than as fuel for simple combustion. **Natural gas processing** is a complex industrial process designed to clean raw natural gas by separating impurities and various nonmethane hydrocarbons and fluids to produce what is known as **pipeline-quality** dry natural gas.

LIQUEFIED NATURAL GAS

LNG is an acronym for Liquefied Natural Gas—the gaseous portion of a natural gas (mostly the methane and a little ethane) in the liquefied state (−160°C) and 1 atmosphere pressure. This is an efficient way to move, transport, and handle large, bulk quantities of natural gas instead of trying to store it in the gaseous state at elevated pressures.

The LNG value chain is depicted in Fig. 38.2.

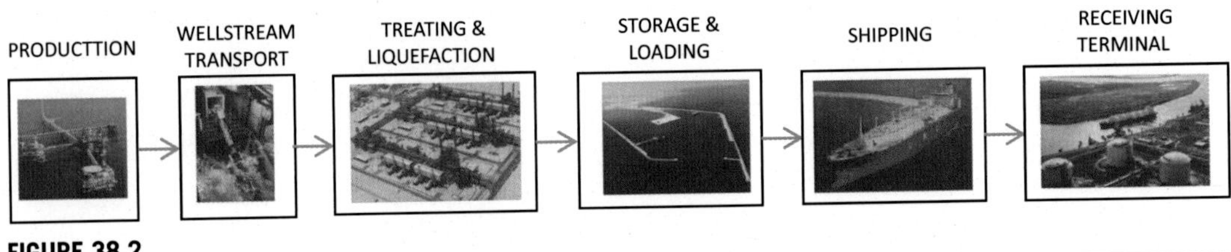

FIGURE 38.2

LNG value chain.

38.2 GAS PLANT–SPECIFIC PERFORMANCE ISSUES COVERED IN PREVIOUS CHAPTERS

Tables, figures, and boxes giving examples related to gas plants are outlined as follows:

PART 1: SYSTEMS

Chapter 2: The Business Unit

Figure 2.3: Hydrocarbon integration.

Chapter 3: Models

Figure 3.4: LNG gas plant mass/energy balance model.
Figure 3.5: Plant-specific power and plant thermal efficiency model.

Chapter 10: Alignment

Figure 10.2: Oil and gas systems alignment.

PART 3: RISK AND PERFORMANCE

Chapter 12: Risk Control Mechanisms

Box 12.8: Gas Plant Risk Assessment

Chapter 13: Merging Risk and Performance

Box 13.3: Shell Pearl GTL Maintenance Contracts

PART 4: PERFORMANCE INDICATOR SELECTION

Chapter 15: Performance Focus

Figure 15.2: Views of a ton of LNG.

Chapter 17: Relationships

Figure 17.2: Gas plant mass balance process.
Figure 17.5: Relationship between Utilization and Availability as per PTAI benchmarking consultants.
Box 17.6: Gas Plant Energy Balance

PART 5: ASSET PERFORMANCE MANAGEMENT

Chapter 20: Human Capital

Box 20.4: Secondees —GTL Plant
Box 20.10: Behavior-Based Safety—STOP Cards

Chapter 21: Finance

Figure 21.2: Gas plant Opex model.

Chapter 22: Physical Asset Performance Management

Box 22.7: System Maturity—Gas Plants

Chapter 23: Turnarounds

Box 23.1: Scope Challenge—Gas Plant Turnaround

PART 6: BENCHMARKING

Chapter 25: Common Denominators

Box 25.3: Equivalent Gas Processing Capacity (EGPC) Application
Box 25.4: Example of Cost Comparison Using RV as Denominator

Chapter 26: Data Verification

Box 26.1: Exclusions—Body Count
Box 26.2: Interpretation of Definitions—Maximum Sustainable Capacity

Chapter 27: Using Benchmarking Results

Box 27.1: Benchmarking Envelopes—Gas Plant Example
Box 27.3: Summary Bar Graph—Gas Plant Benchmarking Study Summary

38.3 GAS PLANT PERFORMANCE INITIATIVES

Some performance initiatives are listed next.

ENERGY CONSERVATION

Reduction in greenhouse gases (GHG) is a balancing act between increased liquid gas production and utilization of waste gas so as to reduce flaring. LNG plants typically use 6%–9% of weight on intake (woi) gas for the process. GTL plants use typically 25% woi. Energy conservation is, therefore, a good place to start for performance improvement.

The following are some performance improvement examples:

- Modifications to heat recovery systems
  - Waste heat recovery
  - Rearrangement of heat exchanger network (pinch technology)
- Improvement in operation
  - Reduction in excess combustion air
  - Application of Advanced Process Control (APC)
- Reduction in electric power consumption
  - Power recover turbine
  - Switching pump operation from motor to steam turbine driver
- Reduction of heat radiation loss
  - Insulation of bare pipe upstream of steam traps
  - Replace steam traps with proper type
  - Apply optimal thickness insulation on piping and vessels

New technologies that can help are as follows:

- High-efficiency burners
- Co-generation of power
- Gas recovery systems for loading LNG
- Helium recovery
- Installation of Coanda flares[1]
  - The Coanda effect proves that a gas passing over a carefully profiled, curved surface will adhere to that surface, creating a near vacuum that pulls in substantial amounts of air. The air turbulently mixes with the gas flow, resulting in highly efficient combustion.
- Membrane tanks for storage and shipping of LNG (see Fig. 38.3 for view of membrane tank)

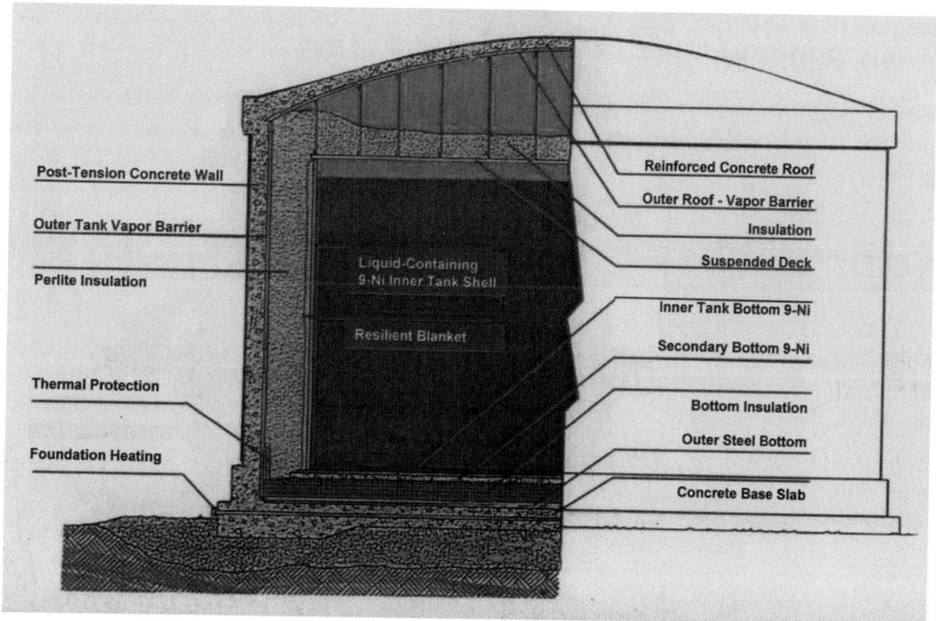

FIGURE 38.3

LNG membrane tank.

PROCESS FIELD DATA COLLECTION AND MONITORING

Honeywell's IntelaTrac process knowledge system (PKS) uses mobile technology to deliver reliable field data faster to the right people, resulting in lower costs, better return on plant assets, and improved profits.

PROCESS SAFETY PROGRAM (PSP) INITIATIVE[3]

This is a structured approach to improving operational process safety. It is of major importance, specifically for gas plants.

LNG SHIPPING

During shipping LNG to customers around the world, a 14-day journey can result in a 2% cargo loss from Boil-Off Gas (BOG). Fuel for the ships engines can also account for 10%–15% of the LNG shipping costs. Previously, boilers and steam turbine propulsion units (at 30% efficiency) disposed of the BOG. The move is now toward using slow-speed diesels (at 50% efficiency) for propulsion and having a package reliquefaction unit for BOG, giving 100% cargo delivery. In the future, dual fuel engines may be introduced, should the price of oil rise above that of gas.

LNG ships have evolved from the conventional spherical tank LNG ships, which restrained the increase in capacity, to membrane LNG ships, allowing much bigger capacities.

Examples:

Q-Flex and Q-Max membrane LNG ships are the largest LNG carriers in the world.

38.4 THE PROCESS SAFETY PROGRAM

The implementation of a PSP is of particular importance for gas plants. It relates directly to the operation of the process plant. Primary elements are the management of the shift cycle and improvement in alarm management.

SHIFT CYCLE

The typical cycle is as follows:

1. Shift handover
2. Start of shift orientation
3. Shift team meeting planning
4. **Proactive monitoring**
5. End of shift review
6. **Report shift performance.**

Proactive monitoring and **shift performance reporting** are discussed in more detail as follows.

Proactive monitoring entails the following:

1. Console operator proactive monitoring
2. Outside operator proactive monitoring including front line maintenance activities
3. Technical proactive monitoring through the use of APC and other tools
4. Managing abnormal situations with the use of active **alarm management** and metrics.

The initial setup of the operating limits entails the following:

1. Product quality limits
2. Equipment integrity limits
3. Reliability limits
4. Environmental permit limits
5. Other operating limits.

Operating targets need to be determined and validated. These include the following:

- Product quality targets: determined by industry standards
- Optimization targets: determined by optimization of the process (e.g., APC)
- Throughput and yield targets: determined by design of the plant, simulation, and benchmarking
- Reliability targets: determined by the production cycle (see Section 23.7: Strategic Planning and Run Length Determination)
- Energy targets: determined by regulatory authorities and benchmarking
- Chemical/catalyst targets: determined by process operational severity and chemical/catalyst cost
- Other targets.

Shift reporting needs to be structured. An electronic log book ensures better record keeping. Improvement to the process includes the following:

1. Process performance metrics
2. Process outcomes metrics
3. Investigation results and learning
4. Other learning

Outcomes entail the following:

- Safe operations
- Environmental compliance
- Meeting production plans
- Reports on equipment and process conditions
- Threat mitigation.

The cycle is depicted in Fig. 38.4.

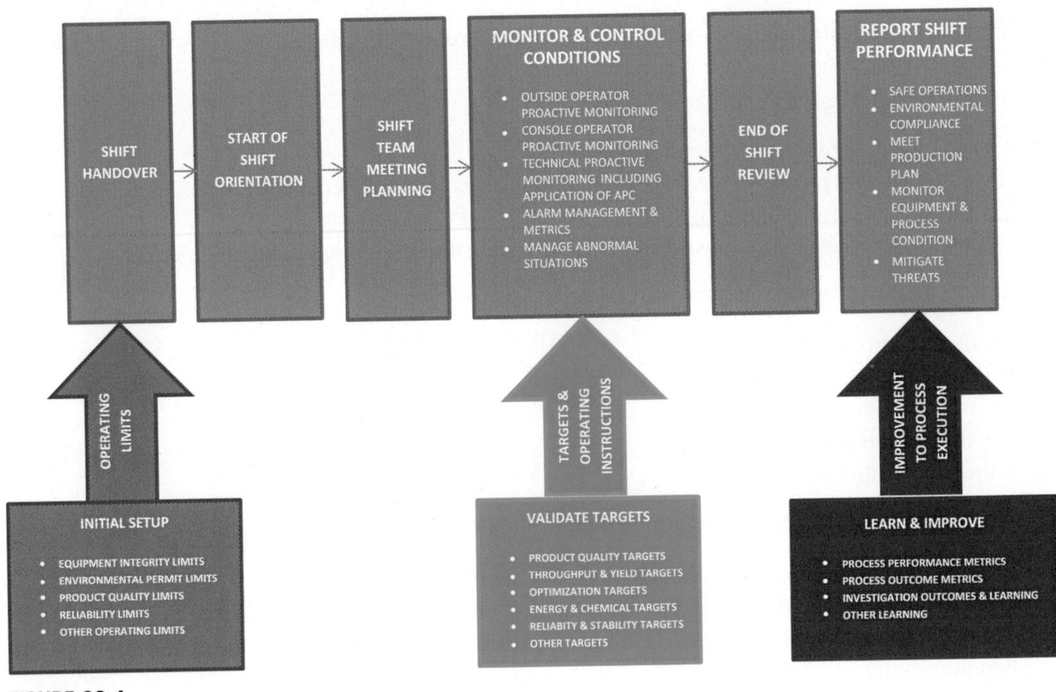

FIGURE 38.4

Typical shift cycle.

ALARM MANAGEMENT

Alarm management entails the following:

- The scrutiny of all alarms to determine their relative importance and urgency
- The establishment of prioritization hierarchies to assist the operator to manage an emergency situation.

Alarm management is of particular importance so as to avoid "alarm flood." A study of all alarms is required, redundancy determined, and prioritization established. Also, a Key Performance Indicator (KPI), reported to top management, is **alarm frequency**, or the number of alarms per panel per hour. A typical industry target is five.

Safe limits for operation have to be clearly identified for operators to work within. Fig. 38.5 depicts limits of operation where an alert is given when target upper and lower limits are exceeded. Management of change (MoC) is required for going beyond the Standard upper and lower limits.

The MoC process helps ensure that changes to a process plant do not inadvertently introduce new hazards or unknowingly increase risk of existing hazards. It includes a review and authorization process for evaluating proposed adjustments to facility design, operations, organization, or activities before implementation to make certain that no unforeseen new hazards are introduced and that the risk of existing hazards to employees, the public, or the environment is not unknowingly increased. It also includes steps to help ensure that potentially affected personnel are notified of the change and that pertinent documents, such as procedures, process safety knowledge, and so forth, are kept up-to-date.

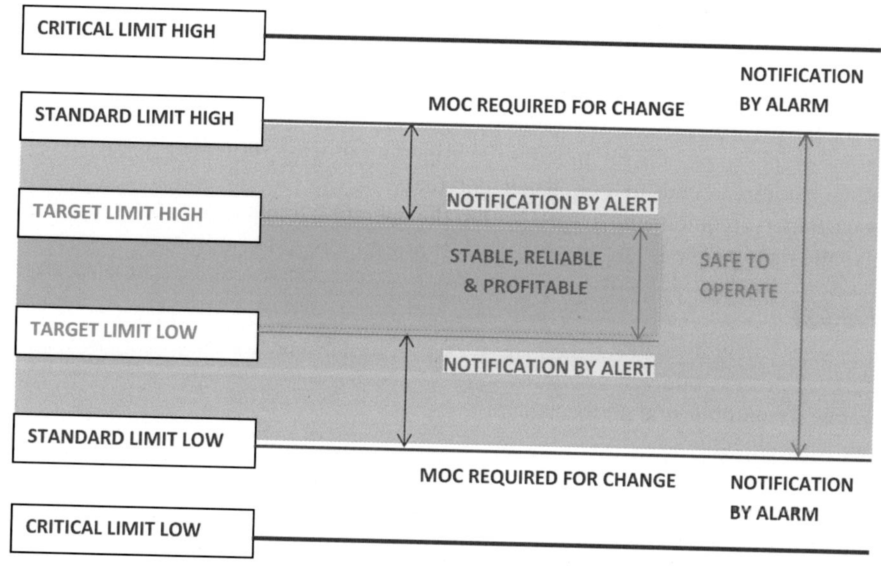

FIGURE 38.5

Alarm limits.

Box 38.1 gives an overview of a sample gas plant PSP.

BOX 38.1 SAMPLE PSP OVERVIEW

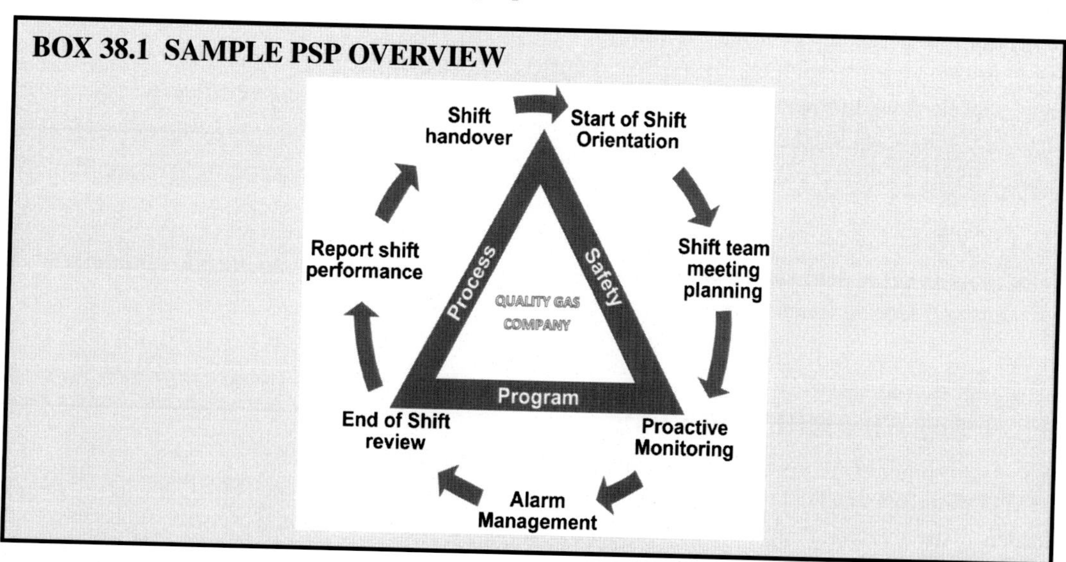

PSP Tenets for Incident Free Operation

Always:

1. Operate within limits and know your limits.
2. Comply with all applicable rules and regulations.

3. Address abnormal conditions via triple S—stabilize, slowdown, and shutdown.
4. Ensure safety devices are in place and functioning.
5. Follow safe work practices and procedures.
6. Maintain integrity of dedicated systems.
7. Use trend graphs to proactively monitor the important process variables.
8. Conduct operating rounds to look, listen, and feel.
9. Meet customers' requirements/targets within the operating limits to avoid abnormal situations.
10. Involve the right people in decisions that affect procedures and equipment.

Benefits of PSP

- Fewer process safety incidents:
  - Improved process safety
  - Improved environmental performance
  - Improved reliability
  - Reduced maintenance and operation costs
- Improved and standardized communication through shift reporting, shift handover, orientation, and shift team meeting and planning
- Proactive monitoring of operations to ensure staying within limits and manage abnormal situations to avoid undesired operating conditions
- Verification that the alarms are aligned with the operating limits
- Operators empowered to take remedial action predefined within the operating limits
- Capturing knowledge of operators and staff of process limitations, operations, and related areas (prior to key operators and staff retiring)
- Consistent prioritization of alarms drives consistent management of abnormal situations:
  - Critical: act now and get help
  - Standard: act to resolve and get help as soon as possible
  - Target: optimization, act to resolve, get help as needed.

Example:

Qatargas has recently been ranked as a pacesetter gas plant by PTAI. Among other initiatives, Qatargas has implemented a PSP.

38.5 GAS-TO-LIQUID PROCESS

The gas-to-liquid (GTL) process is basically depicted as shown in Fig. 38.6.

The proprietary technology is in the synthetic crude production unit of which Shell[4] and Sasol[2] are the leaders in the field. Sasol refined their technology from the production of liquid transport fuels from coal, with a coal gasification process at the front end.

Mass and energy balance is normally undertaken separately in four blocks—feed gas processing (FGP), synthetic crude production (SCP), liquid processing unit (LPU), and utilities.

Box 38.2 gives an overview of Shell Pearl GTL.

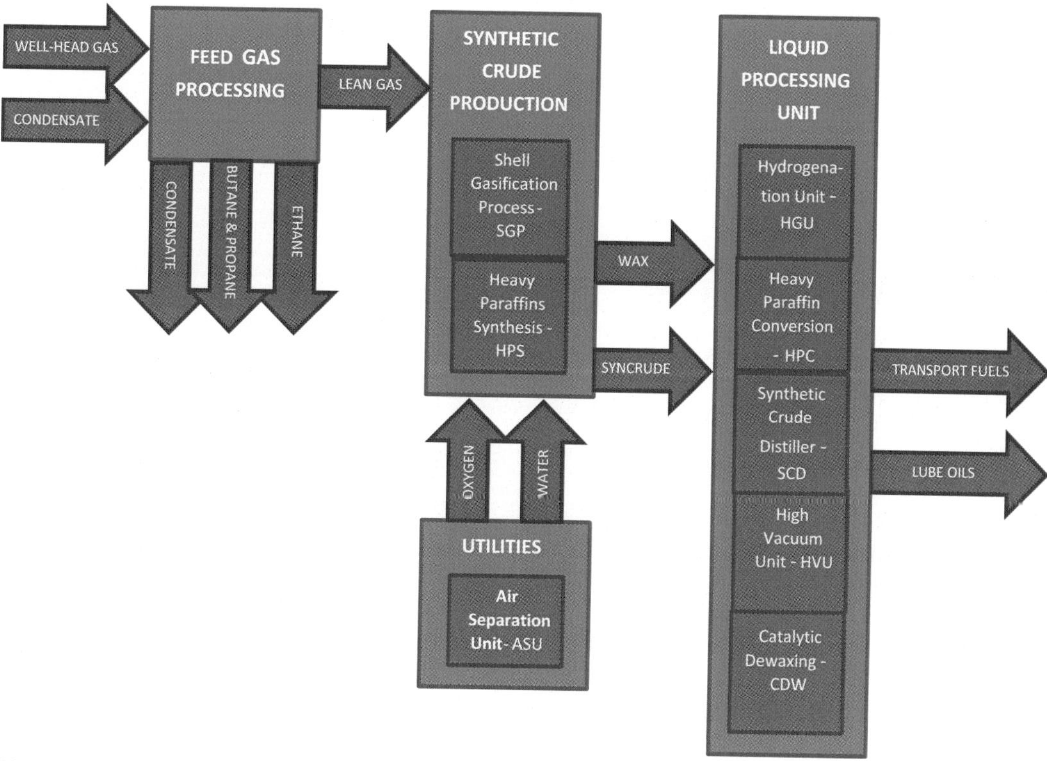

FIGURE 38.6

Basic GTL process—Shell technology.

BOX 38.2 SHELL PEARL GTL

"Pearl GTL is a fully integrated upstream-downstream development capturing in one operating business the full gas value chain from offshore development through onshore gas processing, the conversion of gas to hydrocarbon liquids, and the refining to finished products. Up to 1.6 billion cubic feet per day of wellhead gas from 22 offshore wells is converted to gas-to-liquids using Shell's proprietary Shell Middle Distillate Synthesis (SMDS) process, built on 3500 patents. From this, a range of high-performing GTL products is created, from gasoil, kerosene, and base oil to naphtha and normal paraffins for the petrochemicals industry.

GTL products represent a pioneering innovation to increase the supply of highly-demanded liquid hydrocarbons. Pearl GTL products are virtually sulfur free and have practically no contaminants such as heavy metals or aromatics. Their paraffinic chemical nature ensures that they are highly biodegradable and almost odorless. When GTL gasoil is used in automotive applications, such as buses, it can help improve local air quality, such as soot (particulates) emissions, when compared to the use of conventional diesel fuel.

The Pearl GTL plant has 24 reactors, weighing 1200 tons a piece. They each contain 29,000 tubes full of Shell's cobalt synthesis catalyst, which speeds up the chemical reaction.

Construction of Pearl GTL was completed on schedule with the complex starting up end of Q1 2011. Full ramp-up was achieved toward the end of 2012. Despite the massive number of workers involved and the complexity of Pearl GTL's construction, a strong safety culture helped Qatar and Shell break industry records. In 2010, the project achieved 77 million hours worked without a single lost time injury (LTI) and an overall LTI frequency of 0.04 LTI/million man-hours corresponding to about 1/10th of the industry average."

From Shell website (www.shell.com).

38.6 GAS PLANT TOP KEY PERFORMANCE INDICATORS

The following is a basic list of top KPIs. Others are listed in Appendix B, KPIs.

1. Energy
   a. Energy use on intake: % woi
   b. Fuel and loss: % woi
   c. GHG: % LNG
   d. GHG: % sales/flowing gas
   e. Flaring: % woi
   f. Total LNG loaded
   g. Total sales/flowing gas
2. Financial
   a. Fixed cost per ton of salable product
   b. Operating cost per ton of salable product
3. Physical assets
   a. Reliability
   b. Availability
   c. Fatal Accident Rate (FAR)
   d. Lost Time Incident Frequency (LTIF)
   e. Loss Of Primary Containment (LOPC) level 1 incidents
   f. Loss Of Primary Containment (LOPC) level 2 incidents
   g. Loss Of Primary Containment (LOPC) level 3 selected leading indicators
4. Human assets
   a. Overall Personnel Index
   b. Employee turnover

38.7 SUMMARY

Gas plants are complex process plants. Loss Of Primary Containment is the primary focus. Also, reduction in greenhouse gases is a balancing act between increased liquid gas production and utilization of waste gas so as to reduce flaring.

For LNG, BOG loss is a continuing challenge. Recycling BOG during loading of LNG tankers considerably reduces flaring. Membrane tank design for storage tanks and LNG ships reduces the losses and increases delivered product.

Homing in on the few strategies that will make a big difference to performance entails a lot of analysis and discussion with interested parties. There are many strategies to choose from, but again the trick is to choose only a few. Once these few strategies have been decided, the change process needs to be structured to ensure a step improvement to the next level of performance.

Gas plant–specific performance issues covered in previous chapters are listed for review. A structured process safety program is essential for safe operation of a gas plant.

Gas-to-liquid plants complement conventional LNG plants where there is an abundant supply of gas.

For convenience, gas plant top KPIs are grouped: energy, finance, physical asset, and human asset.

REFERENCES

1. Coanda Flares. http://www.johnzink.com/products/flare-systems/products-2/coanda-flares.
2. Sasol GTL technology. http://www.sasol.com/innovation/gas-liquids/technology.
3. Qatargas process safety program. http://www.qatargas.com/English/MediaCenter/news/Pages/QGDeliversKey.aspx.
4. Shell. www.shell.com.

FURTHER READING

1. Carlsson A. NLNG maximises LNG production in supply constraint. *Oil and Gas Journal* July 2010.
2. Software defuses demographic time-bomb. University of Cambridge. Lifetrack software. http://www.cam.ac.uk/news/software-defuses-demographic-time-bomb.
3. *The outlook for energy: a view to 2040.* ExxonMobil. http://corporate.exxonmobil.com/en/energy/energy-outlook.
4. Castel J, Gadelle P, Hagyard P, Ould-Bamba M. LNG and GTL drive 50 years of technology evolution in the gas industry. *Hydrocarbon Processing* July 2012.

PART

CONCLUSION

10

OVERVIEW

HOW DO YOU KNOW IF THE PERFORMANCE OF THE BUSINESS IS MOVING IN THE RIGHT DIRECTION?

- You know where you are relative to the competition (benchmarking).
- You have a good set of rules for management of the business (governance).
- You have a structured improvement system based on ISO 9001 (systems).
- You have motivated people who have an inherent urge to improve (thinking organization).
- You have high business ethics for long-term sustainability of the business.

OUTLINE OF PART 10

- Awarding excellence: two approaches—organizational excellence and business performance excellence
- ISO standards: ensuring the continuous improvement loop
- People: creating a learning organization
- Sustainability: ensuring long-term survival of the business

CHAPTER

ALIGNMENT TO ACHIEVE RECOGNITION FOR EXCELLENCE

39

39.1 INTRODUCTION

This book has covered all aspects that affect the performance of a business in the process industry. The "black box" modular Business Unit approach makes it possible to address complex issues related to the industry. This chapter sums up performance in the process industry and relates it to businesses in general.

39.2 EXCELLENCE DEFINITIONS AND CRITERIA

"Performance excellence" and "business excellence" converged into "organizational excellence" resulting in the formation of the Multinational Alliance for Advancement of Organizational Excellence (MAAOE) in November 1998. However, in the context of this book, I wish to differentiate between Organizational Excellence (OE) and Business Performance Excellence (BPE).

DEFINITIONS

Organizational Excellence *(MAAOE Definition)*

Organizational Excellence is the overall way of working that balances stakeholder concerns and increases the probability of long-term organizational success through operational, customer-related, financial, and marketplace performance excellence.

OE could be identified by the receipt of an award or by carrying out a self-assessment. This is applicable to any business.

Business Performance Excellence *(Hey Definition)*

Business Performance Excellence is the outcome of the comparative performance of a business with its competitors in similar industries or functions.

Top performers are determined by quantitative evaluation relative to their peers.

BPE is normally the measurement of core operational practices against similar businesses. The best in that industry are regarded as "world class" or "pacesetters."

Performance Management for the Oil, Gas, and Process Industries. http://dx.doi.org/10.1016/B978-0-12-810446-0.00039-6

OE is based on core business practices, whereas BPE is based on core operational practices. Performance criteria are thus different.

The awards are also different. OE is based on the criteria set by the specific organization giving the award. BPE is based on objective performance indicators used for comparison to similar industries or functions. BPE awards are based on the benchmarked outcomes of a particular consultant's study.

Organizational maturity gives an indication of progression toward excellence.

OE and BPE approaches are differentiated as shown in Fig. 39.1.

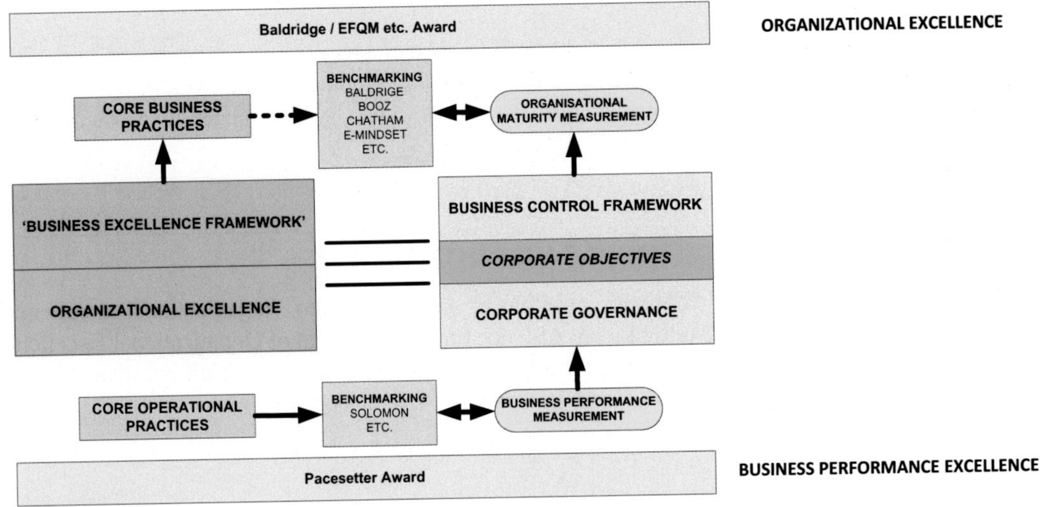

FIGURE 39.1

OE and BPE.

OE CRITERIA

The following generic criteria have been derived from the international quality prize models:

1. Resources
2. **Leadership**
3. Policy and strategy
4. **People** and knowledge
5. Partners/partnerships
6. **Customer**-focused processes
7. Organizational performance
8. Information and analysis

9. Innovation and learning
10. Business results
11. **Customers**
12. Society

The highlighted items are common to the eight principles of ISO 9001:2008 as discussed in Section 39.5, Excellence and Certification Standards.

BPE CRITERIA

These are set by the particular benchmarking consultant. For operational benchmarking, they can be grouped as follows:

- Energy
- Finance
- Physical assets
- Human assets

39.3 OE AWARDS

The following are the most popular awards:

1. Baldrige National Quality Program–US—A "performance excellence" model[1]
2. European Foundation for Quality Management (EFQM)–Europe—A "business excellence" model[2]
3. Business Excellence Framework (SAI Global)[3]—Australia
4. Singapore Quality Awards
5. Japan Quality Award
6. Dubai Quality Award

Companies submit survey details to these organizations for the annual awards for OE.

Box 39.1 outlines the European Foundation for Quality Management (EFQM) model.

BOX 39.1 THE EFQM EXCELLENCE AWARD FOR BUSINESS EXCELLENCE

EFQM postulates that "excellent organizations achieve and sustain superior level of performance that meet or exceed the expectations of all stakeholders."

The model consists of **enablers** and **results**.

Enablers include leadership, strategy, people, partnerships and resources, and processes/products/services.

Results are categorized as people, customer, society, and business.

Feedback entails learning, creativity, and innovation.

The U.S. Baldrige model is shown in Box 39.2.

BOX 39.2 THE BALDRIGE MODEL

The Baldrige Criteria for Performance Excellence are a set of questions in seven interrelated areas (known as categories) that guide you in assessing your organization's performance. This is depicted next.

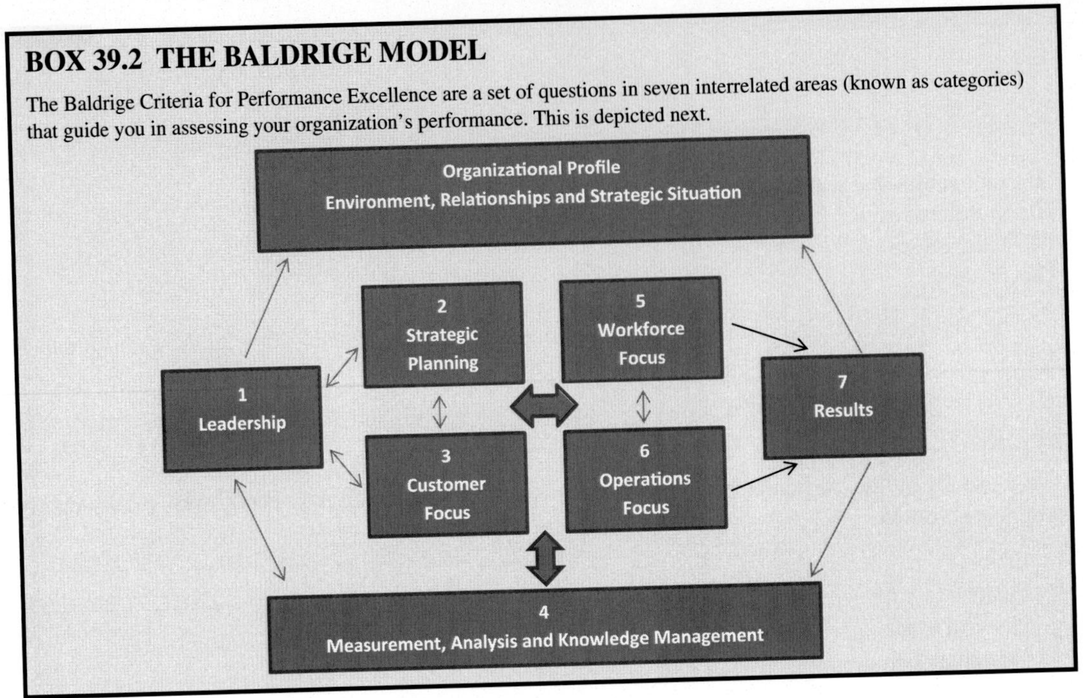

Chevron's approach to comply with Baldrige is shown in Box 39.3.

BOX 39.3 CHEVRON "BEST PRACTICE" RESOURCE MAP—BALDRIGE AWARD CRITERIA

The Chevron approach to achieving Organizational Excellence is outlined next.

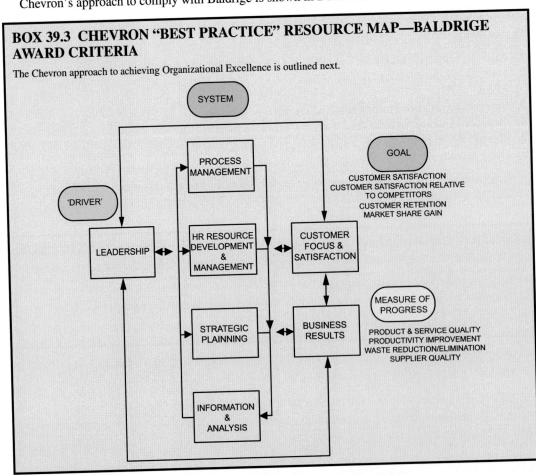

SYSTEM

GOAL

CUSTOMER SATISFACTION
CUSTOMER SATISFACTION RELATIVE TO COMPETITORS
CUSTOMER RETENTION
MARKET SHARE GAIN

'DRIVER'

LEADERSHIP

PROCESS MANAGEMENT

HR RESOURCE DEVELOPMENT & MANAGEMENT

STRATEGIC PLAINNING

INFORMATION & ANALYSIS

CUSTOMER FOCUS & SATISFACTION

BUSINESS RESULTS

MEASURE OF PROGRESS

PRODUCT & SERVICE QUALITY
PRODUCTIVITY IMPROVEMENT
WASTE REDUCTION/ELIMINATION
SUPPLIER QUALITY

39.4 SELF-ASSESSMENT

OE comparisons could be carried out internally. The South African Business Excellence Model is based on the European and U.S. models and is used for self-assessment. Box 39.4 describes the criteria.

BOX 39.4 THE SOUTH AFRICAN BUSINESS EXCELLENCE MODEL

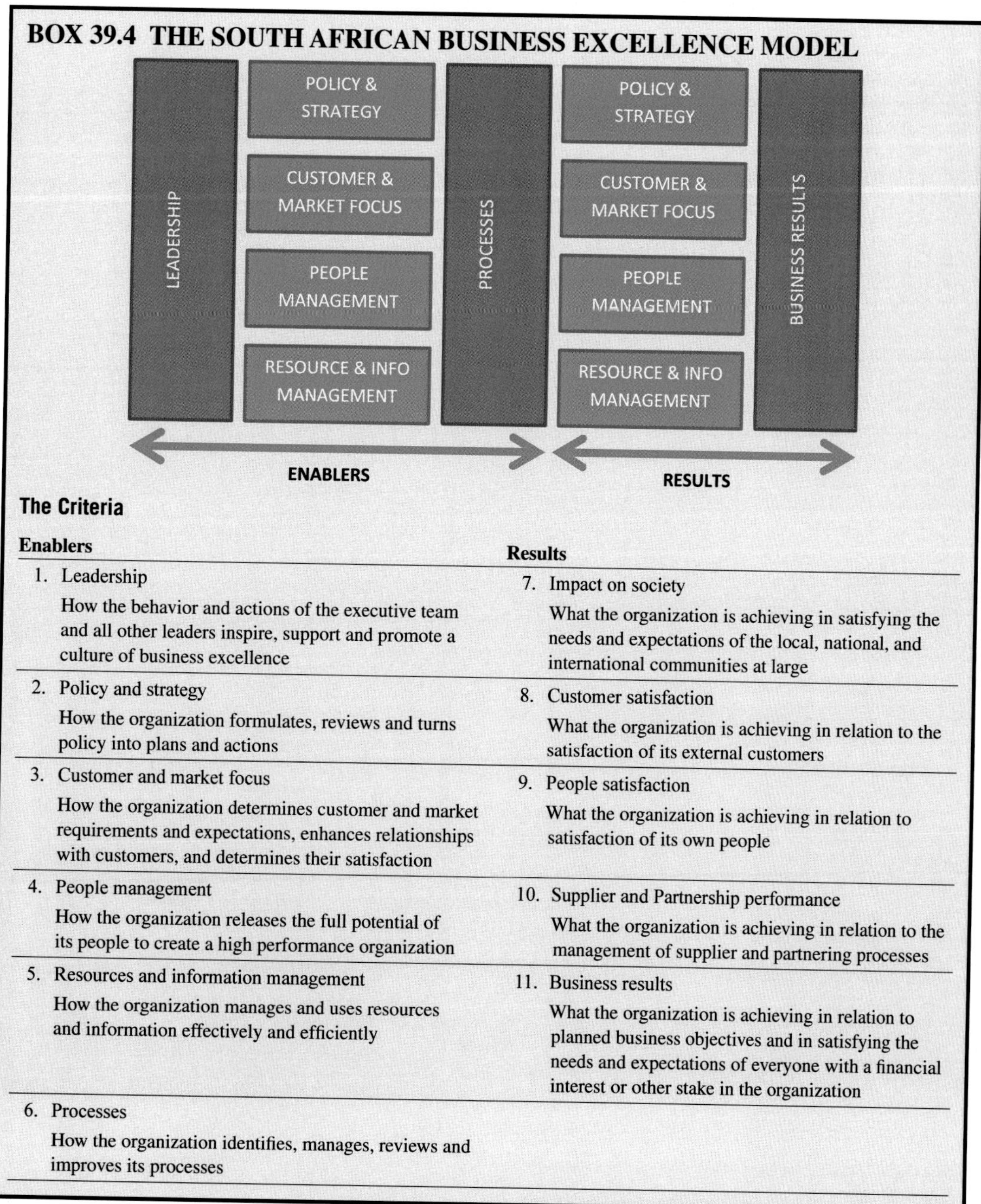

The Criteria

Enablers	Results
1. Leadership How the behavior and actions of the executive team and all other leaders inspire, support and promote a culture of business excellence	7. Impact on society What the organization is achieving in satisfying the needs and expectations of the local, national, and international communities at large
2. Policy and strategy How the organization formulates, reviews and turns policy into plans and actions	8. Customer satisfaction What the organization is achieving in relation to the satisfaction of its external customers
3. Customer and market focus How the organization determines customer and market requirements and expectations, enhances relationships with customers, and determines their satisfaction	9. People satisfaction What the organization is achieving in relation to satisfaction of its own people
4. People management How the organization releases the full potential of its people to create a high performance organization	10. Supplier and Partnership performance What the organization is achieving in relation to the management of supplier and partnering processes
5. Resources and information management How the organization manages and uses resources and information effectively and efficiently	11. Business results What the organization is achieving in relation to planned business objectives and in satisfying the needs and expectations of everyone with a financial interest or other stake in the organization
6. Processes How the organization identifies, manages, reviews and improves its processes	

Box 39.5 depicts an internal assessment/survey carried out by an International Oil Company (IOC). SAI Global Business Excellence Framework[3] can also be used for self-assessment.

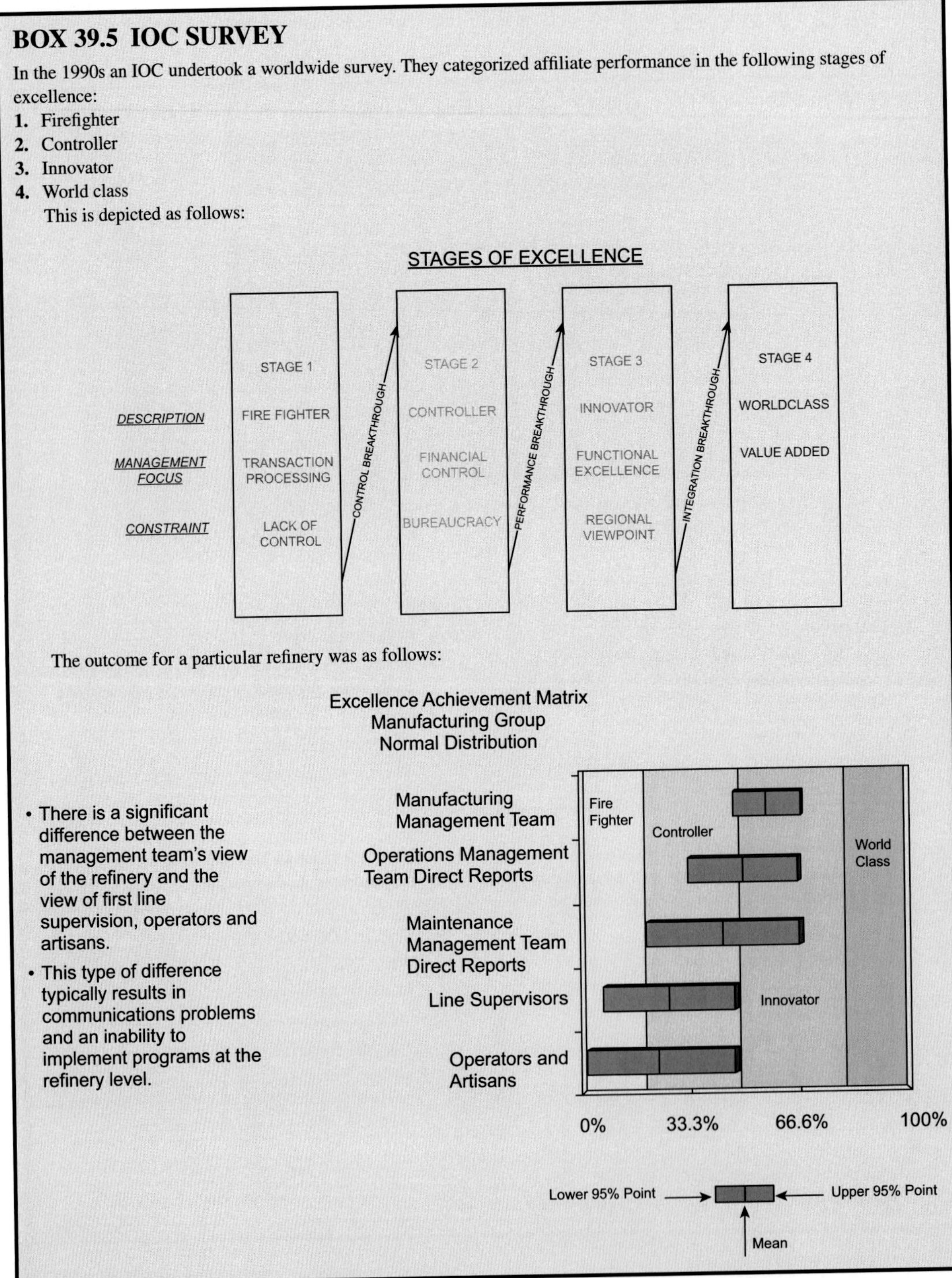

BOX 39.5 IOC SURVEY
In the 1990s an IOC undertook a worldwide survey. They categorized affiliate performance in the following stages of excellence:
1. Firefighter
2. Controller
3. Innovator
4. World class
This is depicted as follows:
STAGES OF EXCELLENCE
DESCRIPTION
MANAGEMENT FOCUS
CONSTRAINT
STAGE 1
FIRE FIGHTER
TRANSACTION PROCESSING
LACK OF CONTROL
CONTROL BREAKTHROUGH
STAGE 2
CONTROLLER
FINANCIAL CONTROL
BUREAUCRACY
PERFORMANCE BREAKTHROUGH
STAGE 3
INNOVATOR
FUNCTIONAL EXCELLENCE
REGIONAL VIEWPOINT
INTEGRATION BREAKTHROUGH
STAGE 4
WORLDCLASS
VALUE ADDED
The outcome for a particular refinery was as follows:
Excellence Achievement Matrix
Manufacturing Group
Normal Distribution
• There is a significant difference between the management team's view of the refinery and the view of first line supervision, operators and artisans.
• This type of difference typically results in communications problems and an inability to implement programs at the refinery level.
Manufacturing Management Team
Operations Management Team Direct Reports
Maintenance Management Team Direct Reports
Line Supervisors
Operators and Artisans
Fire Fighter
Controller
Innovator
World Class
0%
33.3%
66.6%
100%
Lower 95% Point
Upper 95% Point
Mean

BOX 39.5 IOC SURVEY—cont'd

Discussion

The differences in perception indicated in this example needed the process shown in Fig. 30.4, Moving From Blame Culture, to identify the real problems to be overcome to get to the next stage of excellence. This particular refinery actually became more unionized as a result of the differing perceptions.

39.5 EXCELLENCE AND CERTIFICATION STANDARDS

EXCELLENCE AND IMPROVEMENT

The Certification Standards are only tools for a structured approach in the quest for excellence. An example may help differentiate between excellence at different quality levels.

Example

The name Rolls Royce has always been associated with excellence. Yet, the average person cannot hope to buy a Rolls Royce car because it's far beyond the means of the average person. However, many people are buying Hyundais, which are far more affordable. Hyundai produced its first car in 1968 and is now the fifth largest vehicle manufacturer in the world.

Excellence can be achieved on different levels. Hyundai has had an aggressive performance improvement program to achieve its leadership position, yet its quality level is related to its targeted market segment, which is not the same as that of Rolls Royce. It has a market penetration based on value for money.

THE IMPROVEMENT CYCLE

The eight principles of ISO 9001:2008 are the fundamentals for continual improvement. These are as follows:

1. Customer focus
2. Leadership
3. Involvement of people
4. Process approach
5. System approach to management
6. Continual improvement
7. Factual approach to decision-making
8. Mutually beneficial supplier relationships

How Do They Relate to This Book?

Customer Focus

Horizontal alignment of the Systems approach ensures customer satisfaction.

Leadership

Vertical alignment with strong committed leadership from the top is essential.

Involvement of People

Competent and committed human assets make it happen.

Process Approach

Well-developed Processes, as parts of Systems, are essential.

System Approach to Management

Strategic, Core, and Support Systems ensure optimum value addition over the life of the investment.

Continual Improvement

The improvement cycle as part of each System is embedded everywhere in the Business Unit.

Factual Approach to Decision-Making

Facts must be based on benchmarking, and the use of quantitative performance indicators that have a significant impact on the bottom line of the business.

Mutually Beneficial Supplier Relationships

Suppliers have to have the incentive to improve service. Thus, well thought out, mutually beneficial contractual arrangements are essential to achieve common objectives.

The continual improvement cycle of ISO 9001 is relevant throughout the business. It is depicted in Fig. 39.2.

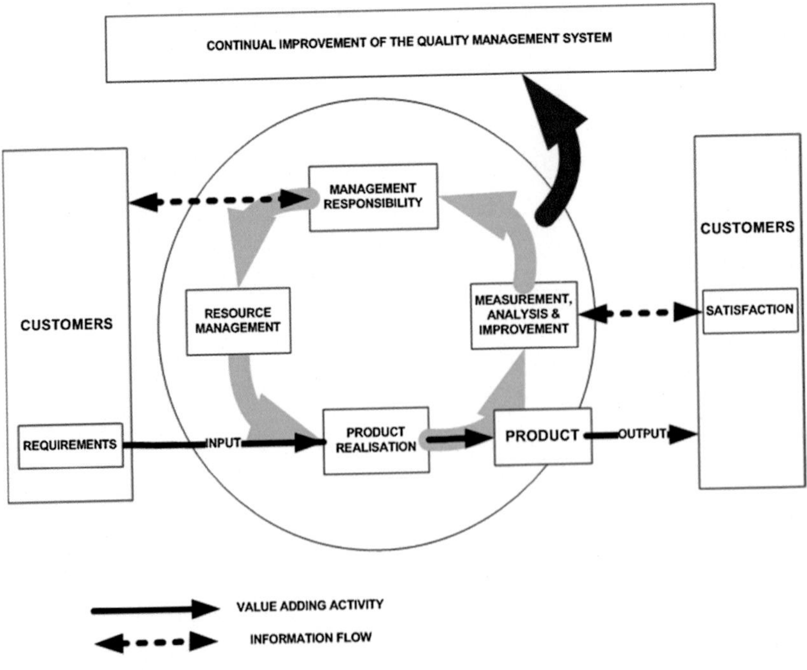

FIGURE 39.2

The continual improvement cycle.

COMMON FORMATS OF INTERNATIONAL STANDARDS

The standard paragraphs applicable to ISO standards requiring certification are as follows:

1. Introduction
2. Scope
3. Normative references
4. Terms and definitions
5. Management System requirements
6. Management responsibility
7. Internal audits
8. Management review
9. Improvement

The following standards require certification.

- ISO 9001 Quality Management Systems
- ISO 14001 Environment Management Systems
- ISO 45001 Occupational Health and Safety Systems
- ISO 27001 Information Security Management Systems
- ISO 50001 Energy Management Systems
- ISO 55001 Asset Management Systems

39.6 ALIGNMENT AND CONTINUOUS IMPROVEMENT

Alignment entails vertical and horizontal alignment as discussed in Chapter 10, Alignment. However, Fig. 39.3 summarizes key aspects of alignment and continuous improvement.

39.7 THE LEARNING ORGANIZATION

DEFINITION

> The learning organization is capable of continual regeneration from the variety of knowledge, experience and skills of individuals within a culture that encourages mutual questioning and challenge around a shared purpose or vision.

Advocates of the learning organization point out that the collective knowledge of all the individuals in an organization usually exceeds what the organization itself "knows" and is capable of doing; the formal structures of organizations typically stifle organizational knowledge and creativity. They argue that the aim of management should be to encourage processes that unlock the knowledge of individuals, and encourage the sharing of information and knowledge, so that every individual becomes sensitive to changes occurring around them and contributes to the identification of opportunities, risks, and required changes.

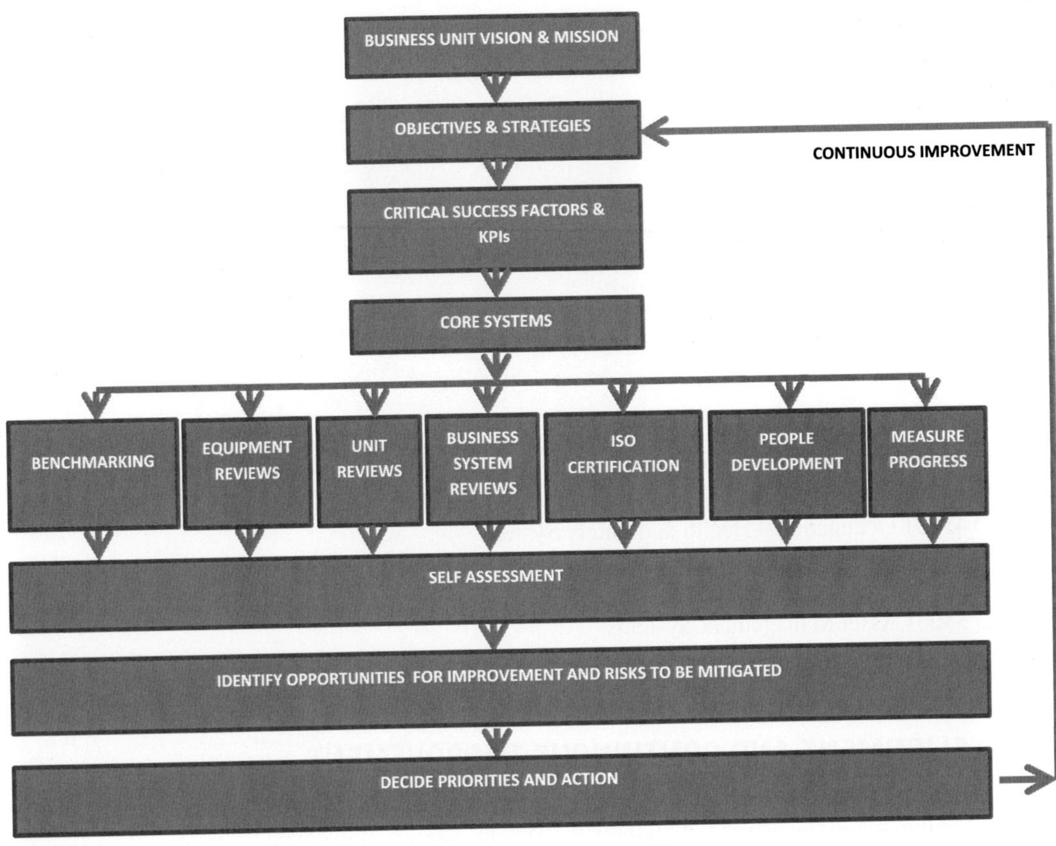

FIGURE 39.3

Continuous improvement model.

BUSINESS MATURITY

Fig. 8.1, Before and after applying a governance framework and Systems approach, gives the extremes of maturity. Section 19.8, The Maturity of Information Systems, discusses system maturity.

Business maturity, however, is the combination of the 3 Ps—Processes (and Systems), People, and Performance. The evolution of maturity is driven by People—competency, experience, and motivation. The cultivation of a learning organization is thus critical for real Performance improvement.

STAFF RECOGNITION

As mentioned in Chapter 20, Human Capital, a well-structured Personnel Performance Management System (PPMS) is required to evaluate each staff member's contribution and calculate an annual performance bonus. The line of sight to the company vision is essential.

LEADERSHIP[4]

Leadership should take place at all levels of the organization. Situational leadership is a key to success.

Example

An engineering manager can be trained as an emergency coordinator so as to be able to take charge in an emergency situation.

The difference between leaders and managers is generally that:

1. leaders produce change by establishing direction, aligning team members, and motivating and inspiring people;
2. managers produce a degree of predictability and order by planning and budgeting, organizing and staffing, and controlling and problem solving.

39.8 SUSTAINABILITY

As discussed in Chapter 7, What Are the Principles of Good Governance? there is now a greater focus than ever on sustainability. Various initiatives ensure that sustainability is given greater prominence.

DOW JONES SUSTAINABILITY INDEX[5]

The Dow Jones Sustainability World Index was launched in 1999 as the first global sustainability benchmark. The DJSWI family is offered cooperatively by RobecoSAM and S&P Dow Jones indices. The family tracks the stock performance of the world's leading companies in terms of economic, environmental, and social criteria. The indices serve as benchmarks for investors who integrate sustainability considerations into their portfolios and provide an effective engagement platform for companies who want to adopt sustainable best practices.

Only the top ranked companies in terms of corporate sustainability within each industry are selected for inclusion in the Dow Jones Sustainability Index family. No industries are excluded from this process.

Selected criteria are as follows:

1. Corporate governance
2. Supply chain management
3. Risk and crisis management
4. Operational eco-efficiency
5. Climate strategy
6. Releases to the environment
7. Social Impacts on communities
8. Occupational health and safety
9. Labor practice indicators and human rights

Thai Oil PLC was Energy Industry Group Leader in 2015.

UN GLOBAL COMPACT[6]

The United Nations (UN) Global Compact is a UN initiative to encourage businesses worldwide to adopt sustainable and social responsibility policies, and to report on their implementation. The UN Global Compact is a principle-based framework for businesses, stating 10 principles in the areas of human rights, labor, the environment, and anticorruption. Under the Global Compact, companies are brought together with UN agencies, labor groups, and civil society.

The 10 principles are as follows:

Human Rights

Businesses should:

- Principle 1: support and respect the protection of internationally proclaimed human rights; and
- Principle 2: make sure that they are not complicit in human rights abuses.

Labor Standards

Businesses should uphold:

- Principle 3: the freedom of association and the effective recognition of the right to collective bargaining;
- Principle 4: the elimination of all forms of forced and compulsory labor;
- Principle 5: the effective abolition of child labor; and
- Principle 6: the elimination of discrimination in employment and occupation.

Environment

Businesses should:

- Principle 7: support a precautionary approach to environmental challenges;
- Principle 8: undertake initiatives to promote environmental responsibility; and
- Principle 9: encourage the development and diffusion of environmentally friendly technologies.

Anticorruption

- Principle 10: Businesses should work against corruption in all its forms, including extortion and bribery.

39.9 SUMMARY

Excellence has evolved on two fronts: comparison with others in the same business, and compliance with a "business excellence framework" established by governments or national quality organizations.

Excellence determined by comparison clearly identifies whether the company is a leader in its industry, thus obtaining the status of "pacesetter" or "world class". On the other hand, the "business excellence framework" is generic and applicable across all commercial and industrial sectors in a country, and could result in an award from an independent committee.

The benchmarking approach focuses on core operational practices and business performance measurement. Corporate governance, corporate objectives and corporate control framework link the benchmarking approach to organizational maturity, core business practices and the resulting "business excellence framework."

All aspects of performance management discussed in this book contribute to both fronts.

A number of ISO standards, with similar mechanisms to ensure improvement, have been established. They all include a **Systems** approach to management, **System** requirements, management responsibility, a requirement for auditing and management review, which ensures an ongoing improvement cycle. Compliance with these International Quality Standards ensures that a performance improvement framework is in place. ISO 9001 lists eight elements for the improvement cycle to be successful.

People are the heart of any business. The cultivation of a learning organization is critical for real performance improvement. The business then matures as the staff experience is built up. The line of sight from the shareholder aspirations, all the way to the motivated staff members, must always be clear to ensure that individual employees can be effective.

The focus on sustainability has become more acute with issues such as global warming. Initiatives such as the Dow Jones Sustainability World Index and UN Global Compact give focus on sustainability.

We hope that this book has helped the reader to:

- apply a "**Systems approach**" to managing the company;
- use **best practice principles of good governance** for long-term performance enhancement;
- identify the most **significant performance indicators** for overall business improvement;
- set appropriate and realistic **short-term** and **long-term targets**;
- apply **strategies** to ensure that targets are met in agreed time frames;
- use appropriate **reporting tools** to influence **effective decision making**.

The bottom line is: **never** give up on performance improvement.

> Excellence is a skill conquered through training and practice. We are what we repeatedly do. Excellence, therefore, is not an act, but a habit.
>
> **Aristotle (384–322 BC)**

REFERENCES

1. Baldrige National Quality Award Program. 1999. http://www.nist.gov/baldrige/index.cfm.
2. European Foundation for Quality Management. *EFQM model for business excellence*. www.efqm.org .
3. *The Business Excellence Framework*. SAI. https://www.saiglobal.com/business-improvement/process/framework/excellence.htm.
4. Kotter JP. *A force for change: how leadership differs from management*. The Free Press; 1990.
5. *Dow Jones Sustainability Index*. http://www.sustainability-indices.com/review/annual-review-2015.jsp.
6. UN Global Compact. *Guide to corporate sustainability*. UN; 2014.

FURTHER READING

1. *BP energy outlook 2016 edition.* https://www.bp.com/content/dam/bp/pdf/energy-economics/energy-outlook-2016/bp-energy-outlook-2016.pdf.
2. BP Statistical Review of World Energy. *2015-all energy production and consumption by country.* http://www.bp.com/en/global/corporate/energy-economics/statistical-review-of-world-energy.html.
3. *The outlook for energy: a view to 2040.* ExxonMobil. http://corporate.exxonmobil.com/en/energy/energy-outlook.
4. *Implement—from quality to organizational excellence.* http://www.businessballs.com/dtiresources/TQM_implementation_blueprint.pdf.
5. Rigby D.K. *Management tools 2013: an executive's guide*. Bain and Company. www.bain.com.
6. American Society for Quality (ASQ). http://asq.org/index.aspx.
7. *Chevron operational excellence.* www.chevron.com/about/operationalexcellence/.
8. Williams JC. *A retrospective view of the South African excellence model* [MBA thesis]. Stellenbosch University; 2008. http://scholar.sun.ac.za/handle/10019.1/783.
9. Senge P, et al. *The fifth discipline fieldbook*. London: Nicholas Brealey; 1994.
10. Branson R. *The Virgin way – how to listen, learn and lead.* Penguin Random House; 2014.
11. Steil L, Bommelje R. *Listening leaders: the ten golden rules to listen, lead and succeed.* Beaver's Pond Press; 2004.
12. ISO/IEC 26000:2010. *Guidance on social responsibility.*

Appendix A: Glossary of Terms

The terms listed are common terms used in the oil and gas industry.

A

Accounting	
EBIT	Earnings before interest and tax
LE	Latest estimate
Accident	See incident
ACHE	Air cooled heat exchanger
ADNOC	Abu Dhabi National Oil Company
Affiliate	Any company in which a shareholder has a direct or indirect interest. This includes all joint ventures (JVs), subsidiaries, and production sharing agreement (PSA) companies
AFE	Approval for expenditure
AGE	Acid gas enrichment; part of sulfur recovery to enrich the feed to sulfur recovery unit (SRU)
AGR	Acid gas recovery
AHT	Anchor handling tug
AIM	Asset integrity management
Agreements	
Investment life cycle	
LoI	Letter of intent; normally signed prior to a prefeasibility study
MoU	Memorandum of understanding; normally signed at start of feasibility study
CA	Confidentiality agreement; normally between a process licensor and operator
HOA	Heads of agreement; similar to MoU
JVA	Joint venture agreement; normally signed at the start of front end engineering (FEED)
ARDPSA	Amended and restated development and production sharing agreement
AT&AA	Accession transfer and amendment agreement
DFA	Development and fiscal agreement
DPSA	Development and production sharing agreement
EPSA	Exploration and production sharing agreement
EGSPA	Export gas sales and purchase agreement
ESPA	Ethane sales and purchase agreement
FSA	Facilities sharing agreement
FSDSA	Facilities services and data sharing agreement
LTSA	Long-term service agreement
MSA	Master swop agreement
OMSA	Operations maintenance sharing agreement
S&LA	Storage and loading agreement
SPA	Sales/supply and purchases agreement
TSA	Technical services agreement

AMS — Alarm management system

An effective alarm management system is one where:
(1) nuisance alarms have been identified and eliminated;
(2) standing alarms are managed; and
(3) the impact of alarm flooding is minimized through alarm prioritization and application of tools, such as alarm suppression.

ALARP — As low as reasonably practical

Amine treatment — H_2S, CO_2, and mercaptan removal

Annualization

Shutdowns — Actual costs and time of last shutdown (SD) spread over each year in the interval (turnaround cycle) to the next shutdown. Planned future costs and interval used where insufficient SD history is available.

Nonshutdown — Maintenance costs and time summed for last 2 years and averaged for a year (Solomon)

ANSI — American National Standards Institute

AOT — Asset oversight team

AP — Air Products; a major cryogenic equipment manufacturer (Lurgi is another)

APCI — Air Products and Chemicals Industries (same as previous)

APC — Advanced process control

API — American Petroleum Institute

APMS — Asset performance management system

APW — Appraisal and data well

ARC — Advanced regulatory control

ARIS — Software product for business process mapping (management); BPM

ASD — Aspirating smoke detector (also see VESDA)

ASME — American Society of Mechanical Engineers

ASV — Automatic safety valve

ATG — Automatic tank gauging: max allowable tolerance 4 mm confirmed once per quarter (typical gas storage tank)

A&P — Allocation and planning

A&R plan — Allocation and reconciliation plan

Availability and utilization

Generic

Utilization = actual production/(calendar days*capacity)*100

Availability = (total production days – shutdown days for maintenance)/total prod days*100

PTAI

Utilization measures the fraction of the capacity used for processing. It is expressed as the ratio of actual production to maximum sustainable capacity over the period.

Reliability: 100% reliability is the ability of a plant/train to have no unplanned or consequential lost capacity.

Availability: as per generic

AWOY — Average without you; PTAI terminology for comparative analysis

B

BA — Business architecture

BAHE/BAHX — Brazed aluminum heat exchanger (cold box heat exchangers)

Battery limits	Isolation of a specific process unit by means of a double block and bleed
IBL	Inside battery limits
OBL	Outside battery limits
BAT	Best available techniques/technology
BATNA	Best alternative to a negotiated agreement
BBL	Barrel
BBS	Behavior-based safety
BBSP	Behavior-based safety program
BCF	Billion cubic feet
BCM	Business continuity management (see also IPOCM)
BCP	Business continuity planning
BIDI pig	Bi-directional pipeline pig
BIP	Business improvement priority
Black Box	A "black box" is a clearly defined modular entity that helps us manage complex systems, processes, and their interrelationships, which we may not fully understand
BoD	Board of directors
BOE	Barrel of oil equivalent
BOM	Business opportunity manager
BOPD	Barrels of oil per day
Boundaries	For performance evaluation, these need to be carefully scrutinized for adjustment of data so as to compare "apples" with "apples."
BOG	Boil off gas
BOL	Bill of loading
BP&B	Business plan and budget; sometimes referred to as work plan and budget (WP&B)
BPM	Business process management or mapping
BPMN	Business process mapping notation
BPO	Bypass overrides
Break-ins	Jobs added to a maintenance schedule without proper lead time
BTU	British thermal units
Business Management System	See System
Butane slipping	Spiking condensate with butane (adding butane to condensate)

C

CA	Criticality analysis; part of reliability-centered maintenance (RCM)
Capacity	Continuous sustainable throughput

- Basis
  - Field development plan (FDP) capacity
  - Design capacity
  - Engineer procure construct (EPC) acceptance test (built) capacity
  - Corrected design capacity (based on actual feed gas composition)

Capex	Capital expense (investment expenditure)
CAR	Corrective action request

CAR insurance	Construction all risks insurance
CBA	Cost benefit analysis
CBS	Cost breakdown structure
CDDR	Contingency drawdown requests
CDM	Clean development mechanism
CDP	Capacity and distribution planner
CECs	Certified emission credits
CEMS	Continuous emission monitoring system GTs >25 MW: NO_x monitoring; required by many operators/authorities SRUs: SO_2 monitoring; required by many operators/authorities RATA: relative accuracy test audits
CEO	Chief executive officer (of a company)
CER	Cold eyes review
CEL	Corrected energy and loss index (PTAI defined)
CERs	Certified emission reductions
CF	Configuration factor
CGR	Condensate gas recovery/condensate to gas ratio: weighted average field condensate to gas ratio
Chemical loss	Includes water produced from H_2S burning in air
CI	Corrosion inhibitor Nalco/Baker Petrolite (see KHI)
CICA	Critical instruments, controls, and alarms
CIRT	Computer incident response team
CLSCSEF	Common low sulfur condensate storage and export facilities
COMAH	Control of major accident hazards; UK Regulation 1999
Core Management System	Core or primary business of the organization
CM	Corrective maintenance
CMMS	Computerized maintenance management system
CMS	Condition monitoring system (Bentley Navada or other)
Composition (gas)	
Methane	CH_4
Methanol	CH_3OH used as refrigerated solvent in Selexol AGR (UOP licensed)
Ethane	C_2H_6
Ethylene	C_2H_4; the simplest olefin or alkene
Propane	C_3H_8
Butane	C_4H_{10}
Isobutane methyl propane	Used as a refrigerant
N-Pentane C_5H_{12}	Normal pentane
Propane refrigerant	Used for LNG cooling instead of mixed refrigerant (MR)
LNG refrigerant	Propane
Isopentane	Methylbutane; used as a refrigerant
Mercaptans	CH_3SH (methanethiol) and C_2H_5SH (ethanethiol)
NOX	Nitrous oxides
DSO	Disulfide oil

N-paraffins	C_nH_{2n+2}: methane, octane, mineral oil, paraffin wax
Naphtha	Feedstock for gasoline and petrochemicals
Gasoil	Feedstock for diesel
DFC	Deodorized field condensate
Components of hydrocarbons	
Component Balance:	Molecules of different gases make up 100% of the gas
• C, H, O, N, He, S	
Component Recovery:	Molecules "in" balanced against molecules "out"
• Water	
• Sulfur	
• Nitrogen	
• Methane	
• Ethane	
• Propane	
• Butane	
• C5+	
CM	Corrective maintenance
CMOF	Common marine offloading facility
COE	Cash operation expense (see opex)
COP	Critical operating parameter is a plant operating condition (e.g., pressure, level, flow, or temperature composition), which, if exceeded, could result in a major event, such as uncontrolled emission, fire, or explosion, that could pose serious danger to people, assets, or the environment, and for which either: no automated protection, e.g., process shutdown, exists to automatically guard against such a major event or where the automated protection system does not cover all potential scenarios, and the possibility exists for the process conditions to exceed the design conditions. Instead, there is complete reliance on the correct manual intervention to return the process condition to a safe state or normal operating window.
Corruption	Abuse of entrusted power for private gain
CP	Corrosion protection
CPC	China Petroleum Corporation
CPI	Cost performance index
CPM	Critical path method/model
Crisis management	Level 1: managed in plant Level 2: industrial area involvement Level 3: regional/national services involvement
Criticality	**A**: Primary process stream equipment shutdown causes production loss *Example: single heat exchanger in primary process stream* **B**: Primary process stream equipment with parallel full-capacity equipment *Example: electrically powered crude feed pump with identical parallel steam-powered crude feed pump* **C**: Equipment that could cause production loss *Example: failure of more than one air compressor supplying the plant air system*

CSE	Confined space entry
CSFs	Critical success factors
CSS&L	Common sulfur storage and loading
CSP	Common sulfur plant
CSR	Corporate social responsibility
CSU	Commissioning and startup
CTO	Consent to operate; government environmental agency approval required to operate (annual renewal)
CTO	Coal to olefins
CV	Cost variance (for projects)
CVOC	Common volatile organic compound
CWI	Completion and well intervention

D

DCS	Distributed control systems
DEPs	Design engineering practices
DFC	Deodorized field condensate
DGS	Downgraded situations
DOSS	Disturbance of safety systems
DPSA	Development and production sharing agreement
DRA	Design risk assessment
DRC	Decision review committee (for projects)
DRS	Data reconciliation system
DSO	Disulfide oil
DT	Destructive testing
Due Diligence	A fact-finding exercise designed to evaluate risks, including testing integration plans and confirming valuation assumptions. It normally includes business principles, regulatory and compliance, financial, legal, intellectual property, HSE, technical, and other specialist due diligence.

E

EAC	Estimate at completion (for projects)
EBIT	Earnings before interest and tax
EBITDA	Earnings before interest, tax, depreciation, and amortization
ECG	Export gas compressor
EDC	Equivalent distillation capacity (Solomon Associates defined term)
EFPH	Energy, finance, physical asset, and human asset categories for assessment
EFQM	European Foundation for Quality Management
EGPC	Equivalent gas processing capacity (Solomon Associates defined term)
EI	Energy index
EII	Energy intensity index (Solomon Associates defined term)
EITI	Extractive Industries Transparency Initiative
ELT	Enterprise leadership team

EMR	Emissions monitoring and reduction
CEMS	Continuous emission monitoring systems
RATA	Relative accuracy testing and auditing
FEM	Fugitive emissions monitoring
LDAR	Leak detection and repair
GHG	Greenhouse gas
EA&R	Emissions Accounting and Reporting
Energy efficiency	
CEL	Corrected energy loss index (PTAI)
NEI	Normalized energy indices
LHV	Lower heating value
Environment	
Acid gas	CO_2 and H_2S; amine treating removes these gases from the process gas
CEMS	Continuous emission monitoring system: control for oxides of nitrogen
GHG	Greenhouse gases: CO_2, CH_4, N_2O reported as CO_2 equivalent
NO_x	Nitrous oxides
SCMI	Site carbon management index (PTAI index)
VOC	Volatile organic compounds
EOR	Enhanced oil recovery
EPC	Engineer and procure contract
EPC	Equivalent processing capacity (for benchmarking)
EPCM	Engineering procurement construction maintenance (contract)
EPIC	Engineering Procurement Installation and Commissioning (contract)
EPL	Equipment parts lists
EPM	Event potential matrix
EPMS	Employee performance management system
EPOCM	Emergency planning and operational continuity management (see IPOCM)
EPRS	Emergency pipeline repair system
EPS	Expanded polystyrene
Equipment criticality	See criticality
ERC	Employee-related cost
ERM	Enterprise risk management or enterprise resource management
ERP	Enterprise resource planning (SAP, Oracle, etc.) or emergency response plan
Escalation indices	Nelson-Farrar Cost Indexes: refinery construction and operating published in Oil & Gas Journal
	Chemical Engineering Plant Cost Index (CEPCI): published in Chemical Engineering Journal
	HIS CERA Indices: Operating Cost Analysis Forum (OCAF) "Moor" zone (Middle East)
ESD	Emergency shutdown
ESDVs	Emergency shutdown valves
ESTS	Emergency ship-to-shore global network
ESR	Engineering service request
ESU	Expansion startup

ETP	Effluent treatment plant
Event	See incident
E&Y	Ernst and Young consultants

F

F&G	Fire and gas, as in process systems
FAC	First aid case
FAR	Fatality accident rate
FAT	Factory acceptance tests
FC	Field condensate/flow computer
FCC	Fluidic catalytic cracking process
FCCU	Fluidic catalytic cracking unit (physical process unit)
FDA	Future development assumptions
FDP	Field development plan/final development plan
FEED	Front end engineering and design
FEL	Front end loading (in preparation for a turnaround)
FERA	Fire and explosion risk analysis
FFF	Fuel from feed (gas)
FFS	Fitness for service
FG	Fuel gas
FGP	Feed gas processing/preparation
FI	Facilities integrity
FIAR	Functional integrity assurance review (Shell)
FICV	Fast intervention crew vessel
FID	Final investment decision
FIDIC	Fédération Internationale Des Ingénieurs-Conseils (International Federation of Consulting Engineers)
FIMS	Facilities integrity management system
KEIs	Key equipment issues
EKEIs	Emerging key equipment issues
FIN EKEIs	Financial emerging key equipment issues
CDFDs	Critical device function test
PRV/CDFDs	Pressure relief valve/critical device function test
PPM	Planned plant maintenance
FIMS exception approval process	Used to postpone critical inspection or maintenance work
CICA	Critical instrument and control alarms
Finance	
EBIT	Earnings before interest and tax
EBITDA	Earnings before interest and tax, depreciation, and amortization
FCF	Free cash flow
MOD	Money of the day
NRE	Nonrecurring expense (turnarounds, etc.)
OCF	Operating cash flow
RT	Real term

FLT	Flare low temp
FM200	Fire extinguishing agent to replace CO_2 (Dupont product)
FMECA	Failure modes and effects and criticality analysis
FOB	Free on board
Framework	An all-encompassing mechanism for control of a strategic activity documented in a "framework" or "manual" (see also **manuals and guidelines**).
Fuel gas makeup	
AGR flash gas	Acid gas recovery flash gas
EFG	End flash gas
FFF	Fresh feed fuel (gas)
BOG	Boil off gas (may be excluded from fuel gas pool): tankage and loading (JBOG – Jetty BOG)
Fuel gas analysis	Fuel gas produced; used for mass balance
	Fuel gas consumed; used for energy efficiency
Function	Functions have executive roles with delegated authorities and accountabilities. They assist the MD by providing functional direction, support, and leadership to the group of companies and provide services to the Business Units and other functions.
FPU	Feed gas processing unit (Shell GTL)
FSA	Formal safety assessment
FSDSA	Facilities services and data sharing agreement
FSG	Flare sour gas
FTEs	Full-time employees
FTNIR analyzer	Fourier transformation near infrared analyzer: online, real-time, continuous, multiple, and simultaneous measurement for properties and component concentration of various processes (APC analyzer)
FWL	Fire water line
FWS	Full wellhead stream

G

G&A	General and admin (in budgets)
Gas pricing	
HH	Henry Hub US ($/mmbtu)
NBP	National Balancing Point UK ($/mmbtu)
JCC	Japan Crude Cocktail ($/bbl)
JKM	Japan Korea Marker Platts ($/mmbtu)
Brent	North Sea ($/bbl)
FDA	Future development (pricing) assumptions
Gas streams	
AGR flash gas	Acid gas recovery flash gas
BOG	Boil off gas (may be excluded from fuel gas pool): Tankage and loading (JBOG – Jetty BOG)
DISO OH	Disulfide oil overheads

Dry gas	Contains no water; after dehydration process
EFG	End flash gas
ERG	Ethane rich gas
FFF	Fresh feed fuel (gas)
Feed gas	Gas entering the battery limits of a process unit
FWS	Full well stream; maximum well output
GHG	Greenhouse gases
Khuff gas	From a specific gas reservoir layer
Lean gas	Out of NGL recovery unit
RAG	Rich associated gas/OFFRAG: offshore rich associated gas
RG	Royalty gas
SAG	Stripped associated gas/OFFSAG offshore/ONSAG onshore
Sales gas	Gas exported to a customer through a fiscal meter to an agreed specification
Sour gas	Contains more than 4 ppm by volume of H_2S
Sweet gas	Contains less than 4 ppm by volume of H_2S
SWG	Sweet dry gas; out of dehydration/mercury removal unit
Tank boil off	
Wellhead gas	
Wet gas	
GC	Gas chromatographs (key inline analyzer for royalty/fiscal gas measurement)
GCC	Gulf Cooperative Council
GHG	Greenhouse gases
GIIGNL	International group of LNG importers
GIT	Goods in transit
GPA	Gas Producers Association
GPM	Gross profit margin
GRC	Governance risk and compliance (or controls)
GT	Gas turbine
GTL	Gas to liquid (Pearl, Oryx, Bintulu, Mossgas)
GTW	Gap troubleshooting workshops

H

HAS	Hydrocarbon accounting system
HC	Hydrocarbon
HDPE	High-density polyethylene
Health and safety (H&S)	
API 754	Process safety performance indicators
DRA	Design risk assessment
FAC	Fatal accident case
FAR	Fatal accident rate
FIR	Fatal incident rate
HAZOP	Hazard operability study
KP4	UK Health and Safety Exec Aging and Life Inspection Program
LTIF	Lost time injury frequency

LOPC	Loss of primary containment
Tier 1: events of greater consequence	
Tier 2: events of lesser consequence	
Tier 3: challenges to safety systems	
MUPE	Major unplanned event
TRIF	Total reportable incident/injury frequency
TROIF	Total reportable occupational injury frequency
LTA	Lost time accident
LTI	Lost time incident
LTIR	Lost time incident rate
LTIF	Lost time injury frequency
TRIR	Total recordable injury rate
TRCF	Total reportable case frequency
RCFA	Root cause failure analysis: Taproot, Tripod, Proact
RWDC	Restricted work day case
RWC	Restricted work case
SHE case	Safety, health, and environment scenario; UK Health and Safety Executive
SIL	Safety integrity level
STOP	HSE unsafe act cards
FMEA	Failure modes and effects analysis
FTA	Fault tree analysis
ETA	Event tree analysis
HRA	Human reliability analysis
POD	Probability of detection
Heat exchangers	
BAHE	Brazed aluminum heat exchanger
HRSG	Heat recovery steam generator
MCHE	Main cryogenic heat exchangers
HEMP	Hazard evaluation and management process (Shell)
HEMS	Helicopter emergency medical services
HeRU	Helium recovery unit
HeXU	Helium extraction unit
HHV	Higher heating value versus LLV (lower heating value)
HIAs	High impact actions; as a result of an assessment
HIPS	High integrity protection systems
HIPPS	High integrity pressure protection systems; function to avoid overpressure through automatic isolation
HMI	Human machine interface
HRSG	Heat recovery steam generator
HSE	Health, safety, and environment
HSE	Health and Safety Executive; a UK government agency
HSSE	Health, safety, security, and environment
HTF	Heat transfer fluid
HYSYS	Process simulation software by AspenTech
Hydrate	Sold ice/methane blocking gas pipelines; see also KHI

I	
IAP	Integrity assurance program
IASB	International accounting standards board
ICP	Indonesian crude price
ICSR	Industrial control system requirements
ICS	Instrument control system
ICSS	Integrated control and safety systems
IDPs	Individual development programs
IEC	International Electrotechnical Commission
IET	International emissions trading
IFRS	International Financial Reporting Standards
IID	Initial investment decision
IMO	International Maritime Organization
IMS	Integrated Management System
Incident/threat	Condition/event that may prevent us from meeting desired objectives
Information systems: process	
LIMS	Laboratory information management system
RTIS	Real-time information system
PI/PHD	Process information system/process historian database
Inspection	
Acoustic emission	
CUI	Corrosion under insulation
HGPI	Hot gas path inspection
MFL	Magnetic flux leakage
NDT	Nondestructive testing
SLOFEC	Saturation low-frequency eddy current
Thermography	Infrared photography
TOFD	Time of flight diffraction
UT	Ultrasonic testing (including phased array)
Instruments	
FT	Flow transmitter
FI	Flow indicator
FC	Flow controller
FIC	Flow indicator and controller
PIC	Pressure indicator and controller
PC	Pressure controller
Integrity	
SIMS	Structural integrity management system
IOC	International oil company BP, Shell, ExxonMobil, Chevron, Total, etc.
IPMS	Integrated project management system
IPOCM	Incident preparedness and operational continuity management
IPR	Independent project review
IPS	Instrument protection system
TMR	Triple modular redundant logic design

IS	Information system
ISGOTT	International Safety Guide for Oil Tankers and Terminals
ISO	International Standards Organization
ISP	Infrastructure service provider
ISPS	International Ship and Port Security Code
IT	Information technology
ITD	Information technology department
ITR	Instrument technical/termination room
ITT	Invitation to tender
IVMS	In-vehicle monitoring system

J

JADT	Joint asset development team
JBOG	Jetty boil off gas
JFA	Joint facilities agreement
JCF	Juran complexity factor
JHI	Juran hydrocarbon index
JSA	Job safety analysis
JV	Joint venture
JVA	Joint venture agreement

K

KBBL	Thousand barrels
KEIs	Key equipment issues and EKEIs (emerging key equipment issues) Categories: HSE and FIN
Kelton	Consultant for product measurement in oil and gas industry
KHI	Kinetic hydrate inhibitor; corrosion inhibitor injected into the gas stream offshore (Nalco or Baker)
KPC	Kuwait Petroleum Company
KPI	The words "**KPIs**," "**key performance indicators**," and "**performance indicators**" are synonymous. They are used generically in all parts of the business. They could be applied to a **Process**, **System,** or the complete **Business Unit**.
KRIs	Key results areas

L

LC	Letter of credit
LDAR	Leak detection and repair
LDPE	Low-density polyethylene
LEP	Limited equity project
LGR	Lean gas recovery
LIMS	Laboratory information management system
LLDPE	Linear low-density polyethylene
LOPA	Layers of protection analysis
LOPC	Loss of primary containment

LNG	Liquefied natural gas
Rich	Including ethane
Lean	Excluding ethane
LPB	Liquid product berth
LPG	Liquefied petroleum gas
LR	Loading rack
LSC	Low sulfur condensate
LTI	Lost time incident
LTIF	Lost time incident frequency
LTR	Lost time rate: total fatalities plus lost work days incident rate
LTSA	Long-term service agreement

M

MAAOE	Multinational Alliance for Advancement of Organizational Excellence
MaAoA	Memorandum and articles of association (for formation of a company)
Maintenance	
Break-in	Jobs added to a maintenance schedule without proper lead time
Emergency maintenance (EM)	
Planned maintenance (PM)	
Corrective maintenance (CM)	
Criticality	See criticality
Management Systems	See Systems
HAS	Hydrocarbon accounting System
SCMS	Supply chain Management System
RPMS	Resource planning Management System
Mancom	Management Committee: in place of a Board of Directors for PSAs and unincorporated companies
Manuals and Guidelines	More detailed documents provided to help staff comply with legal and regulatory requirements and group policies, standards, and regulations, and to follow best practices. Manuals largely contain instructions, which are mandatory, on how to implement group policies regulations, standards, and frameworks. Guidelines contain more nonmandatory guidelines on good practice to assist staff to carry out their duties properly in areas where specialist expertise is needed.
Mass Balance Software	
Sigmafine	Mass balance model (OSIsoft product)
AspenTech	"aspenONE" Aspen Operations Reconciliation and Accounting
PV	Process value
RV	Reconciled value
MBR	Management Board review
MCHE	Main cryogenic heat exchanger
MCR	Main control room
MD	Managing Director (of a company)
Measurement	
Accuracy	The closeness of the agreement between the result of the measurement and the true value of the quantity (ISO 4006)
	Manufacturer's stated accuracy based on method of metering

Tolerance	
Allocated	Assigned to specific ventures/cost centers
ATG	Automatic tank gauging: max tolerance 4 mm confirmed once per quarter
Calculated	Measurement derived based on calculations
Condensate sales measurement:	
Field condensate: barrels	
Plant condensate: tons/density	
ISO 5168:2005	Fluid flow uncertainties
Least Squares	Square root of sum of squares
MRMR	Monthly royalty measurement system reading report
MSMP	Measurement surveillance and monitoring process
Metering device	Coriolis, Ultrasonic (US), Orifice, etc.
Mbtu	Million BTUs
Mmscfm	Million standard cubic feet per minute (typical for gas)
Reconciled	Calculated based on measurements and/or assumptions/data sheets
VEF	Vessel experience factor; measurement of product on board a vessel
Mechanical availability	See availability
Meters	
Royalty (feed from wells); based on volume	
Fiscal	
Custody transfer	
Key allocation	
Allocation	
GC	Online Gas chromatograph (analyzer); analyzes C2, C3, IC4, NC4, IC5, NC5, C6+
Density	
Pressure and temperature compensation	
FTNIR	Fourier transformation near infrared
USM	Ultrasonic meter
MEG	Monoethylene glycol; injected into wet gas streams to remove water
Meridium	Asset performance management system from GE Digital used by Rasgas, Qatargas, Bapco, etc.
Meridium QIS	Reliability incident reporting tool
MI	Maintenance index (Solomon defined)
MICC	Main instrument and controls contractor
MIPs	Management improvement projects
MM	Million
MMS	Machine monitoring system
MMSCF	Million standard cubic feet
MMSCM	Million standard cubic meters
MOC	Management of change
MOD	Money of the day: the actual value of money at the time of the transaction, as opposed to real terms (RT)
RT	Real terms: money corrected to the value of constant purchase power by taking inflation into account
MOF	Marine offloading facility

MPC	Multivariable process/predictive control
MPP	Maximum production potential
MSCR	Maximum sustained capacity rate
MSF	Maintenance scaling factor
MTA	Million tons per annum
MTC	Medical treatment case
MUc	Maintenance units corrected for size of the equipment

N

NBS	Neutron back scanner for checking corrosion rate
NDT	Nondestructive testing
NGL	Natural gas liquefaction
NOB	Net on board
NOC	National Oil Company: Saudi Aramco, Bapco, KPC, QP, Petronas Malaysia, PDVSA, Petrobras, etc.
Normalization	Technique used to compare multiple sites of various sizes and complexity (benchmarking) or annualizing turnaround costs
NPAT	Net profit after tax
NPV	Net present value
NRE	Nonrecurring expenses
NSP	Normalized shift positions (PTAI)

O

OBL	Outside battery limits, as opposed to IBL (inside battery limits)
OBS	Organization breakdown structure
OECD	Organization for Economic Cooperation and Development
OCP	Operations capital projects
OEM	Original equipment manufacturer
OEP	Original equipment producer
O&G	Oil and gas
OGP	International Association of Oil and Gas Producers
OHSAS	Occupational Health and Safety Assessment Series
OIMS	Operations Integrity Management System (ExxonMobil)
OLT	Operations leadership team
OM&S	Oil movement and storage
OPC	Open productivity and connectivity in industrial automation and the enterprise systems that support industry; a series of standards specifications
Operator	On behalf of the **asset owners**, the operator is responsible for: **1.** operating and maintaining the assets **2.** maintaining asset integrity **3.** exercising cost management **4.** monitoring and reporting asset performance
Opex	Operating expense
OPM	Operating profit margin
ORIP	Operational reliability improvement process

OSBL	On-site battery limit
OTS	Operator training simulator

P

PAA	Plant automation applications	
PAC	Project acceptance and completion	
Pacesetter	PTAI definition:	Average of top 2 performers
	Solomon definition for refineries:	Top quartile for x indicators for three successive studies (6 years)
P&ID	Piping and instrument diagram	
PAS	Publicly Available Specification; a standardization document that closely resembles a formal standard in structure and format but which has a different development model. The objective of a PAS is to speed up standardization.	
PC	Plant condensate	
PCS	Process control systems (DCS, SIS, PLC)	
DCS	Distributed control system	
SIS	Safety instrument system	
PLC	Programmable logic controller	
PCD	Production commencement date	
PCN	Proposal to commence negotiations	
PCN	Petrochemical naphtha	
PCR	Process change request	
PDCA	Plan do check act	
PDMS	Pipeline data management system	
PDSA	Plan, do, study, act	
PDVSA	Petróleos de Venezuela, S.A.; Venezuelan state-owned oil and natural gas company	
PE	Poly ethylene	
PET	Poly ethylene terephthalate	
PFD	Process flow diagram	
PGP	Power generation package	
PHA	Process hazard analysis	
PHD	Process historian database (Honeywell)	
PID	Preliminary investment decision/piping and instrument diagram	
PIMS	Plant information management system (AspenTech)	
PIMS	Pipeline integrity management system	
PIN	Project initiation note	
PIPs	Process improvement projects	
PM	Planned, preventive, or predictive maintenance	
PMCRs	Preventive maintenance change requests	
RCM	Reliability-centered management/maintenance reliability change management	
Production (Gas)	Salable production: LNG, sales gas, condensate, propane, butane, sulfur, (ethane, helium)	
	Commodities: condensate, propane, butane, sulfur (ethane, helium)	
PMS	Project management system	
	Pipeline management system	
	Power management system	

PIN	Project initiation note
PIN	Plant information network
PKS	Process Knowledge System (Honeywell Experion)
PM	Planned maintenance
PMCR	Planned maintenance change request
POIS	Plant operations information system
Policy	Set of principles by which a business is managed. Policies are divided as follows: group: relating to governance of the group of companies corporate: relating to governance of the directly managed Business Units HSEQ: relating the corporate HSEQ activities operational: relating to Business Unit operations activities
PP	Polypropylene
PPM	Planned or predictive plant maintenance
PPM	Parts per million
PPPM	Per person per month
PROACT	**Pr**eserving failure data, **o**rdering the analysis, **a**nalyzing the data, **c**ommunicating findings and recommendations, and **t**racking for results; RCA methodology
Procedure	Description of a Process or other business activities
Process	A set of interacting activities that transform inputs into outputs
Project (general)	
CSU	Commissioning and startup
EOT	Extension of time
ESU	Engineering support unit
IPR	Independent project review
LE	Latest estimate
MAC	Main automation contract
PIMS	Project information management system
PMC	Project management contractor/premechanical completion
PMT	Project management team
P&T	Projects and technology
RFSU	Ready for startup
SIMOPS	Simultaneous operations
TSI	Telecommunication system integration
Project drivers	
Financial	
HSE	
Maintenance and repair	
Business interest	
PRV	Pressure relief valve
PS	Polystyrene
PSA	Production sharing agreement: can either be a development and production sharing agreement (DPSA) or an exploration and production sharing agreement (EPSA)
PScan	Corrosion mapping tool
PSD	Process shutdown/process safety device
PSE	Process safety events (ref. API Guide to Reporting Process Safety Events); Tiers 1 to 4

PSE	Process safety equipment
PSM	Process safety management
PSP	Process safety program
PPM	Planned plant maintenance
PTA/DMT	Purified terephthalic acid/Dimethyl terephthalate
PTAI	Philip Townsend Associates International: successor to Shell Global Solutions for benchmarking, covering 75% of worldwide LNG capacity
PTE	Plant thermal efficiency (Shell defined)
PTR	Performance test run
PTW	Permit to work
PVC	Polyvinyl chloride
PX	Polyxylene

Q

QA	Quality assurance
QC	Quality control
QE	Quantitative easing
QPR	Quarterly performance report
QRA	Quantitative risk assessment
LSIR	Location-specific individual risk

R

RAGAGEP	Recognized and generally accepted good engineering practices
RAM	Reliability and maintenance (benchmarking)
RAM	Risk assessment matrix • Severity: 0–5 • Probability of consequence: A–E • Risk: low, medium, high
RAMS	Risk assessment management system
RAPID	Refinery and petrochemical integrated development (Petronas Pengerang Johor)
RATA	Relative accuracy test audits relating to continuous emission monitoring system (CEMS) calibration
RBI	Risk-based inspection
RBIMS	Risk-based inspection management system
RCA	Root cause analysis
RCFA	Root cause failure analysis (using Taproot, Tri-Beta, or other tool)
RCM	Reliability-centered maintenance
R&D	Research and development
Reconciliation Software	**1.** Sigmafine: OSIsoft **2.** Shell Advisor **3.** Honeywell "Production Balance" **4.** Spreadsheets
RDAS	Remote data acquisition system (machinery monitoring data acquisition)

Refrigeration
- APX — 3 cycles: propane, MR, nitrogen (Air Products)
- MR — Mixed refrigerant
- SMR — Single mixed refrigerant
- C3MR — Propane precooled mixed refrigerant (Shell)

Reservoir management
- MPLT — Memory production logging tool
- PLT — Production logging tool
- RST — Reservoir saturation tool

Reserves — Proved developed, proved undeveloped, probable, possible, remaining recoverable reserves

RIMAP — Risk-based inspection and maintenance practices

RIS — Refinery information system

Risk management software
- ARM; active risk management
- @Risk (EPG)
- Easyrisk

RM — Reactive or routine maintenance

RMDD — Routine maintenance down days

ROCE — Return on capital employed

ROE — Return on equity

ROI — Return on investment

RONOA — Return on net operating assets

ROV — Remote operated vehicle/valve

RP — Reserve production ratio

RPT — Rapid phase transition explosion when LNG contacts water

RSPL — Required spare parts list

RT — Real terms: Money corrected to the value of constant purchase power by taking inflation into account

RTDB — Real-time data base

RTIS — Real-time information system (see also PHD)

RTP — Request to pay

RV — Replacement value: Solomon determines from company asset register and "normalized" to present value based on "Chemical Engineering" plant cost indices

RWC — Restricted work case

S

S&W — Sediment and water

Safety
- DOSS — Demands on safety systems
- LOPA — Layers of protection analysis
- PHA — Process hazard analysis
- SIFs — Safety instrument functions
- SIL — Safety integrity level
- SISs — Safety instrument systems
- SOLE — Safe operating limit exceedences
- SRS — Safety requirements specification reports

Safety Case	
QRA	Quantitive risk assessment
FSA	Formal safety assessment
SAE	Society of Automotive Engineers
SAP	A global software company
SAP BW	SAP Business Warehouse
SAP PSO	Projects system optimization
SAP MM	Materials module purchasing and warehousing linked to SAP PM
SAP PM	Planned maintenance module
RBWS	Risk-based work selection
	SAP notification category
	Categories of criticality: criticality A, B, etc.
SAR	Stock amendment request
SAT	Site acceptance test
SC	Slug catcher (in a gas plant)
SC	Stabilized condensate
SCADA	Supervisory control and data acquisition
SCCB	Single common control building
SCF	Standard cubic foot
SCE	Safety critical equipment
SCMI	Site carbon management index: PTAI
SCMS/HAS	Supply chain management system/hydrocarbon accounting system
SCOT	Shell claus off-gas treating
SD	Sustainable development
SD	Shutdown of production, planned or unplanned (see also turnaround)
Service Business Units	These Units supply services in varying degrees to other Business Units.
SGP	Shell gasification process as part of gas to liquid (GTL)
SGS	Shell Global Solutions
SGS	Environmental consultant (not Shell Global Solutions)
SIH	Satellite instrument house
SIL	Safety integrity level
SIMS	Structure integrity management system
SIPs	System improvement programs
SIS	Safety instrument systems
SMART	Simple, measurable, aligned, repeatable, timely
SME	Subject matter expert
SOPA	Statement of petroleum account
SOPC	Statement of petroleum costs
SPI	Site personnel index (PTAI defined)
SRU	Sulfur recovery unit
Stewardship report	Report from affiliate on performance of affiliate
SPA	Supply purchase agreement
SPC	Statistical process control
SPC	Statement of petroleum costs
SPI	Schedule performance index (projects)

SPI	Site personnel index (PTAI defined term)
SPM	Single point mooring
CALM	Catenary anchor leg mooring
SRU	Sulfur recovery unit
Standard Activities	PTAI: activities relating to the operation, maintenance, and support of assets normally found in LNG sites (but not necessarily installed at all sites)
SPI	Schedule performance index
STG	Steam turbine generators
Subsidiary	This is a generally a shareholding company (also known as a joint stock company). It may take two different forms: **I.** a public or open shareholding company (usually listed on a stock exchange) **II.** a private or closed shareholding company
SV	Schedule variance (related to projects)
SWS	Saltwater system/sour water system
System	"A structured and documented set of interdependent practices, Processes, and procedures used by the managers and the workforce at every level in a company to plan, direct, and execute activities" A **System** must contain all of the key elements described in Section 4.5, The Key Elements of a System. "**system**" with a lowercase "s" is generic and does not necessarily comply with the previously mentioned definition. For example, it could refer to a computer system.

T

TA	Turnaround: process plant shutdown planned at least a year in advance and budgeted for
TADD	Turnaround annualized down days
TARP	Troubled assets relief program/turn-a-round process
TCR	Total case incident rate (safety)
TDMS	Technical document management system
TEG	Triethylene glycol
Tendering RTA	Recommendation to award
TF	Total flaring
TGT	Tail gas treater
TI	Turnaround index (Solomon defined)
TI	Turnaround and inspection
TIW	Treated industrial waste/water
TML	Thickness measurement location
TO	Terminal operations
TOFA	Table of financial authorities
TOP	Take or pay power agreement
TPA	Tons per annum
TPP	Technical proficiency program
TRCF	Total recordable case frequency
Trend group	A group of similar businesses used for comparison purposes for at least 5 consecutive years (used in benchmarking)
TRIR	Total recordable incident rate

TROIF	Total recordable occupational injury frequency
TSA	Technical services agreement: an agreement between two or more parties for the provision of specified technical advice and services
TTF	Time to failure
TTR	Time to repair
Turnaround	Planned shutdown of production for maintenance and internal inspection
Turnarounds: Gas Plants	
HGPI	Hot gas path inspection
COI	Mid-cycle combustion inspection
MOI	Major overhaul and inspection

U

UEDC	Utilized equivalent distillation capacity (Solomon term)

Unplanned downtime

Unplanned downtime is caused by unplanned shutdowns that lead to an interruption of production.

Unplanned downtime is defined as the time (in days or parts of days) required for shutdowns that are not specified in the yearly shutdown plan.

Unplanned downtime related to unreliability can be caused by maintenance, process upsets (trips, heat exchanger fouling, etc.); consequential downtime caused by unplanned outages of upstream equipment and equipment failure or breakdown (leakage, fire, etc.) and by external events such as power failure. Shutdowns resulting from exceedances of flaring or emission levels as defined in the permit to operate also fall under this category.

In reporting unplanned downtime, the following rules should apply:

1. Downtime caused by an unforeseen event that can be deferred and scheduled in the next monthly or quarterly shutdown schedule should still be reported as unplanned downtime. Only when it can be deferred and included into the next year annual plan, it can be classified as planned downtime.
2. **Anytime that a turnaround takes longer than defined in the annual plan for causes mentioned previously shall be classified as unplanned downtime.**

Reliability

Reliability = (Total Days – Shutdown Days – Unplanned downtime) × 100/(Total Days – Shutdown Days)

Shutdown days in the above formula shall be defined as those days that are regarded as turnaround days in the WP&B/BP&B.

UNFCCC	United Nations Framework Convention on Climate Change
UOP	Union Oil Products Inc.; a process technology contractor
UT	Ultrasonic testing

V

Value addition	The enhanced “change of state” of a material or the supply of a service of value
Value chain	The sequence of “changes of state” of a material from the source of raw material to end products
Valves	
ESDV	Emergency shutdown value
MOV	Motor operated valve
ROV	Remote operated valve
VDT	Volume downtime (bscfpa) due to facility integrity incidents (SHE, FIN KEIs, or EKEIs)
VEF	Vessel experience factor; measurement of product on board a vessel

VEI	Volumetric expansion index (Solomon defined)
VESDA	Very early smoke detection apparatus
VFD	Variable frequency drive
VOC	Volatile organic compounds
VOWD	Value of work done
VOWF	Value of work forecast
VRS	Vapor recovery system

W

WBS	Work breakdown structure
WCN	Work change notifications
WOI	Weight on intake
World class	See pacesetter
WDI	Weld diameter inches (estimating method)
WDU	Waste disposal unit
WHB	Waste heat boiler
WHG	Wellhead gas
WHS	Wellhead stream (also see FWS; full wellhead stream)
WMD	Waste management department/weapons of mass destruction (US terminology)
WOI	Weight on intake
WP&B	Work plan and budget; upstream annual submission to Board/Management Committee
WWD	Wastewater disposal
WWDSF	Wastewater disposal surface facility
WWW	Wastewater wells

Y

YEO	Year end outlook
YTD	Year to date

Z

ZLD	Zero liquid discharge
Zyme-Ox	Slug catcher decontamination chemical

Appendix B: KPIs

Nine focus areas are defined in this Appendix.

Primary requirement areas are defined as:

1. Corporate (including shareholder)
2. Management of Business Unit
3. Benchmarking (BM)

1. STRATEGY

Requirement	KPI	Unit	Definition	Comments
Corporate	Risk 1 action	Traffic light red amber green	Access to reserves	Ernst and Young top 10 risks 2011 number 1
Corporate	Risk 2 action	Traffic light red amber green	Uncertain energy policy	Ernst and Young top 10 risks 2011 number 2
Corporate	Risk 3 action	Traffic light red amber green	Cost containment	Ernst and Young top 10 risks 2011 number 3
Corporate	Risk 4 action	Traffic light red amber green	Worsening fiscal terms	Ernst and Young top 10 risks 2011 number 4
Corporate	Risk 5 action	Traffic light red amber green	Health, safety, and environmental risks	Ernst and Young top 10 risks 2011 number 5
Corporate	Risk 6 action	Traffic light red amber green	Human capital deficit	Ernst and Young top 10 risks 2011 number 6
Corporate	Risk 7 action	Traffic light red amber green	New operational challenges	Ernst and Young top 10 risks 2011 number 7
Corporate	Risk 8 action	Traffic light red amber green	Climate change concerns	Ernst and Young top 10 risks 2011 number 8
Corporate	Risk 9 action	Traffic light red amber green	Price volatility	Ernst and Young top 10 risks 2011 number 9
Corporate	Risk 10 action	Traffic light red amber green	Competition from new technologies	Ernst and Young top 10 risks 2011 number 10
Corporate	Opp 1 action	Traffic light red amber green	Frontier acreage	Ernst and Young top 10 opportunities 2011 opp 1
Corporate	Opp 2 action	Traffic light red amber green	Unconventional sources	Ernst and Young top 10 opportunities 2011 opp 2
Corporate	Opp 3 action	Traffic light red amber green	Conventional reserves in challenging areas	Ernst and Young top 10 opportunities 2011 opp 3
Corporate	Opp 4 action	Traffic light red amber green	Rising emerging market demands	Ernst and Young top 10 opportunities 2011 opp 4
Corporate	Opp 5 action	Traffic light red amber green	NOC-IOC partnerships	Ernst and Young top 10 opportunities 2011 opp 5

Continued

Requirement	KPI	Unit	Definition	Comments
Corporate	Opp 6 action	Traffic light red amber green	Investing in innovation and R&D	Ernst and Young top 10 opportunities 2011 opp 6
Corporate	Opp 7 action	Traffic light red amber green	Alternative fuels	Ernst and Young top 10 opportunities 2011 opp 7
Corporate	Opp 8 action	Traffic light red amber green	Cross-sector strategic partnerships	Ernst and Young top 10 opportunities 2011 opp 8
Corporate	Opp 9 action	Traffic light red amber green	Building regulatory confidence	Ernst and Young top 10 opportunities 2011 opp 9
Corporate	Opp 10 action	Traffic light red amber green	Acquisitions or alliances to gain new capabilities	Ernst and Young top 10 opportunities 2011 opp 10
Corporate	SD projects	Number	Sustainability development projects	Focus area: Management System projects—governance climate change and energy
Corporate	CSR commitments	%	Corporate social responsibility commitments for year achieved	Focus area: Management System projects—governance
Corporate	M S trans	Number	Management System transformations	Focus area: Management System projects
Corporate	Closed out by due date	%	Shareholder audit action status	Focus area: audit
Corporate	Closed out by due date	%	Internal audit action status	Focus area: audit
Corporate	Open audit items	Index	Open audit items	Focus area: audit
Corporate	Open audit items after due date	Number	Open audit items after due date	Focus area: audit

2. HSSE

HEALTH AND SAFETY KPIs—LINKED TO HUMAN CAPITAL

Requirement	KPI	Unit	Definition	Comments
Corporate/management/BM	FAR	Number	Fatal accident rate—number of fatalities	Shareholder requirement
Corporate/management/BM	LTIF	Index	Lost time injury—frequency per million hours	
Management	TRIR	Index	Total recordable incident rate	
Management	Consecutive injury-free days	Number		
Management	Total injury-free days	Number		
Management	Near misses	Number		

Requirement	KPI	Unit	Definition	Comments
Management	RWC	Number	Restricted work case	
Management	MTC	Number	Medical treatment cases	
Management	FAC	Number	First aid cases	
Management	Number of STOP cards issued	pppm	Per person per month	Behavior-based safety initiative: STOP card prevents/notifies unsafe acts
Management	Mandatory HSE training	% of staff	Minimum % required	
Management	Mandatory HSE training	% Complete	Progress for year	
Management	Fire and explosions: significant	Number	Categorized	
Management	Fire and explosions	Number	Total	
Management	Fires	Number	Electrical equipment, IC engines, welding/cutting and grinding, hot equipment, pyrophoric, etc.	
Management	PTW compliance	%	Number of permit to work actions (PTWs) in full compliance/number of PTWs audited	
Management	PTW compliance critical infringements	Number	Deviation from procedures, inadequate documentation, site verification and hazard id, equipment certification, lack of supervision	
Management	Implement e-PTW	%	Conversion from manual permit to work (PTW) increases efficiency	
Management	Improve SD PTW system	%	SD PTW streamlining is critical to success of a turnaround	
Management	Implement PSP	%	Process safety program (PSP) implementation	
Management	PSP implementation	Date	Shift cycle (e-log and shift management system), alarm management, proactive monitoring	
Management	Alarm management	Number/panel/h	After alarm management program implementation	
Management	Life-saving rules violations	Number		Examples: Under suspended load, PTW not valid, no CSE certificate, no seat belt, no fall protection, using cell phone while driving
Management/BM	TRCF	Index	Total reportable case frequency per million hours	

Continued

Requirement	KPI	Unit	Definition	Comments
Management/ BM	TROIF	Index	Total occupational injury frequency per million hours	
BM	Total recordable incidents	TRI/million exp hours	Offshore: total recordable incidents per million exposure hours	

ENVIRONMENTAL KPIs—LINKED TO HUMAN CAPITAL

Requirement	KPI	Unit	Definition	Comments
Corporate	Water disposal quality	ppm	Oil in water measured as ppm for produced water before disposed	Consent to operate (CTO) requirement
Corporate/ management/BM	Total GHG emissions (direct and indirect)	Tons CO_2e	Other GHG gases converted to CO_2	UNFCCC requirement
Management	Unplanned events flaring	Volume	Reporting for CTO for duration exceeding an agreed time	CTO requirement
Management	Unplanned events flaring	Duration	Reporting for CTO for duration exceeding an agreed time	CTO requirement
Management	Total flaring	mmscm	Million standard cubic meters	Focus area: climate change and energy
Management	Direct GHG emissions	% woi	Weight % of total intake	Applicable to all HC plants
Management	Purge flaring	Volume		Applicable to all HC plants
Management	Tankage flaring	Volume		Applicable to a gas plant
Management	Jetty flaring	Volume	For each loading	Applicable to a gas plant
Management	Jetty flaring	Duration	For each loading	Applicable to a gas plant
Management	Flaring during planned shutdown	Volume	For each shutdown	Applicable to a gas plant
Management	Flaring during planned shutdown	Duration	For each shutdown	Applicable to a gas plant
Management	Restart flaring	Volume	After each shutdown	Applicable to a gas plant
Management	Restart flaring	Duration	After each shutdown	Applicable to a gas plant
Management	Major gas turbine emissions	Volume		Selected GTs greater than 25 MW
Management/BM	GHG emissions	Ratio	Ton GHG/ton prod (CO_2 equivalent)	Applicable to a gas plant PTAI

Requirement	KPI	Unit	Definition	Comments
Management/BM	Waste disposal	Ratio	Ton waste/ton intake	Applicable to a gas plant PTAI reference target 90T/MMT intake
Management/BM	Waste recycled	%	% Waste recycled	Applicable to a gas plant PTAI Reference target 70%
BM	SCMI	Index	Site carbon management index	Applicable to a gas plant PTAI

OTHER HEALTH, SAFETY, ENVIRONMENTAL, AND ASSET INTEGRITY KPIs

Requirement	KPI	Unit	Definition	Comments
Corporate	Flaring	% woi		Shareholder requirement for gas plants
Corporate	Flaring	% wt on sweet gas, e.g., AGR		Applicable to a gas plant
Corporate	Flaring	MMSCF/day	Volume of flaring recorded	
Corporate	Total flaring	mmscm		Focus area: climate change and energy
Corporate	Unplanned events	Number	All unplanned events affecting production.	
Corporate/BM	Direct GHG emissions	% woi	Weight % of total intake	PTAI
Corporate/ management/BM	LOPC1	Number	Tier 1	
Corporate/ management/BM	LOPC2	Number	Tier 2	
Management	LOPC3	Number	Tier 3 challenges to safety systems	
Management	LOPC1 to LOPC3 total	Number	Flange/fitting/packing failure, isolation failure, material quality, overfill, seal failure, corrosion, fatigue (wear and tear)	
Management	LOPC3 any loss other than 1 or 2	Number		Leading indicator
Management	Other fires not related to process units	Number		Leading indicator
Management	Central operator panel alarm priority 1 exceedances	Number		Leading indicator

Continued

Requirement	KPI	Unit	Definition	Comments
Management	Demand on safeguarding system	Number		Leading indicator
Management	Overdue inspection recommendations	Number		Leading indicator
Management	Near misses	Number		Leading indicator
Management	Number. STOP cards initiated	pppm	Per person per month	An STOP card is initiated when a safety violation is observed
Management	Consecutive days without RAM3+LOPC	Days		RAM3: risk assessment matrix category 3
Management	Purge flaring	Vol		Applicable to a gas plant
Management	Major GT emissions	Number		
Management	Emergency drills tier 1	%	Actual versus planned	IPOCM managed by the operating staff
Management	Emergency drills tier 2	%	Actual versus planned	IPOCM Fire department called out
Management	Emergency drills tier 2	Number		
Management	Emergency drills tier 3	%	Actual versus planned	IPOCM support required from civil defense and other industries in the area
Management	Emergency drills tier 3	Number		
Management	IPOCM risk actions beyond the agreed implementation date	Number		IPOCM—Incident preparedness and operational continuity management
BM	TRI	TRI/million exp hours	Offshore: Total recordable incidents per million exposure hours	McKinsey
Management	SHE case implementation for onshore	%		UK HSE regulation requirement
Management	SHE case implementation for offshore	%		UK HSE regulation requirement
Management	SRU efficiency <99%	Days	SRU: sulfur recovery unit	
Management	LDAR program implementation	%	LDAR: Leak Detection and Repair	
Management	Tanks emissions inventory project	%		
Management	Unplanned events: CA implementation within prescribed time	%	CA: Corrective Action. Target should be 100%	

Requirement	KPI	Unit	Definition	Comments
Management/BM	TRCF	Index	Total reportable case frequency per million hours	
Management/BM	TROIF	Index	Total reportable occupational injury frequency per million hours	
BM	LOPC/NSP	Index	Normalized loss of primary containment	PTAI
BM	Normalized number of fires	Index	Fires and explosions per NSP	PTAI
BM	GHG emissions	Ratio	Ton GHG/ton prod (CO_2 equivalent)	PTAI

3. HYDRO CARBON MANAGEMENT

Requirement	KPI	Unit	Definition	Comments
Corporate	Production	BOPD MMSCF/D	Actual production	Shareholder requirement
Corporate	Flaring production ratio	MMSCF/BBL	Gas flared/oil production	Shareholder requirement. Oil well
Corporate/BM	Energy usage	% Energy on intake		PTAI
Corporate	Generation performance index	%	Steam and power (electrical)—Boilers, STGs, GTs: generation efficiency actual/generation efficiency design	
Corporate	Unaccounted loss/ gain	% wt on slugcatcher over head (oh) gas	HC mass balance by venture	Measured values required. Applicable to a gas plant
Management	Known losses	% wt on slugcatcher oh gas	Including flaring, CO_2 to AGR: "chemical," and other	Shareholder requirement. Applicable to a gas plant
Management	Well head gas flow rate	mmscfd Year on year	Current year to date versus previous year to date	Applicable to a gas plant
Management	Yields of heavy products	bbl/mmscf wh gas year to date	Field condensate Total condensate Butane and condensates	Applicable to a gas plant
Management	Well head gas composition	H_2S C5+ C4+	Basis: SC OH gas analysis	Applicable to a gas plant
Management	Plant yields	Ethane (year to date)		Applicable to a gas plant

Continued

Requirement	KPI	Unit	Definition	Comments
Management	Plant yields	Component recoveries (year to date)	Propane Butane C5+ in plant condensate GTL prods	Applicable to a gas plant
Management	Plant yields	Component recoveries (versus EPC test run)	Propane Butane C5+ in plant condensate Sulfur	Applicable to a gas plant
Management	GTL yield efficiency	bbl/mmscf		Applicable to a GTL plant
Management	Total natural gas used	Million M3		Focus area: climate change and energy
Management	Total energy used (direct and indirect)	GJ		Focus area: climate change and energy
Management	Consumption performance index	%	Process blocks, incl. Utilities: energy consumption actual/energy consumption design	Applicable to a gas plant
Management	Loss performance index	%	Steam and condensate—mass and energy: loss performance actual/loss performance design	Applicable to a gas plant
Management	FG own use	% woi		Applicable to a gas plant
Management	FG nonenergy use	% woi		Applicable to a gas plant
Management	Generation efficiency: GT1 etc.	%	Generation efficiencies (boilers, WH boilers, GTs)	For each item of equipment
Management	Transportation loss (steam+condensate)	%	Deduced from model	
Management	Dump condensing as % primary steam produced	%		
Management	Mass balance per steam grade	%		
Management	Total energy consumption in process	Ton NRG	Ton national reference gas	Applicable to a gas plant
Management	Energy consumption per production block	Ton NRG	Ton national reference gas	Applicable to a gas plant
Management	PTE	%	Plant thermal efficiency—mmbtu out/mmbtu in	Applicable to a gas plant
Management	TF	wt%	Total flaring/woi	Applicable to a gas plant

Requirement	KPI	Unit	Definition	Comments
Management	PSP	KW/tpd	Plant specific power = total shaft power/tons LNG rundown per day	Applicable to a gas plant
Management	Own fuel use	% woi		Applicable to a gas plant
Management	Royalty/fiscal/custody transfer mismeasurement reports	Number		Highlights imbalance issues
Management	Energy exposed to royalty/energy out of well	%		Applicable to a gas plant
Management	Fuel and loss	% woi	Weight % of intake	Applicable to a gas plant
Management	GHG % LNG	%	GHG × 100/LNG produced	Applicable to a gas plant
Management	GHG % sales/flowing gas	%	GHG × 100/flowing gas produced	Applicable to a gas plant
BM	Energy on intake	% woi	Weight % of intake	Applicable to a gas plant
BM	Losses	% woi	Weight % of intake	PTAI
BM	EI	Index	Energy index	PTAI
BM	CEL	Index	Corrected energy and loss index	PTAI
BM	Total identified losses (excl. chemical losses)	% Weight on intake		PTAI
BM	VEI	Index	Volumetric expansion index	Solomon. Applicable to a refinery
BM	EII	Index	Energy intensity index	Solomon. Applicable to a refinery
BM	CEI	Index	Carbon emission index	Solomon
BM	JHI	Index	Juran hydrocarbon Index	Juran
BM	Overall energy usage	Btu/bbl		Applicable to a gas plant
BM	Purchased energy consumption	Btu/bbl		Applicable to a gas plant and refinery
BM	Refinery produced energy consumption	Btu/bbl		Applicable to a refinery
BM	Electricity consumption	kWh/bbl		Applicable to a refinery
BM	EII	Index	Energy intensity index	Solomon: applicable to a refinery

4. PRODUCTION AND DOWNTIME

Requirement	KPI	Unit	Definition	Comments
Corporate	Unplanned SD days	Actual/planned	Unplanned SD days deviation from the estimate in the WP&B	Should be stated in the BP&B/WP&B
Corporate	Facilities downtime	Days		Applicable to an oil well
Corporate	Equipment uptime—gas compressors	Days		Applicable to an oil well
Corporate	Equipment uptime—water injection	Days		Applicable to an oil well
Corporate	Equipment uptime—power generation	Days		Applicable to an oil well
Corporate	Pigging compliance	%		Applicable to an oil well
Management	LNG loaded	TPA loaded/mmscf feed gas		Range: 16 to 17 t/mmscf. Applicable to a gas plant
Management	LNG loaded	Actual YTD TPA loaded/planned YTD TPA loaded		Applicable to a gas plant
Management	Yield efficiency	kbbl/mmscf	Offshore gas to bbls conversion ratio	Applicable to a GTL plant
Management	Total sales/flowing gas	mmscf/d		Applicable to a gas plant
Management	Annual GTL production	kbbls		Applicable to a GTL plant
Management	Planned SD days actioned	Days	Actual planned SD days actioned to date versus BP&B per unit/stream	
Management	Planned SD days	Actual/planned	Turnaround over/underrun	
Management	Unplanned SD days actioned	Days	Actual unplanned SD days actioned to date versus BP&B per unit/stream	Direct reliability indicator for each process unit/stream
Management	Process unit availability	%		
Management	LNG availability	% of year		Planned and unplanned in the year. Applicable to a gas plant
Management/BM	Lost capacity tech	% of capacity		PTAI
Management/BM	Lost capacity other	% of capacity		PTAI

Requirement	KPI	Unit	Definition	Comments
BM	Utilization	% of capacity		Max sustainable capacity definition to be agreed and applied as per PTAI
BM	Reliability	% of year	Days on stream × 100/365	PTAI
	Unplanned maintenance	Days in year	365 days on stream	Excluding planned shutdowns and other outages
BM	Operational availability	% of year		PTAI—planned days annualized
BM	Lost capacity	% of total capacity		PTAI
BM	Production efficiency	% MPP	Prod eff = throughput/ MPP (maximum production potential), MPP = structural MPP − market losses	Applicable to an oil well
BM	Opportunity to increase production (%)	%		PTAI
BM	Annualized lost capacity for maintenance (%)	%		PTAI
BM	Annualized unplanned lost capacity (%)	%		PTAI

5. FINANCE

Requirement	KPI	Unit	Definition	Comments
Corporate	ROCE	EBIT/total capital employed	Return on capital employed	EM primary measure of O&G business performance
Corporate	NPAT	$	Net profit after tax	
Corporate	NPAT	%	Net profit after Tax	
Corporate	GPM	%	Gross profit margin (%)	
Corporate	OPM	%	Operating profit margin (%)	
Corporate	Royalty	$		
Corporate	Royalty	%		
Corporate/BM	LNG unit costs	$/ton LNG		Applicable to a gas plant
Corporate	Opex	$/t LNG	Actual opex/actual production	Applicable to a gas plant
Corporate	Opex	$	Actual operating expense	Shareholder requirement

Continued

Requirement	KPI	Unit	Definition	Comments
Corporate	Capex	$	Actual capital expense	Shareholder requirement
Corporate	Unit operating cost	$/bbl	Actual opex/actual production	Shareholder requirement
Corporate/BM	Operator staff costs	$/NSP		Opex template PTAI
Corporate/BM	Support staff costs	$/NSP		Opex template PTAI
Corporate/BM	Maintenance and engineering costs	$ %/RV		Opex template
Corporate/BM	Support staff costs	% of total work force		Opex template
Corporate/BM	Maint and eng costs	$/NSP		Opex template PTAI
Management	EBIT	$	Earnings before interest and tax (EBIT) $	
Management	EBITDA	$	Earnings before interest, tax, depreciation and amortization (EBITDA) $	
Management	Net income	Billion $		
Management	Net income price variance	Ratio		
Management	Net income—volume variance	Ratio		
Management	Total opex	$		
Management	Total expenditure	Million $		
Management	Operating cash flow			
Management	Ship cost	$		Capex, shareholder advances, repayment of debt, undistributed cash
Management	Optimization gross revenue uplift	Billion $		
Management	Deviation from predicted cost	%		New investment
Management	Deviation from planned startup date	Days		New investment
Management	gross margin	$/barrel	profit	Applicable to a refinery
Management	Net (cash) margin	$/barrel	opex	Refineries—profit per barrel
Management	Opex per ton	$/ton	opex	Applicable to a gas plant
Management	Opex per mmbtus	$/million BTUs	opex	Applicable to a gas plant
Management	Opex per ton	$/ton	opex	Petrochemical plants

Requirement	KPI	Unit	Definition	Comments
Management	Opex per MWH	$/MWH	opex	Applicable to a power plant
Management	Opex var	%	Deviation of actual from planned	
Capex/project management	Actual versus budget	%		
Capex/project management	Cost variance	%		
Capex/project management	Estimate at completion	$		
Capex/project management	Actual CERs versus total CERs	%	Actual certified emission reductions (CERs) versus total certified emission reductions (CERs)	Total certified emission reductions (CERs) = actual CERs + potential CERs
Capex/project management	Actual versus budget	%		
Capex/project management	Return on investment	ROI		
BM				
BM	Fixed cost	$/MUc		PTAI
BM	Unit opex	$/ton salable prod	Onshore cost	PTAI
BM	Unit fixed costs	$/ton salable prod	Onshore cost	PTAI
BM	Unit opex	$/ton salable prod	Total (offshore and onshore) cost	WP&B
BM	Unit fixed costs	$/ton salable prod	Total (offshore and onshore) cost	WP&B
BM	Unit variable costs	$/ton salable prod		PTAI
BM	Operating cost index	Index		PTAI
BM	Maint costs	$/MUc		PTAI
BM	Maint labor costs	$/MUc		PTAI
BM	Maint material costs	$/MUc		PTAI
BM	Maint activity costs	$/MUc		PTAI
BM	Fixed costs (adjusted for maintenance)	$/MU		PTAI
BM	Annualized maintenance costs	$/MUc		PTAI
BM	Annualized maintenance material costs	$/MUc		PTAI
BM	Average personnel costs	$/hr		PTAI

Continued

Requirement	KPI	Unit	Definition	Comments
BM	Unit lifting cost	$/boe	ULC=total opex partner billings less processing tariff costs divided by throughout including third party processed volumes	McKinsey
BM	Cost/topside weight	$/t		McKinsey
BM	Logistics costs/ topside weight	$/t		McKinsey
BM	Gross margin index (GMI)			Solomon
BM	(cash) opex per EDC	$/EDC	opex	Solomon
BM	Nonenergy (cost efficiency) cost	$/EDC		Solomon
BM	Maintenance (cost efficiency) index	$/EDC		Solomon
BM	Capital invest-ment index	$/EDC		Solomon
BM	Maint cost/opex	%	Percentage cost of main-tenance as part of opex	Opex template
BM	$ per normal shift position (NSPs)	$/NSP		PTAI
BM	$ per equivalent distillation capac-ity (EDC)	$/EDC		Solomon: Applicable to a refinery
BM	$ per equivalent processing capac-ity (EPC)	$/EPC		Solomon: applicable to a gas plant
BM	Production cost	$/KWH		Applicable to a power plant
BM	Production cost	$/mmbtu	mmbtu of salable product	Applicable to a gas plant
BM	Production cost	$/ton	Ton of salable product	Applicable to a gas plant

6. ASSET MANAGEMENT

Requirement	KPI	Unit	Definition	Comments
Management	Mechanical availability	%	Mechanical availability (MA) is calculated from total lost maintenance days for a particular year MA=(days in current year−TADD—RMDD)×100/days in current year	Very important KPI

Requirement	KPI	Unit	Definition	Comments
Management	Reliability	%	Reliability = (total days – shutdown days – unplanned down time) × 100/(total days – shutdown days)	Very important KPI
Management	Alarms	Number	Number per panel per hour	Target 5
Management	Critical (A) equipment PPM compliance	%	PPM—planned plant maintenance. Critical A causes production loss	Target 100%
Management	Critical (B) equipment PPM compliance	%	Critical B needs immediate attention to prevent production loss	Target 100%
Management	Essential equipment (C) PPM compliance	%	If not attended to timeously, critical C will escalate to critical B or even critical A	Target >95%
Management	Schedule compliance	%	Actual versus planned work order completion	Target >90%
Management	Work order backlog	Weeks	Larger backlog needs attention wrt the cause: restrained manpower, availability of spares, access to plant, etc.	Target 3–6
Management	Maintenance notification denied	%	Reasons need to be given and analyzed	Target <5%
Management	Break in	%	Jobs added to a maintenance schedule without proper lead time	Target <5%
Management	Emergency	%		More emergency work orders indicates potential for major incidents Target <1%
Management	Critical PPMs overdue >30 days	Number		Longer backlog of critical PPMs indicates potential for major incidents/loss of production. Target 0%
Management	Maintenance rework (within 3 months)	%		Indicates poor workmanship or root cause of issue not addressed Target <3%
Management	Number of exceptions endorsed, not released by ops	%	Critical PPM compliance	
Management	Number of exceptions endorsed, SD required	%	Critical PPM compliance	

Continued

Requirement	KPI	Unit	Definition	Comments
Management	Number of exceptions endorsed, equipment not running	%	Critical PPM compliance	
Management	Number of exceptions endorsed, other	%	Critical PPM compliance	
Management	Number of exceptions endorsed, awaiting spares	%	Critical PPM compliance	
Management	Monthly planned maintenance (PM) compliance for safety critical elements	%		
Management	Monthly planned maintenance (PM) compliance	%		
Management	Maintenance overtime	%		
Management	KEIs	Number	KEIs: key equipment issues	
Management	EKEIs	Days	EKEIs: emerging key equipment issues	
	PRV testing compliance backlog		PRV: pressure relief valve	Important issue—high backlog indicates a higher risk of loss of primary containment
Management	CDFT backlog	Days	Critical device function test (CDFT)	Important issue—high backlog indicates a higher risk of loss of primary containment
Management	Facilities integrity (FI) incidents: vol down time (VDT)	Total hours lost production		
Management	FI incidents: VDT	Number per year		
Management	Risk adjusted potential losses (million $)	million $		
Management	Number of process SD demands per unit	Number		
Management	Fire and gas demands per unit	Number		
Management	ESD demands per unit	Number		
Management	Past due actions process safety equipment	Number		
Management	Past due actions process hazard analysis	Number		
Management	Past due actions DGS	Number	DGS: down graded situations	
Management	Inhibits/bypasses of safety systems	Number		
Management Maintenance	Number of safety critical work requests outstanding for approval and implementation	Number		

Requirement	KPI	Unit	Definition	Comments
Management Maintenance	% Completion of testing of SCE	%	SCE: safety critical equipment	
Management Maintenance	Number of failure to function of SCE	Number		
Management Maintenance	Maintenance notifications priority 1 and 2 within capacity	%		Resourcing capacity limit should exceed priority 1 and 2 maintenance notifications
Management Maintenance	PTW prep later than 8 a.m.	Number	PTW: permit to work	
Management Maintenance	Maintenance backlog >6 months	%		
Management Maintenance	PM compliance to schedule	%		Target >85%
Management Maintenance	% work orders classified as urgent	%		Excellent leading indicator Target <5%
Management	% of PM versus all work orders	%		Target >80%
Management Maintenance	Safety critical elements PM compliance	%		
Management Maintenance	PM/CM ratio	ratio	Planned maintenance/corrective maintenance ratio	Target 80:20
Management Maintenance	Backlog >6 months	%	Total work order backlog	
Management Maintenance	Outstanding work order (backlog)	Weeks		
Management Maintenance	PRV, ESDV, FG PM compliance	%	Pressure relief valves (PRVs), emergency shutdown valves (ESDVs), fuel and gas systems (FG)	
Management Maintenance	Metering PM compliance	%	All metering especially royalty, fiscal, allocation	
Management Inspection	Number of overdue inspections	Number	Inspections of vessels, pipelines, structures, and subsea	
Management Inspection	Static equipment inspection	%	Actual versus planned	
Management Inspection	Piping inspection	%	Actual versus planned	
Management Inspection	PRV inspection	%	Actual versus planned	
Management Inspection	Corrosion control and monitoring compliance	%	Actual versus planned	

Continued

Requirement	KPI	Unit	Definition	Comments
Management Inspection	Coating survey	%	Actual versus planned	
Management Inspection	CP surveys	%	Corrosion protection surveys: actual versus planned	
Management Inspection	Lifting equipment inspection	%	Actual versus planned	
Management Inspection	QA/QC audits	Number	Actual versus planned	
Management Inspection	High priority recommendations overdue	Number	Actual versus planned	
Management Inspection	PMCRs	%	Preventative maintenance change requests: implemented versus accepted	
Management Inspection	High criticality RCFAs completed	Ratio	Highly critical root cause failure analysis: actual versus planned	
Management Inspection	Receiving material inspection	%	% of total	
Management	Bypass override (min number)	Number		LOPC tier 4
Management	Alarm management	Number/hr/panel		LOPC tier 4
Management Maintenance	PM compliance	%		LOPC tier 4
Management Maintenance	Work order completions fully documented	%		Target 100%
Management Maintenance	Safety critical elements PM compliance	%		LOPC tier 4
Management Maintenance	Safety critical elements test exceedances min number	Number		LOPC tier 4
Management Inspection	Inspection schedule compliance	%	Equipment, piping, corrosion inhibitor program, etc.	LOPC tier 4
Management Inspection	Compliance to process safety audit schedule	%		LOPC tier 4
Management Inspection	Overdue audit items	Number	Process safety field observations	LOPC tier 4
Management Maintenance	Overdue maintenance audit items	Number		
Management Maintenance	Metering compliance	%		
Management Maintenance	ESDV, PRV, FG, ESC, PM compliance	%		
Management Maintenance	Rolling backlog	Weeks		

Requirement	KPI	Unit	Definition	Comments
Management Maintenance	Overall PM:CM ratio	%		
Management Maintenance	Wrench time measurement			
Management Inspection	Internal inspection	%	Static equipment	
Management Inspection	External inspection	%	Static equipment	
Management Inspection	UT inspection	%	Static equipment	
Management Inspection	Thermography survey	%	Static equipment	
Management Inspection	On-line UT inspection	%	Pipelines	
Management Inspection	External inspection	%	Pipelines	
Management Inspection	Safety critical items due	%	(Number of inspections of safety critical items of plant and equipment due during the measurement period and completed on time/ total number of inspections of safety critical items of plant and equipment due during the measurement period) × 100%	Maintenance of mechanical integrity, CCPS
Management Inspection	Time plant in operation with safety critical items in failed state	%	(Length of time plant is in production with items of safety critical plant or equipment in a failed state, as identified by inspection or as a result of breakdown/ length of time plant is in production) × 100%	Maintenance of mechanical integrity, CCPS
Management Inspection	Past due safety action items	%	(Number of past due and/or having approved extension of process safety action items/total number of active or open action items) × 100%	Action items follow-up, CCPS
Management	Audited MOCs	%	% of audited MOCs that satisfied all aspects of the site's MOC procedure	Management of change, CCPS
Management	Audited changes	%	% of audited changes that used the site's MOC procedure prior to making the change	Management of change, CCPS

Continued

Requirement	KPI	Unit	Definition	Comments
Management turnaround	Trouble free startup	%	% start-ups following plant changes where no safety problems related to the changes were encountered during recommissioning or startup	Management of change, CCPS
Management	IT back-up planned versus actual	%		Information asset
Management	High-risk IT issues outstanding	Number		Information asset
Management	IT audit finding outstanding beyond agreed implementation date	Number		Information asset
Management Maintenance	Scheduling compliance	%	% Scheduled with respect to total actual work	SAP maintenance module
Management Maintenance	Planning estimation compliance	%	Man-hour estimate versus actual	SAP maintenance module
Management Maintenance	Emergency work	%	Maintenance hours spent on priority "A" jobs × 100/total man-hours worked	SAP maintenance module
Management Maintenance	Total backlog	%	Man-hours planned on pending orders with status "AW**" × 100/total man-hours of confirmed for the period of analysis	SAP maintenance module: work orders pending categorized by delay codes ("awaiting")
Management Maintenance	Maintenance planned work performance	Ratio	Planned versus unplanned	SAP maintenance module
Management Maintenance	Maintenance work types ratio: corrective	%	Actual CORR × 100/actual total	SAP maintenance module
Management Maintenance	Maintenance work types ratio: preventive	%	Actual PREV × 100/actual total	SAP maintenance module
Management Maintenance	Maintenance work types ratio: inspection	%	Actual INSP × 100/actual total	SAP maintenance module
Management Maintenance	Maintenance work types ratio: PCR	%	Actual PCRO × 100/actual total	SAP maintenance module
Management Maintenance	Maintenance work types ratio: preventive + inspection	%	Actual (PREV + INSP) × 100/actual total	SAP maintenance module
Management Maintenance	Manpower utilization ratio	Ratio	Actual completed work for the period divided by the available capacities	SAP maintenance module
Management Maintenance	Safety equipment program effectiveness	%	Number of CORR orders actually started for func. loc. category "X"* 100/total number of corrective orders started for the same period	SAP maintenance module: safety equipment functional location categorized as "X"

Requirement	KPI	Unit	Definition	Comments
Management Maintenance	Safety system inspection compliance	Ratio	Compare the completed "safety" preventive orders against the scheduled "safety" preventive orders	SAP maintenance module
Management turnaround	Behind/ahead of critical path: hours	Hours		Next turnaround
Management turnaround	Schedule variance (SV): actual/planned duration	%		Next turnaround
Management turnaround	Schedule performance index (SPI)	index	Deviation from planned to date: earned value/planned value	Next turnaround—alternative to schedule variance (SV)
Management turnaround	Cost variance (CV): actual/planned costs	%		Next turnaround
Management turnaround	Cost performance index (CPI)	index	Deviation from budget to date: earned value/actual cost	Next turnaround—alternative to cost variance (CV)
Management turnaround	Deviation from budget: estimate at completion/budget at completion	%		Next turnaround
Management turnaround	Additional work cost: actual versus contingency	%		Next turnaround
Management turnaround	Emergent work: man-hours as % of total TA man-hours	%		Next turnaround: minimize
Management turnaround	Lost time incidents (LTIs): number	Number		Next turnaround: target: "zero"—monitor lower level HSE indicators to prevent even one incident
Management turnaround	Flawless startup (no incidents and no leaks)	Number		Next turnaround: target: "zero" measure from "hand back" to "on-stream at required quality"
Management turnaround	Schedule efficiency	Days versus days	Shutdown duration relative to others with similar scope and complexity	Future TAs
Management turnaround	Plant availability	%		Future TAs: max interval between turnarounds with min unplanned down days and min planned down days—annualized

Continued

Requirement	KPI	Unit	Definition	Comments
Management turnaround	Runtime	Months	Planned based on analysis	Future TAs: target based on benchmarking
Management turnaround	Front end loading (FEL)	Index	Preparedness indicator	Future TAs
BM	Maintenance index (MI)	Index	Sum of the nonturnaround (routine) maintenance index (RI) and the turnaround maintenance index (TI)	Solomon
BM	Nonturnaround (routine) maintenance index (RI)	Index		Solomon
BM	Turnaround maintenance index (TI)	Index		Solomon
BM	Asset management cycle av. lost cap (% on cap)	%		PTAI
BM	Lost cap: LNG treating and liquefaction planned (%)	%		Applicable to a gas plant. PTAI
BM	Lost cap: LNG treating and liquefaction unplanned (%)	%		Applicable to a gas plant. PTAI
BM	Lost cap: sales gas treating and compression planned (%)	%		Applicable to a gas plant. PTAI
BM	Lost cap: sales gas treating and compression unplanned (%)	%		Applicable to a gas plant. PTAI
BM	Lost cap: condensate stab planned (%)	%		Applicable to a gas plant. PTAI
BM	Lost cap: condensate stab unplanned (%)	%		Applicable to a gas plant. PTAI
BM	Lost cap: ethane and LPG fract planned (%)	%		Applicable to a gas plant. PTAI
BM	Lost cap: ethane and LPG fract unplanned (%)	%		Applicable to a gas plant. PTAI
BM	Lost cap: sulfur recovery planned (%)	%		Applicable to a gas plant. PTAI
BM	Lost cap: sulfur recovery unplanned (%)	%		Applicable to a gas plant. PTAI
BM	Lost cap: fired boilers and WHBs planned (%)	%		PTAI
BM	Lost cap: fired boilers and WHBs unplanned (%)	%		PTAI

7. PERSONNEL (HUMAN CAPITAL)

Requirement	KPI	Unit	Definition	Comments
Corporate/BM	Staff turnover	%		A very good indicator that could reflect in a number of other indicators
Corporate	Key positions vacant for more than 3 months	Number		Indication of strategic risk
Management	Assessment of technical competencies	%		Includes all operations departments (Ops, Maint, Eng) excludes mandatory training (HSE etc.)
Management	Technical competencies development plans	%		An indication of training being on track
Management	Implement BBSP program	%		Implement a behavior-based safety program (BBSP)
Management	Direct hire overall attrition	%		A good indication of retention
Management	Vacancies	%		Question: can we do without the positions or do we need to improve offers/ change recruiting tactics?
Management	Employee awards	Number	Number. per employee per year	
Management	EPMS objective roll down?	%		Employee Performance Management System (EPMS) SMART objective management
Management	Completed midyear reviews	%		EPMS
Management	Completed annual appraisals	%		EPMS
Management	Overtime	%		An indication of emergency maintenance work—raised risk
Management	Direct hire overall attrition	%		Good leading indicator

Continued

Requirement	KPI	Unit	Definition	Comments
Management	Key positions vacant for more than 3 months	Number		Good leading indicator
Management	Assessment of technical competencies	%		Includes all operations departments (Ops, Maint, Eng) excludes mandatory training (HSE, etc.)
Management	Technical competencies development plans	%		
Management	Training for PSM critical positions	%		Process safety training and competency, CCPS
Management	Training competency assessment	%		Process safety training and competency, CCPS
Management	Failure to follow procedures/safe working practices	Number		Process safety training and competency, CCPS
Management	Complete formal training for maintenance planners and schedulers	%		
NOC management	% Total nationals	%	Nationalization action plan	
NOC management	% Nationals in established positions	%	Nationalization action plan	
NOC management	Increase % nationals in established positions	%	Nationalization action plan	
NOC management	Nationals total head count	Number	Nationalization action plan	
NOC management	IDPs for Nationals	%	Individual development programs for nationals (actual versus plan)	
BM	Overall PI	Index	Overall personnel index	Solomon
BM	Operations PI	Index	Operations personnel index	Solomon
BM	Maintenance PI	Index	Maintenance personnel index	Solomon
BM	Support PI	Index	Support personnel index	Solomon
BM	Normalized shift positions (NSP)	Index		PTAI
BM	Site personnel index (SPI)	Index		PTAI

Requirement	KPI	Unit	Definition	Comments
BM	Maintenance effort	Maint hours/MUc	Extent of maintenance required related to complexity of plant	PTAI
BM	FTEs per topside weight	FTE/1000 tons	Full-time employees per topside weight	McKinsey
BM	Overall PI	Index	Overall personnel index	PTAI
BM	Operations PI	Index	Operations personnel index	PTAI
BM	Maint PI	Index	Maintenance personnel index	PTAI
BM	Support PI	Index	Support personnel index	PTAI
BM	Maint effort	Maint hours/MUc	Maint effort	PTAI
BM	Staff turnover	%		PTAI
BM	Personnel efficiency index (PI)	index		Solomon
BM	FTEs per topside wt	FTE/1000 tons		McKinsey

8. SUBSURFACE

Requirement	KPI	Unit	Definition	Comments
Corporate	RP	Ratio	Reserve replacement ratio	Assessment of number of years of production at the current rate

9. COMMERCIAL AND SHIPPING (INCLUDING CUSTOMER)

Requirement	KPI	Unit	Definition	Comments
Management	Customer satisfaction index (CSI)	Index		Required for ISO 9001 certification. Customer satisfaction tailored for industry
Management	Late deliveries	Number		Customer satisfaction
Management	Off spec	Number		Customer satisfaction
Management	Reliable deliveries	%		Customer satisfaction
Management	Terminal tanks <2.5 days	Number		Applicable to a gas plant
Management	LNG loading zero debris	Number of months		Applicable to a gas plant

Continued

Requirement	KPI	Unit	Definition	Comments
Management	LNG prices	$/mmbtu		Applicable to a gas plant
Management	Condensate prices	$/bbl		Applicable to a gas plant
Management	LNG volumes by region	million tons		Applicable to a gas plant
Management	LNG volumes by region	%		Applicable to a gas plant
Management	Shipping costs	$/mmbtu		Applicable to a gas plant
Management	Bunker average price variance from plan	%		Applicable to a gas plant

Appendix C: Oil and Gas Rules of Thumb

1. GENERAL

- Refinery-produced fuel consumed in a fuels refinery: 6.05 MMBtu per fuel oil equivalent (FOE) barrel (Solomon).
- One barrel of oil has approximately six times the energy content of 1 MMBtu of natural gas.
- One ton of liquefied natural gas (LNG) is equivalent to 8.5 Oil Equivalent Barrels.
- One barrel per day is equivalent to 49.8 tons per year production.
- 1 Btu = 1055 J.
- 1 kWH = 9.09 Btu = 9590 J [average net or low thermal heat input that the power industry uses to generate electricity using a combination of different types of generating equipment (Solomon)].

2. HEALTH, SAFETY, AND ENVIRONMENT

Liquefied Natural Gas

• Lost time Injury Frequency (LTIF)	0.2 per million hours
• Total Reportable Case Frequency (TRCF)	1.0 per million hours
• No. of fires per Normal Shift Position (NSP)	0.09
• Greenhouse gas (GHG) emissions	28% woi average
• GHG emissions	19.6% woi pacesetter
• GHG intensity (total GHG)	0.33–0.5 ton CO_2e/ton LNG
• GHG intensity (total GHG)	0.45 CO_2e/ton LNG

3. HYDROCARBON MANAGEMENT

Mass Balance

• Refining: Volume gain/Mass loss	+0%/−0.2%
• Gas Plants: Mass loss/gain less than	±0.6%

Gas Plant Yield

• Total Condensate	40 bbl/MMscf WHG (Wellhead Gas)
• Field Condensate Factor	35.6 bbl/MMscf RG (Royalty Gas)
• Plant Condensate Factor	4.1 bbl/MMscf SDG (Sweet Dry Gas)
• Helium Factor	24.8 kg/MMscf SDG

- Sulfur Factor — 151.5 kg/MMscf RG
- Deisopentanizer Factor — 1.4 bbl/MMscf SDG
- Storage LNG Factor — 40.6 m^3/MMscf SDG
- Free on board LNG Factor — 40.5 m^3/MMscf SDG
- Gas to Liquid (GTL) design yield — 105 bbl/MMscf

Energy

Energy Consumption (Including Imported or Generated Electricity)

- LNG: small train (3.3 MTA/C3MR) — to 9.7% woi
- LNG: medium train (4.7 MTA/C3MR) — 6%–9% of woi
- LNG: mega-train (7.8 MTA/AP-X) — 7.6%–8.6% of woi
- Sales gas train — 2.5%–4.5% of woi
- Refinery (hydroskimming) — 4% woi
- Refinery energy cost (Europe) — 60% of total cash opex
- Refinery energy cost (Solomon) — 45%–55% of total cash operating expenses
- Gas to Liquid (GTL) — 25% woi

Known Losses

- Sales gas — 5% wt on SC OH gas
- LNG — 0.5%–1.5% woi gas and condensate

Power Generation Efficiency

- Condensing Steam Turbine — 31%–33%
- Gas Turbines — 51.1%
- Boilers — 85%–87%
- Fired Heaters — 92%
- Sulfur Recovery Unit (SRU) waste heat boilers — 66%
- GE 7FA Combined Cycle — 56%
- National Reference Gas (1 NRG) — 1000 Btu/scf
- 1 ton NRG — 4.44 MWH
- Average net or low thermal heat input that the power industry uses to generate electricity using a combination of different types of generating equipment (Solomon) — 9590 kJ/kWH (9090 Btu/kWH)
- Fuel Oil Equivalent — 6383 MJ/FOE barrel (6.05 MBtu/FOE barrel) (Solomon)

4. PRODUCTION AND DOWNTIME

- Unplanned shutdowns per year:
  - LNG: 2%–3% of available capacity; 7–11 days per year
  - LNG mega-trains design: 8 days per year
  - Refinery: 2%–4% of available capacity; 7–15 days

- LNG initial startup:
  - Dry out 26 days
  - Cool down 14 days
- LNG mega-train: 12–16 days annualized, based on:
  - GE Frame 9E
    - Interval 6-year cycle
  - Hot Gas Path Inspection (HGPI)
    - Interval every 3 years
    - Duration 27–34 days (17+ days plus 10 days sd/su)
  - Major Overhaul and Inspection (MOI)
    - Interval every 6 years
    - Duration 42–44 days (30 days plus 12+ days sd/su)
  - Combustion Inspection (COI)
    - Interval every 12,000 h
    - Duration 7 days plus 10 days sd/su (could be eliminated)

5. FINANCE

Profit

- Refinery skimming 0 $/bbl gross margin (2005–15)
- Refinery complex
  - Average 5 $/bbl gross margin (2005–15)

Opex

- LNG train cost per ton of salable product (2012 on- and offshore): 16–24 $/Te (Te: equivalent LNG tons)
- LNG train cost per ton of salable product (2012 onshore, World): 15–20 $/Te
- LNG train fixed cost per ton of salable product (2012 onshore, World): 10–15 $/Te
- Maintenance cost less than x% of Replacement Value (RV)
  - Oil and Gas Industry 1.2%–1.9% (middle third of 50 plants)
  - Refinery 1.7%
  - Fluidic Catalytic Cracking (FCC) 1.8%
  - Sulfur Recovery Unit (SRU) 3.2%

Note: BP Texas refinery apparently drove their costs down to 1% and had a major disaster in March 2005. Some people blamed the setting of stringent benchmarking targets.

- Maintenance costs for LNG $1900/corrected Mechanical Units (Muc) to $3000/MUc

Capex

- Return On Investment (ROI) 20%

Shipping and Commercial

- LNG
  - Overall shipping cost $1.30 to $1.60/MMBtu

6. ASSET MANAGEMENT

Operations

- Panel Alarm Rates
  - Target 5 per panel per hour
  - Pacesetter Average 12 per panel per hour

Maintenance and Inspection

- Intelligent Pigging of undersea gas transport lines: 5-year cycle
- Turnarounds Duration:
  - Crude Unit 30 days
  - Fluidic Catalytic Cracking (FCC) 42 days
  - LNG Gas Plant Train (MOI) 42–44 days
  - Pipeline/Sales Gas Plant Train (MOI) 22–28 days
- Turnarounds Interval
  - Refining Crude Units >4 years
  - FCC >3.5 years
  - LNG Gas Trains (major TA) 6 years
  - LNG Gas trains (turbines) 5.5–6 years
  - LNG Gas trains (combustion) 18 months
  - LNG Gas trains (hot gas pass) 3 years
  - LNG Gas trains (mole sieve):
  - Sales gas plants 3–3.5 years
  - Gas to Liquid (GTL) 6 years (planned P)
  - Benchmarking annualization
    - Philip Townsend Associates International (PTAI): last turnaround duration divided by previous interval
    - Solomon: based on actual average intervals and last turnaround duration of the particular company
    - Internal: predicted turnaround interval and duration
- Maintenance
  - Break-ins (interruption of production or utility service):
  - Emergency Maintenance Work Orders: Less than 3%–5% of total Work Orders
  - Planned work versus all work 80%
  - Actual work versus all work 70%
  - Planned Maintenance/Corrective Maintenance Ratio 65% (pacesetter effort based)
  - Planned Maintenance/Corrective Maintenance Ratio 80:20 (pacesetter equipment objects based)
  - Compliance to schedule >85%
  - Planned Maintenance (PM) Compliance target 90%
  - Man-hours utilized 90%
  - Safety equipment effectiveness ratio 5%
  - Safety Critical Elements PM Compliance target 100%
  - Benchmarking annualization
    - PTAI and Solomon: last 2 years nonturnaround average
    - Internal: Nonturnaround actual for year

- Reliability
  - LNG Gas Train Typical 97%–98% (7–11 unplanned maintenance days per annum)
  - GE 7FA
    - Reliability (GE stats) 98%
    - DLN2.6 combustor (Dry low NOx) 5–9 ppm NOx
    - Simple Cycle Heat Rate (171.7 MW) 9360 Btu/kWh (9873 kJ/kWh)
    - Combined Cycle Heat Rate (262.6 MW) 6090 Btu/kWh (6424 kJ/kWh)
      - Net Plant Efficiency 56%
  - GE 9E
    - DLN2.6 combustor (Dry low NOx) 5–15 ppm NOx
    - Simple Cycle Heat Rate 9210–9860 Btu/kWh (9717–10403 kJ/kWh)
      - Net Plant Efficiency 34.6%–37%
    - Combined Cycle Heat Rate 6270–6460 Btu/kWh (6615–6816 kJ/kWh)
      - Net Plant Efficiency 52.8%–54.4%

7. PERSONNEL

- Gas Plants
  - Staff turnover (Pacesetter) 2.5%
  - Site Personnel Index (Pacesetter) 220
  - Operations Personnel Index (Pacesetter) 160
  - Maintenance Personnel Index (Pacesetter) 340
  - Support Personnel Index (Pacesetter) 300

8. SUBSURFACE

- Replacement/Production (RP) ratio: greater than the life of the investment of 20–30 years

9. SHIPPING AND COMMERCIAL (INCLUDING CUSTOMER)

- LNG
  - Shipping cost $1.33/MMBtu
  - Fuel cost 10%–15% of shipping cost
  - Cargo loss 2% for 14-day journey
  - Customer service Zero complaints and zero late deliveries

Appendix D: Operational Performance Health Checker Example

Operational Performance Checklist
Company/Business Unit:

Date:

rev 130103	Applied: Y/N/?	Level of Maturity	Comments
FA1-STRATEGY			
ERM with corporate risk register.			
Integrated Management Systems (IMS)			
Top 10 risks / opportunities			
Elements of good corporate governance: objectives, core values, responsibilities, principles & behaviors, sustainable development commitment, disclosure & transparency.			
Statement of Code of Conduct: Ethics & Conflict of Interest			
KPI scorecard			
Shareholder audit rights routine			
ISO 9001 Quality Management Systems certification			
Business Operations benchmarked with application of full improvement cycle			
Use of ARIS or similar Business Process Management tool to map / align business processes with the organizational structure			
FA2- HSSE			
Health & Safety			
OHSAS 18001 Occupational Health & Safety Management systems certification			
RCA process using Tripod Beta, Taproot, or other.			
SHE case implementation - offshore			
SHE case implementation - onshore			
Process Safety Program (PSP) implementation including Shift handover, Start of shift orientation, Shift team meeting planning, Proactive monitoring, Alarm management, End of shift review and Report shift performance.			
Implementation of ePTW			
Application of Energy Institute Process Safety Management Framework or similar.			
Communication & Collaboration such as shareholders network, contractors safety forum			
Environment			
ISO 18001 Environmental Management Systems certification			
GHG Compliance with UNFCCC			
FA3- Hydro Carbon Management			
Hydro Carbon Mass Balance Modeling			
Hydro Carbon Mass Balance Modeling with component balance			
Royalty / Fiscal/ Custody Transfer Metering as per Company Guidelines for the measurement of hydrocarbon fluids.			

Integrated HC Accounting System for Reconciliation & Allocation including Economic Optimization Modelling			
Energy Balance Modeling			
Consent to Operate (CTO) status			
CEMS (Continuous Emission Monitoring System) Code of Practice (CoP)			
ISO 50001 Energy Management certification			
FA4 - Production & Downtime			
Electronic Permit To Work (ePTW) process			
APC application to LNG trains			
Use of linear programing models			
PTW process for shutdowns			
FA 5 - Finance			
Application of International Financial Reporting Standards (IFRS)			
Institute of Internal Auditors (IIA) Audit Process			
Application of 9 Best Practices of Cash Operating Cost Management: Management Stewardship, Five Year Plan, Chart of Standard Accounts, Cost Driver Analysis, Cost Optimization Initiatives, Best Practice Business Processes, Training & Capability, Enterprise Systems, Engagement at all levels.			
Maintenance costs: routine, turnaround, overheads, cost of unreliability / lost profit opportunity.			
FA 6 - Asset Management			
Application of Maintenance Excellence Dashboard			
Maintenance: The right work at the right time in the most cost effective manner including systems & practices for work identification, methods & practices for work investigation & scoping, use of risk based decision making for work screening, work prioritization practices & methods, job planning, work scheduling, CMMS use & effectiveness, backlog management tools & techniques, materials management & spare parts, operator maintenance & reliability programs, contracting issues, routine maintenance organization, general assessment of craft skills/ number/ organization/ tools, use of latest industry tools & techniques for work execution, use of MoC in the maintenance process, overall mapping of routine maintenance work process.			
ERM Maintenance Module (Computerized Maintenance Management System - CMMS: SAP, Maximo, Oracle or other)			
Maintenance contractors performance workshops			
Maintenance risk register			
Reliability: Preventive & Predictive Maintenance programs, Reliability Bad Actor programs & RCFA, use of equipment criticality analysis, rotating equipment monitoring & improvement programs, inspection programs & corrosion control practices, operational reliability practices & programs, use of equipment plans & policies, RBI implementation, use of RCM techniques, reliability considerations in new capital projects.			
Facilities Integrity Management System (FIMS)/ Asset Performance Management System (APMS): Meridium or other, including Reliability Incident Reporting Tool.			
Application of Reliability Centered Maintenance as per RCM: SAE standard JA1011, "Evaluation Criteria for Reliability-Centered Maintenance (RCM) Processes," or equal.			Criticality reviews prior to implementation of RCM
Application of Risk Based Inspection (RBI) as per API 580 or equal.			
Safety Instrumented Systems (SIS) as per ISA 84/IEC 61511 and IEC 61508 or equal.			
Implementation of HAZOP/HAZAN studies as per IEC 61882 Hazards and Operability Studies Application Guide or equal.			
Thickness Measurement Location (TML) Management including monitoring & thickness calculation according to ASME and API industry standards for piping, tanks and pressure vessels or equal.			

Addressing & Reporting of Potential and Actual Volume Downtime (Key Equipment Issues - KEIs and Emerging Key Equipment Issues - EKEIs)			
Application of Meridium QIS Reliability Incident Reporting Tool or similar tool			
RCFA process application			
Bad actor process implementation			
Turnarounds: work scope generation practices, use of risk based work selection for scope optimization, control procedures for scope creep, philosophy for routine vs. turnaround work, turnaround cycle optimization, turnaround planning lead times, operations effectiveness in equipment preparation & start-up, integration of process & mechanical turnaround plans, efficiency improvements in work execution, use of latest industry tools & techniques for work execution, critical path minimization practices, TA contracting issues, TA organization & controls, TA budgeting & cost control.			
Use of Turnaround Critical Path Modelling Tool: Primavera etc.			
Application of Turnaround spares expediting processes			
Application of structured phased approach to managing turnarounds including the use of dedicated staff to manage planned activities			
The application of specific turnaround workshops as follows: 1. turnaround scope challenge, 2. turnaround schedule optimization, 3. flawless turnaround review, 4. turnaround readiness review.			
Turnarounds benchmarked			
Application of safety critical equipment materials planning process			
Pipeline Management System (PMS) built using eDPM® simulator functions provided by Scandpower PT (or similar) including Real-time monitoring • Pipeline Operating Condition Monitoring • Liquid Management • Pig Tracking • Hydrate Formation Monitoring • Pipeline Blockage Detection • Corrosion/ Erosion Monitoring • Pipeline Leakage Detection • Look-ahead monitoring • Production forecasting in the near future • What-if monitoring • Check of operational procedures			
OLGA®transient multi-phase flow simulator and the APIS® industrial IT platform (or similar).			
Implementation of electronic Management Of Change (eMOC) using consistent risk based approach & hurdle criteria implementation			
Instrument Control Systems (ICS) Security Standards			
ICS intrusion, detection & protection			
ICS Disaster Recovery Plan			
FA 7 - Personnel			
Employee Performance Management System (EPMS) including Behavioral & Technical competencies			
Special recognition awards system implemented			
Evidence of staff focus on core functions & contractor supervision and the use of contractors for non-core functions.			
Evidence of task force approach to problem resolution			
Behavior based safety program implementation			
FA 8- Sub-surface			
Reservoir Modeling			
FA 9 - Marketing & Shipping			
Logistics modeling			

MATURITY ASSESSMENT

Policy / Strategy	Level Of Maturity	CLASSIFICATION	
No explicit strategy. No allocation of resources. Information management rather than knowledge management.	1 No formal System	Laissez faire	Importance & manageability of knowledge assets is recognized: spontaneity
Designing policies for entry, retrieval, storage sharing & utilization of information. Focus on knowledge transfer among sub units.	2 Under development	Content based	Designing organized efforts to leverage knowledge. Building up formal systems for managing knowledge assets
Leveraging not only internal but also external knowledge. Recognizing subsidiaries as centers of excellence & global innovation.	3 Well established	Network based	Move towards a global knowledge management system. Active knowledge management. Shift to knowledge creation & innovation
Systematic design of intellectual capital rights. Designing polices that allow subsidiaries to gain power by exchanging knowledge.	4 Fully integrated	Process based	True integration of entire value system. Continuous innovativeness is a key asset.

Index

'*Note*: Page numbers followed by "f" indicate figures, "t" indicate tables and "b" indicate boxes.'

B

D

F

G

M

N

O

S

U

V

W

Y